102

Inorganic Solid Fluorides
Chemistry and Physics

This is a volume in the
Materials Science and Technology series.
Editors: *A. S. Nowick and G. G. Libowitz*

A complete list of the books in the series appears at the end of the volume.

Inorganic Solid Fluorides

Chemistry and Physics

Edited by

Paul Hagenmuller

Laboratoire de Chimie du Solide du Centre
   National de la Recherche Scientifique
Université de Bordeaux
Talence, France

1985

ACADEMIC PRESS, INC.

(Harcourt Brace Jovanovich, Publishers)

Orlando   San Diego   New York   London
Toronto   Montreal   Sydney   Tokyo

ACADEMIC PRESS, INC.
Orlando, Florida 32887

United Kingdom Edition published by
ACADEMIC PRESS INC. (LONDON) LTD.
24–28 Oval Road, London NW1 7DX

Library of Congress Cataloging in Publication Data

Main entry under title:

Inorganic solid fluorides.

    (Materials science and technology series)
    Includes bibliographies and index.
    1. Fluorides--Handbooks, manuals, etc. I. Hagenmuller,
Paul. II. Series: Materials science and technology.
QD181.F1I56   1985   546'.731   84-18591
ISBN 0-12-313370-X (alk. paper)

PRINTED IN THE UNITED STATES OF AMERICA

85 86 87 88   9 8 7 6 5 4 3 2 1

Contents

3. Crystal Chemistry of Fluorides

DIETRICH BABEL and ALAIN TRESSAUD

4. The Crystal Chemistry of Transition Metal Oxyfluorides

B. L. CHAMBERLAND

5. Defects in Solid Fluorides

C. R. A. CATLOW

6. High Oxidation States in Fluorine Chemistry

RUDOLF HOPPE

11. Electronic Conduction in Fluorides

ALAIN TRESSAUD

12. Fast Fluorine Ion Conductors

JEAN-MAURICE RÉAU and JEAN GRANNEC

13. Nonlinear Properties of Fluorides

J. RAVEZ

14. Optical Properties of Fluorides

C. FOUASSIER

Contributors

Numbers in parentheses indicate the pages on which the authors' contributions begin.

Dietrich Babel (77), Fachbereich Chemie, Philipps-Universität, D-3550 Marburg, Federal Republic of Germany

Neil Bartlett (331), Department of Chemistry, University of California, Berkeley, California 94720

Charles A. Baud (545), Institute of Morphology, University Medical Center, 1211 Geneva 4, Switzerland

C. R. A. Catlow (259), Department of Chemistry, University College London, London WC1E 6BT, England

B. L. Chamberland (205), Department of Chemistry and Institute of Materials Science, University of Connecticut, Storrs, Connecticut 06268

Bernard Cochet-Muchy (565), Atochem, Centre de Recherches de Lyon, 69310 Pierre Bénite, France

Jean-Michel Dance (371, 489, 525), Laboratoire de Chimie du Solide du CNRS, Université de Bordeaux I, 33405 Talence, France

P. B. Fabritchnyi (519), Department of Chemistry, Moscow State University, Moscow 117234, USSR

G. Ferey (395), Laboratoire des Fluorures et Oxyfluorures Ioniques, Faculté des Sciences, 72017 Le Mans, France

C. Fouassier (477), Laboratoire de Chimie du Solide du CNRS, Université de Bordeaux I, 33405 Talence, France

Jean Grannec (17, 423), Laboratoire de Chimie du Solide du CNRS, Université de Bordeaux I, 33405 Talence, France

Paul Hagenmuller (1), Laboratoire de Chimie du Solide du CNRS, Université de Bordeaux I, 33405 Talence, France

Rudolf Hoppe (275), Institut für Anorganische und Analytische Chemie, Justus Liebig–Universität, 6300 Giessen, Federal Republic of Germany

M. Leblanc (395), Laboratoire des Fluorures et Oxyfluorures Ioniques, Faculté des Sciences, 72017 Le Mans, France

Lucien Lozano (17), Laboratoire de Chimie du Solide du CNRS, Université de Bordeaux I, 33405 Talence, France

Thomas Mallouk (331), Materials and Molecular Research Division, Lawrence Berkeley Laboratory, Berkeley, California 94720

Katsuki Miyauchi (489), Central Research Laboratory, Hitachi, Ltd., Tokyo 185, Japan

Tsuyoshi Nakajima (331), Department of Industrial Chemistry and Division of Molecular Engineering, Kyoto University, Kyoto 606, Japan

Tetsu Oi (489), Central Research Laboratory, Hitachi, Ltd., Tokyo 185, Japan

J. Pannetier (395), ILL Grenoble, 38042 Grenoble, France

R. de Pape (395), Laboratoire des Fluorures et Oxyfluorures Ioniques, Faculté des Sciences, 72017 Le Mans, France

Josik Portier (309, 553, 565), Laboratoire de Chimie du Solide du CNRS, Université de Bordeaux I, 33405 Talence, France

J. Ravez (469), Laboratoire de Chimie du Solide du CNRS, Université de Bordeaux I, 33405 Talence, France

Jean-Maurice Réau (423), Laboratoire de Chimie du Solide du CNRS, Université de Bordeaux I, 33405 Talence, France

Dirk Reinen (525), Fachbereich Chemie, Universität Marburg, D-3550 Marburg, Federal Republic of Germany

H. Selig (331), Department of Inorganic and Analytical Chemistry, The Hebrew University, 91904 Jerusalem, Israel

Hidekazu Touhara (331), Department of Industrial Chemistry and Division of Molecular Engineering, Kyoto University, Kyoto 606, Japan

Alain Tressaud (77, 371, 415), Laboratoire de Chimie du Solide du CNRS, Université de Bordeaux I, 33405 Talence, France

Jean Jacques Videau (309), Laboratoire de Chimie du Solide du CNRS, Université de Bordeaux I, 33405 Talence, France

Gérard Villeneuve (493), Laboratoire de Chimie du Solide du CNRS, and École Nationale Supérieure de Chimie et de Physique de Bordeaux, 33405 Talence, France

Nobuatsu Watanabe (331), Department of Industrial Chemistry and Division of Molecular Engineering, Kyoto University, Kyoto 606, Japan

Foreword

If solid-state science is to be carried forward in an exciting and effective way, there should be a cooperative interplay of the skills and knowledge of synthesizers, theorists, and those who measure physical properties. In its treatment of solid-state fluorides, this book encourages that broad interdisciplinary approach.

Although many topics, such as synthesis, compositional and structural relationships, defect materials, and magnetic, electronic, and optical properties, are common to most solid-state texts, there is much, even in the coverage of these topics, that is special to fluorine chemistry. Indeed, the characteristic influences of the fluorine ligand are everywhere evident.

The compactness of the fluorine ligand derives from the perfect-octet configuration of neon, the electron configuration of which atom the fluorine ligand always approximates to. Only hydrogen or oxygen compete with fluorine in smallness, and the last can, as is shown, replace those ligands in some circumstances. Substitution of fluorine for hydrogen in biologically active molecules can be valuable in structural work (in which the nuclear magnetic properties of fluorine are exploited). Such substitution can also profoundly influence the role of such molecules in their biological action. Also, it is the smallness and "hardness" of the fluoride ion that contribute importantly to the fast fluoride ion conductivity of some fluorides. Low polarizability of the fluorine ligands also has a major influence on the optical properties of fluorides, and on the low-friction character of fluoride surfaces.

At a more detailed level of appreciation, we perceive that it is the fluorine ligand's π-donor character, and the absence of any π-acceptor character, that determines its low-field (high-spin) character. It is the latter that is crucial to magnetic and optical behavior of transition metal fluorides. Fluorine is almost alone in its stabilization of high-spin configuration in some second- and third-transition-series species. Fluorine can also facilitate superexchange interactions between magnetic centers, much as oxygen can do. Like oxygen it also acts as bridging ligand, and fluorine–ligand bridges are known to exist even in polymeric gaseous species. The bridging habit, as will be appreciated particularly from the crystal chemistry surveys, is tied to coordination number and oxidation state. It is subtly dependent on fluoro-acid and fluoro-base properties. These acid–base properties determine the limits of complex fluoride formation. The variety that those limits permit is vast indeed.

The extraordinary range of fluorine chemistry (only three elements, helium, neon, and argon, are not known to form fluorides) is associated with the great strength of the bonds made by the small, highly electronegative fluorine ligand. The extraordinary thermodynamic stability of fluorides is also due to the weakness of the bond in the F_2 molecule itself. As pointed out in the chapter on Fluorine Chemistry and Energy, it is the high thermodynamic stability of the compounds of this lightweight ligand that makes fluorides so attractive as agents of energy storage. This is a matter of growing importance and opportunity.

It is the same properties that account for the high thermodynamic stability of fluorides, which also account for the exceptional capability fluorine has of exciting oxidation states otherwise unattainable. Such features are provocatively discussed in the survey of High Oxidation States in Fluorine Chemistry. Associated with this (and a consequence of the monovalence of the ligand), is the production of species of unusually high coordination number (such as IF_7 and ReF_7). Moreover, as we observe in the case of the KrF_2 molecule (which is less well bound than the F_2 molecule itself), some of the chemical bonds that the fluorine ligand makes are among the weakest known to us.

This book provides an excellent introduction to the chemical and physical world of fluorides, a world that, we can be sure, will continue to provide challenges and rewards for those who seek to explore its mysteries.

NEIL BARTLETT

Preface

The specificity of fluorine as the most electronegative element of Mendeleev's periodic table leads to a number of uncommon properties, for example, large optical transmission domain, high resistivity, electron acceptor behavior, or anionic conductivity. On the other hand, the proximity of the F^- and O^{2-} anions connects fluorides with oxides, whose features may often be improved by oxygen–fluorine substitution. Fluorine is also a close neighbor of chlorine, allowing a modulation of the physical properties: chlorides for instance have a nearer absorption threshold in the ultraviolet, but a further multiphonon transparency in the infrared. A counterpart is often higher hygroscopicity of the chlorides: they are easier to prepare, but more difficult to handle.

The singularity of fluorine accounts for its prevailing role in novel materials such as graphite intercalation compounds, materials for power lasers or low-loss optical fibers, high-permittivity thin layers, etc. Most of these materials are thoroughly discussed in this book.

Solid inorganic fluorides constitute a challenge for both chemists and physicists, as they often require very specific preparation methods—not to mention the difficulties of spinning fibers—and they raise questions regarding largely accepted physical models. Nevertheless, despite relatively high prices and sometimes difficult elaboration conditions, they will probably lead to many advanced applications in fields as different as electrochemistry, microelectronics, or optical transmission.

We hope that this book constitutes a valuable step in this ambition.

PAUL HAGENMULLER

1

General Trends

PAUL HAGENMULLER

Laboratoire de Chimie du Solide du CNRS
Université de Bordeaux
Talence, France

I. Introduction

Fluorine is a cornerstone of Mendeleev's periodic table. This odd position, which explains why, despite its relative abundance, it was isolated only in 1886 by Moissan, results in *high electronegativity*: in the Allred and Rochow scale its value is 4.10 compared with those of its neighbors, 3.50 for oxygen and 2.83 for chlorine. High electronegativity leads to small polarizability of the fluorine anion: 0.81×10^{-24} cm^3 according to Pauling, compared with 3×10^{-24} cm^3 for O^{2-} and 2.98×10^{-24} cm^3 for Cl$^-$. Its electronic configuration accounts for many other aspects of fluorine chemistry.

Due to the high redox potential of the F_2/F^- systems (2.87 V using a standard hydrogen reference), elemental fluorine can be prepared only electrochemically.

II. Bonding Problems

In the absence of available d orbitals, diatomic gaseous fluorine has a dissociation energy that is much smaller than that of other halogens (37.8 kcal/mol versus 57.1 kcal/mol for Cl_2). The resulting high reactivity, strong electronegativity, and small size of the fluorine anion obtained account for the *existence of fluorides for most elements*, including the heavy rare gases (contrary to the prejudice largely accepted until 1962, certain rare-gas fluorides such as XeF_4 or XeF_6 are even exothermal). Nevertheless, due to the higher electrical charge of O^{2-} and π back bonding enhanced by smaller electronegativity, oxides are generally thermodynamically more stable than fluorides. Otherwise, we would probably live in a world of fluorides in supposing our own body preserved enough from an oxygen–fluorine substitution! Nevertheless, the strength of fluorine bonding is sufficient to explain the stability of widely used materials such as Freon and Teflons.

Its specificity probably easily accounts for the role played by fluorine in molecular chemistry. In *solid-state chemistry*, if many properties result directly from the unique electronic configuration of fluorine, its behavior is not contradiction free: Because of the small size of the fluorine anion, the lattice energy largely exceeds that of homologous halogenides, leading to higher melting points. Nevertheless, due to the weaker covalency of fluorine bonds, but also to appropriate structures, many fluorides are excellent ionic conductors, the F^- sublattice being below melting temperature in a "quasi-liquid" state.

Higher stability than that of similar halogenides combined with weak covalency also accounts for the volatility without decomposition of some high-oxidation-state fluorides (e.g., WF_6 and UF_6). This feature leads to unique industrial applications.

Most fluorides have *large energy gaps*, resulting in insulating properties and often in a white color despite a small covalent contribution to chemical bonding (e.g., MgF_2 has a 12-eV gap). Such wide gaps are indeed the consequence of the high electronegativity of fluorine, which reduces wavelength function admixture.

Due to its position in the periodic table, F^- is expected to be the smallest anion of the $n = 2$ row, in agreement with the high value of the nuclear charge. In solid oxyfluorides the M—O bonds are actually often stronger than the M—F bonds because of π bonding favored by higher charge of oxygen, resulting in smaller M—O distances.

Such complex features emphasize the fact that fluorine is simultaneously an utmost element and an element with properties that often cannot be predicted by simple extrapolation from the behavior of its neighbors in Mendeleev's classification. Hence, in fields such as solid-state physics and chemistry, which are expanding more and more but where predicting structures and properties is a fundamental way of thinking, the introduction of fluorine justifies careful investigation before any systematization is undertaken. Such cautious behavior is all the more justified because solid fluorides have a growing importance in advanced industrial applications due to specific properties.

In this volume a large place is devoted to *preparative chemistry*, both because it is a traditional bottleneck preventing many physicists from using high-purity fluorides as often they would be willing to and because of the large variety of reactants and techniques used. Elaboration of fluorides cannot be separated from such problems as *contamination, corrosion,* and *safety.*

High-pressure fluorination is becoming, for instance, a powerful technique for attaining *very high oxidation states,* sometimes exceeding those obtained for the same cations with oxygen despite smaller lattice energy. The high electroaffinity of fluorine may compensate for the strong ionization potentials of fully occupied d shells such as those of copper, silver, gold, krypton, and xenon. The highest oxidation states are often obtained in ternary compounds, where the bonds with fluorine are strengthened by the highly ionic character of competing bonds (e.g., with cesium or rubidium).

III. Structural Features

A large chapter is devoted to the *intercalation of fluorine in graphite*, which is favored by the availability in graphite of delocalized π electrons but leads to partially unsolved bonding problems. Some of these intercalation compounds are already being used as electrodes in primary batteries, the counterelectrode as a rule being lithium. Intercalation of high-oxidation fluorides, which act as electron acceptors, may lead to reduction reactions such as $AsF_5 + 2\,e^- \rightleftarrows AsF_3 + 2\,F^-$, where the formation of F^- is enhanced by the electrochemical chain (see Chapter 8).

Due to weak layer bonding, fluorine–graphite intercalates are excellent lubricants that compete with layer sulfides or Teflon. They can be used at higher temperatures.

Fluoroglasses prepared either by conventional glass techniques or by splat-cooling are of growing importance owing to increasing industrial demand. Glasses containing a beryllium or aluminum former network, which contain tetrahedral groups analogous to those of silicates or strongly bound octahedra, are excellent host lattices for neodymium power lasers, a weak Nd—F bond competing with rather strong Be—F or Al—F bonds and giving rise to long lifetimes in the excited neodymium levels (this advance is enhanced by low refractive index and high Abbe number). Such vitreous materials are also appropriate for utilization as ultraviolet-transparent windows due to large energy gaps. In contrast, heavy-element fluoride glasses have a dis-ordered structure resulting from the various coordination possibilities of those elements whose bonding has a very strong ionic contribution; they are ideal materials for far-infrared applications, as fibers, for instance, when they can be spun in spite of the small interval between melting and vitreous transition temperatures and low viscosity at melting point. Their theoretical optical attenuation falls to 10^{-3} dB/km compared with 0.2 dB/km for silica glasses. Due to strong electronegativity, fluorides have a low tendency to give rise to Schottky defects, which probably require too much energy, but may easily give rise in appropriate lattices to Frenkel defects (e.g., in a fluorite-type structure where 50% of the cationic sites are theoretically empty). High electronegativity also accounts for an *absence of measurable homogeneity ranges in binary fluorides* except in some very peculiar structures such as AgF_2. This property clearly distinguishes the fluorides from the homologous oxides, the existence domains of which are often very wide. Ternary fluorides may show large homogeneity domains due to a Frenkel defect type of insertion. On the basis of x-ray diffraction data, they seem to form solid solutions in which extra fluorine anions are randomly distributed in the available sites, but vacancies and intercalated anions actually tend to group in *short-range clusters* similar to those of isostructural oxides. These clusters gradually lead to long-range ordering with increasing concentration, con-stituting new crystallographic lattices.

From the standpoint of *crystal chemistry*, fluorides can be classified like the oxides in terms of anionic packing and occupation of the interstitial positions, but as a result of smaller covalency the coordination number of the cations is higher (e.g., 6 in Fe^{3+} fluorides rather than 4 or 6 for the corresponding oxides; 9 or 11 in binary trivalent rare-earth fluorides rather than 6 or 7). Oxide–fluorides sometimes have a strongly distorted surrounding due to the difference in behavior between oxygen and fluorine (e.g., InOF; Vlasse *et al.*, 1973). Indeed, covalent binary groups such as VO_2^+ or UO_2^{2+} do not exist in pure fluorides owing to the essentially ionic character of the bonds in the fluoride lattices. However, ordering of MF_6 octahedra into cis- or trans-connected chains or into layers can lead to one- or two-dimensional-type properties analogous to those of the oxides. The bonds are often strong enough to give rise to homologous 3D frameworks as well. A significant example is given by the tungsten bronze structures that were reproduced, for instance, for the A_xFeF_3 and later the A_xVF_3 series (A = alkali element) (de Pape, 1965; Tressaud *et al.*, 1970, 1972; Cros *et al.*, 1975). Due to weakening of the bonds in the frameworks, they very easily adapt to the size and concentration of the inserted A^+ cations, as shown in Fig. 1.

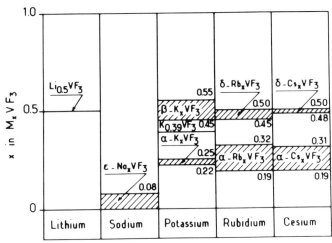

Fig. 1. Various $A_x VF_3$ phases as a function of x and size of the A^+ cation. $Li_{0.5}VF_3$: trirutile type; ϵ: VF_3 type; α: hexagonal tungsten bronze; β: tetragonal tungsten bronze; δ: pyrochlore.

IV. Influence of Oxygen–Fluorine Substitution

A very common feature of fluorine in solid-state chemistry is *easy substitution of a large amount of oxygen by fluorine in oxides* due to the similar size of the O^{2-} and F^- anions. This substitution may occur in structures as different as rutile, perovskite, pyrochlore, garnet, or fluorite. It has to be compensated for either by reduction of the cationic charge (e.g., in $Fe_3O_{4-x}F_x$) or by appropriate cationic replacements (e.g., in $Pb_{2-x}Na_xNb_2O_{7-x}F_x$).

The physical properties are often strongly modified by such an oxygen–fluorine substitution. A good example is the marked decrease in the Curie temperature T_C in ferroelectric oxides due to the smaller covalency of the fluorine bonds, which lowers the thermal energy required for breaking the ferroelectric distortion in the lattice. Table I illustrates the fact that oxygen–fluorine substitution is much more efficient for decreasing the Curie point than more classical replacement of niobium by tantalum, titanium, or tungsten (Ravez and Dabadie, 1973; Ravez *et al.*, 1976).

Oxygen–fluorine substitution in ferroelectric oxides not only diminishes the Curie temperature but also, probably as a result of the formation of composition domains, gives rise in a large temperature range to materials of high permittivity that are practically temperature and frequency independent.

TABLE I

Influence of Some Cationic and Anionic Substitutions on the Ferroelectric Curie Temperature T_C of Tetragonal Tungsten Bronze-Type $Sr_2KNb_5O_{15}$

Phase	T_C (K)	Substitution Reaction	ΔT_C (K)
$Sr_2KNb_5O_{15}$ ↓ $SrK_2Nb_4WO_{15}$	437 ↓ 338	$Sr^{2+} + Nb^{5+} \longrightarrow K^+ + W^{6+}$	-99
$Sr_2KNb_5O_{15}$ ↓ $Sr_3Nb_4TiO_{15}$	437 ↓ 403	$K^+ + Nb^{5+} \longrightarrow Sr^{2+} + Ti^{4+}$	-34
$Sr_2KNb_5O_{15}$ ↓ $SrK_2Nb_5O_{14}F$	437 ↓ 95	$Sr^{2+} + O^{2-} \longrightarrow K^+ + F^-$	-342
$Sr_2KNb_5O_{15}$ ↓ $Sr_2KNb_4TiO_{14}F$	437 ↓ 80	$Nb^{5+} + O^{2-} \longrightarrow Ti^{4+} + F^-$	-357

By selection of the oxide and choice of the substitution rate, one can adjust the temperature region to fit the needs (Campet *et al.*, 1974; 1979), for example, down to ambient temperature.

Substitution of fluorine for oxygen in ferrimagnetic oxides decreases the magnetic Curie temperature due to weaker magnetic couplings resulting from the decrease of anion–cation orbital mixing in fluorine bonds. Because anionic substitution is, of course, coupled to appropriate cationic substitution, the evolution of the magnetic properties can provide information on site occupation (e.g., octahedral Ni^{2+} sites in $Gd_3Fe_{5-x}Ni_xO_{12-x}F_x$ or tetrahedral Zn^{2+} sites in $Gd_3Fe_{5-x}Zn_xO_{12-x}F_x$ garnets) (Claverie *et al.*, 1972; Tanguy *et al.*, 1979).

Replacement of oxygen by fluorine also has a strong influence on metal–nonmetal transitions in oxides either in increasing the electronic localization ($Na_xWO_{3-y}F_y$; Doumerc, 1974) or in lowering the conduction band ($VO_{2-x}F_x$; Bayard, 1974; Bayard *et al.*, 1975).

Figure 2 shows, for example, how the anionic substitution in VO_2 decreases the temperature at which the monoclinic distortion due to d–d bonding disappears. Figure 3 illustrates the influence on the thermal behavior of conductivity of the appearance of a V^{3+}–V^{4+} hopping mechanism in $VO_{2-x}F_x$ and of the gap narrowing. It also shows that fluorine, in large amounts, plays the role of electron trapper in the rutile phase. The influence of fluorine on the behavior of VO_2 has been, in fact, much less thoroughly studied than cationic substitutions ($V_{1-x}Cr_xO_2$, $V_{1-x}Nb_xO_2$) owing to the bad reputation of fluorine as an odd element both in elaboration processes and in theoretical physical models.

Curiously enough, reciprocally oxygen can replace only very small amounts

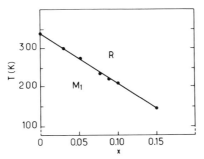

Fig. 2. Variation with composition of the metal–nonmetal transition temperature of $VO_{2-x}F_x$ (R is the rutile phase, metallic for low x values but becoming gradually semiconducting for high x values; M_1 is the monoclinic distorted rutile phase, which is semiconducting throughout the whole existence range).

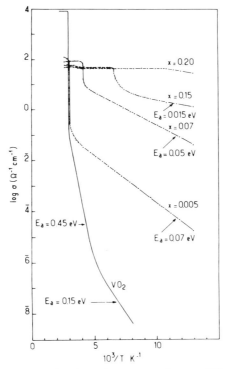

Fig. 3. Variation of log σ with reciprocal temperature for some $VO_{2-x}F_x$ single crystals.

of fluorine in fluorides. Such substitutions, however, may represent a harmful contamination of fluoride preparations when traces of water vapor are present: despite low concentration, they strongly influence certain physical properties in creating anionic vacancies—for example, the intensity of optical emission spectra, which is very sensitive to impurities ($2 F^- = O^{2-} + \square$).

In contrast, large existence domains with fluorine–hydroxyl substitutions have been detected for rare-earth $Ln(OH)_{3-x}F_x$ compounds prepared by hydrofluoric synthesis under pressure. The nephelauxetic effect on the optical spectra clearly illustrates the higher covalency of the OH bonds, which leads to shifting toward longer wavelengths (Marbeuf et al., 1971).

Substitution of oxygen by fluorine in bones, which often has a pathological aspect, is discussed in detail in Chapter 19.

Attempts have also been made to substitute fluorine for nitrogen. New materials have been obtained, but their number is limited by the overly large covalency difference between M—N and M—F bonds, and they tend to lead to ordered structures (Juza et al., 1966; Pezat et al., 1976).

V. Magnetic Properties

The magnetic properties of the fluorides are highly analogous to those of most oxides, but in fluorides strong electronic localization excludes phenomena such as narrow-band or itinerant electron magnetism. As a consequence of lower covalency the magnetic interactions are weaker. At first approximation they often concern only the nearest-neighbor magnetic cations because couplings with the next-nearest neighbors may be neglected. This is why Néel's molecular field theory is more applicable to the fluorides than to the oxides, though they constituted the basis of its first successes; in many transition element series [e.g., the iron(III) fluorides] the ordering temperature is thus roughly proportional to the number of nearest cations (Tressaud and Portier, 1971).

In fluorides, however, the number of single electrons is often higher than in oxides, which may raise the intensity of the couplings. *High-spin configurations* are much more common for fluorides than for oxides due to smaller d-orbital splitting, at least when the temperature is sufficiently high. Let us quote the unusual high-spin configurations in octahedral sites of Co^{3+} $(t_{2g}^4 e_g^2)$ or Ni^{3+} $(t_{2g}^5 e_g^2)$, without mentioning the high-spin configurations found for $4d$ and $5d$ cations and inaccessible in other materials.

There are probably relatively more families of ferrimagnetic fluorides than oxides despite the absence of ferrimagnetic fluoride spinels or garnets (they are dia- or paramagnetic) (Dance et al., 1977). Some of them were discovered because of their structural analogy with diamagnetic oxides containing different cationic sublattices (chiolites and weberites) (McKinzie et al., 1972; Cosier et al., 1970).

The possibilities offered by the site diversity of hexagonal perovskites can be extended to new families in the case of fluorine compounds. Figure 4 shows an example for $Cs_{1-x}Rb_xMF_3$ compounds, the structures of which can be modified according to the nature of the M^{2+} ion, the x substitution rate, and eventually the pressure used (Goodenough et al., 1972; Dance et al., 1979). Some isolated clusters of two or three face-sharing magnetic octahedra, obtained by a relevant choice of the concerned cations, allow worthwhile theoretical studies to be performed on magnetic interactions involving a small number of isolated atoms grouped in single geometric units. These studies can be extended to *spin frustration problems* in which fluorides offer particularly relevant examples of networks of triangular platelets.

There are few examples of ferromagnetic fluorides. The concerned rare-earth fluorides have strong anisotropic magnetizations, leading to significant theoretical work, but the Curie points are very low. Ferrodistortive ordering of Jahn–Teller ions in layer compounds is often the origin of ferromagnetic behavior. The ferromagnetic Pd_2F_6-type compounds are a good illustration of the Goodenough–Kanamori rules, as only a limited number of ions can be involved here in ferromagnetic behavior ($Ni^{2+}Mn^{4+}F_6$ is a particularly good example because it contains a $t_{2g}^6 e_g^2$ and a $t_{2g}^3 e_g^0$ cation).

Unfortunately, the hope that spontaneous magnetization in fluorides could be used in magnetic bubble or magneto-optical applications owing to transparency in the visible region has so far hardly been fulfilled (Wolfe et al., 1970). The highest Curie temperatures for ferrimagnetic fluorides do not exceed 150 K. The only exception with unstable CoF_3 is FeF_3. The spontaneous magnetization that holds up to 363 K is due to simple spin canting of antiferromagnetic interactions but leads to strong Faraday rotation at room temperature and involves strong anisotropy. Unfortunately, large

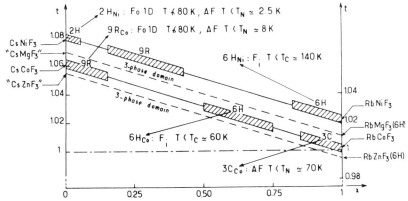

Fig. 4. Some $Cs_{1-x}Rb_xMF_3$ systems at 800°C and under ordinary pressure (M = Mg, Co, Ni, or Zn; AF, antiferromagnetic).

birefringence effects and the existence of a low temperature transition (690 K), which make it difficult either to grow FeF_3 crystals even of small size or to depose thin layers, do not allow competition with more classical materials such as ytlrium iron garnet (Bertaut, 1963). Applications of fluorides as ferrofluid particles are under consideration owing to their good resistance to corrosion, but they require higher Curie temperatures.

VI. Electric Properties

We already have stressed the *electronic insulating character* of most fluorides due to the strong electronegativity of fluorine and large energy gap. Exceptions are the intercalation compounds in graphite and layer structure materials such as Ag_2F, the metallic character of which allows even the insertion of organic acceptors (Koshkin *et al.*, 1976). Metallic chain compounds such as Hg_3AsF_6 or Hg_3SbF_6 may also be mentioned as odd exceptions due to structural singularities (Cutforth *et al.*, 1977).

Chapter 13 is devoted to *nonlinear properties.* They are obviously less common in fluorides than in oxides due to the weaker covalency of the former, which does not favor noncentrosymmetric structures. However, there are worthwhile exceptions, such as ferroelectric $SrAlF_5$, which fits very well with the rules proposed by Abrahams relating ferroelectric Curie temperature, spontaneous polarization, and out-of-symmetry-center shift at low temperature in oxides (Abrahams *et al.*, 1981). Use at moderate temperatures is often recommended because of high dielectric losses related to easy diffusion of F^- (Claverie *et al.*, 1974).

In contrast, there are a relatively large number of oxyfluoride families of nonlinear materials; they are often characterized by ferroelectric–ferroelastic couplings (e.g., chiolites and $Rb_3MoO_3F_3$-type materials) (Ravez *et al.*, 1979, 1980; Simon and Ravez, 1980). Owing to a very large field of application, such materials will probably undergo extensive development in the future—for instance, for signal conversion, as transmission devices, or in electro-optic utilizations (Ravez and Micheron, 1979).

Since the early 1970s research on *fluorine ion solid electrolytes* has developed to a great extent partially because of potential industrial valorization, but also as a result of the existence of many fluorite-type fluorides that are electronic insulators, contain available vacancies, and hence allow large-scale comparative investigations to be undertaken. Analogous oxides deriving from zirconia were already known and even utilized as oxygen sensors in the industry, but their conductivity is much lower because of the stronger bonds in oxides, so that they are usable only at relatively high temperatures. It was found that the mobility of F^- is favored by the high polarizability of the cations, especially when they have a lone pair, and by the disordered distribution of fluorine in ternary fluorides. Large size difference between the cations is also a favorable factor. The existence of defect clustering and the easy diffusion of fluorine, which below melting point is in a quasi-liquid state as shown by high heat capacity, lead to complex structural investigations requiring simultaneous use of appropriate models and careful experimental work (Chapters 5 and 12). Figure 5 shows the position of these materials between high-performance conductors such as α-AgI or β''-alumina, on one hand, and poor electrolytes such as stabilized zirconia, on the other.

Some of these materials are used for measuring F^- concentrations in aqueous solutions (mostly using an Eu^{2+}-doped LaF_3 membrane); others have been proposed as elements of "all-solid" F^- sensor batteries for determining traces of reducing gases or HF pressures (Fig. 6). Investigations are in progress to integrate F^- solid electrolytes into electrochemical chains using intercalation graphite compounds, but the difficulty to surmount is to avoid the formation of a stable insolating layer at the negative electrode (LiF, for instance, when lithium is selected). Fluorine fast ion conductors have also

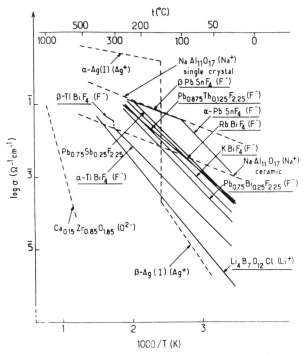

Fig. 5. Variation of the logarithm of the conductivity versus reciprocal temperature of some fluorite-type or derived materials.

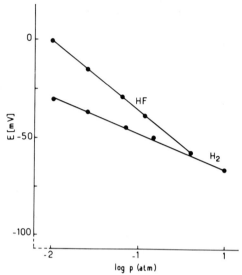

Fig. 6. Variation of the voltage measured with a fluorine Ag|PbSnF$_4$|Pt battery as a function of p_{H_2} or p_{HF}.

been used on an experimental scale for microelectronic devices requiring high reliability (Schoonman and Bottelberghs, 1978). Some other fluorides have found an application in electrochromic display devices, but they are actually based on lithium ion conductivity rather than on that of fluorine ions (Chapter 15). Materials that are simultaneously Li^+ and F^- conductors probably have a future in electrochemistry due to high voltage and mass capacity, but so far they remain a challenge to solid-state chemists.

VII. Spectroscopic Aspects

The odd *optical properties* of the fluorides described in Chapter 14 result largely from the specific features of fluorine: high electronegativity, small polarizability, and weak covalency of the M—F bonds. They explain, for instance, low refractive index, wide transmission domain of the absorption spectra, and shift of the $4f$ levels to lower wavelengths. The charge-transfer spectra are in the far ultraviolet owing to small orbital admixture in valence and conduction bands. However, relatively low melting points, making crystal growth easier than in the case of oxides, and strong chemical stability also have to be considered, for instance, in power laser applications.

Because of the weak crystal field and small covalent mixing, the decay times resulting from *d–d* transitions may be very long, resulting in appropriate applications (e.g., radar screens). In a similar way the $4f-4f$ transitions show weak concentration quenching even at high temperature, especially when the fluorine coordination octahedra are isolated from each other.

For divalent rare-earth ions such as Eu^{2+} and Sm^{2+}, the high energy of the $5d$ band allows strong monochromatic emissions due to internal transitions within the $4f$ configuration. This phenomenon is all the more pronounced as the rare-earth–fluorine distance is larger (e.g., due to substitution for Ba^{2+}), the coordination number of the rare earth is higher, and the competing bonds are stronger (Fouassier et al., 1976) (Fig. 7). The intense monochromatic emission of Eu^{2+} may induce strong emission in the visible region of other rare-earth ions by energy transfer due to overlapping of the Eu^{2+} emission spectrum and corresponding absorption spectra of the second rare-earth ion (e.g., in BaY_2F_8: Eu^{2+} with Er^{3+}, Ho^{3+}, or Tb^{3+}).

In contrast, weak phonon energies, with respect to oxides, reduce the nonradiative decay rates. Such a feature as a counterpart favors radiative emission and eventual Auzel-type up-conversion by cross-relaxation (e.g., from infrared to visible; Fig. 8).

For the fluorides with lone-pair cations (Tl^+, Pb^{2+}, etc.) the $s \rightarrow p$ absorption bands are shifted to high energies with respect to oxides or other halogenides as a result of the small covalency.

The optical properties of the fluorides are used largely for energy conversion, signal transmission, display devices, information storage, or other sophisticated applications such as tunable lasers.

A large place is given in this volume to studies of the fluorides by various *resonance techniques* owing to the precise information they provide concerning the nature of the surrounding, coordination number, bonding, electronic transfer, etc. Whereas epr and Mössbauer spectroscopy characterize primarily the cation and can be considered only as sound methods able to be used with other materials, nmr is specific for ^{19}F. The intense signal obtained and the absence of quadrupole effects due to an $I = \frac{1}{2}$ nuclear spin make this technique quite powerful for solving all short-range problems arising in fluorides. Determination of nmr can be used as well for solving static problems

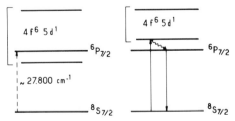

Fig. 7. Representation of both Eu^{2+} emission possibilities and their dependence on bond strength with the fluorine surrounding.

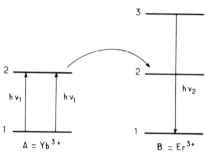

Fig. 8. Conversion of infrared to visible light.

(involving, e.g., fine structures of glasses or orbital admixtures with fluorine) as for studying dynamic behaviors such as diffusion mechanisms of F^- in fast ion conductors. Second moment analysis can often allow one to distinguish the oxygen and fluorine positions in oxide–fluorides.

VIII. Economic Implications of Fluorides

At a time when advanced science is becoming more and more related to applications, it is important to consider the *industrial aspects* of fluorine solid-state chemistry. These are discussed in the chapters devoted to the specialized aspects of fluorides or, in Chapters 20 and 21, from the viewpoint of the user.

The role of fluorides in *energy conversion* or *storage* is emphasized in a broad review devoted to application fields as various as efficient utilization of solar energy, thermal storage involving fluorides of high melting enthalpy, electrochemical facilities. Nd^{3+} high-peak power lasers and isotopic separation of uranium by gaseous diffusion of UF_6.

The specific role of materials such as graphite fluorides, C_2F for instance, as high-performance lubricants and antifriction agents due to low surface energies has been developed in several chapters.

Emphasis is also given to the sensitivity of the electronic conductivity of $M^{II}M^{IV}F_6$ fluorides such as Pd_2F_6 for piezoresistive applications. The logarithm of the resistivity at room temperature varies linearly with pressure over a large domain (Langlais *et al.*, 1979).

Passivation of metallic surfaces by fluorophosphates is another field of interest (Cot *et al.*, 1976).

The industrial applications considered include relatively classical problems such as electrolytic elaboration of aluminum metal, fluorine elaboration, and production of gases such as HF or BF_3; the processes used and the difficulties encountered are described. The applications under consideration also involve more advanced utilizations of fluorides—for example, those required for defense necessities or the growing integration of electronic devices into daily life (e.g., use as insulating films on III–V semiconductors). For obvious reasons, one being the rapid development of applications of this type, they have not always been reported in great detail, the risk of rapid absolescence being rather high.

The first task of this volume has been to fill a vacuum in a developing field where general information is needed to accelerate scientific progress. The nature of the field involves not only chemists and physicists but biologists and specialists of mechanical properties as well: pluridisciplinary work is becoming a daily necessity in all advanced fields of materials science. Solid-state fluorides, however, is probably one of the areas where scientists interested in chemical and physical properties are confronted with the greatest difficulties in elaboration and characterization, as well as in safety requirements and where chemists mastering new preparation techniques feel the strongest call for new openings and breakthroughs.

References

Abrahams, S., Ravez, J., Simon, A., and Chaminade, J. P. (1981). *J. Appl. Phys.* **52**, 4740.

Bayard, M. (1974). Thèse de Doctorat-ès-Sciences, Université de Bordeaux I.

Bayard, M., Pouchard, M., Hagenmuller, P., and Wold, A. (1975). *J. Solid State Chem.* **12**, 41.

Bertaut, E. F. (1963). *In* "Magnetism" (G. T. Rado and H. Suhl, eds.), Vol. 3. Academic Press, New York.

Campet, G., Claverie, J., Perigord, M., Ravez, J., Portier, J., and Hagenmuller, P. (1974). *Mater. Res. Bull.* **9**, 1589.

Campet, G., Claverie, J., Hagenmuller, P., and Périgord, M. (1979). *Rev. Phys. Appl.* **14**, 415.

Claverie, J., Portier, J., and Hagenmuller, P. (1972). *Z. Anorg. Allg. Chem.* **393**, 314.

Claverie, J., Campet, G., Périgord, M., Portier, J., and Ravez, J. (1974). *Mater. Res. Bull.* **9**, 585.

Cosier, R., Wise, A., Tressaud, A., Grannec, J., Olazcuaga, R., and Hagenmuller, P. (1970). *C. R. Hebd. Seances Acad. Sci.* **271**, 142.

Cot, L. (1978). U.S. Patent 789,658.

Cros, C., Feurer, R., Pouchard, M., and Hagenmuller, P. (1975). *Mater. Res. Bull.* **10**, 383.

Cutforth, B. D., Datars, W. R., Van Schyndel, A., and Gillespie, R. J. (1977). *Solid State Commun.* **21**, 377.

Dance, J. M., Tressaud, A., Portier, J., and Hagenmuller, P. (1977). *Rev. Roum. Chim.* **22** (4), 587.

Dance, J. M., Kerkouri, N., and Tressaud, A. (1979). *Mater. Res. Bull.* **14**, 869.

de Pape, R. (1965). *C. R. Hebd. Seances Acad. Sci.* **260**, 4527.

Doumerc, J. P. (1974). Thèse de Doctorat-ès-Sciences, Université de Bordeaux I.

Fouassier, C., Latourrette, B., Portier, J., and Hagenmuller, P. (1976). *Mater. Res. Bull.* **11**, 953.

Goodenough, J. B., Longo, J. M., and Kafalas, J. A. (1972). *In* "Preparative Methods in Solid State Chemistry" (P. Hagenmuller, ed.), p. 2. Academic Press, New York.

Juza, R., Sievers, R., and Juny, W. (1966). *Naturwissenschaften* **53**, 551.

Koshkin, V. M., Yakubshii, E. B., Milorer, A. P., and Zabrodskii, Y. B. (1976). *JETP Lett.* (*Engl. Transl.*) **24**, 110.

Langlais, F., Demazeau, G., Portier, J., Tressaud, A., and Hagenmuller, P. (1979). *Solid State Commun.* **29**, 473.

McKinzie, H., Dance, J. M., Tressaud, A., and Portier, J. (1972). *Mater. Res. Bull.* **7**, 673.

Marbeuf, A., Demazeau, G., Turrell, S., and Hagenmuller, P. (1971). *Solid State Chem.* **3**, 677.

Pezat, M., Tanguy, B., Vlasse, M., Portier, J., and Hagenmuller, P. (1976). *J. Solid State Chem.* **18**, 381.

Ravez, J., and Dabadie, M. (1973). *Rev. Chim. Miner.* **10**, 765.

Ravez, J., and Micheron, F. (1979). *Actual. Chim.* **9**, 9.

Ravez, J., Perron-Simon, A., and Hagenmuller, P. (1976). *Ann. Chim.* (*Peris*) **1**, 251.

Ravez, J., Elaatmani, M., von der Mühll, R., and Hagenmuller, P. (1979). *Mater. Res. Bull.* **14**, 1083.

Ravez, J., Perandeau, G., Arend, H., Abrahams, S. C., and Hagenmuller, P. (1980). *Ferroelectrics* **26**, 767.

Schoonman, J., and Bottelberghs, P. H. (1978). *In* "Solid Electrolytes" (P. Hagenmuller and W. van Gool, eds.), p. 335. Academic Press, New York.

Simon, A., and Ravez, J. (1980). *Ferroelectrics* **24**, 305.

Tanguy, B., Portier, J., Morell, A., Olazcuaga, R., Francillon, M., Pauthenet, R., and Hagenmuller, P. (1979). *Mater. Res. Bull.* **7**, 1339.

Tressaud, A., and Portier, J. (1971). "Appareillages et Techniques de Caractérisation des Composés Minéraux Solides." Masson, Paris.

Tressaud, A., de Pape, R., Portier, J., and Hagenmuller, P. (1970). *Bull. Soc. Chim. Fr.* **10**, 3411.

Tressaud, A., Ménil, F., Georges, R., Portier, J., and Hagenmuller, P. (1972). *Mater. Res. Bull.* **7**, 1339.

Vlasse, M., Massies, J. C., and Chamberland, B. (1973). *Acta Crystallogr., Sect. B* **B29**, 627.

Wolfe, R., Kurtzig, A. J., and Le Craw, R. C. (1970). *J. Appl. Phys.* **41**, 1218.

2

Preparative Methods

JEAN GRANNEC and LUCIEN LOZANO
Laboratoire de Chimie du Solide du CNRS
Université de Bordeaux
Talence, France

INORGANIC
SOLID FLUORIDES

I. Introduction

As emphasized in the preceding chapter, one of the main characteristics of fluorine is its very high electronegativity. This property, connected with the low dissociation energy of the molecule and the relatively large bond energies in many fluorine compounds, explains why the chemistry of fluorine differs appreciably from that of other halogens.

In recent years an increasing number of physical properties of solid inorganic fluorides have been studied. Indeed, the electronic localization that characterizes the fluorides allows better insight into solid-state research, but generally only so far as single crystals can be grown. Furthermore, although applications of solid fluorides have remained relatively restricted, many future developments can be predicted in certain fields (e.g., related to optical, electrical, and energetic properties) based on such specific features as low refractive index, high ionic conductivity, and low boiling or melting points. This new interest in fluorides therefore justifies acquiring better knowledge and increasing the availability of appropriate synthetic techniques. Indeed, the specific nature of the preparative methods must not be neglected. Because of their strong oxidizing power, fluorine and some highly reactive derivatives react spontaneously with many compounds and give very exothermic reactions.

The objective of this chapter is not to review all the techniques of fluoride synthesis that have been reported; some preparative volumes on fluorine chemistry have already been written in this field and are recommended (Simons, 1950, 1954, 1964; Sharpe, 1960; Peacock, 1960; Bartlett, 1965). Furthermore, we shall attempt to describe the various types of fluorination procedures rather than the synthesis of individual fluorides. We shall deal, of course, essentially with inorganic compounds that are solid at room temperature.

We would like to point out that the selection of a procedure may be strongly influenced by the equipment available; in addition, the choice of a material often depends on the working conditions. Indeed, temperature, pressure, concentration of fluorine or derivatives, purity, physical state (gaseous or liquid), dynamic or static use, and contact time are factors that can change the behavior of a given material.

Most experiments require the use of a metallic high-vacuum line; such a tool was promoted by Bartlett (1965) and later developed in many other fields. An example of a fluorination line is given in Fig. 1 (Tressaud *et al.*, 1976a); this system was adapted from the original one. The use of separate pumps is recommended: a mechanical pump for destroying waste fluorine (or derivatives) through a soda lime tower, a large mechanical pump for rough pumping, and in some cases an oil diffusion pump for high-vacuum capacity. Working pressures can be measured with different kinds of gauges, depending on the pressure range.

The different parts of the line must be easy to dismantle. All newly assembled parts must be carefully tested for leaks and retested when connections are broken. A passivation of the whole system (line and apparatus) with elemental fluorine is imperative after each operation. The small percentage of hydrogen fluoride contained in fluorine is removed by passing the gas through sodium fluoride pellets or through a coil immersed in liquid nitrogen.

At moderate temperatures, reactions can be carried out in quartz and even in Pyrex glass vessels. Proper drying by thorough flaming is then necessary to avoid the formation of HF. In the same way reaction vessels and small systems built up with Teflon (polytetrafluorethylene), Kel'f (monochlorotrifluoro-ethylene, and FEP (fluorinated ethylene–propylene copolymer) are suitable for static use. Most of these materials are attractive because they allow one to follow the reactions visually, but they are not recommended for dynamic applications or for pressure working.

On the other hand, metallic equipment is necessary for work at relatively high temperatures. The most convenient metals and alloys employed are aluminum, copper, stainless steel, nickel, brass, Inconel, and Monel. High-

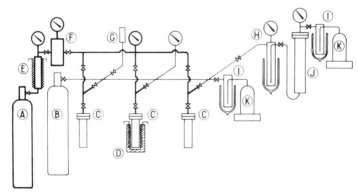

Fig. 1. Fluorination line: A, fluorine tank; B, nitrogen tank; C, reactors; D, standardized resistance; E, trap for HF; F, fluorine "reserve"; G, vacuum gauge; H, pressure-reducer container; I, liquid N_2 traps; J, fluorine destruction tower; K, pumps.

TABLE I

Main Materials Used for Handling Fluorine[a]

Material	Highest Working Temperature (°C)	Gaseous Fluorine Service	Liquid Fluorine Service
Nickel	650	Lines, fittings, flanges, storage tanks	Lines, fittings, flanges, storage tanks
Monel	550	Lines, fittings, flanges, valves, storage tanks, valve bellows	Lines, fittings, flanges, valves, storage tanks, valve bellows
Copper	400	Lines, fittings, flanges, valve seats, gaskets	Lines, fittings, flanges, valve seats, gaskets
Brass	200, low pressure	Lines, fittings, flanges, valve bodies	—
Stainless steel	200	Lines, fittings, flanges, storage tanks, valve bodies	Lines, fittings, flanges, storage tanks, valve bodies
Aluminum	400	Lines, fittings (Al 2017, 2024), gaskets (Al 1100)	Lines, fittings (Al 2017, 2024), gaskets (Al 1100)
Teflon	80	Valve packing (low static pressure), gaskets	Valve packing (low static pressure), gaskets

[a] See Gakle (1960).

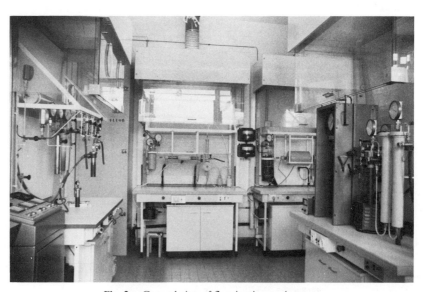

Fig. 2. General view of fluorinating equipment.

pressure/high-temperature reactions must be carried out in Monel or nickel bombs.

The general subject of handling elemental fluorine and reactive fluorides has been discussed by several reseachers. The interested reader is directed for more details to the experimental sections of corresponding articles (Cady, 1950; Hyman and Katz, 1965; Kirk and Othmer, 1966; Canterford and O'Donnell, 1968). A list of commonly used materials is given in Table I.

Finally, we must remind the reader that most fluorides are moisture sensitive, some of them being very corrosive. Hence, their manipulation requires the rigorous maintenance of a dry environment; the use of a very proof dry box is necessary. Some special devices for identification [e.g., x-ray powder and infrared (are) spectra] are also essential. All experiments must be carried out under perfectly ventilated hoods. Figure 2 shows a general view of various types of equipment that can be used in a laboratory.

II. Gas-Phase Reactions

In this section, we present some details of gas–gas interactions and describe methods requiring very low temperature work. In the latter conditions, fluorides, which are gaseous at room temperature, sometimes react in the liquid or even in the solid state.

The remarkable oxidizing properties of platinum hexafluoride, discovered by Bartlett and Lohmann (1962a,b), led to rare-gas chemistry. The first xenon compound ($XePtF_6$) was synthesized by Bartlett (1962) by spontaneous interaction of xenon and PtF_6 vapor at room temperature. Soon after, binary xenon fluorides were prepared by heating mixtures of xenon and fluorine in nickel or Monel vessels. High fluorine gas pressures and lower temperatures favor XeF_6 formation; an excess of xenon over fluorine gives a good yield of XeF_2 (Claassen et al., 1962; Hoppe et al., 1962; Weaver et al., 1963; Bartlett and Jha, 1963; Malm et al., 1963; Dudley et al., 1963; Falconer and Sunder, 1967). The latter can also easily be obtained in Pyrex glass vessels by exposure to sunlight of xenon/fluorine mixtures at room temperature (Streng and Streng, 1965; Holloway, 1966). Several other xenon fluorides have been isolated; one can mention, for example, $XeF_5^+PtF_6^-$ characterized after reaction between xenon and PtF_5 vapor with fluorine gas under 6 bars pressure at 180 to 220°C (Bartlett et al., 1966b).

It was found later that, like PtF_6, other fluorides were capable of oxidizing molecular oxygen. Shamir and Binenboym (1968) described a photochemical synthesis of O_2AsF_6 in a Pyrex glass system; the O_2^+ was also stabilized by GeF_4 (Christe et al., 1976a). Furthermore, the hexafluorides of the third transition series oxidize NO to yield nitrosonium salts $NO^+MF_6^-$. In the reaction with nitrosyl fluoride ONF, ReF_6 readily forms $(NO)_2ReF_8$ at $\sim 20°C$. From a comparative study, the electron affinity was seen to increase regularly from WF_6 to PtF_6 (Bartlett et al., 1966a; Bartlett, 1968). In the second transition series, MoF_6 was shown to react instantaneously with NO by mixing the two gases at temperatures between 30 and 60°C either in a glass apparatus or in a metallic container (Gleichmann et al., 1962).

We may also mention the interaction of nonmetal fluorides with strong F^- acceptors, such as BF_3 or metal pentafluorides: $SF_3^+BF_4^-$ and $SF_3^+AsF_6^-$ were prepared by co-condensation of SF_4 in excess on BF_3 or AsF_5 in a Monel can fitted with a Teflon-gasketed lid (Bartlett and Robinson, 1956; Gibler et al., 1972).

Many studies have been carried out on highly energetic NF_4^+ salts because of their potential application to high-detonation-pressure explosives (Christe, 1980). Low-temperature glow discharge was used for the synthesis of $NF_4^+AsF_6^-$ from NF_3, AsF_5, and fluorine (Christe et al., 1966). A high yield of NF_4BF_4 can also be prepared; this involves ultraviolet photolysis at $-196°C$, using BF_3 instead of AsF_5. The compound NF_4PF_6 was obtained by an exchange reaction from a mixture of NF_4BF_4 and PF_5 (Christe et al., 1976b). The thermochemistry of these NF_4^+ salts has been studied (Bougon et al., 1982).

Finally, the preparation of KrF_2 must be pointed out, although it takes place from a liquefied mixture. This compound was first characterized by Turner and Pimentel (1963). It can be synthesized in several ways. All methods make use of krypton/fluorine mixtures held at liquid-nitrogen temperature. They involve, for instance, irradiation with γ rays or uv light (Slivnik et al., 1975). A procedure using the electrical discharge of the gaseous mixture also gives excellent results (Schreiner et al., 1965). This fluoride is thermodynamically unstable and is therefore potentially an extremely strong oxidation agent. It must be stored under a positive pressure of high-purity argon in Dry Ice conditions.

III. Reactions in Solution

A.   REACTIONS IN HYDROFLUORIC ACID SOLUTIONS

1.   Atmospheric Pressure

Atmospheric-pressure synthesis in hydrofluoric acid solutions is the most simple method because it does not require elaborate devices. Binary or ternary fluorides can be prepared in a platinum crucible or a Teflon vessel by reacting solids (or solid and liquid) with an aqueous solution of HF. Nevertheless, the compounds isolated in these conditions sometimes contain a small amount of oxygen or hydrogen. Because F^-, OH^-, and O^{2-} have very similar ionic radii, they are easily interchangeable in a crystal lattice. An example of a mistake that can be made because of such substitutions is that of rubidium iron tetrafluoride. This compound, prepared in aqueous acid solution, was first claimed to crystallize with a tetragonal unit cell. Tressaud *et al.* (1969) later proved that after careful preparation it was orthorhombic and that the compound prepared in solution was, in fact, $RbFeF_{4-x}(OH)_x$.

The method allows one to obtain hydrated fluorides, as in the following reaction (Tedenac *et al.*, 1969):

$$Be(OH)_2 + 4\,HF + MO \longrightarrow MBeF_4 \cdot xH_2O$$

$$(x = 5 \text{ if } M = Cu, x = 6 \text{ if } M = Ni \text{ or } Zn)$$

Many anhydrous compounds, however, can be isolated from HF solutions. Various starting materials are used for such syntheses, for instance:

1. Mixture of NaF and hydrated indium fluoride for Na_3InF_6 (Ensslin and Dreyer, 1942)
2. Mixture of alkali metal fluoride and TiO_2 for M_2TiF_6 (Cox and Sharpe, 1953)
3. Mixture of thallium(I) carbonate and germanium oxide for Tl_2GeF_6 (Hájek and Benda, 1972)
4. Mixture of alkali metal carbonate and lead acetate in BeF_2 solution for $M_2Pb(BeF_4)_2$ (Le Fur and Aleonard, 1970)
5. Mixture of ammonium hydrogen fluoride and manganese oxide hydroxide for $(NH_4)_2MnF_5$ (Emori *et al.*, 1969)

Therefore, further treatment in a stream of gaseous HF is recommended to dry the products and to avoid the presence of oxygen or hydroxides. Infrared spectroscopy is a powerful method for detecting the presence of water molecules or hydroxyl groups.

An original method for synthesizing divalent chromium compounds may be pointed out here, although HF solutions are not used: $KCrF_3$ can be prepared by interaction of an aqueous solution of KHF_2 with divalent chromium acetate. The valence state $+2$ is preserved during the experiment by protecting the compound against atmospheric oxygen with an overflowing bed of ether (Dumora *et al.*, 1970; Dumora, 1971).

2. High Pressure: Hydrofluorothermal Syntheses

For hydroflurothermal syntheses the starting products (oxide and fluoride) are placed in a thin gold tube in the presence of an HF solution. After sealing, the tube is subjected to pressure either by a liquid or by a gas in an autoclave. Such an apparatus can provide 5–6 kbars pressure at temperatures close to 500°C.

More details are given later because this technique can be used to obtain crystals. In some cases, hydroxyfluorides can also be prepared. A phase with composition $In[(OH)_{1-x}F_x]_3$ ($0.53 \leq x \leq 0.67$) was isolated after reaction of HF solution on In_2O_3 at 300 to 500°C (Grannec, 1970). The pyrochlore $M_8[(OH)_{1-x}F_x]_{24} \cdot 3H_2O$ (M = Cr or Ga) was prepared from the oxides at 3.5 kbars in a very narrow temperature range ($300 \pm 10°C$) (Rault et al., 1970).

B. REACTIONS IN NONAQUEOUS SOLVENTS

Several solvents are used, some of them remaining inert, some being fluorinating agents. The reactions can be carried out in metallic vessels, but fluorine-containing polymers such as Kel'f or FEP are also usually employed, especially for experiments involving liquid HF. In some cases quartz may also be a useful material, even with a very reactive solvent.

1. Reactions in Halogen Fluorides

The main solvents used for syntheses are BrF_3, BrF_5, and IF_5.

a. BrF_3. Bromine trifluoride is a good fluorinating agent. The high dielectric constant and the appreciable conductivity ($8 \times 10^{-3} \ \Omega^{-1} \ cm^{-1}$) of the liquid lead to excellent solvent properties. Autoionization of BrF_3 has been proposed:

$$2 \, BrF_3 \rightleftharpoons BrF_2^+ + BrF_4^-$$

In this solvent compounds formed with F^- donors such as $MBrF_4$ (M = alkali metal) are bases, and salts obtained with F^- acceptors such as $BrF_2^+AsF_6^-$ behave as acids. Solutions of bases and acids are electrically conducting, and many reactions can be followed by conductivity measurements.

Bromine trifluoride can be used in the preparation of binary fluorides. For example, it reacts exothermally with gold powder placed in a quartz reactor to yield $AuF_3 \cdot BrF_3$ (most likely $AuF_4^- BrF_2^+$). The recovery of AuF_3 is then realized by pyrolysis at 180°C (Sharpe, 1949).

In the same way, palladium(II) hexafluoropalladate(IV) was obtained by the following sequence of reactions (Sharpe, 1950):

$$2 \, PdBr_2 + 4 \, BrF_3 \longrightarrow 2[PdF_3 \cdot BrF_3] + 3 \, Br_2$$

$$2 \, PdF_3 \cdot BrF_3 \xrightarrow{180°C} Pd^{II}Pd^{IV}F_6 + 2 \, BrF_3$$

The same type of reaction has been used to isolate Pb_2F_6 (Charpin et al., 1972). Manganese trifluoride has been synthesized from $Mn(IO_3)_2$ without intermediate complex formation (Nyholm and Sharpe, 1952).

Bromine trifluoride is also an excellent reagent and solvent for the preparation of ternary fluorides. The following reactions show the trends:

$$2\,CsBr + 2\,Ru + 4\,BrF_3 \longrightarrow 2\,CsRuF_6 + 3\,Br_2 \qquad \text{(Hepworth } et\ al.\text{, 1954)}$$

$$6\,CsBr + 6\,IrBr_4 + 12\,BrF_3 \longrightarrow 6\,CsIrF_6 + 21\,Br_2 \qquad \text{(Hepworth } et\ al.\text{, 1956)}$$

$$3\,PdBr_2 + 3\,GeO_2 + 6\,BrF_3 \longrightarrow 3\,PdGeF_6 + 3\,O_2 + 6\,Br_2 \qquad \text{(Bartlett and Rao, 1964)}$$

$(NO)_2PdF_6$ is synthesized by dissolving NOCl and $PdBr_2$ in BrF_3 in a quartz bulb (Leary, 1975).

b. BrF_5. Bromine pentafluoride is also often used for solvolysis reactions, for instance, in the preparation of hexafluorobromates of alkali metals (Whitney $et\ al.$, 1964):

$$MF + BrF_5 \longrightarrow MBrF_6 \qquad (M = K,\,Rb,\,or\,Cs)$$

It is a good solvent for XeF_2 and F^- acceptors; in liquid BrF_5, xenon fluoride forms with AsF_5 two adducts that can actually be formulated $Xe_2F_3{}^+AsF_6{}^-$ and $XeF^+AsF_6{}^-$ (Sladky $et\ al.$, 1968). Xenon difluoride also reacts with many other pentafluorides that dissolve easily in BrF_5 but that do not form with the solvent adducts stable above $0°C$; BrF_5 can therefore be readily removed by vacuum distillation. A series of compounds of general formula $2XeF_2 \cdot MF_5$ and $XeF_2 \cdot MF_5$(M = Ru, Os, Ir, or Pt) have been characterized (Sladky $et\ al.$, 1969).

Adducts involving XeF_6 in combination with F^- acceptors have also been reported in this connection. For instance, $XeF_5{}^+AsF_6{}^-$ was prepared by dissolving XeF_6 in BrF_5 contained in a Kel'f weighing bottle provided with a Kel'f valve. After AsF_5 was condensed, the contents were mixed thoroughly by shaking the container at room temperature. All volatiles (BrF_5 and AsF_5 in excess) were removed in dynamic vacuum at $0°C$, the solid adduct remaining (Selig, 1964; Bartlett and Wechsberg, 1971).

c. IF_5. Iodine pentafluoride is not as efficient a solvent as BrF_3. It can be used in conjunction with iodine, which serves as reducing agent, in the preparation of low-valence fluorides (Holloway and Peacock, 1963b):

$$10\,RuF_5 + I_2 \xrightarrow{\;IF_5\ \text{solution}\;} 10\,RuF_4 + 2\,IF_5$$

It is also useful for the synthesis of potassium hexafluoroplatinate(V) or cesium hexafluororhodate(V). The former is made by reaction of dioxygenyl hexafluoroplatinate(V) on KF (Bartlett and Lohmann, 1964), the latter according to (Holloway $et\ al.$, 1965)

$$Cs^+IF_6{}^- + IF_4{}^+RhF_6{}^- \longrightarrow CsRhF_6 + 2\,IF_5$$

In contrast with its behavior toward BrF_5, xenon difluoride forms with IF_5 the adduct $XeF_2 \cdot IF_5$ (Sladky and Bartlett, 1969). Other adducts of IF_5 have been isolated with SbF_5 (Woolf, 1964) or with alkali metal fluorides (Christe, 1972).

Finally, mention must be made of the spontaneous interaction at ambient temperature in a quartz bulb of IF_5 with O_2AuF_6, which contains two powerful oxidizing species, O_2^+ and Au(V), to give $IF_6^+AuF_6^-$ (Bartlett and Leary, 1976):

$$4\,O_2AuF_6 + 3\,IF_5 \longrightarrow 3\,IF_6^+AuF_6^- + AuF_3 + 4\,O_2$$

2. Reactions in Selenium, Antimony, and Vanadium Fluorides

Selenium tetrafluoride is a good ionizing solvent. It has been used to convert oxides to fluorides, for example, $KMnF_5$ from $KMnO_4$ (Peacock, 1953). It can form adducts with fluorides, and its reducing behavior allowed the preparation of very pure PdF_2 (Bartlett and Quail, 1961):

$$6\,PdF_3\cdot BrF_3 + 12\,SeF_4 \longrightarrow 6\,(SeF_4)_2PdF_4 + Br_2 + 4\,BrF_3$$
$$(SeF_4)_2PdF_4 \xrightarrow{155°C} PdF_2 + SeF_4 + SeF_6$$

Antimony pentafluoride is a very strong F^- acceptor. It can react with several fluorides to give adducts.

The compound $XeF_2\cdot 2SbF_5$ was first isolated by Edwards et al. (1963); it may be represented by the formal notation $XeF^+Sb_2F_{11}^-$. Complexes of this general type have since been identified for krypton: $KrF^+Sb_2F_{11}^-$, $KrF^+SbF_6^-$, and $Kr_2F_3^+SbF_6^-$ (Selig and Peacock, 1964; Frlec and Holloway, 1973; Gillespie and Schrobilgen, 1976). Xenon tetrafluoride also behaves as an F^- donor and the $XeF_3^+SbF_6^-$ and $XeF_3^+Sb_2F_{11}^-$ salts have been established (McKee et al., 1973). All these compounds are prepared by treating the rare-gas fluorides with SbF_5 in Kel'f, quartz, or Pyrex glass containers at appropriate temperatures.

The synthesis of $NF_4^+SbF_6^-$ and $O_2^+SbF_6^-$ was performed according to the following reactions:

$$NF_3 + F_2 + SbF_5 \xrightarrow[200°C,\,F_2\,pressure]{in\,Monel\,vessel} NF_4^+SbF_6^- \qquad \text{(Tolberg et al., 1967)}$$

or

$$NF_3 + 2\,F_2 + SbF_3 \xrightarrow[30-70\,atm]{250°C} NF_4^+SbF_6^- \qquad \text{(Wilson and Christe, 1980)}$$

$$2\,O_2 + F_2 + 2\,SbF_5 \xrightarrow[uv\,irradiation]{in\,Pyrex\,bulb} 2\,O_2^+SbF_6^- \qquad \text{(McKee and Bartlett, 1973)}$$

The adduct $UF_5\cdot 2\,SbF_5$ was obtained either from the interaction between the two components or from the reaction of UF_6 with SbF_5 in the presence of Freon 114 (Bougon and Charpin, 1979).

Vanadium pentafluoride seems also to exhibit a capacity to accept F^- in solution, but there is little published evidence of this property. The existence of

$XeF_2 \cdot VF_5$ has been established (Žemva and Slivnik, 1971). The better F^- donor capacity of XeF_6 in comparison with XeF_2 has made it possible to prepare $2XeF_6 \cdot VF_5$ (Moody and Selig, 1966), $XeF_6 \cdot VF_5$, and $XeF_6 \cdot 2VF_5$ (Jesih et al., 1982).

These reactions are performed in a nickel vessel. For instance, the 1:2 compound is prepared by introducing XeF_6 to the reactor by sublimation. An excess of VF_5 is also added by sublimation, and the mixture is warmed to room temperature, so that all the XeF_6 dissolves in the VF_5. The excess of VF_5 is then pumped away at $-35°C$.

Mention may also be made of another adduct, $KrF_2 \cdot VF_5$, although it melts at 5°C (Žemva et al., 1975).

3. Reactions in Liquid Hydrogen Fluoride

Liquid HF is a solvent for F^- acceptors or donors, and therefore it can be used as a medium for the preparation of ternary fluorides.

As an example, we shall consider the preparation of heptafluouranates(V) mentioned by Frlec et al. (1966). The reaction is carried out in fluorothene tubes sealed at one end and attached at the other end through a flare connection to an all-stainless steel valve. Uranium hexafluoride is sublimed onto HF in one of the reaction vessels, and the mixture is then condensed on hydrazinium difluoride in a second one. The reaction occurs at $0°C$, the mixture being shaken several times. After 6 to 10 hr, unreacted UF_6 and HF are pumped off. The overall reaction might be described by the equation

$$4\,UF_6 + 5\,N_2H_6F_2 \longrightarrow 4\,N_2H_6UF_7 + N_2 + 6\,HF$$

If an excess of UF_6 is used, the reaction proceeds according to the following equation (Frlec and Hyman, 1967):

$$4\,UF_6 + 3\,N_2H_6F_2 \longrightarrow 2\,N_2H_6(UF_6)_2 + N_2 + 6\,HF$$

Some reactions that cannot be realized by heating two components have been carried out by dissolving the components in liquid HF. This is the case for the interaction between XeF_2 and AuF_5 leading to $Xe_2F_3^+AuF_6^-$ (Holloway and Schrobilgen, 1975; Vasile et al., 1976) or between XeF_6 and Pd_2F_6 giving a compound formulated $4XeF_6 \cdot PdF_4$ (Leary et al., 1973; Leary, 1975). In the same way the reaction $2\,AgF + UF_6 \rightarrow Ag_2UF_8$ is very fast in anhydrous HF if the concentration of AgF is higher than 2 M (Malm, 1983).

Many perfluoroammonium salts can be prepared in HF solutions by metathesis reactions between Cs_2MF_6 (M = Si, Ti, Mn, Ni, or Sn) and NF_4SbF_6 (Christe et al., 1977; Wilson and Christe, 1982):

$$2\,NF_4SbF_6 + Cs_2MF_6 \longrightarrow 2\,CsSbF_6\downarrow + (NF_4)_2MF_6$$

These reactions are carried out in an apparatus consisting of three interconnected Teflon–FEP U traps, which allows for easy separation of the $CsSbF_6$ precipitate. The compound NF_4GeF_5 was also isolated from the interaction of $(NF_4)_2GeF_6$ with an excess of GeF_4 (Christe et al., 1976b).

Liquid HF is also useful for reactions between compounds that have very different volatilities. By successive cooling and heating it is possible to ensure that two reactants will come in contact in the condensed form. This method has been used to prepare a complex between SbF_5 and ClF (Christe and Sawodny, 1969) and to isolate adducts of the type $MF_2 \cdot 2AsF_5$ (M = Mg, Co, Ni,..., Ba) (Frlec et al., 1982). Nitrosyl fluorometallate salts $NOMF_6$ (M = Nb, Sb, Ta, or U) have also been obtained by reaction of N_2O_4/HF mixtures on metals (Sato and Kigoshi, 1982).

Binary fluorides can also be synthesized. Antimony pentafluoride reacts with readily available hexafluorometallate(V) anions when dissolved in anhydrous HF to produce the weaker Lewis acids MF_5(M = Nb, Mo, W, or Os) (O'Donnell and Peel, 1976). In the same way β-UF_5 was prepared by reaction of a saturated solution of UF_6 in anhydrous HF with HBr (Jacob, 1973). Bougon and Lance (1983) synthesized AgF_3 after interaction of AgF_2 with KrF_2 in HF solution.

4.   Reactions in Sulfur Dioxide

Liquid SO_2 may be a suitable solvent for reactions between fluorides. For instance, when SO_2 solutions of ReF_6 and KI are brought together, iodine is immediately liberated, and the complex fluoride is precipitated (Peacock, 1957b):

$$2\,ReF_6 + 2\,KI \longrightarrow I_2 + 2\,KReF_6$$

The salt is obtained in a pure form by removal of volatile materials under vacuum at 200°C.

The F^- acceptor property of arsenic and antimony pentafluorides also led to the preparation, in liquid SO_2, of such compounds as $SnF_2 \cdot SbF_5$ and $MF_2 \cdot 2SbF_5$ (M = Mn, Fe, or Ni). Antimony pentafluoride reacts either with metal difluorides (Birchall et al., 1971) or with a powdered metal (Dean, 1975).

IV.   Gas–Solid Reactions

Many fluorides must be prepared at high temperature. Thus, fluorinations by gaseous HF or by fluorine constitute excellent synthesis methods. The use of other fluorinating agents can also be taken into consideration.

A. REACTIONS USING GASEOUS
   HYDROGEN FLUORIDE

The most common technique consists in using streaming gaseous HF over metals, halides, or oxides heated at high temperature. Figure 3 shows an apparatus for carrying out fluorinations at low pressures up to 700°C (Grannec, 1970). The sample is placed in a gold crucible introduced into a nickel or Inconel tube under a flow of nitrogen. All the other tubing is made of copper, which thoroughly resists HF at room temperature. Controlled fluorination can be obtained by using a mixture of HF and either nitrogen or hydrogen that has been previously dried and freed of oxygen. The flow rate is regulated by the use of valves. An internal mobile tube allows collection of possible volatile products. A summary of such fluorinations that have been described in the literature is given in Table II.

Intermediate hydride phases may be worthwhile for some elements. It is not surprising that an H_2/HF mixture was shown to convert uranium metal to UF_4 at lower temperature than does pure HF (Meister, 1954). Sometimes two different fluorides are obtained. If an H_2/HF mixture is passed over TaH at 570°C, solid TaF_3 is formed in the crucible, whereas TaF_5 crystals lay on the cooled walls of the tube (Emeleus and Gutmann, 1950).

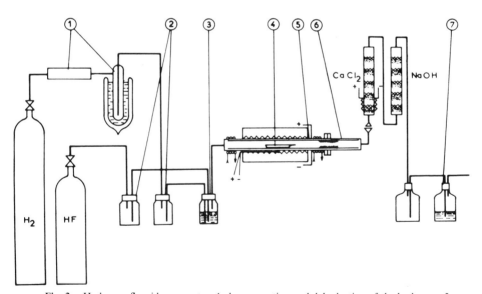

Fig. 3. Hydrogen fluoride apparatus: 1, deoxygenation and dehydration of the hydrogen; 2, gas mixers; 3, gas flow control; 4, gold crucible; 5, nickel tube; 6, mobile nickel tube; 7, flow control for end of reaction.

TABLE II

Action of Hydrogen Fluoride on Various Metals, Halides, Oxides, and Hydrides

Starting Material	Experimental Conditions, °C	Compound Formed	References
Mn or Co	180	MnF_2 or CoF_2	Muetterties and Castle (1961)
Sb	300	SbF_3	Muetterties and Castle (1961)
U	500	UF_4	Meister (1954)
$TiH_{0.7}$ or $TiCl_3$	700	TiF_3	Ehrlich and Pietzka (1954); Siegel (1956)
$TiCl_4$	275	TiF_4	Haendler et al. (1954)
VCl_3	300–500	VF_3	Jack and Gutmann (1951); Nyholm and Sharpe (1952)
VCl_4	20–30	VF_4	Ruff and Lickfett (1911); Cavell and Clark (1962)
$CrCl_3$	550	CrF_3	Nyholm and Sharpe (1952); Hepworth et al. (1957)
$FeCl_2$	500	FeF_2	Henkel and Klemm (1935)
$FeCl_3$	200–400	FeF_3	Tressaud (1969)
$CoCl_2 \cdot 6H_2O$	550	CoF_2	Baur (1958)
$CoCl_2$	600	CoF_2	Lozano and Grannec (1969)
UO_2	570	UF_4	Smiley and Brater (1958)
ThO_2	600	ThF_4	L. Lozano and J. Grannec (private communication, 1969)
Y_2O_3	HF/H_2; 700	YF_3	Zalkin and Templeton (1953)
ZrH	HF/H_2; 750	ZrF_4	Ehrlich et al. (1964)
Ta or TaH	HF/H_2; 300	TaF_3 and TaF_5	Emeleus and Gutmann (1950)
VF_3	HF/H_2; 1150	VF_2	Stout and Boo (1966)
V	HF/H_2; 1200	VF_2	Shafer (1969)
VCl_2	HF/H_2; 650	VF_2	Cros et al. (1975)

Ternary fluorides can also be synthesized by action of gaseous HF on mixtures of halides or on ternary halides:

$$CsF + M(II)Cl_2 \xrightarrow[600°C]{HF} CsM(II)F_3 \quad \text{(Dance, 1974)}$$

$$(M = Fe, Co, Ni)$$

$$K_2ReBr_6 \xrightarrow[200°C]{HF} K_2ReF_6 \quad \text{(Weise, 1956)}$$

In such cases, however, especially if experiments involve alkali metal fluorides, it is necessary to pump over the compound at about 100°C to avoid absorption of HF by the product.

We may recall here the role of gaseous HF in preparing very pure fluorides after synthesis in aqueous solutions. Treatment of the binary or ternary fluorides under a stream of HF eliminates traces of oxygen or hydroxides.

B. REACTIONS USING FLUORINE

1. Low-Pressure Fluorine

Fluorine handling at low pressures is not very dangerous, provided that a few precautions are taken. The fluorine source may be either an electrolytic cell or a commercial cylinder. It must be located in a well-ventilated, independent structure, and all pieces of the apparatus must be built under hoods. Two kinds of fluorination units may be used, depending on the working pressure.

For fluorinating materials at pressures not exceeding 2 bars a system similar to that shown in Fig. 3 is very useful, but it must be slightly modified, as follows (Fig. 4; Grannec, 1970; Grannec et al., 1971). Just next to the nitrogen and fluorine tanks (which replace the hydrogen and HF cylinders) a container is filled under 2 or 3 bars, a precaution that makes it possible to turn out the main fluorine source during the fluorinating process. This reserve is also used to dilute fluorine with nitrogen; the relative amounts of the two gases are controlled by a gauge. Any traces of HF are trapped by passage through a column filled with NaF pellets. The reaction tube is made of nickel or Monel; a gauge allows one to check the pressure during the experiments. The samples are placed in a nickel crucible, but sintered alumina can also be used at moderate temperatures (Haszeldine and Sharpe, 1951). The appropriate temperature is obtained by a resistance that is wound around the nickel tube. It is controlled by a thermocouple inside a nickel sheath placed just under the

Fig. 4. Low-pressure fluorine apparatus.

reaction crucible. The unreacted fluorine is destroyed in a tower containing anhydrous alumina or calcium oxide.

For fluorinations under pressures of up to 10 to 12 bars it is preferable to use another kind of apparatus consisting of vessels connected to a high-vacuum line (see reactors C in Fig. 1) by Whitey IKS4 brass valves. Such devices are detailed in Fig. 5 (Tressaud *et al.*, 1976b). The reactor is made of nickel or Monel. The starting material is loaded into a small nickel can in a dry box. The lid of the reactor is provided with a Teflon gasket, which ensures a very good tightness. The bottom is heated with a standardized resistance. The upper walls are kept cold with a strong flow of coolant.

Palladium tetrafluoride has been prepared in such a way by fluorinating Pd_2F_6 under 7 bars F_2 at 300°C for several days (Rao *et al.*, 1976). This method also allows one to collect very simply volatile fluorides after fluorination either of the corresponding metal (Ir → IrF_5; Bartlett and Rao, 1965) or of the oxide–fluoride ($TiOF_2$ → TiF_4) or even of a lower-valency fluoride (BiF_3 → BiF_5) (Tressaud *et al.*, 1976b). The separation of fluorides having different volatilities also becomes easier: Tressaud *et al.* (1976a) succeeded thus in isolating $Pt^{II}Pt^{IV}F_6$ from PtF_5.

Fluorine reacts with most metals and gives rise to compounds generally having a high oxidation state. A mixture of fluorides is sometimes obtained, however. Syntheses with transition metals are listed in Table III. To facilitate reactions it is preferable to use a powdered form of the metal. Control of the

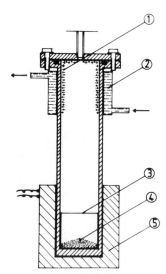

Fig. 5.   Fluorinating reactor (F_2 up to 12 bars): 1, volatile fluoride deposit; 2, coolant; 3, nickel can; 4, starting material; 5, standardized resistance.

TABLE III

Action of Fluorine on Some Transition Metals

Metal	Experimental Conditions, °C	Compound Formed	References
Ti	350	TiF_4	Haendler et al. (1954)
V	300–350	VF_5	Clark and Emeleus (1957); Trevorrow et al. (1957)
Cr	300–600	CrF_3, CrF_4, CrF_5	Wartenberg (1941); Clark and Sadana (1964)
Mn	300–600	MnF_3, MnF_4	Hepworth and Jack (1957); Roesky et al. (1965)
Co	300–600	CoF_3	Nyholm and Sharpe (1952); Hepworth et al. (1957)
Nb	300–500	NbF_5	Fairbrother and Frith (1951)
Mo	Ar/F_2; 105 min	MoF_3	Hoppe and Lehr (1975)
Mo	400	MoF_5, MoF_6	Barber and Cady (1956); Iwasaki et al. (1962)
Ru	300	RuF_5	Holloway and Peacock (1963a)
Rh	400	RhF_4, RhF_5, RhF_6	Holloway et al. (1965)
Pd	500	Pd_2F_6	Tressaud et al. (1976c)
Ag	80	AgF	Portier et al. (1970); Grannec et al. (1983)
Ag	200	AgF_2	Gruner and Klemm (1937)
Ir	350–380	IrF_5	Bartlett and Rao (1965)
		IrF_6	Robinson and Westland (1956)
Pt	700	PtF_5, PtF_6	Bartlett and Lohmann (1964)
Platinides	Various Temperatures	Hexafluorides	Weinstock (1962)

reaction must be maintained, however, to prevent the formation of a hot spot, which may lead to fusion of the powder. This control can be achieved by the dilution of fluorine with nitrogen. The process yields very pure silver(I) fluoride (Portier *et al.*, 1970; Grannec *et al.*, 1983). Another problem in the fluorination of metals is that thin fluoride films may form and block the reaction; layers of compounds having different compositions are then observed (Gillardeau, 1970).

To eliminate these drawbacks, better starting materials are often oxides, sulfides, halides, or lower-valency fluorides. Sometimes the reaction with oxides yields oxide–fluorides (Table IV).

Many ternary fluorides can also be synthesized under fluorine. Very different mixtures of salts may be used, just as ternary oxides or ternary fluorides of lower valence. Sometimes dehydration of hydrated fluorides under fluorine is necessary to avoid hydrolysis (Table V).

2. High-Pressure Fluorine

The handling of fluorine at high pressures requires taking many more precautions than for reactions described previously. Safety considerations are therefore very important, but if a strict procedure is carefully followed, this technique can be applied more easily. It is necessary to describe briefly the safety aspects and technology before giving some typical examples of compounds prepared by this method.

a. ADVANTAGES OF HIGH-PRESSURE TECHNIQUES.   Because of the strong oxidizing power of fluorine, autoclaves usually make it possible to work between 20 and ~ 4000 bars. In this zone the pressure has a strong effect on chemical reactions involving gaseous products. Indeed, these gases are very compressible, and the pressure produces a large increase in their concentrations in the reaction vessel. In most cases, the law of mass action explains quantitatively the effects produced in the system.

A beneficial effect of pressure occurs for reactions involving a decrease in volume, as predicted by the Le Châtelier principle. For a quantitative appreciation of this phenomenon, however, it is necessary to take into account the fugacity of the gases under synthesis conditions. Such reactions can also be considered as a means of activating fluorine, which can be achieved in another connection by heating, irradiation, or electrical discharge.

A final advantage of this technique is that it allows the reactions to be followed by the pressure changes in the reactor. It is the appropriate method for obtaining, among other products, fluorine compounds in high oxidation states.

b. SAFETY ASPECTS.   Elementary precautions in handling fluorine are always absolutely necessary, but further safeguards must also be observed.

TABLE IV

Action of Fluorine on Oxides, Sulfides, Halides, and Fluorides

Initial Product	Experimental Conditions, °C	Compound Formed	References
V_2O_5	475	VOF_3	Haendler et al. (1954); Trevorrow (1958)
CeO_2	500	CeF_4	Asker and Wylie (1965)
Co_2O_3	300	CoF_3	Brauer (1960)
Tl_2O_3	300	TlF_3	Hannebohn and Klemm (1936)
MnO or Mn_3O_4	100	MnF_3	Aynsley et al. (1950)
SnO	300	SnF_4	Hoppe and Dähne (1962)
$Mo(CO)_6$	$-70 \xrightarrow{} Mo_2F_9$ 170 (decomp.)	MoF_4	Peacock (1957a)
ZrO_2	525	ZrF_4	Haendler et al. (1954); Sense et al. (1954)
U_3O_8	—	UF_6	Brčič and Slivnik (1955)
CdS	300	CdF_2	Haendler and Bernard (1951)
CuO or $CuCl$	400–500	CuF_2	Billy and Haendler (1957)
RhI_3	400	RhF_3	Hepworth et al. (1957)
$AuCl_3$	300–600	AuF_3	Jack (1959)
CoF_2	250	CoF_3	Belmore et al. (1947)
PbF_2	300	PbF_4	Wartenberg (1940)
UF_4	275	UF_6	Katz and Rabinowitch (1951)
MnF_2	300	MnF_3	Lorin (1980)

TABLE V

Action of Fluorine on Various Products

Starting Material	Experimental Conditions, °C	Compound Formed	Reference
$MPbO_3$ [M(II) = Ca, Sr, or Ba]	400–500	$MPbF_6$	Hoppe and Blinne (1958)
$BaFeO_3$	—	$BaFeF_5$	Peacock (1960)
$2\ LiCl + MnO_2$	400–500	Li_2MnF_6	Hoppe et al. (1961a)
$3\ NaF + 3\ LiF + 2\ CoO$	500	$Na_3Li_3Co_2F_{12}$	de Pape et al. (1967)
$2\ (NH_4)_2PbCl_6 + Ag_2SO_4$	450	$Ag^{II}Pb^{IV}F_6$	Hoppe and Müller (1969)
$2\ MF + NaF + \frac{1}{2}\ Tl_2O_3$ [M(I) = K, Rb, or Cs]	400–500	M_2NaTlF_6	Grannec et al. (1970)
$M_2CO_3 + Bi_2O_3$ [M(I) = Rb or Cs]	400–500	$M^IBi^VF_6$	Hebecker (1970)
$2\ KF + M_2O_3$ [M(III) = In or Tl]	400–500	KMF_4	Grannec (1970)
$M_2CO_3 + 2\ AuCl$ [M(I) = Li or Na]	400	$MAuF_4$	Hoppe and Homann (1970)
$MCl + CeO_2$ [M(I) = Li, Rb, or Cs]	300–600	$MCeF_5$	Delaigue and Cousseins (1972)
$2\ Ag + NiF_2 + MF_3$ [M(III) = Al, Cr, or Fe]	80 and refiring	Ag_2NiMF_7	Dance et al. (1974)
$5\ NaF + 3\ MCl_2$ [M(II) = 3d transition elements]	500	$Na_5M^{III}_3F_{14}$	Dance (1974)
$2\ RbF + MF + CuF_2$ [M(I) = Li or K]	150–350	Rb_2MCuF_6	Grannec et al. (1975a)
$2\ NaF + LiF + MF_2$ [M(II) = Ni or Cu]	500 P_{F_2} for Cu	$Na_2LiM^{III}F_6$	Sorbe (1977)
$MCl + Ag + Me_2O_3$ [M(I) = Rb or Cs; Me(III) = Al, Fe, or Ga]	F_2/Ar; >500	$M^IAg^{II}MeF_6$	Müller (1981)
$0.2\ Cs_2CO_3 + 0.4\ MgCO_3 + 1.6\ [Co(NH_3)_6]Cl_3$	400–500	$Cs_{0.4}Mg_{0.4}Co_{1.6}F_6$	Hartung and Babel (1982)
$CuMF_6 \cdot 4H_2O$ [M(IV) = Ti, Zr, Sn, or Hf]	300–500	$CuMF_6$	Friebel et al. (1983)

Indeed, a compressed gas represents, at the same pressure and volume, a much greater risk than a liquid, because the energy contained also represents the work necessary to compress the fluid; it is higher, of course, for a gas than for a liquid. Therefore, the reactor must be used in an isolated and very well ventilated hood. It has to be adequately shielded to protect the operators against possible projections of metallic plates in the event of an explosion of some weaker part of the apparatus.

To limit these drawbacks, it is recommended that thick-walled bombs machined without welding be used. The presence of oil or organic compounds must also be absolutely avoided because these products can initiate a combustion reaction with fluorine. In that case the protective fluorine layer could be destroyed and the combustion might spread rapidly with the risk of an explosion due to increasing pressure. The strength of the materials of which the reactor is made and the size of the reactor are also important safety criteria.

c. CHOICE OF VESSEL MATERIAL AND ACCESSORY EQUIP-MENT. Nickel and Monel are suitable metals for building autoclave bombs. With these materials a reactor can be designed to stand up to the maximum pressures produced during a reaction. Several factors must be considered in choosing the reactor's dimensions; help may be obtained in this connection from the theories concerning the maximum principal stress, the maximum deformation, the maximum shear stress, etc. An appropriate cylindrical reactor may have external and internal radii of about 40 and 20 mm, respectively, the net volumes being 30–50 ml.

The tightness must be ensured by a copper gasket. It is essential to pretest the bomb in a hydraulic press or by using nitrogen under high pressure. Reactions involving relatively small quantities of product may be carried out in devices made from metal tubing, from which it is easy to obtain fairly large wall thicknesses for low internal diameters. This type of reactor also has the advantage of being easy to weigh on precision balances.

VALVES. Valves made of Monel are necessary for working with fluorine under pressure. For example, those built by Autoclave Engineering Inc. are quite suitable for such experiments. Some of them may be used for higher temperatures because the Teflon packing is separated from the main body of the valve.

MANOMETERS. Gauges made of Monel or nickel are the most suitable. A good sensitivity in a wide pressure range can be found with gauges provided by Bourdon Company Inc. They can be used up to 5000 bars, and some of them allow the recording of pressure:

Autoclave Engineering Monel compression fittings are also very useful. Nickel liners are recommended to minimize corrosion. Further details about this technology can be found in other reviews (Dodge, 1950; Comings, 1956; Kelly *et al.*, 1963; Bougon *et al.*, 1972).

d. EXAMPLES OF FLUORINATION UNDER F_2 PRESSURE. We detail here the preparation of lithium hexafluoronickelate. The overall reaction (Grannec *et al.*, 1975b) can be written

$$6 \, LiF + 2 \, NiF_2 + F_2 \qquad 2 \, Li_3NiF_6$$

It occurs with a reduction in the mass of gas and is therefore favored by high pressure.

Figure 6 shows the apparatus used for this synthesis. The reactor is made of Monel. The joint is a lens–ring type made of copper, strongly tightened. The head of the reactor is made up of a threaded portion into which an Autoclave Engineering Monel valve is adapted. This bomb is pretreated under fluorine pressure of a few bars at 300 to 400°C. After this operation it is dismantled in a dry glove box. The LiF/NiF_2 mixture is placed in a Monel finger tube (sintered alumina is also suitable), which is introduced in the reactor. Before every experiment, the apparatus must be put under high vacuum. After purification by passing over NaF pellets, fluorine is introduced under a determined pressure into a calibrated copper container. It is then condensed into the bomb by means of liquid nitrogen; the volume of the bomb is much smaller than that of the container, which allows high pressures to be reached at room temperature. The reactor is then isolated by closing the Monel valve and

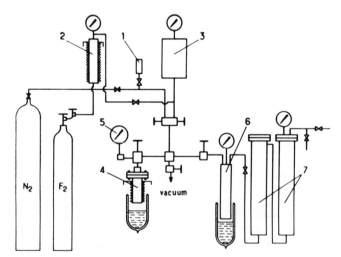

Fig. 6. High-pressure apparatus: 1, vacuum gauge; 2, trap for HF; 3, calibrated fluorine "reserve"; 4, high-pressure bomb; 5, Bourdon pressure gauge; 6, pressure-reducer container; 7, fluorine destruction towers.

heated with a standardized resistance. A shield is placed in front of the apparatus. The appropriate pressure is obtained by heating; it is controlled by a Bourdon gauge.

When the reaction is considered to be over, the reactor is cooled again at $-196°C$. The unreacted fluorine is then allowed to distill slowly into a large container in order to lower its pressure before its destruction on soda lime or calcium oxide. The sample is removed under dry argon and handled in a dry glove box. Very pure Li_3NiF_6 has been synthesized under 80 bars at 500°C, but such autoclaves enable one to use pressures of up to 400 bars at 500°C.

Other examples of fluorination under pressure may be mentioned. For instance, $KClF_4$ and $RbClF_4$ have been isolated from KCl and RbCl under 4000 bars of fluorine at $\sim 300°C$ (Kelly et al., 1963). The synthesis of $CsClF_4$ was carried out in a nickel tube having a volume of ~ 9 ml on which was fixed a manometer and a nickel guard disk designed to yield 5000 bars (Asprey et al., 1961). A description of the appropriate devices and guidelines for their use in the preparation of some binary fluorides such as CrF_6 or the perfluorides of rhenium, osmium, and platinum have been given by Glemser et al. (1963, 1966) and Slivnik et al. (1965).

Several ternary fluorides containing rare-earth or transition elements in high oxidation states have been synthesized under different conditions of temperature and pressure (Table VI). Some other ternary fluorides of transition elements can also be prepared under pressure in the presence of xenon fluorides. Leary and Bartlett (1972) characterized a gold(V) salt formulated $Xe_2F_{11}{}^+AuF_6{}^-$ by heating in a Monel bomb a mixture of AuF_3 and XeF_2 in gaseous fluorine (70 mmol at 65 bars). $[XeF_5{}^+]_2PdF_6{}^{2-}$ was isolated from Pd_2F_6 and XeF_2 under approximately similar conditions (Leary et al., 1973).

The combined action of oxygen and fluorine under pressure also leads to high oxidation states. This is the case for the preparation of $O_2{}^+MnF_5{}^-$ (Hoppe, 1978) or $O_2{}^+AuF_6{}^-$ (Bartlett and Leary, 1976); the decomposition of the former gives reportedly pure MnF_4 (Hoppe et al., 1961b). By heating MnO_2 at about 380 to 420°C under $PF_2/O_2 > 3500$ bars, Müller (1980) succeeded in crystallizing $O_2{}^+Mn_2F_9{}^-$.

Finally, dioxygenyl hexafluoropalladate(V), $O_2{}^+PdF_6{}^-$ must be mentioned. Falconer et al. (1976) stabilized this palladium(V) compound as follows: A free volume of 5 cm³ high-pressure Monel reactor containing 0.4 g of powdered palladium metal was filled with a $2:5$ O_2/F_2 liquid mixture at $-196°C$. The reactor was heated to 320°C for 133 hr, generating an internal pressure of about 4 kbars. After removal of the unreacted gases, the product was ground in a helium-filled dry box, reintroduced into the high-pressure autoclave, and oxyfluorinated under similar conditions for 94 hr. A brown-gold free-flowing powder was then obtained.

TABLE VI

Ternary Fluorides Synthesized under F_2 Pressure

Starting Material	Experimental Conditions, P, bars (°C)	Compound Formed	References
$2 NaF + NiF_2$	350 (350)	Na_2NiF_6	Bougon (1968)
Li_2CrO_4	300 (300)	Li_2CrF_6	Siebert and Hoppe (1972a)
$MCO_3 + CrCl_3$	300 (300)	$MCrF_6$	Siebert and Hoppe (1972b)
[M(II) = Ca, Ni, Zn, Sr, Cd,....]			
$2 MF + CuF_2$ [M(I) = Rb or Cs]	230–300 (500)	M_2CuF_6	Sorbe et al. (1976)
M_2CO_3 or $2 MCl + 2 RhCl_3$	300–400	$M^ICl^IRh^VF_6$	Wilhelm and Hoppe (1976)
[M(I) = Na, K, Rb, or Cs]	(400–500)		
$2 MF + CoF_2$ [M(I) = K, Rb, or Cs]	60–250 (500)	M_2CoF_6	Grannec et al. (1976)
$2 CsF + AgF_2$	100 (400)	Cs_2AgF_6	Sorbe et al. (1977)
MF/PrF_3, various compositions	400 (350)	MF/PrF_4 systems	Avignant and Cousseins (1978)
[M(I) = Na, K, Rb, or Cs]			
$4 CsF + AgF_2 + GaF_3$	100 (500)	$Cs_2Ag_{0.5}Ga_{0.5}F_6$	Sorbe et al. (1978)
$Ba[Ni(CN)_4]\cdot 4H_2O$	140 (280)	$BaNiF_6$	Fleischer and Hoppe (1982a)
$Li_2[Ni(CN)_4]\cdot 3H_2O$	300 (400)	Li_2NiF_5	Fleischer and Hoppe (1982b)
$CsCuCl_3$	350 (400)	$CsCuF_4$	Fleischer and Hoppe (1982c)
$CaCO_3$(or CdO) $+ 2 LiCl + \frac{1}{6} Pr_6O_{11}$	400–500 (500)	$M^{II}Pr^{IV}Li_2F_8$	Feldner and Hoppe (1983)
$6 MCO_3 + Pr_6O_{11}$	400–500 (500)	$MPrF_6$	Feldner and Hoppe (1983)
[M(II) = Sr or Ba]			

C.   REACTIONS USING OTHER FLUORINATING GASES

Two types of reactions are considered in this section: Fluorinating gases may give rise to binary fluorides, but sometimes adducts are also isolated.

1.   Chlorine Trifluoride

Chlorine trifluoride is a strong fluorinating agent. It reacts with metals or metal halides, usually leading to the highest oxidation state. With oxides it has the disadvantage of sometimes giving chlorine oxides or oxide–fluorides. An advantage, however, is that it can be transported in liquid form.

The strong oxidizing power of ClF_3 often allows it to react at lower temperature than fluorine. Thus, AgF_2 can be prepared from AgF at 100°C instead of 200°C with fluorine (Charpin et al., 1966). To avoid excessively violent fluorinations it is recommended that the gas be diluted in a dry nitrogen stream. The reactions may be carried out in metallic vessels (copper or nickel), but at low temperature Kel'f reactors are also useful because the reactions can be followed visually.

Other examples of fluorinations with ClF_3 include the preparation of cobalt(III) fluoride from $CoCl_2$ at 250°C (Rochow and Kuken, 1952) and cerium(IV) fluoride from the trifluoride at 350°C (Batsonova et al, 1973). Chlorine trifluoride has much less tendency to form complexes with binary fluorides than does BrF_3.

2.   Sulfur Tetrafluoride

Sulfur tetrafluoride can be a convenient fluorinating agent of oxides or sulfides because it reacts in a moderate way. With inorganic oxides, it usually leads to thionyl fluoride, as shown in the following reactions:

$$MO_3 + 3\,SF_4 \xrightarrow{350°C} MF_6 + 3\,SOF_2 \qquad \text{(Oppegard et al., 1960)}$$

$$(M \ = Mo, W)$$

$$\left.\begin{array}{l} UO_2 + 2\,SF_4 \xrightarrow{500°C} UF_4 + 2\,SOF_2 \\[4pt] UO_3 + 3\,SF_4 \xrightarrow{300°C} UF_6 + 3\,SOF_2 \end{array}\right\} \quad \text{(Johnson et al., 1961)}$$

In the latter experiment, the conversion is different if the reaction occurs in a sealed reactor or under gas flow.

Sulfur tetrafluoride forms complexes with F^- acceptors such as SbF_5 or AsF_5. Such adducts can be the starting point for the preparation of ternary fluorides (Smith, 1962):

$$AsF_5 \cdot SF_4 + CsF \longrightarrow CsAsF_6 + SF_4$$

Adducts have also been isolated with UF_5 (Rediess and Bougon, 1983).

3. Boron, Germanium, and Arsenic Fluorides

Boron, germanium, and arsenic fluorides give exclusively adducts with binary fluorides.

Boron trifluoride reacts at 350°C with alkaline-earth fluorides. The reactions are carried out in an 18/8 Mo stainless steel tube under 1 bar BF_3 pressure. The compounds $MF_2 \cdot 2BF_3$ (M = Ca, Sr, or Ba), which can be formulated $M(BF_4)_2$, are formed (de Pape and Ravez, 1962; Ravez and Hagenmuller, 1964). Boron trifluoride is also known to give adducts by combining with PbF_2 (Ravez et al., 1966) and XeF_6 (Selig, 1964).

It has been found that the reaction of excess XeF_6 with GeF_4 proceeds readily at room temperature to yield the solid $4XeF_6 \cdot GeF_4$. In contrast, an excess of GeF_4 leads to $XeF_6 \cdot GeF_4$ (Pullen and Cady, 1967).

Finally, AsF_5 reacts as a liquid in an FEP tube in intimate contact with KrF_2 at $-63°C$, before being allowed to expand at room temperature, to give salts containing the KrF^+ cation (Gillespie and Schrobilgen, 1976).

V. Partial or All Solid-State Reactions

Most methods described here involve syntheses in the solid state, but all types of reactions between inorganic solids must be quoted, even when the reactions occur above the melting point of one of the starting materials.

A. REACTIONS USING AMMONIUM
AND POTASSIUM DIFLUORIDES
OR HYDRAZINIUM FLUORIDE

1. Ammonium and Potassium Difluorides

Ammonium and potassium difluorides can be used to prepare ternary fluorides from oxides, halides, or carbonates. The reactions occur simply by heating together the bifluoride and the appropriate salt in an open crucible. $(NH_4)_2MgF_4$ was obtained from an appropriate mixture of the difluoride and $MgCO_3$ (Roux et al., 1968). This technique can be used with transition elements to stabilize low valence states:

$$RuI_3 + 3\,KHF_2 \longrightarrow K_3RuF_6 + 3\,HI \qquad \text{(Peacock, 1956)}$$

$$V_2O_3 + 6\,NH_4HF_2 \longrightarrow 2\,(NH_4)_3VF_6 + 3\,H_2O \qquad \text{(Sturm, 1960)}$$

Binary fluorides can be synthesized in the same way, but further treatment under gaseous HF is usually recommended to eliminate traces of oxygen.

A similar method is used for the preparation of fluoride glasses. These materials are handled basically in the same way as classical glasses. The chemical behavior of fluoride melts, however, may cause difficulties in glass-melting operations, such as volatilization (Hightower and McNeese, 1972), hydrolysis (Poulain et al., 1977), vessel corrosion (Mackenzie, 1978), and premature nucleation (Poulain, 1983). Crystallization may be limited by specific processes involving a reactive atmosphere (Robinson et al., 1980) or ammonium difluoride (Poulain and Lucas, 1978; Stević et al., 1978). Many details on such materials can be found in Chapter 7.

2. Hydrazinium Fluoride $N_2H_6F_2$

Hydrazinium(II) fluoride can be used as a fluorinating agent. Its handling is easy because it is not hygroscopic. Several ternary fluorides have been isolated; they contain either $N_2H_6{}^{2+}$ ions, for instance, in $N_2H_6MF_6$ (M = Ti or Zr) (Slivnik et al., 1964), or $N_2H_5{}^+$ ions in $(N_2H_5)_3CrF_6$ (Slivnik et al., 1967) and $N_2H_5CoF_3$ (Slivnik et al., 1976).

In the same way, treatment of aluminum or AlF_3 with molten $N_2H_6F_2$ at 210°C in a copper cylinder equipped with a loose copper stopper leads to ammonium hexafluoroaluminate (Volavšek et al., 1972).

B. LOW-PRESSURE REACTIONS

1. Reactions in Solid State

Ternary or quaternary fluorides can be prepared from binary fluorides in a direct solid-phase synthesis. Although this method seems very much like conventional chemical techniques, some precautions should be taken: most of the binary fluorides are sensitive to moisture, and their mixtures must be ground in a dry atmosphere.

Two techniques are used. First, the appropriate mixture is crushed in a dry box and put in a nickel, gold or platinum tube, which is then argon sealed. This prevents hydrolysis of the compounds when they are heated to the required temperature. Second, if such tubes are not available the reaction can be performed in an open crucible; however, the working system must be preliminarily evacuated with care and outgassed before the products are heated under dry argon. In order to hasten the reactions, it may be advisable to press the mixture and heat it as pellets. Some examples of such syntheses are given in Table VII.

2. Reactions Involving Rare-Gas Fluorides

The preparation of adducts and binary fluorides (e.g., AgF_3) by using rare-gas fluorides in nonaqueous solvents is described is Section III,B. Such complexes can also be synthesized by simple interaction in solid state or more often in molten XeF_2 or XeF_6.

TABLE VII

Reactions in Solid State

Starting Mixture	Experimental Conditions, °C	Compound Formed	References
Fluoride/fluoride			
x MF + x FeF$_2$ + (1 − x) FeF$_3$ [M(I) = alkali metal]	300–1100 300–700	— M$_x$FeF$_3$	Babel (1967) de Pape (1965); Tressaud et al. (1970)
NaF + CdF$_2$ + 2 MF$_2$ [M(II) = Mg, Ni, Cu, or Zn]	700	NaCdM$_2$F$_7$	Hänsler and Rüdorff (1970)
TlF/TlF$_3$ system	300	Tl$_3$TlF$_6$, Tl$_2$TlF$_5$ TlTlF$_4$, TlTl$_2$F$_7$	Grannec et al. (1971)
2 LiF + MF$_2$ + M'F$_4$ [M(II) = Ca, Mn, or Cd; M'(IV) = Zr, Hf, Ce, Th, or U]	550	M'MLi$_2$F$_8$	Vedrine et al. (1973)
2 AF + BF + MF$_3$ [A, B = alkali metal; M(III) = Al, V, Cr, Fe, or Ga]	600–900	A$_2$BMF$_6$	Babel et al. (1973)
MF$_2$ + M'F$_3$ [M(II) = Ca, Cr, Mn, Cd, Sr, Ba, or Eu; M'(III) = Al, Ti, V, Cr,..., Tl]	600–800	MM'F$_5$	Von der Mühll and Ravez (1974)
2 CsF + MF$_2$ + M'F$_2$ [M(II), M'(II) = Mg, Mn, Co, Ni, Zn, or Cd]	700	Cs$_2$MM'F$_6$	Dance et al. (1975)
5 NaF + 3 MF$_3$ [M(III) = Al, V, Cr, or Fe]	600–700	Na$_5$M$_3$F$_{14}$	Tressaud and Dance (1977)
MF + InF$_3$ [M(I) = alkali metal, Tl, or NH$_4$]	300–700	M$_3$InF$_6$, MInF$_4$, MIn$_2$F$_7$,...	Champarnaud-Mesjard et al. (1978)
2 AgF + MF$_2$ + M'F$_3$ [M(II) = Mg, Mn, Co, or Ni; M'(III) = Al, Sc, Ga, or In]	500–700	Ag$_2$MM'F$_7$	Koch et al. (1982)

Most reactions are carried out in metallic reactors. For instance, argon-arc-welded copper or nickel pressure vessels (90–120 ml) equipped with Teflon-packed brass valves are very useful for such experiments. After an appropriate mass of solid fluoride is weighed in the vessel, an excess of XeF_2 or XeF_6 is added by sublimation in high vacuum. The reaction vessel is then heated for several hours, usually at 120°C in the case of XeF_2 and 60°C in the case of XeF_6 in a thermostated bath. The products are then separately pumped off on the vacuum line from liquid-nitrogen temperature up to 20°C.

Slivnik and Žemva (1971) and Žemva et al. (1973) studied the reactions between Xenon and chromium fluorides. $XeCr_2F_{10}$ was prepared according to

$$2 \, CrF_5 + n \, XeF_2 \longrightarrow XeCr_2F_{10} + XeF_4 + (n-2) \, XeF_2$$

After interaction of an excess of XeF_6 with CrF_2, they isolated a $XeF_6 \cdot CrF_4$ compound. Hydrazinium(II) fluoroferrate(III) and hydrazinium(I) fluoro-cobaltate(II) reacting on XeF_6 at room temperature yield $XeF_6 \cdot FeF_3$ and $XeF_6 \cdot CoF_3$, respectively (Slivnik et al., 1976).

In the same way, the following sets of compounds were synthesized from manganese difluoride: $nXeF_2 \cdot MnF_4$ ($n = 1$, 0.5) and $nXeF_6 \cdot MnF_4$ ($n = 4, 2, 1, 0.5$) (Bohinc et al., 1976). Pullen and Cady (1966) obtained the adduct $4XeF_6 \cdot SnF_4$, whereas Bartlett et al. (1973) isolated crystals of $XeF^+RuF_6^-$ after having melted XeF_2 and RuF_5 at 120°C. The oxidizing capability of XeF_2 allows it to oxidize PtF_4 or mixed-valency Pd_2F_6 as follows (Bartlett et al., 1976):

$$5 \, XeF_2 + 2 \, PtF_4 \longrightarrow 2 \, Xe_2F_3PtF_6 + Xe$$
$$3 \, XeF_2 + Pd_2F_6 \longrightarrow 2 \, XePdF_6 + Xe$$

It has also been used to convert TbF_3 to the tetrafluoride at 300 to 350°C (Spitsyn et al., 1973).

Sometimes metallic reactors are not necessary. Kel'f containers have been conveniently used to prepare the 1:2 adduct of XeF_2 with $XeF_5^+AsF_6^-$ at 85°C (Bartlett and Wechsberg, 1971).

Finally, mention must be made of the adducts formed by combination of XeF_6 with alkali fluorides (Peacock et al., 1964). In particular, the reversible formation of the complex $NaF \cdot XeF_6$ provides a reliable method for the purification of XeF_6 because the other xenon fluorides do not give such adducts (Sheft et al., 1964).

C. HIGH-PRESSURE REACTIONS

High-pressure techniques are used in a large range of preparations, but they are of major interest in the field of fluoride elaboration.

Pressure can be applied through a fluid, either by compression of a gas or by hydrothermal synthesis. With such equipment, pressures of 5 kbars can be reached (Demazeau, 1973). Very high pressures are obtained by using a "belt"-type apparatus, a tetrahedral anvil press, etc. (Hall, 1960; Kennedy and La Mori, 1961; Bradley, 1966). The reactants must be loaded into gold- or platinum-sealed capsules to protect the materials from hydration and to avoid corrosion of the apparatus.

The role of pressure it twofold. It ensures intimate contact between the solids, which has a great influence on the reaction kinetics. It also makes more difficult the decomposition of unstable phases or the sublimation of volatile compounds.

The influence of pressure can be considered in three ways. First, it has been found by thermodynamic considerations that some preparations require high pressures. When reactions do not take place at atmospheric pressure, high pressure may induce a synthesis by lowering the free energy of the compound relative to that of the components (in fact, by increasing the density). In the study of the series $CsMF_3$ [M(II) = Mg, Mn, Fe, Co, Ni, or Zn], Longo and Kafalas (1969) showed that, whereas a mixture $(Cs_4Mg_3F_{10} + MgF_2)$ is obtained at atmospheric pressure, $CsMgF_3$ forms only at pressures and temperatures exceeding 30 kbars and 700°C, respectively. Likewise, the cubic perovskite $Tl^ITl^{III}OF_2$ was prepared under 4.5 kbars pressure at 500°C by reaction of either TlOF and TlF or TlF and O_2 (Demazeau et al., 1969).

Second, this method may induce allotropic transformations due to the higher density of the high-pressure forms. For instance, $CsMnF_3$ and $TlNiF_3$ change under high pressure (20–50 kbars) into cubic perovskites (Syono et al., 1969). In the same way, the influence of pressure on the crystallographic properties of AMF_3 compounds has been studied by Longo and Kafalas (1969) and that of some cryolites by Pistorius (1975). Demazeau et al. (1971) found the following structural evolution under high pressure in the hexafluorides Li_2MF_6 [M(IV) = Ti, Ge, Zr, or Sn]:

$$Li_2GeF_6 \, \beta \text{ type} \longrightarrow \text{trirutile} \longrightarrow Li_2ZrF_6 \text{ type}$$

The rutile form of PdF_2 is transformed at 50 kbars and 400°C into a structure derived from the fluorite type with a coordination number of $6 + 2$ instead of 6 for the cations (Tressaud et al., 1981; Müller, 1982).

The third effect of high pressures is the stabilization of unusual oxidation states. An electronic transition linked to the equilibrium

$$Pd(II) + Pd(IV) \Longleftrightarrow 2 Pd(III)$$

has been found in Pd_2F_6 from resistivity measurements between 1 bar and 80 kbars (Langlais et al., 1979). More recently, oxidation state 3 of palladium has been stabilized after reaction at 70 kbars in $NaPdF_4$ (Tressaud et al., 1982) and in the elpasolite-like structure of K_2NaPdF_6 (Khaïroun et al., 1983).

VI. Other Preparation Methods

A. DECOMPOSITION OF FLUORIDES

The thermal degradation of binary or ternary fluorides may lead to phases that are sometimes difficult to obtain by the previously mentioned techniques. This method is used for compounds that are in an unstable oxidation state or that decompose at relatively low temperature. In some cases the decomposition is only a photochemical effect (Hargreaves and Peacock, 1960) or occurs when a solvent is used (Soriano et al., 1966) (Table VIII).

B. DECOMPOSITION OF HYDRATES

Some anhydrous fluorides can be prepared by dehydration of the corresponding hydrates. For instance, $CsMnF_4$ was isolated after heating $CsMnF_4 \cdot 2H_2O$ to $100°C$ (Dubler et al., 1977). Günter et al. (1978) investigated the reversible dehydration of $Rb_2MnF_5 \cdot H_2O$, which led to Rb_2MnF_5. They clearly showed the chain-controlled topotactic reaction mechanism.

In some cases it is possible to find intermediate hydrates. In this connection, the behavior of $Fe_2F_5 \cdot 2H_2O$ is quite characteristic. The thermal decomposition scheme proposed by Charpin and Macheteau (1975),

$$Fe_2F_5 \cdot 2H_2O \xrightarrow[-H_2O]{170°C} Fe_2F_5 \cdot H_2O \xrightarrow[230°C]{200°C} FeF_3 \cdot H_2O + FeF_2 \xrightarrow[-H_2O]{250°C} FeF_3 + FeF_2$$

was fully confirmed by Ferey et al. (1981), who succeeded in preparing single crystals of the intermediate phases. They also proposed a mechanism for the intermediate formation of the pyrochlore $Fe_2F_5 \cdot H_2O$.

Another example worth noting is that of $(H_2O)_{0.33}FeF_3$, which was synthesized by Leblanc et al. (1983). Dehydration of this compound leads to a new form of FeF_3, the structure of which is related to that of the hexagonal tungsten bronzes (HTB).

C. REDUCING PROCESS

Some reducing processes have already been mentioned, for instance, the use of SeF_4 in the preparation of PdF_2 (Section III,B,2). Most of the reduction methods involving a gas or a solid are regrouped here.

1. Solid–Gas Reduction

Beyond its important role when mixed with gaseous HF (see Table II), hydrogen itself is one of the most common reducing agents. It must be thoroughly dried and freed of oxygen traces, however, to avoid hydrolysis and to prevent formation of oxide–fluorides. The following are examples of such a

TABLE VIII

Decomposition of Some Binary and Ternary Fluorides

Starting Material	Experimental Conditions	Compound Formed	Reference
Mo_2F_9	170°C	$MoF_4 + MoF_5$	Peacock (1957a)
OsF_6	uv, 20°C	OsF_5	Hargreaves and Peacock (1960)
$(NH_4)_3AlF_6 \xrightarrow{170°C}$	$NH_4AlF_4 + 2\,NH_4F \xrightarrow{720°C} \alpha\text{-AlF}_3$		Shinn et al. (1966)
$\xrightarrow{300°C}$	$\gamma\text{-AlF}_3 + NH_4F$		
Na_2PrF_6	Liquid HF; 20°C	$PrF_4 + 2\,NaHF_2$ (solv)	Soriano et al. (1966)
NH_4CrF_3	300°C; vacuum	CrF_2	Dumora and Ravez (1969)
$(NH_4)_3GaF_6 \xrightarrow{210°C}$	$NH_4GaF_4 + 2\,NH_4F \xrightarrow{455°C} \alpha\text{-GaF}_3$		Beck et al. (1973)
$\xrightarrow{335°C}$	$\gamma\text{-GaF}_3 + NH_4F$		
$(N_2H_5)_3CrF_6$	$(NH_4)_3CrF_6 + NH_3 + N_2 \xrightarrow{350-500°C} CrF_3$		Bukovec (1974)
$NO^+AuF_6^- \xrightarrow{230°C}$	400°C	$NO^+AuF_4^- \uparrow$	Leary (1975)
$2\,CsUF_7$	180–220°C; vacuum	$Cs_2UF_8 + UF_6 \uparrow$	Bougon et al. (1976)
$KrF^+AuF_6^-$	60–65°C; 8 hr	$AuF_5 \;(+ Kr + F_2)\uparrow$	Holloway and Schrobilgen (1975)
O_2AuF_6	180°C; vacuum	$AuF_5 \;(+ O_2 + \tfrac{1}{2}F_2)\uparrow$	Vasile et al. (1976)

route to lower valence states:

$$EuF_3 \xrightarrow{1100°C} EuF_2 \qquad \text{(Klemm and Döll, 1939)}$$

$$UF_4 \xrightarrow{1000°C} UF_3 \qquad \text{(Katz and Rabinowitch, 1951)}$$

$$AMnF_5 \xrightarrow{250°C} AMnF_4 \qquad \text{(Hoppe et al., 1959)}$$

$$(A = \text{alkali metal})$$

Other reducing gases can be used; these include carbon monoxide (or metal carbonyls), phosphorus trifluoride, and even SF_4 (although this gas reacts sometimes as a fluorinating agent; see Section IV,C,2):

$$RuF_5 + CO \xrightarrow{149°C} RuF_3 + COF_2 \qquad \text{(Peacock, 1960)}$$

$$2\,UF_6 + PF_3 \xrightarrow{-78\text{ to }20°C} 2\,\beta\text{-}UF_5 + PF_5 \qquad \text{(O'Donnell et al., 1983)}$$

$$2\,IrF_4 + SF_4 \xrightarrow{400°C} 2\,IrF_3 + SF_6 \qquad \text{(Robinson and Westland, 1956)}$$

2. Reaction Involving a Solid as Reducing Agent

Most of the methods involve the reduction of a fluoride by the corresponding metal:

$$3\,MoF_5 + 2\,Mo \xrightarrow{400°C} 5\,MoF_3 \qquad \text{(La Valle et al., 1960)}$$

$$GeF_4 + Ge \xrightarrow{300°C} 2\,GeF_2 \qquad \text{(Bartlett and Yu, 1961)}$$

$$4\,NbF_5 + Nb \xrightarrow{300°C} 5\,NbF_4 \qquad \text{(Gortsema and Didchenko, 1965)}$$

$$\text{(also possible with Si, P, etc.)} \qquad \text{(Schäfer et al., 1965)}$$

$$2\,SmF_3 + Sm \xrightarrow{1600°C} 3\,SmF_2 \qquad \text{(Stezowski and Eick, 1970)}$$

$$3\,NbF_5 + 2\,Nb \xrightarrow[3\text{ kbars}]{750°C} 5\,NbF_3 \qquad \text{(Pouchard et al., 1971)}$$

$$Pd_2F_6 + Pd \xrightarrow{980°C} 3\,PdF_2 \qquad \text{(Müller and Hoppe, 1972)}$$

$$2\,YbF_3 + Yb \xrightarrow{700°C} 3\,YbF_2 \qquad \text{(Petzel and Greis, 1973)}$$

$$4\,IrF_5 + Ir \xrightarrow{350-400°C} 5\,IrF_4 \qquad \text{(Bartlett and Tressaud, 1974; Rao et al., 1976)}$$

$$(IrF_5 \text{ used in excess})$$

$$2\,VF_3 + V \xrightarrow{\text{at red heat}} 3\,VF_2 \qquad \text{(Cros, 1978)}$$

Sulfur and iodine have also proved to be useful, for instance, in reducing RuF_5 at 200°C to give the trifluoride (Aynsley et al., 1952).

Ternary or quaternary fluorides can be prepared by a similar process. Zirconium has been used to stabilize compounds of trivalent uranium (Fonteneau and Lucas, 1974) and silicon to synthesize a fluorogarnet with trivalent titanium (de Pape et al., 1967).

Electrolytic reductions may also be pointed out: K_2NaTiF_6 was isolated after electrolysis of a mixture containing KCl and $NaCl$ with either K_2TiF_6 or Na_2TiF_6 (Bright and Wurm, 1958).

VII. Crystal Growth

It is not the aim of this section to give a detailed review of all the methods of growing single crystals. Most of the techniques utilized for fluorides are the same as those developed for other materials. These methods have been surveyed by many authors, and the main characteristics of various processes can be found in a large number of books or conference proceedings, for example, Lawson and Nielsen (1958), Gilman (1962), Schäfer (1964, 1971), White (1964), Peiser (1967), Laudise (1970), and Anthony and Collongues (1972).

It is difficult to specify in advance whether a technique will be preferable to another because each material has its own particularities according to the mechanical or physical properties desired. Among all the methods that will be described, several may sometimes be applied to the same compound. In such cases the choice is not limited, but the most suitable technique will be found only after appropriate investigation of all the problems: crucible, working atmosphere, regulating system of the furnace, melting point of the components, inclusion of doping agents etc.

A.   GROWTH IN ANHYDROUS HYDROGEN FLUORIDE

Small single crystals for x-ray diffraction studies can be obtained by using anhydrous HF as a solvent. The reactions are usually carried out in Kel'f or Teflon containers. For instance, crystals of uranium(V) complexes were isolated by Sturgeon et al. (1965) as follows. Weighed amounts of UF_5 and the desired alkali fluoride were introduced in an inert-atmosphere box into a Kel'f tube, the top of which was equipped with a plug of Teflon wool to allow filtration. The tube was then connected to the vacuum line, and HF was condensed by cooling with liquid nitrogen. After the mixture was stirred at room temperature, the HF solution was cooled to about 0°C and decanted through the Teflon wool into a second cooled Kel'f tube. The HF was slowly evaporated, and crystals formed, usually within 12 hr. They were then removed from the mother liquor and dried.

Oxonium pentafluorotitanate was prepared by Cohen et al. (1982) according to the reaction

$$TiF_4 + HF + H_2O \longrightarrow H_3O^+TiF_5^-$$

Water was added with a syringe to a cooled suspension of TiF_4 in anhydrous HF contained in a Teflon bottle. A clear, saturated solution resulted at 20°C after addition of anhydrous HF. Crystals remained after evaporation of anhydrous HF at 25°C.

The adduct $XeF_4 \cdot SbF_5$ was prepared by directly combining and melting together equimolar quantities of the two fluorides (McKee et al., 1973); single crystals were obtained from a solution in anhydrous HF by slow removal of the solvent at $-10°C$.

B. GROWTH IN HYDROFLUORIC ACID

1. Atmospheric Pressure

Growing crystals in hydrofluoric acid under atmospheric pressure is very simple, but it usually yields small crystals. A large number of ternary or quaternary fluorides can be obtained by dissolving various components (oxides, hydroxides, carbonates, fluorides, or difluorides) in aqueous solutions of HF. Crystals can sometimes be ground by slowly evaporating the solution at low temperature (20–60°C); this is the case for K_2TiF_6 (Siegel, 1952) and $(NH_4)_2MnF_5$ (Sears and Hoard, 1969). If several phases may form from the same mixture, the appropriate crystal must be isolated under accurate conditions; for example, crystals of $Na_3Li(BeF_4)_2$ are obtained only in the beginning of crystallization of concentrated and hot solutions containing lithium and sodium fluoroberyllates. At the end of crystallization, a mixture of $NaLiBeF_4$ and $Na_2LiBe_2F_7$ precipitates (Pontonnier and Aleonard, 1972).

Under certain conditions the elaborated compound is not an anhydrous fluoride, but a hydrate. This is the case for the alums $MCr(BeF_4)_2 \cdot 12H_2O$ (M = Rb, Cs, Tl, or NH_4) (Lari-Lavassani et al., 1969) and manganese(III) derivatives such as $Cs_2MnF_5 \cdot H_2O$ (Riss and Vituhnovskaja, 1958), $K_2MnF_5 \cdot H_2O$ (Edwards, 1971), and $Cs[MnF_4(H_2O)_2]$ (Bukovec and Kaučič, 1977).

2. Hydrothermal Syntheses

If hydroxyfluorides are prepared by hydrothermal synthesis (see Section III,A,2), single crystals either of hydrated fluorides or of anhydrous compounds can also be grown. The various possibilities of this technique have been investigated by de Pape at Le Mans University (France) and developed by Ferey et al. (1975) and Plet et al. (1979). Their apparatus is built as follows. The starting materials are introduced into a platinum capsule (9 mm in diameter and ~100 mm in length), which is then sealed and put into a chromium–steel autoclave. The filling rate in the tube is 70%, and the solvent may be water, a mixture NH_4HF_2 and H_2O, or HF of various concentrations. When the temperature increases, the internal pressure is balanced by an external pressure provided by a compressor. The upper part of the autoclave remains outside the furnace in order to create a good temperature gradient inside the tube. The working temperature is 300–400°C under 1 or 2 kbars.

This method is very suitable for growing crystals of compounds that decompose before melting, such as NH_4AlF_4, or whose formation is very slow, such as $CsAlF_4$. This technique is especially appropriate for ferric fluoride crystallization, because it can be used at temperatures below 410°C, at which the structure changes from the rhombohedral to the cubic type. If $FeF_3 \cdot H_2O$ and $FeF_3 \cdot 3H_2O$ may be isolated, depending on the nature of the solution and the working conditions, single crystals of anhydrous FeF_3 are also produced by the use of 49% HF at ~ 400°C under 1200 to 2000 bars pressure (Passaret et al., 1974).

When the structure of a compound is especially stable, the fluoride can hold out against hydrolysis. Crystals of $Na_3Li_3Al_2F_{12}$ garnet were ground under hydrothermal conditions at 300°C and 300 bars with a garnet/water ratio of 0.15 (Naka et al., 1979).

Other hydrates also crystallize under pressure. Fourquet et al. (1981) isolated $Hg_2AlF_5 \cdot 2H_2O$ in this way, although attempts to prepare the compound directly from HF solutions under atmospheric pressure failed. Single crystals of $Fe_3F_8 \cdot 2H_2O$ have been obtained in aqueous HF under 30 kbars at 250°C (Herdtweck, 1983).

C. BRIDGMAN–STÖCKBARGER METHOD

The Bridgman–Stöckbarger method consists in melting a fluoride or a mixture of fluorides in a crucible and in cooling the liquid under particular conditions, leading to the formation of a single crystal (Bridgman, 1925; Stöckbarger, 1936). The oriented solidification of the melt can be achieved in three ways: (1) lowering the crucible through a temperature gradient, (2) moving the furnace over the crucible, (3) keeping both furnace and crucible stationary and cooling the furnace along a thermal gradient.

This technique may be used with either a vertical or a horizontal tube. Figure 7 shows a Bridgman furnace in which single crystals of rare-earth trifluorides can be grown in a sealed platinum crucible (Garton and Walker, 1978). After being transferred to the crucible, the sample is evacuated at 300°C for ~ 24 hr. The crucible is then sealed off under vacuum and put into a platinum tube inside the furnace; this tube protects the furnace elements in the event of a crucible fracture. In this system the crucible, its holder, and the platinum tube remain stationary and the furnace moves upward at a constant

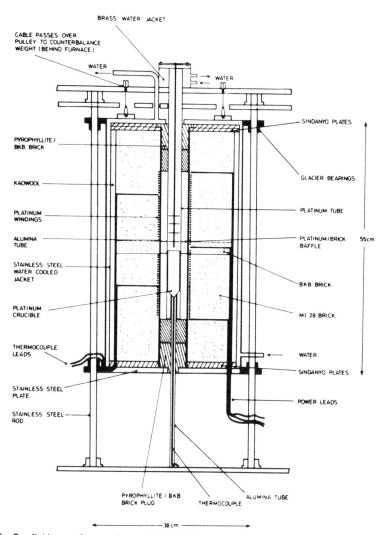

Fig. 7. Bridgman furnace for sealed crucible crystal growth (Garton and Walker, 1978). Copyright by *Materials Research Bulletin*; used by permission of the copyright owner.

speed in the range 1–10 mm hr^{-1}. The heating elements consist of two zones of resistance controlled independently. A temperature gradient of $\sim 30°$C cm^{-1} is normally used. The temperature of the sample is monitored by a thermocouple inserted into the platinum tube near the tip of the crucible. The sample is soaked at 50°C above its melting point for 24 hr, and growth rates of 1 to 5 mm hr^{-1} are used, the best crystals being obtained for 1 mm hr^{-1}.

The shape of the crucible plays a great part in crystallization. A pointed tip favors the formation of only one crystal, the cone apex angle being an important factor in the crystal quality. The principal characteristic is that some parts of the solid–liquid interface are in contact with the crucible, which ensures complete control of the shape of the crystal. Sometimes, however, this contact may result in strained crystals or in nucleation of new, differently aligned crystals.

If sealed platinum crucibles are very suitable, opened graphite crucibles are also highly satisfactory to work either in an inert atmosphere or under vacuum ($Ba_{1-x}La_xF_{2+x}$ solid solutions by Wapenaar et al., 1981). Special devices are used to improve the crystallization: a double graphite crucible to produce Ho^{3+}-doped $Er_{1-x}Y_xLiF_4$ crystals (Jones et al., 1975) and a specially designed crucible for materials having a high vapor pressure such as PbF_2 (Jones, 1976).

Table IX gives other examples of crystal growth by the Bridgman–Stöckbarger method.

To avoid hydrolysis, experiments are sometimes carried out under a flow of HF. The HF atmosphere can also be provided by decomposition of KHF_2; such a procedure is used in the Laboratoire de Chimie du Solide du CNRS, Université de Bordeaux (Fig. 8). The fluoride is introduced into a platinum crucible set on an alumina holder protected by a platinum tube. A flanged bored lid is filled with KHF_2. The crucible and its lid are capped with a closed platinum tube, which prevents gaseous HF from attacking the external alumina tube during the experiment. The whole system is brought up into the furnace, and the fluoride is kept several hours under vacuum at 110°C. The furnace is then heated and the compound is soaked for 4 or 5 hr at 50°C above its melting point under the reactive atmosphere of HF resulting from the decomposition of KHF_2. Growth is induced either by keeping the crucible stationary and cooling the furnace very slowly or by lowering the whole system containing the compound through the temperature gradient. Lithium fluoride crystals of $30 \times 30 \times 20$ mm size obtained by this method are shown in Fig. 9.

Finally, the possibility of growing small crystals by simple melting in gold or platinum tubes and slow cooling in a furnace must be pointed out. Many fluorides lead to crystals that can be used for x-ray diffraction studies, but no other details will be given on this very common technique.

D. CZOCHRALSKI METHOD

The Czochralski method (1917) is a process for single-crystal pulling, starting with a small oriented seed. The crucible contains the melt, which is

TABLE IX

Crystal Growth by the Bridgman–Stöckbarger Method

Compound	Maximal Temperature and Experimental Conditions	Shape and Dimensions (mm)	References
CsMnF₃	900°C; HF atm	$3 \times 3 \times 1$	Zalkin et al. (1962)
REF₃ [RE (rare earth) = La, Ce, Pr, or Nd]	1430–1550°C; HF atm multiport crucible	—	Robinson and Cripe (1966)
RbNiF₃	1055°C	$15 \times 5 \times 5$	Shafer et al. (1967)
	960°C	Elliptical, 10 cm³	Smolensky et al. (1967)
	—		Guggenheim (1963); Als-Nielsen (1972)
REF₃	—	$10 \times 3 \times 5$	Jones and Shand (1968)
LiYF₄	900°C; 48% YF₃ and 52% LiF	$50 \times 18 \times 18$	Shand (1969)
CaF₂-doped NdOF	Induction heater	$40 \times 12 \times 12$	Petit Le Du (1970)
RbNi₁₋ₓCoₓF₃	900°C		Elbinger et al. (1971)
CsFeF₃	800°C; HF atm	$0.5 \times 0.5 \times 0.5$	McGuire and Shafer (1971)
		$0.5 \times 0.5 \times 2$	Kestigian et al. (1966)
Sr₁₋ₓEuₓ²⁺AlF₅	Graphite cell sealed in an evacuated silica capsule		Meehan and Wilson (1972)
MF₂ (M = Ca, Sr, or Ba)	In HF diluted with He	Ingots 50 mm diameter	Pastor and Arita (1975)
MMeF₃ (M = Na or K; Me = Mg, Zn, or Cd)	Induction furnace	$40 \times 8 \times 10$	Orlov (1975)
MF₂:UF₄:CeF₃ (M = Ca, Sr, or Ba)	1450–1500°C		Catalano and Wrenn (1975)
HoF₃ and ErF₃	—		Sobolev et al. (1976)
K₂CuF₄	20°C above mp	$40 \times 15 \times 15$	Walker (1979)
Cs₂KNdF₆	820°C	50 mm³	Merchant et al. (1980)
KMn₁₋ₓMₓF₃ (M = Mg, Fe, Co, or Ni)	1330–1420°C, horizontal version	3–10 cm³	Skrzypek et al. (1980)

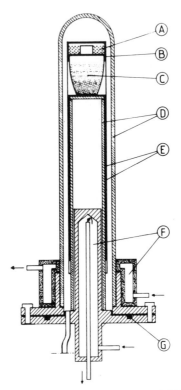

Fig. 8. Apparatus for crystal growth under HF atmosphere: A, KHF_2 powder; B, crucible and its flanged bored lid; C, melted fluoride; D, alumina tubes (holder and external tube); E, platinum sheaths; F, coolant; G, Teflon gasket.

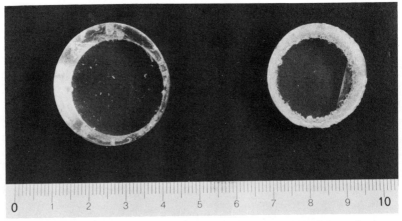

Fig. 9. Lithium fluoride crystals (growth under HF atmosphere). Scale in centimeters.

maintained at a temperature slightly higher than the melting point. The seed is put in contact with the melt, and the crystal is grown at the free top surface, so that there is no contact between the crystal and the crucible. The crystal is slowly pulled upward as it grows.

There is another version of this method, known as the Kyropoulos method (1926). Here, a strongly cooled crystal seed is brought to the surface of the liquid. The melt temperature is lowered slowly; crystallization then proceeds within the melt.

The crucible and the seed holder can be rotated in reverse directions to provide a better homogeneity of liquid concentration and temperature. An advantage of this method is that it does not produce strains due to crystal–crucible contact.

Such a technique was used by Plovnick and Camobreco (1972) for growing $RbMnF_3$ and $KMnF_3$ single crystals of high optical quality. In a typical experiment, after the material is transferred from the hydrofluorination apparatus to a platinum crucible, the furnace chamber is continuously evacuated and gradually heated up to about 300°C over a period of 24 hr. The system is then backfilled with sufficient ultra-high-purity argon gas to generate 0.1 bar overpressure at the crystal growth temperature (about 1000–1050°C). In general a (100)-cleaved crystal seed fastened with a platinum wire to the stainless steel water-cooled pulling rod is used. The crystal is pulled from the melt at a speed of about 10 mm hr^{-1} with 3 rpm rotation. Typical boule dimensions were $60 \times 20 \times 20$ mm.

Several other examples of crystal growth by the Czochralski method are given in Table X.

Mention may be made here of the use of CF_4 or C_2F_4 instead of HF as an agent of reactive atmosphere processing in the crystal growth of alkaline-earth fluorides (Pastor and Robinson, 1976).

E. MELTING-ZONE TECHNIQUE

The melting-zone technique, a well-known method of purifying many compounds, may also be suitable for growing crystals. Figure 10 shows the apparatus used in the Laboratoire de Chimie du Solide du CNRS, Université de Bordeaux. The binary or ternary fluoride is placed in a platinum boat, which is introduced into a platinum tube connected to the HF container. An external silica tube keeps this platinum tube absolutely straight when the temperature is very high. The soldered parts are kept cold with a strong flow of coolant. All other tubing is made of copper, the different parts being assembled by Monel unions equipped with Teflon ferrules. The flow rate of HF is regulated by Monel valves and controlled by passing through a perfluorated oil. The appropriate temperature is obtained by an induction coil connected to

TABLE X

Crystal Growth by the Czochralski Method

Compound	Maximal Temperature, Experimental Conditions	Shape and Dimensions (mm)	Crystal-Pulling Rate	Reference
LiYF$_4$	—	10 mm length	0.8–6 mm hr^{-1}	Gabbe and Harmer (1968)
Sr$_{1-x}$Eu$_x$$^{2+}AlF_5$	Flow N$_2$/H$_2$	Boule 5–10 mm $\varnothing$	3–6 mm hr^{-1}	Meehan and Wilson (1972)
REF$_3$ [RE (rare earth) = Gd, Dy, or Ho]	Flow HF			Pastor and Robinson (1974)
ThF$_4$	Flow HF; 1120°C	~50 × 8 × 8	—	Pastor and Arita (1974)
Ca$_{1-x}$RE$_x$F$_{2+x}$	Flow HF	~80 × 6 × 6	—	Pastor et al. (1974)
LiREF$_4$ (RE = Y, Ho, Er, or Tm)	Flow HF/He; 950°C	60 × 12 × 12	3–12 mm hr^{-1}	Pastor et al. (1975)
TmF$_3$, YbF$_3$, or LuF$_3$	—	—	—	Sobolev et al. (1976)
LiHoF$_4$	HF atm; rotation, 50 rpm	Boules up to 2 cm^3	1 mm hr^{-1}	Walker (1980)
KZnF$_3$	Crystal rotation speed, 15 rpm; crucible rotation speed, 5 rpm	50–80 mm length; 30–40 mm $\varnothing$	10 mm hr^{-1}	Gesland (1980)

Fig. 10.   Apparatus for crystal growth by the melting-zone technique.

a high-frequency generator. The compound is brought to melt at one end, and the melting zone is then moved along the boat at a speed of 0.2 to 1.6 mm min^{-1}. The quality of the crystal is controlled mainly by the rate of displacement and by the frequency of passing over. This setting makes it possible to work at 1200 to 1300°C. A crystal of $RbMnF_3$ (dimensions: 60 × 10 × 6 mm) obtained by this method is shown in Fig. 11.

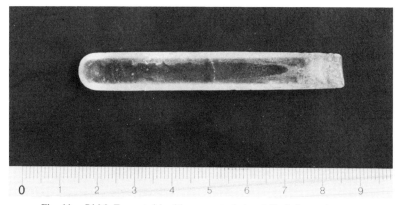

Fig. 11. $RbMnF_3$ crystal (melting-zone technique). Scale in centimeters.

F.   TRANSPORT METHOD

The transport technique involves the transport of a material as a chemical compound (frequently a halide) that decomposes in the growth area. Many details on the criteria for the choice of transport reactions have been given by Schäfer (1962). In brief, this method is possible only when, aside from the transported solid material, only gases participate in the reaction, so that all the components are mobile throughout the gas phase. The reaction must also be reversible.

Although this technique has been developed for many compounds, it has very seldom been used for fluorides, probably because it is difficult to find the appropriate transport agent. For instance, Cl_2 and HCl are not very suitable if the reactions are carried out in quartz ampoules because the gaseous phase resulting from the chemical attack of the fluorides contains either fluorine or HF.

The method is interesting in that it enables one to obtain compounds in their low-temperature form. Bonnamy et al. (1978) succeeded, for instance, in isolating crystals of AlF_3, TiF_3, and FeF_3 by using $SiCl_4$ as transport agent. The chemical transport was performed in silica tubes 15 mm in diameter and 200 mm in length placed in a temperature gradient of $\sim 10°C\ cm^{-1}$, the experiment duration varying from 200 to 600 hr. Kestigian et al. (1966) have also demonstrated the crystallization of FeF_2 by heating up to 950 to 1000°C either $FeF_2 \cdot xH_2O$ or a mixture of Fe_2O_3 and C in an HF/Ar carrier gas transport system.

G.   GROWTH FROM THE VAPOR PHASE

The principle of growth from the vapor phase is very simple. It involves the direct transport of material by sublimation from a hot source and condensation on a cool growth region. It can be used either in vacuum or with a moving gas stream and sometimes under pressure. For instance, if an HF/H_2 mixture is passed over NbH at 570°C in an Inconel tube, whereas NbF_3 is formed in the hot part, NbF_5 crystals are condensed on the cooled walls of the tube (Ehrlich et al., 1964).

Needles of AuF_3 are formed after 12 to 14 hr of treating freshly precipitated gold powder with fluorine gas at 10 bars and 350°C in a Monel reaction vessel, the top of which is cooled to 20°C (Einstein et al., 1967). Iron trifluoride is obtained from the vapor phase by heating a hydrated fluoride in HF at 1000°C, but most of the crystals are multiply twinned (Kurtzig and Guggenheim, 1970).

Crystals of many compounds can be obtained by this method, but most of them are small and useful only for crystal structure determination. The following examples may be mentioned:

$BrF_4{}^+Sb_2F_{11}{}^-$: By slow sublimation at $30°C$ in a Teflon–FEP system (Lind and Christe, 1972)

$[XeF_5{}^+]_2PdF_6{}^{2-}$: By heating $4XeF_6·PdF_4$ up to $400°C$ in a bomb under 66 bars F_2, then cooling slowly down to room temperature, excess F_2 and XeF_6 being removed under vacuum (Leary et al., 1973)

RhF_5: By vacuum sublimation in a quartz tube, the parent sample at $\sim100°C$ (Morrell et al., 1973)

Sometimes large single crystals can be obtained; this is the case for AlF_3 doped with transition metal ions, which crystallizes from $AlF_3·xH_2O$ heated rapidly in a 500-cm^3 platinum crucible up to $1200°C$ and held at this temperature for 20 hr (Grecu and Wanklyn, 1968).

H. FLUX GROWTH TECHNIQUE

Growth from a melting flux provides a simple way of preparing crystals. This method has been developed for several years in the fluorides field largely by B. M. Wanklyn at Oxford University and by R. de Pape in Le Mans. The method is particularly suitable for preparing doped crystals when various dopant concentrations are required. It also allows compounds that melt incongruently to grow at relatively low temperatures.

Some essential properties are required for a good solvent: low melting point far removed from the boiling point, high dissolving capacity, low viscosity and volatility, low toxicity, ease of removal of the crystals after growth, compatibility with the crucible, and no formation of solid solutions with the material that is grown.

It is sometimes difficult to find a solvent that complies with all these criteria, but a large number have many of the desired attributes. For instance, PbF_2 has been used in the crystallization of K_3CrF_6 or K_2NaCrF_6 (Garton and Wanklyn, 1967) but it has not proved to be very satisfactory as a flux for $KNiF_3$ or NiF_2. The latter gives clear rods in KCl or $PbCl_2$ melt; the addition of a small percentage of PbF_2 to $PbCl_2$ is very useful to reduce nucleation (Wanklyn, 1969). Traces of oxides or adsorbed water, which forms oxides by hydrolysis at higher temperatures, can be removed by incorporating ammonium or potassium difluorides to the flux. This technique proved to be effective for many ternary fluorides such as K_2FeF_5, $RbFeF_4$, K_2NiF_4, and $LiRF_4$ (R = Gd to Er) (Wanklyn, 1975).

The use of chloride mixtures has become more common, however, due to a number of possible advantages:

1. Chlorides melt at relatively low temperatures and may easily dissolve fluorides. Their crystallization temperatures are therefore largely

lowered, and the loss of volatile components is reduced. This improves the crystalline quality.

2. Flux composition can be easily modified and will be chosen as near as possible to that of the eutectic mixture.
3. Traces of oxides are trapped owing to the formation of possible stable oxide–chlorides.
4. Chlorides and fluorides do not form solid solutions.
5. Crystals are protected by the flux against a contingent external pollution.
6. Chlorides are usually soluble in water and can therefore be easily removed from the melt.
7. The crystals are faceted, and this feature makes their orientation easier.

To obtain a maximum yield of crystals and to avoid side crystallization, the phase diagrams of chloride mixtures were used by Nouet *et al.* (1971), who succeeded in preparing fluorinated perovskites and pyrochlores as follows:

$$5\,RbCl + 3\,CoF_2 \longrightarrow 2\,RbCoF_3 + 3\,RbCl + CoCl_2$$

$$8\,RbCl + CoCl_2 + 3\,CoF_2 + 2\,CrF_3 \longrightarrow 2\,RbCoCrF_6 + 6\,RbCl + 2\,CoCl_2$$

These reactions show that exchange occurs between the solvent and the solute. In many cases, however, the crystals isolated are those of the starting fluoride; for instance, CrF_3 crystallizes from a $PbCl_2 + NH_4HF_2$ flux with CrF_3 as starting material (Wanklyn, 1969).

Other examples of crystal growth by the flux technique are given in Table XI.

Most of the experiments are carried out in platinum crucibles, but platinum alloyed with 5 or 10% rhodium also gives good devices, the incorporation of rhodium increasing the mechanical and thermal strengths. A flanged lid may closely fit the rim of the crucible to avoid sublimation of flux, but it is better to use a welded lid topped by a small-section tube, allowing the crucible to be filled and sealed easily under argon. Molybdenum or graphite crucibles provided with graphite lids are also highly satisfactory for work under high vacuum or under controlled atmosphere in a nickel or Monel reactor. The furnace must possess a great thermal inertia. Crystal growth takes place as follows. After heating in order to melt the flux, the furnace is kept at uniform temperature for several hours so that complete dissolution occurs. Then the furnace is slowly cooled at the rate of 0.5 to $10°C\,hr^{-1}$. When the flux becomes solid, the temperature is allowed to decrease at a faster rate (at $\sim 50°C\,hr^{-1}$) down to ambient temperature. Crystals are separated from the flux by dissolution in water or methanol.

A simple apparatus for flux growth under reducing conditions has been set up by Garrard *et al.* (1974) (Fig. 12). A nickel tube is mounted coaxially within a mullite tube and held in position by pyrophyllite spacers; these are loosely

TABLE XI

Examples of Crystal Growth by the Flux Technique

Compound	Composition of the Mixtures (Maximal Temperature, °C)	Shape and Dimensions (mm)	Color	Reference
Na_2NiCrF_7	$7\ NaCl + NiCl_2 + 4\ NiF_2 + 2\ CrF_3 \longrightarrow$ $2\ Na_2NiCrF_7 + 3\ NaCl + 3\ NiCl_2$ (800)	Prismatic, $\sim 1\ mm^3$	Dark green	Miranday (1972)
$CsNa_{0.5}Al_{1.5}F_6$	In 25% NaCl/75% CsCl (in moles) flux (800)	Truncated rhombohedron, $0.2\ mm^3$	Colorless	Courbion et al. (1974)
α-$Na_5Cr_3F_{14}$	$5\ NaF + 3\ CrF_3 + 5\ NaCl + 5\ CoCl_2$ (750)	Prismatic, $2 \times 2 \times 2$	Dark green	Miranday et al. (1975)
Na_2MFeF_7 [M(II) = Mn or Ni]	$7\ NaCl + MCl_2 + 4\ MF_2 + 2\ FeF_3 \longrightarrow$ $2\ Na_2MFeF_7 + 3\ NaCl + 3\ MCl_2$ (600)	Parallelepipedic, $2 \times 3 \times 2$ Prismatic, $2 \times 2 \times 1$	Light brown Light green	Tressaud et al. (1976d)
$K_5V_3F_{14}$	$5.3\ g\ VF_3 + 1.3\ g\ V + 15.6\ g\ KF + 65\ g\ PbCl_2$ $(H_2/N_2\ atm)$ (850)	Rectangular, $6 \times 4 \times 1$	Bright green	Wanklyn et al. (1976)
α-$NaTiF_4$	$3.34\ NaF + 3.34\ TiF_3 + 6\ NaCl + 6\ MnCl_2$ (800)	Small plates	Light brown	Omaly et al. (1976)
$KNiF_3$	$11.6\ g\ KF + 9.7\ g\ NiF_2 + 38.5\ g\ PbCl_2 +$ $3.5\ g\ NH_4HF_2$ (900)	5 mm on edge	—	Safa et al. (1977)
$LiTh_4F_{17}$	$LiF + 4\ ThF_4 + 2\ LiCl + 2\ ZnCl_2$ (600)	Rectangular prism	Colorless	Cousson et al. (1977)
K_2FeF_5	$2\ KF + FeF_3 + excess\ PbCl_2$ (650)	$4.7 \times 1.4 \times 1.2$ $4 \times 3 \times 3$	Colorless	Vlasse et al. (1977)
$KTiF_4$	$KTiF_4 + 3\ KCl + 3\ ZnCl_2$	Small plates, $0.28 \times 0.26 \times 0.07$	Red-brown	Sabatier et al. (1979)
NdF_3	27.7% NdF_3 + 28.9% KCl + 43.4% $NdCl_3$ (in moles) (910)	Plates, $20 \times 7 \times 0.6$	Pink	Aride et al. (1980)
$Cs_4CoCr_4F_{18}$	$18\ CsCl + 3\ CoCl_2 + 2\ CsF + 2\ CoF_2 + 4\ CrF_3$ $\longrightarrow Cs_4CoCr_4F_{18} + 4\ CoCl_2 + 16\ CsCl$	Plates	Greenish	Courbion et al. (1983)

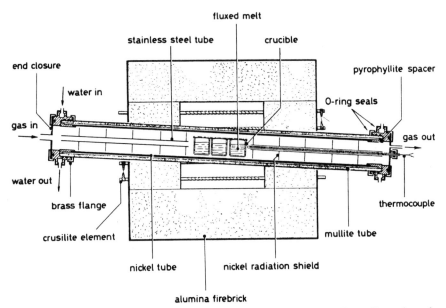

Fig. 12. Controlled-atmosphere furnace for flux growth technique. From Garrard *et al.* (1974). Copyright by the *Journal of Crystal Growth*; used by permission of the copyright owner.

fitted so that nitrogen can enter the space between the tubes. Water-cooled brass flanges with simple O-ring seals are fitted to the end of the mullite tube, and the joints are made gas tight. Nickel radiation shields are mounted on stainless steel tubes and inserted from either end. The incline ensures that the volatile materials that are swept along the tube by a stream of nitrogen to the lower end remain there after condensation. The temperature distribution in the central region is suitable for flux growth in several crucibles simultaneously.

An improved apparatus constructed by the same authors differs from that just described in that the assembly is vertical and up to six crucibles, each 30 mm in height, can be heated simultaneously under conditions that are satisfactory for crystal growth (Garrard *et al.*, 1975).

I. GROWTH IN GELS

Gel growth conventionally involves the diffusion of two soluble reactants through a gel to a region in which they react to form a relatively insoluble product in crystalline form. Typical gels are 1–5% agar or silica gel.

This technique has been investigated by Henisch *et al.* (1965) and Dennis and Henisch (1967). The apparatus tends to be as simple as possible. The gel, with one of the reactants incorporated, is formed in a test tube, and a solution of the second reactant is added after setting. In an alternative method, a neutral gel is set at the base of a U tube, and solutions of two reactants are added via the two arms. A particular advantage of the method is that the grown crystal is held in the gel in a strain-free manner. The growth is usually limited by diffusion, however, and a period of 2 or 3 weeks is necessary to obtain crystals with dimensions of several millimeters.

This method has very seldom been used for growing crystals of fluorides. Leckebusch (1974), however, succeeded in isolating crystals of NH_4MnF_3 and analogous $AMnF_3$ compounds (A = K, Rb, Cs, or Tl). Silica gel cannot be used owing to the formation of Na_2SiF_6 in the presence of F^-. The appropriate gel is agar-agar, which is mixed with a solution containing $MnCl_2$. After 5 mins of heating until the gel becomes hard, a solution of alkali fluoride or alkali difluoride is placed above so that the following reaction occurs:

$$A^+ + Mn^{2+} + 3 F^- \longrightarrow AMnF_3$$

The best results are obtained after 4 weeks at a growth temperature of $35°C$.

VIII. Thin Films and Ceramics

Thin-film ion conductors have potential applications in solid-state electrochromic devices for information processing and display. Multilayer circuits, however, are more and more frequently used in the microelectronics field. The excellent optical properties of fluorides, such as light emission or transmission, a good transparency in the visible and near-ir range, and their ionic conductivity properties have led to attempts at their elaboration in thin films.

There are two techniques for making such deposits: sputtering and vapor deposition. In the former the fluoride is pressed to give a target of 50 mm in diameter, fastened on a holder kept cold by a flow of coolant; atomization is realized with dynamic argon plasma under 8×10^{-3} torr at various rates depending on incident power. In the latter the powdered sample is placed into a molybdenum, tantalum, or platinum crucible that is resistance heated at an appropriate temperature. The evaporation is realized under vacuum (10^{-6} torr), and the vapors may be condensed on various substrates: carbon, glass, mica, etc. Several parameters have a prominent part in thin-film formation: vacuum; evaporation rate, depending on the source temperature; substrate temperature; and cooling rate of the substrate after layer deposition.

Salardenne et al. (1975) made a thin-film preparation of MgF_2 and AlF_3. Barrière (1976) showed that MgF_2 deposits are crystallized but may be amorphous when the substrate temperature is maintained below $180°C$. Thin films of FeF_3 were obtained by Lachter et al. (1978) by fluorination of iron thin films or by sublimation of FeF_3 in a fluorinating atmosphere. Several methods were investigated by Gevers et al. (1980) to obtain thin films of NiF_2: fluorination of nickel plates and sublimation of fluoride powder under vacuum or under a partial pressure of fluorine. They found that the growth rate of thin films obtained by fluorination was highly dependent on the temperature and that sublimation of the fluoride led to a granulous texture. Epitascial solid solutions of fluorides have been grown as passivating layers onto III–V semiconductors (Barrière et al., 1985a,b).

Thin films with m/n varying between 3:1 and 1:3 for $mLiF \cdot nAlF_3$ have been evaporated under vacuum onto glass substrates kept at room temperature. The colorless transparent films are amorphous even after annealing at 600°C (Oi and Miyauchi, 1981). Similar experiments have been carried out in the $mNaF \cdot nMF_3$ systems (M = Al or Cr), and the ionic conductivity of the amorphous thin films has been studied (Dance and Oi, 1983). More details on such deposits for electrochromic devices are be given in Chapter 15.

Finally, we note that fluorine compounds have useful refractory and nuclear applications. For instance, calcium fluoride has been slip cast in ethyl alcohol to avoid erosion problems (Masson et al., 1963). The influence of various fluorides on densification and dielectric properties of polycrystalline barium titanate has also been determined (Beauger et al., 1982); so far, only LiF or a combination of lithium with a fluoride has allowed densification of $BaTiO_3$ at low temperature. The dielectric properties of ceramics result mainly from the role played by fluorine and its associate cation.

IX. Some Applications of Fluorides

It is not the aim of this section to list all the uses of fluorides; Chapter 21 is devoted to industrial applications and prospects. One can briefly recall the importance of these compounds in nuclear industry and siderurgy, their use in surface treatment of metals, and their use as abrasives and flux for brazing. The particular properties of many fluorides will lead to many applications in optics and electrochemistry: lasers, electrochromic devices, solid galvanic cells, etc.

Mention can also be made of the synthesis of organic fluoride derivatives using inorganic fluorides; such fluorinating agents can be used either in solid state or as molten salts. For instance, Booth (1935) replaced chlorine by fluorine in chloride derivatives by means of SbF_3 in a four-center reaction, whereas Klauke et al. (1983) succeeded in analogous substitution by inter-action of tetrachloropyridazine with NaF or KF.

The most significant results have been obtained with fluorides containing elements in an unstable oxidation state, such as MnF_3 and PbF_4 (Stacey and Tatlow, 1960) and especially CoF_3 (Fowler et al., 1951) and AgF_2 (Glemser and Richert, 1961; Sheppard, 1962). Fluorination of various organic compounds thanks to ternary fluorides also gives significant yields. $CsCoF_4$ is a good reagent for fluorinating arenes (Edwards *et al.*, 1972; Bailey *et al.*, 1975) or pyridine and related compounds (Plevey *et al.*, 1982a,b). $KCoF_4$ reacts in a similar way toward pyridine (Coe *et al.*, 1982). The introduction of fluorine into organic molecules has also been carried out by the reaction of XeF_2 in various solvents (Nikolenko *et al.*, 1970; Shaw *et al.*, 1971; Anand and Filler, 1976; Zupan and Pollak, 1976).

References

Als-Nielsen, J., Birgeneau, R. J., and Guggenheim, H. J. (1972). *Phys. Rev. B: Solid State* [3] **6**, 2030.
Anand, S. P., and Filler, R. (1976). *J. Fluorine Chem.* **7**, 179.
Anthony, A. M., and Collongues, R. (1972), *In* "Preparative Methods in Solid State Chemistry" (P. Hagenmuller, ed.), p. 147. Academic Press, New York.
Aride, J., Chaminade, J. P., and Pouchard, M. (1980). *J. Fluorine Chem.* **15**, 117.
Asker, W. J., and Wylie, A. W. (1965). *Aust. J. Chem.* **18**, 959.
Asprey, L. B., Margrave, J. L., and Silverthorn, M. E. (1961). *J. Am. Chem. Soc.* **83**, 2955.
Avignant, D., and Cousseins, J. C. (1978). *Rev. Chim. Miner.* **15**, 360.
Aynsley, E. E., Peacock, R. D., and Robinson, P. L. (1950). *J. Chem. Soc.* p. 1622.
Aynsley, E. E., Peacock, R. D., and Robinson, P. L. (1952). *Chem. Ind (London)* p. 1002.
Babel, D. (1967). *Struct. Bonding (Berlin)* **3**, 1.
Babel, D., Haegele, R., Pausewang, G., and Wall, F. (1973). *Mater. Res. Bull.* **8**, 1371.
Bailey, J., Plevey, R. G., and Tatlow, J. C. (1975). *Tetrahedron Lett.* p. 869.
Barber, E. J., and Cady, G. H. (1956). *J. Phys. Chem.* **60**, 505.
Barrière, A. S. (1976). Thesis, University of Bordeaux, France.
Barrière, A. S., Couturier, G., Grannec, J., Ricard, H., and Sribi, C. (1985a). *Commun. Int. Conf. Format. Semicond. Interf., Marseille.*
Barrière, A. S., Couturier, G., Ricard, H., Sribi, C., and Tressaud, A. (1985b). *Commun. Europ. Symp. Molec. Beam Epitascy, Aussois.*
Bartlett, N. (1962). *Proc. Chem. Soc., London* p. 218.
Bartlett, N. (1965). *Prep. Inorg. React.* **2**, 301.
Bartlett, N. (1968). *Angew. Chem.* **7**(6), 433.
Bartlett, N., and Jha, N. K. (1963). *In* "Noble Gas Compounds" (H. H. Hyman, ed.), p. 23. Univ. of Chicago Press, Chicago, Illinois.
Bartlett, N., and Leary, K. (1976). *Rev. Chim. Miner.* **13**, 82.
Bartlett, N., and Lohmann, D. H. (1962a). *Proc. Chem. Soc., London* p. 115.
Bartlett, N., and Lohmann, D. H. (1962b). *Proc. Chem. Soc., London* p. 5253.
Bartlett, N., and Lohmann, D. H. (1964). *J. Chem. Soc.* p. 619.
Bartlett, N., and Quail, J. W. (1961). *J. Chem. Soc.* p. 3728.
Bartlett, N., and Rao, P. R. (1964). *Proc. Chem. Soc., London* p. 393.
Bartlett, N., and Rao, P. R. (1965). *Chem. Commun.* p. 252.
Bartlett, N., and Robinson, P. L. (1956). *Chem. Ind. (London)* p. 1351.
Bartlett, N., and Tressaud, A. (1974). *C. R. Hebd. Seances Acad. Sci.* **278**, 1501.
Bartlett, N., and Wechsberg, M. (1971). *Z. Anorg. Allg. Chem.* **385**, 5.
Bartlett, N., and Yu, K. C. (1961). *Can. J. Chem.* **39**, 80.
Bartlett, N., Beaton, S. P., and Jha, N. K. (1966a). *Chem. Commun.* p. 168.
Bartlett, N., Einstein, F., Stewart, D. F., and Trotter, J. (1966b). *Chem. Commun.* p. 550.

Bartlett, N., Gennis, M., Gibler, D. D., Morrell, B. K., and Zalkin, A. (1973). *Inorg. Chem.* **12,** 1717.
Bartlett, N., Žemva, B., and Graham, L. (1976). *J. Fluorine Chem.* **7,** 301.
Batsanova, L. R., Zakharev, Y. V., and Opalovskii, A. A. (1973). *Zh Neorg. Khim.* **18,** 905.
Baur, W. H. (1958). *Acta Crystallogr.* **11,** 488.
Beauger, A., Lagrange, A., Houttemane, C., and Ravez, J. (1982). *Proc. Int. Conf. New Trends Passive Components,* p. 10.
Beck, L. K., Haendler Kugler, B., and Haendler, H. M. (1973). *J. Solid State Chem.* **8,** 312.
Belmore, E. A., Ewalt, W, M., and Wodjcik, B. H. (1947). *Ind. Eng. Chem.* **39,** 340.
Billy, C., and Haendler, H. M. (1957). *J. Am. Chem. Soc.* **79,** 1049.
Birchall, T., Dean, P. A. W., and Gillespie, R. J. (1971). *J. Chem. Soc. A* p. 1777.
Bohinc, M., Grannec, J., Slivnik, J., and Žemva, B. (1976). *J. Inorg. Nucl. Chem.* **38,** 75.
Bonnamy, C., Launay, J. C., and Pouchard, M. (1978). *Rev. Chim. Miner.* **15,** 178.
Booth, H. S. (1935). *J. Am. Chem. Soc.* **57,** 1333.
Bougon, R. (1968). *C. R. Hebd. Seances Acad. Sci.* **267,** 681.
Bougon, R., and Charpin, P. (1979). *J. Fluorine Chem.* **14,** 235.
Bougon, R., and Lance, M. (1983). *C. R. Hebd. Seances Acad. Sci.* **297,** 117.
Bougon, R., Ehretsmann, J., Portier, J., and Tressaud, A. (1972). *In* "Preparative Methods in Solid State Chemistry" (P. Hagenmuller, ed.), p. 401. Academic Press, New York.
Bougon. R., Costes, R. M., Desmoulin, J. P., Michel, J., and Person, J. L. (1976). *J. Inorg. Nucl. Chem.,* Suppl. p. 99.
Bougon, R., Bui Huy, T., Burgess, J , Christe, K. O., and Peacock, R. D. (1982). *J. Fluorine Chem.* **19,** 263.
Bradley, R. S. (1966). "Advances in High Pressure Research," Vol. 1, Chapters 1–4. Academic Press, New York.
Brauer, G. (1960). *In* "Handbuch der Präparativen Anorganishen Chemie," 2nd ed., Vol. 1, p. 246. Enke, Stuttgart.
Brčič, B. S., and Slivnik, J. (1955). "*J. Stefan*" *Inst., Rep.* **2,** 541.
Bridgman, P. W. (1925). *Proc. Am. Acad. Arts Sci.* **60,** 305.
Bright, N. F. H., and Wurm, J. G. (1958). *Can. J. Chem.* **36,** 615.
Bukovec, P. (1974). *Monatsh. Chem.* **105,** 517.
Bukovec, P., and Kaučič, V. (1977). *J. Chem. Soc., Dalton Trans.* p. 945.
Cady, G. H. (1950) *In* "Fluorine Chemistry" (J. H. Simons, ed.), Vol. 1, p. 311. Academic Press, New York.
Canterford, J. H., and O'Donnell, T. A. (1968). *Tech. Inorg. Chem.* **7,** 273.
Catalano, E., and Wrenn, E. W. (1975). *J. Cryst. Growth* **30,** 54.
Cavell, R. G., and Clark, H. C. (1962). *J. Chem. Soc.* p. 2692.
Champarnaud-Mesjard, J. C., Frit, B., and Gaudreau, B. (1978). *Rev. Chim. Miner.* **15,** 328.
Charpin, P., and Macheteau, Y. (1975). *C. R. Hebd. Seances Acad. Sci.* **280,** 61.
Charpin, P., Dianoux, A. J., Marquet-Ellis, H., and Nguyen-Nghi (1966). *C. R. Hebd. Seances Acad. Sci.* **263,** 1359.
Charpin, P., Marquet-Ellis, H., Nguyen-Nghi, and Plurien, P. (1972). *C. R. Hebd. Seances Acad. Sci.* **275,** 1503.
Christe, K. O. (1972). *Inorg. Chem.* **11,** 1215.
Christe, K. O. (1980). U.S. Patent **4,** 207,124.
Christe, K. O., and Sawodny, W. (1969). *Inorg. Chem.* **8,** 212.
Christe, K. O., Guertin, J. P., and Pavlath, A. E. (1966). *Inorg. Chem.* **5,** 1921.
Christe, K. O., Wilson, R. D., and Goldberg, I. A. (1976a). *Inorg. Chem.* **15,** 1271.
Christe, K. O., Schack, C. J., and Wilson, R. D. (1976b). *Inorg. Chem.* **15,** 1275.
Christe, K. O., Schack, C. J., and Wilson, R. D. (1977). *Inorg. Chem.* **16,** 849.
Claassen, H. H., Selig, H., and Malm, J. G. (1962). *J. Am. Chem. Soc.* **84,** 3593.
Clark, H. C., and Emeleus, H. J. (1957). *J. Chem. Soc.* p. 119.

Clark, H. C., and Sadana, Y. N. (1964). *Can. J. Chem.* **42**, 50.

Coe, P. L., Holton, A. G., and Tatlow, J. C. (1982). *J. Fluorine Chem.* **21**. 171.

Cohen, S., Selig, H., and Gut, R. (1982). *J. Fluorine Chem.* **20**, 349.

Comings, E. W. (1956). "High Pressure Technology." McGraw-Hill, New York.

Courbion, G., Jacoboni, C., and de Pape, R. (1974). *Mater. Res. Bull.* **9**, 425.

Courbion, G., de Pape, R., Knoke, G., and Babel, D. (1983). *J. Solid State Chem.* **49**, 353.

Cousson, A., Pages, M., Cousseins, J. C., and Vedrine, A. (1977). *J. Cryst. Growth* **40**, 157.

Cox, B., and Sharpe, A. G. (1953). *J. Chem. Soc.* p. 1783.

Cros, C., (1978). *Rev. Inorg. Chem.* **1**(2), 163.

Cros, C., Feurer, R., and Pouchard, M. (1975). *J. Fluorine Chem.* **5**, 457.

Czochralski, J. (1917). *Z. Phys. Chem.* **92**, 219.

Dance, J. M. (1974). Thesis, University of Bordeaux, France.

Dance, J. M., and Oi, T. (1983). *Mater. Res. Bull.* **18**, 263.

Dance, J. M., Grannec, J., Jacoboni, C., and Tressaud, A. (1974). *C. R. Hebd. Seances Acad. Sci.* **279**, 601.

Dance, J. M., Grannec, J., and Tressaud, A. (1975). *C. R. Hebd. Seances Acad. Sci.* **281**, 91.

Dean, P. A. W. (1975). *J. Fluorine Chem.* **5**, 499.

Delaigue, A., and Cousseins, J. C. (1972). *Rev. Chim. Miner.* **9**, 789.

Demazeau, G. (1973). Thesis, University of Bordeaux, France.

Demazeau, G., Grannec, J., Marbeuf, A., Portier, J., and Hagenmuller, P. (1969). *C. R. Hebd. Seances Acad. Sci.* **269**, 987.

Demazeau, G., Menil, F., Portier, J., and Hagenmuller, P. (1971). *C. R. Hebd. Seances Acad. Sci.* **273**, 1641.

Dennis, J., and Henisch, H. K. (1967). *J. Electrochem. Soc.* **114**, 263.

de Pape, R. (1965). *C. R. Hebd. Seances Acad. Sci.* **260**, 4527.

de Pape, R., and Ravez, J. (1962). *C. R. Hebd. Seances Acad. Sci.* **254**, 4171.

de Pape, R., Portier, J., Gauthier, G., and Hagenmuller, P. (1967). *C. R. Hebd. Seances Acad. Sci.* **265**, 1244.

Dodge, B. F. (1950). *In* "Chemical Engineers Handbook" (J. H. Perry, ed.), p. 1233. McGraw-Hill, New York.

Dubler, E., Linowsky, L., Matthieu, J. P., and Oswald, H. R. (1977). *Helv. Chim. Acta* **60**, 1589.

Dudley, F. B., Gard, G., and Cady, G. H. (1963). *Inorg. Chem.* **2**, 228.

Dumora, D. (1971). Thesis, University of Bordeaux, France.

Dumora, D., and Ravez, J. (1969). *C. R. Hebd. Seances Acad. Sci.* **268**, 337.

Dumora, D., Ravez, J., and Hagenmuller, P. (1970). *Bull. Soc. Chim. Fr.* **5**, 1751.

Edwards, A. J. (1971). *J. Chem. Soc. A.* p. 2653.

Edwards, A. J., Holloway, J. H., and Peacock, R. D. (1963). *Proc. Chem. Soc., London* p. 275.

Edwards, A. J., Plevey, R. G., Salomi, I. J., and Tatlow, J. C. (1972). *J. Chem. Soc., Chem. Commun.* p. 1028.

Ehrlich, P., and Pietzka, G. (1954). *Z. Anorg. Allg. Chem.* **275**, 121.

Ehrlich, P., Ploeger, F., and Koch, E. (1964). *Z. Anorg. Allg. Chem.* **333**, 209.

Einstein, F. W. B., Rao, P. R., Trotter, J., and Bartlett, N. (1967). *J. Chem. Soc. A* p. 478.

Elbinger, G., Jäger, E., Keilig, W., and Perthel, R. (1971). *J. Phys., Colloq.* **32**, 626.

Emeleus, J. H., and Gutmann, V. (1950). *J. Chem. Soc.* p. 2115.

Emori, S., Inoue, M., Kishita, M., and Kubo, M. (1969). *Inorg. Chem.* **8**, 1385.

Ensslin, F., and Dreyer, H. (1942). *Z. Anorg. Allg. Chem.* **249**, 119.

Fairbrother, F., and Frith, W. C. (1951). *J. Chem. Soc.* p. 3051.

Falconer, W. E., and Sunder, W. A. (1967). *J. Inorg. Nucl. Chem.* **29**, 1380.

Falconer, W. F., Di Salvo, F. J., Edwards, A. J., Griffiths, J. E., Sunder, W. A., and Vasile, M. J. (1976). *J. Inorg. Nucl. Chem., Suppl.* p. 59.

Feldner, K., and Hoppe, R. (1983). *Rev. Chim. Miner.* **20**, 351.

Ferey, G., Leblanc, M., de Pape, R., Passaret, M., and Bothorel-Razazi, M. P. (1975). *J. Cryst. Growth* **29**, 209.

Ferey, G., Leblanc, M., and de Pape, R. (1981). *J. Solid State Chem.* **40**, 1.

Fleischer, T., and Hoppe, R. (1982a). *Z. Anorg. Allg. Chem.* **489**, 7.

Fleischer, T., and Hoppe, R. (1982b). *Z. Anorg. Allg. Chem.* **490**, 7.

Fleischer, T., and Hoppe, R. (1982c). *Z. Anorg. Allg. Chem.* **492**, 76.

Fonteneau, G., and Lucas, J. (1974). *J. Inorg. Nucl. Chem.* **36**, 1515.

Fourquet, J. L., Plet, F., and de Pape, R. (1981). *Acta Crystallogr. Sect. B* **B37**, 2136.

Fowler, R. D., Burford, W. B., Hamilton, J. M., Sweet, R. C., Weber, C. E., Kasper, J. S., and Litant, I. (1951). *In* "Preparation Properties and Technology of Fluorine and Organo Fluoro-compounds" (C. Slesser and S. R. Schram, eds.). McGraw-Hill, New York.

Friebel, C., Pebler, J., Steffens, F., Weber, M., and Reinen, D. (1983). *J. Solid State Chem.* **46**, 253.

Frlec, B., and Holloway, J. H. (1973). *J. Chem. Soc., Chem. Commun.* p. 370.

Frlec, B., and Hyman, H. (1967). *Inorg. Chem.* **6**, 2233.

Frlec, B., Brčič, B. S., and Slivnik, J. (1966). *Inorg. Chem.* **5**, 542.

Frlec, B., Gantar, D., and Holloway, J. H. (1982). *J. Fluorine Chem.* **19**, 485.

Gabbe, D., and Harmer, A. L. (1968). *J. Cryst. Growth* **3-4**, 544.

Gakle, P. S. (1960). "Design Handbook for Fluorine Ground, Handling Equipment, " WADD Tech. Rep. 60159, Sect. 8.

Garrard, B. J., Wanklyn, B. M., and Smith, S. H. (1974). *J. Cryst. Growth* **22**, 169.

Garrard, B. J., Smith, S. H., Wanklyn B. M., and Garton, G. (1975). *J. Cryst. Growth* **29**, 301.

Garton, G., and Walker, P. J. (1978). *Mater. Res. Bull.* **13**, 129.

Garton, G., and Wanklyn, B. M. (1967). *J. Cryst. Growth* **1**, 49.

Gesland, J. Y. (1980). *J. Cryst. Growth* **49**, 771.

Gevers, G., Lachter, A., Salagoïty, M., Barrière, A. S., and Lozano, L. (1980). *J. Cryst. Growth* **49**, 45.

Gibler, D. D., Adams, C. J., Fischer, M., Zalkin, A., and Bartlett, N. (1972). *Inorg. Chem.* **11**, 2325.

Gillardeau, J. (1970). *In* "Oxidation of Metals" (K. Hauffe, ed.), Vol. 2, No. 3. Plenum, New York.

Gillespie, R. J., and Schrobilgen G. J. (1976). *Inorg. Chem.* **15**, 22.

Gilman, J. J. (1963). "The Art and Science of Growing Crystals." Wiley, New York.

Gleichmann, J. R., Smith, E. A., Trond, S. S., and Ogle, P. R. (1962). *Inorg. Chem.* **1**, 661.

Glemser, O., and Richert, H. (1961). *Z. Anorg. Allg. Chem.* **307**, 313.

Glemser, O., Roesky, H., and Hellberg, K. H. (1963). *Angew. Chem.* **75**, 346.

Glemser, O., Roesky, H., Hellberg, K. H., and Werther, H. U. (1966). *Chem. Ber.* **99**, 2652.

Gortsema, F. P., and Didchenko, R. (1965). *Inorg. Chem.* **4**, 182.

Grannec, J. (1970). Thesis, University of Bordeaux, France.

Grannec, J., Tressaud, A., and Portier, J. (1970). *Bull. Soc. Chim. Fr.* **5**, 1719.

Grannec, J., Lozano, L., Portier, J., and Hagenmuller, P. (1971). *Z Anorg. Allg. Chem.* **385**, 26.

Grannec, J., Sorbe, P., Portier, J., and Hagenmuller, P. (1975a). *C. R. Hebd. Seances Acad. Sci.* **280**, 45.

Grannec, J., Lozano, L., Sorbe, P., Portier, J., and Hagenmuller, P. (1975b). *J. Fluorine Chem.* **6**, 267.

Grannec, J., Sorbe, P., and Portier, J. (1976). *C. R. Hebd. Seances Acad. Sci.* **283**, 441.

Grannec, J., Chartier, C., Reau, J. M., and Hagenmuller, P. (1983). *Solid State Ionics* **8**, 73.

Grecu, V. V., and Wanklyn, B. M. (1968). *J. Phys. C* [2] **1**, 387.

Gruner, E., and Klemm, W. (1937). *Naturwissenschaften* **25**, 59.

Guggenheim, H. J. (1963). *J. Appl. Phys.* **34**, 2482.

Günter, J. R., Matthieu, J. P., and Oswald, H. R. (1978). *Helv. Chim. Acta* **61**, 328.

Haendler, H. M., and Bernard, W. J. (1951). *J. Am. Chem. Soc.* **73**, 5218.

Haendler, H. M., Bartram, S. F., Becker, R. S., Bernard, W. J., and Bukata, S. W. (1954). *J. Am. Chem. Soc.* **76**, 2177.

Hájek, B., and Benda, F. (1972). *Collect. Czech. Chem. Commun.* **37**, 2534.
Hall, T. (1960). *Rev. Sci. Instrum.* **31**, 125.
Hannebohn, O., and Klemm, W. (1936). *Z. Anorg. Allg. Chem.* **229**, 337.
Hänsler, R., and Rüdorff, W. (1970). *Z. Naturforsch.*, **25B**, 1306.
Hargreaves, G. B., and Peacock, R. D. (1960). *J. Chem. Soc.* p. 2618.
Hartung, A., and Babel, D. (1982). *J. Fluorine Chem.* **19**, 369.
Haszeldine, R. N., and Sharpe, A. G. (1951). "Fluorine and its Compounds," p. 4. Methuen, London.
Hebecker, C. (1970). *Z. Anorg. Allg. Chem.* **376**, 236.
Henisch, H. K., Dennis, J., and Hanoka, J. I. (1965). *J. Phys. Chem. Solids* **26**, 493.
Henkel, P., and Klemm, W. (1935). *Z. Anorg. Allg. Chem.* **222**, 73.
Hepworth, M. A., and Jack, K, H. (1957). *Acta Crystallogr.* **10**, 345.
Hepworth, M. A., Peacock, R. D., and Robinson, P. L. (1954). *J. Chem. Soc.* p. 1197.
Hepworth, M. A., Jack, K. H., and Westland, G. J. (1956). *J. Inorg. Nucl. Chem.* **2**, 79.
Hepworth, M. A., Jack, K. H., Peacock, R. D., and Westland, G. J. (1957). *Acta Crystallogr.* **10**, 63.
Herdtweck, E. (1983). *Z. Anorg. Allg. Chem.* **501**, 131.
Hightower, J., and McNeese, L. (1972). *J. Chem. Eng. Data* **17**, 342.
Holloway. J. H. (1966). *Chem. Commun.* p. 22.
Holloway, J. H., and Peacock, R. D. (1963a). *J. Chem. Soc.* p. 527.
Holloway, J. H., and Peacock, R. D. (1963b). *J. Chem. Soc.* p. 3892.
Holloway, J. H., and Schrobilgen, G. J. (1975). *J. Chem. Soc., Chem. Commun.* p. 623.
Holloway, J. H., Rao, P. R., and Bartlett, N. (1965). *Chem. Commun.* p. 306.
Hoppe, R. (1978). *Isr. J. Chem.* **17**, 48.
Hoppe, R., and Blinne, K. (1958). *Z. Anorg. Allg. Chem.* **293**, 251.
Hoppe, R., and Dähne, W. (1962). *Naturwissenschaften* **49**, 254.
Hoppe, R., and Homann, R. (1970). *Z. Anorg. Allg. Chem.* **379**, 193.
Hoppe, R., and Lehr, K. (1975). *Z. Anorg. Allg. Chem.* **416**, 240.
Hoppe, R., and Müller, B. (1969). *Naturwissenschaften* **1**, 56.
Hoppe, R., Blinne, K., and Liebe, W. (1959). *Spec. Publ.—Chem. Soc.* **13**, 132.
Hoppe, R., Liebe, W., and Dähne, W. (1961a). *Z. Anorg. Allg. Chem.* **307**, 276.
Hoppe, R., Dähne, W., and Klemm, W. (1961b). *Naturwissenschaften* **48**, 429.
Hoppe, R., Dähne, W., Mattauch, H., and Rödder, K. M. (1962). *Angew. Chem.* **74**, 903.
Hyman, H. H., and Katz, J. J. (1965). *In* "Non-Aqueous Solvent Systems" (T. C. Waddington, ed.), p. 47. Academic Press, New York.
Iwasaki, H., Yahata, T., Suzuki, K., and Oshina, K. (1962). *Kogyo Kagaku Zasshi* **65**, 1165.
Jack, K. H. (1959). *Commun. Int. Symp. Fluorine Chem., Birmingham, England.*
Jack, K. H., and Gutmann, V. (1951). *Acta Crystallogr.* **4**, 246.
Jacob, E. (1973). *Z. Anorg. Allg. Chem.* **400**, 45.
Jesih, A., Žemva, B., and Slivnik, J. (1982). *J. Fluorine Chem.* **19**, 221.
Johnson, C. E., Fischer, J., and Steindler, M. J. (1961). *J. Am. Chem. Soc.* **83**, 1620.
Jones, D. A. (1976). *J. Cryst. Growth* **34**, 149.
Jones, D. A., and Shand, W. A. (1968). *J. Cryst. Growth* **2**, 361.
Jones, D. A., Cockayne, B., Clay, R. A., and Forrester, P. A. (1975). *J. Cryst. Growth* **30**, 21.
Katz, J. J., and Rabinowitch, E. (1951). "The Chemistry of Uranium," p. 352. McGraw-Hill, New York.
Kelly, D. H., Post, B., and Mason, R. W. (1963). *J. Am. Chem. Soc.* **85**, 307.
Kennedy, G. C., and La Mori, P. N. (1961). *In* "Progress in Very High Pressure Research" (F. B. Bundy *et al.*, eds.), p. 304. Wiley, New York.
Kestigian, M., Leipziger, F. D., Croft, W. J., and Guidoboni, R. (1966). *Inorg. Chem.* **5**, 1462.
Khairoun, S., Dance, J. M., Grannec, J., Demazeau, G., and Tressaud, A. (1983). *Rev. Chim. Miner.* **20**, 871.

Kirk, R. E., and Othmer, D. F., eds. (1966). "Encyclopedia of Chemical Technology," Vol. 9, p. 506. Wiley, New York.

Klauke, E., Oehlmann, L., and Baasner, B. (1983). *J. Fluorine Chem.* **23**, 301.

Klemm, W., and Döll, W. (1939), *Z. Anorg. Allg. Chem.* **241**, 233.

Koch, J., Hebecker, C., and John, H. (1982). *Z. Naturforsch., B:* **37B**, 1659.

Kurtzig, A. J., and Guggenheim, H. J. (1970). *Appl. Phys. Lett.* **16**, 43.

Kyropoulos, S. (1926). *Z. Anorg. Chem.* **154**, 308.

Lachter, A., Lascaud, M., Barrière, A. S., Lozano, L., Portier, J., and Saboya, B. (1978). *J. Cryst. Growth* **43**, 621.

Langlais, F., Demazeau, G., Portier, J., Tressaud, A., and Hagenmuller, P. (1979). *Solid State Commun.* **29**, 473.

Lari-Lavassani, A., Avinens, C., and Cot, L. (1969). *C. R. Hebd. Seances Acad. Sci.* **268**, 1782.

Laudise, R. A. (1970). "The Growth of Single Crystals." Prentice-Hall, Englewood Cliffs, New Jersey.

La Valle, D. E., Steele, R. M., Wilkinson, M. K., and Yakel, H. L. (1960). *J. Am. Chem. Soc.* **82**, 2433.

Lawson, W. D., and Nielsen, S. (1958). "Preparation of Single Crystals." Butterworth, London.

Leary, K. (1975). Ph.D. Thesis, University of California (*Lawrence Berkeley Lab.* [*Rep.*] **LBL-3746**).

Leary, K., and Bartlett, N. (1972). *J. Chem. Soc., Chem. Commun.* p. 903.

Leary, K., Templeton, D. H., Zalkin, A., and Bartlett, N. (1973). *Inorg. Chem.* **12**, 1726.

Leblanc, M., Ferey, G., Chevallier, P., Calage, Y., and de Pape, R. (1983). *J. Solid State Chem.* **47**, 53.

Leckebusch, R. (1974). *J. Cryst. Growth* **23**, 74.

Le Fur, Y., and Aleonard, S. (1970). *Bull. Soc. Fr. Mineral. Cristallogr.* **93**, 260.

Lind, M. D., and Christe, K. O. (1972). *Inorg. Chem.* **11**, 608.

Longo, J. M., and Kafalas, J. A. (1969). *J. Solid State Chem.* **1**, 103.

Lorin, D. (1980). 3rd Cycle Thesis, University of Bordeaux, France.

McGuire, T. R., and Shafer, M. W. (1971). *J. Phys. (Colloq.* 2–3) **32**, 627.

McKee, D. E., and Bartlett, N. (1973). *Inorg. Chem.* **12**, 2738.

McKee, D. E., Adams, C. J., and Bartlett, N. (1973). *Inorg. Chem.* **12**, 1722.

Mackenzie, J. D. (1978). *Lawrence Livermore Lab., UCLA Final Tech. Rep.* **8331705**.

Malm, J. G. (1983). *J. Fluorine Chem.* **23**, 267.

Malm, J. G., Sheft, I., and Chernick, C. L. (1963). *J. Am. Chem. Soc.* **85**, 110.

Masson, C. R., Whiteway, S. G., and Collings, C. A. (1963). *Am. Ceram. Soc. Bull.* **42**, 745.

Meehan, J. P., and Wilson, E, J. (1972). *J. Cryst. Growth* **15**, 141.

Meister, G. (1954). *In* "Rare Metals Handbook" (C. A. Hampel, ed.), p. 547. Van Nostrand, Reinhold, Princeton, New Jersey.

Merchant, P., Grannec, J., Chaminade, J. P., and Fouassier, C. (1980), *Mater. Res. Bull.* **15**, 1113.

Miranday, J. P. (1972). 3rd Cycle Thesis, University of Caen, France.

Miranday, J. P., Ferey, G., Jacoboni, C., Dance, J. M., Tressaud, A., and de Pape, R. (1975). *Rev. Chim. Miner.* **12**, 187.

Moody, G. J., and Selig, H. (1966). *J. Inorg. Nucl. Chem.* **28**, 2429.

Morrell, B. K., Zalkin, A., Tressaud, A., and Bartlett, N. (1973). *Inorg. Chem.* **12**, 2640.

Muetterties, E. L., and Castle, J. E. (1961), *J. Inorg. Nucl. Chem.* **18**, 148.

Müller, B. G. (1980). *J. Fluorine Chem.* **16**, 637.

Müller, B. G. (1981). *J. Fluorine Chem.* **17**, 317.

Müller, B. G. (1982). *J. Fluorine Chem.* **20**, 291.

Müller, B. G., and Hoppe, R. (1972). *Mater. Res. Bull.* **7**, 1297.

Naka, S., Takeda, Y., Kawada, K., and Inagaki, M. (1979). *J. Cryst. Growth* **46**, 461.

Nikolenko, L. I., Jurasova, T. I., and Manjko, A. A. (1970). *Zh. Obshch. Khim.* **40**, 938.

Nouet, J., Jacoboni, C., Ferey, G., Gerard, J. Y., and de Pape, R. (1971). *J. Cryst. Growth* **8**, 94.
Nyholm, R. S., and Sharpe, A. G. (1952). *J. Chem. Soc.* p. 3579.
O'Donnell, T. A., and Peel, T. E. (1976). *J. Inorg. Nucl. Chem., Suppl.* p. 61.
O'Donnell, T. A., Rietz, R., and Yeh, S. (1983). *J. Fluorine Chem.* **23**, 97.
Oi, T., and Miyauchi, K. (1981), *Mater. Res. Bull.* **16**, 1281.
Omaly, J., Batail, P., Grandjean, D., Avignant, D., and Cousseins, J. C. (1976). *Acta Crystallogr. Sect. B.* **B32**, 2106.
Oppegard, A. L., Smith, W. C., Muetterties, E. L., and Engelhardt, V. A. (1960), *J. Am. Chem. Soc.* **82**, 3835.
Orlov, M. S. (1975). *Sov. J. Opt. Technol.* **42**, 733.
Passaret, M., Leblanc, M., and de Pape, R. (1974). *High Temp.—High Pressures* **6**, 629.
Pastor, R. C., and Arita, K. (1974). *Mater. Res. Bull.* **9**, 579.
Pastor, R. C., and Arita, K. (1975). *Mater. Res. Bull.* **10**, 493.
Pastor, R. C., and Robinson, M. (1974), *Mater. Res. Bull.* **9**, 569.
Pastor, R. C., and Robinson, M. (1976), *Mater. Res. Bull.* **11**, 1327.
Pastor, R. C., Robinson, M., and Hastings, A. G. (1974). *Mater. Res. Bull.* **9**, 781.
Pastor, R. C., Robinson, M., and Akutagawa, W. M. (1975). *Mater. Res. Bull.* **10**, 501.
Peacock, R. D. (1953). *J. Chem. Soc.* p. 3617.
Peacock, R. D. (1956). *Chem. Ind. (London)* p. 1391.
Peacock, R. D. (1957a). *Proc. Chem. Soc., London* p. 59.
Peacock, R. D. (1957b). *J. Chem. Soc.* p. 467.
Peacock, R. D. (1960). *Prog. Inorg. Chem.* **2**, 193.
Peacock, R. D., Selig, H., and Sheft, I. (1964). *Proc. Chem. Soc., London* p. 285.
Peiser, H. S. (1967). "Crystal Growth." Pergamon, Oxford.
Petit Le Du, G. (1970). *Rev. Int. Hautes Temp. Refract.* **7**, 100.
Petzel, T., and Greis, O. (1973). *Z. Anorg. Allg. Chem.* **396**, 95.
Pistorius, C. W. F. T. (1975). *J. Solid State Chem.* **13**, 208.
Plet, F., Fourquet, J. L., Courbion, G., Leblanc, M., and de Pape, R. (1979). *J. Cryst. Growth* **47**, 699.
Plevey, R. G., Rendell, R. W., and Tatlow, J. C. (1982a). *J. Fluorine Chem.* **21**, 159.
Plevey, R. G., Rendell, R. W., and Tatlow, J. C. (1982b). *J. Fluorine Chem.* **21**, 265.
Plovnick, R. H., and Camobreco, S. J. (1972). *Mater. Res. Bull.* **7**, 573.
Pontonnier, L., and Aleonard, S. (1972). *Bull. Soc. Fr. Mineral. Cristallogr.* **95**, 507.
Portier, J., Tressaud, A., and Dupin, J. L. (1970). *C. R. Hebd. Seances Acad. Sci.* **270**, 216.
Pouchard. M., Torki, M. R., Demazeau, G., and Hagenmuller, P. (1971). *C. R. Hebd. Seances Acad. Sci.* **273**, 1093.
Poulain, M. (1983). *J. Non-Cryst. Solids* **56**, 1.
Poulain, M., and Lucas, J. (1978). *Verres Refract.* **32**, 505.
Poulain, M., Chanthanasinh, M., and Lucas, J. (1977). *Mater. Res. Bull.* **12**, 151.
Pullen, K. E., and Cady, G. H. (1966). *Inorg. Chem.* **5**, 2057.
Pullen, K. E., and Cady, G. H. (1967). *Inorg. Chem.* **6**, 1300.
Rao, P. R., Tressaud, A., and Bartlett, N. (1976). *J. Inorg. Nucl. Chem., Suppl.* p. 23.
Rault, M., Demazeau, G., Portier, J., and Grannec, J. (1970). *Bull. Soc. Chim. Fr.* **1**, 74.
Ravez, J., and Hagenmuller, P. (1964). *Bull. Soc. Chim. Fr.* p. 1811.
Ravez, J., de Pape, R, and Hagenmuller, P. (1966). *Bull. Soc. Chim. Fr.* **1**, 240.
Rediess, K., and Bougon, R. (1983). *J. Fluorine Chem.* **23**. 464.
Riss, I. G., and Vituhnovskaja, B. S. (1958). *Zh. Neorg. Khim.* **3**, 1185.
Robinson, M., and Cripe, D. M. (1966). *J. Appl. Phys.* **37**, 2072.
Robinson, M., Pastor, R. C., Turk, R. R., Devor, D. P., and Braunstein, M. (1980). *Mater. Res. Bull.* **15**, 735.

74 J. Grannec and L. Lozano

Robinson, P. L., and Westland, G. J. (1956). *J. Chem. Soc.* p. 4481.
Rochow, E. G., and Kuken, I. (1952). *J. Am. Chem. Soc.* **74**, 1615.
Roesky, H. W., Glemser, O., and Hellberg, K. H. (1965). *Chem. Ber.* **98**, 2046.
Roux, N., Charpin, P., and Ehretsmann, J. (1968). *C. R. Hebd. Seances Acad. Sci.* **267**, 484.
Ruff, O., and Lickfett, H. (1911). *Ber. Dtsch. Chem. Ges.* **44**, 2539.
Sabatier, R., Charroin, G., Avignant, D., and Cousseins, J. C. (1979). *Acta Crystallogr., Sect. B.* **B35**, 1333.
Safa, M., Tanner, B. K., Garrard, B. J., and Wanklyn, B. M. (1977). *J. Cryst. Growth* **39**, 243.
Salardenne, J., Danto, Y., and Barrière, A. S. (1975). *Ann Chim.* **10**, 201.
Sato, N., and Kigoshi, A. (1982). *J. Fluorine Chem.* **20**, 365.
Schäfer, H. (1962). "Chemische Transportreaktionen." Weinheim, New York.
Schäfer, H. (1964). "Chemical Transport Reactions." Academic Press, New York.
Schäfer, H. (1971). *Angew. Chem.* **83**, 85; *Angew. Chem., Int. Ed. Engl.* **10**, 143.
Schäfer, H., Schnering, H. G., Niehues, K. J., and Nieder-Vahrenholz (1965). *J. Less-Common Met.* **9**, 95.
Schreiner, F., Malm, J. G., and Hindman, J. C. (1965). *J. Am. Chem. Soc.* **87**, 25.
Sears, D. R., and Hoard, J. L. (1969). *J. Chem. Phys.* **50**, 1066.
Selig, H. (1964). *Science* **144**, 537.
Selig, H., and Peacock, R. D. (1964). *J. Am. Chem. Soc.* **86**, 3895.
Sense, K. A., Snyder, M. J., and Filbert, R. B. (1954). *J. Phys. Chem.* **58**, 995.
Shafer, M. W. (1969). *Mater. Res. Bull.* **4**, 905.
Shafer, M. W., McGuire, T. R., Argyle, B. E., and Fan, G. J. (1967). *Appl. Phys. Lett.* **10**, 202.
Shamir, J., and Binenboym, J. (1968). *Inorg. Chim. Acta* **2**, 37.
Shand, W. A. (1969). *J. Cryst. Growth* **5**, 143.
Sharpe, A. G. (1949). *J. Chem. Soc.* p. 2901.
Sharpe, A. G. (1950). *J. Chem. Soc.* p. 3444.
Sharpe, A. G. (1960). *Adv. Fluorine Chem.* **1**, 29.
Shaw, M. J., Hyman, H. H., and Filler, R. (1971). *J. Org. Chem.* **36**, 2917.
Sheft, I., Spittler, T. M., and Martin, F. H. (1964). *Science* **145**, 701.
Sheppard, W. A. (1962). *J. Am. Chem. Soc.* **84**, 3064.
Shinn, D. B., Crocket, D. S., and Haendler, H. M. (1966). *Inorg. Chem.* **5**, 1927.
Siebert, G., and Hoppe, R. (1972a). *Z. Anorg. Allg. Chem.* **391**, 113.
Siebert, G., and Hoppe, R. (1972b). *Z. Anorg. Allg. Chem.* **391**, 126.
Siegel, S. (1952). *Acta Crystallogr.* **5**, 683.
Siegel, S. (1956). *Acta Crystallogr.* **9**, 684.
Simons, J. H., ed. (1950). "Fluorine Chemistry," Vol. 1. Academic Press, New York.
Simons, J. H., ed. (1954). "Fluorine Chemistry," Vol. 2. Academic Press, New York.
Simons, J. H., ed. (1964). "Fluorine Chemistry," Vol. 5. Academic Press, New York.
Skrzypek, D., Jakubowski, P., Ratuszna, A., and Chelkowski, A. (1980). *J. Cryst. Growth* **48**, 475.
Sladky, F. O., and Bartlett, N. (1969). *J. Chem. Soc.* p. 2188.
Sladky, F. O., Bulliner, P. A., Bartlett, N., De Boer, B. G., and Zalkin, A. (1968). *Chem. Commun.* p. 1048.
Sladky, F. O., Bulliner, P. A., and Bartlett, N. (1969). *J. Chem. Soc. A* p. 2179.
Slivnik, J., and Žemva, B. (1971). *Z. Anorg. Allg. Chem.* **385**, 137.
Slivnik, J., Šmalc, A., Sedej, B., and Vilhar, M. (1964). *Inst. "Jožef Stefan," IJS Rep.* **R-430**.
Slivnik, J., Šmalc, A., and Zemljic, A. (1965). *Inst. "Jožef Stefan," IJS Rep.* **R-472**.
Slivnik, J. Pezdič, J., and Sedej, B. (1967). *Monatsh. Chem.* **98**, 204.
Slivnik, J., Šmalc, A., Lutar, K., Žemva, B., and Frlec, B. (1975). *J. Fluorine Chem.* **5**, 273.
Slivnik, J., Žemva, B., Bohinc, M., Hanžel, D., Grannec, J., and Hagenmuller, P. (1976). *J. Inorg. Nucl. Chem.* **38**, 997.

Smiley, S. H., and Brater, D. C. (1958). *Prog. Nucl. Energy, Ser.* **2**, 107.
Smith, W. C. (1962). *Angew. Chem., Int. Ed. Engl.* **1**, 467.
Smolensky, G. A., Yudin, V. M., Syrnikov, P. P., and Sherman, A. B. (1967) *JETP Lett.* (*Engl. Transl.*) **8**, 2368.
Sobolev, B. P., Ratnikova, I. D., and Fedorov, P. P. (1976). *Mater. Res Bull.* **11**, 999.
Sorbe, P. (1977). Thesis, University of Bordeaux, France.
Sorbe, P., Grannec, J., Portier, J., and Hagenmuller, P. (1976). *C. R. Hebd. Seances Acad. Sci.* **282**, 663.
Sorbe, P., Grannec, J., Portier, J., and Hagenmuller, P. (1977). *C. R. Hebd. Seances Acad. Sci.* **284**, 231.
Sorbe, P., Grannec, J., Portier, J., and Hagenmuller, P. (1978). *J. Fluorine Chem.* **11**, 243.
Soriano, J., Givon, M., and Shamir, J. (1966). *Inorg. Nucl. Chem. Lett.* **2**, 13.
Spitsyn, V. I., Kiselev, Y. M., and Martynenko, L. I. (1973). *Zh. Neorg. Khim.* **18**, 1125.
Stacey, M., and Tatlow, J. C. (1960). *Adv. Fluorine Chem.* **1**, 166.
Stević, S., Videau, J. J., and Portier, J. (1978). *Rev. Chim. Miner.* **15**, 529.
Stezowski, J. J., and Eick, H. A. (1970). *Inorg. Chem.* **9**, 1102.
Stöckbarger, D. C. (1936). *Rev. Sci. Instrum.* **7**, 133.
Stout, J. W., and Boo, W. O. J. (1966). *J. Appl. Phys.* **37**(3), 966.
Streng, L. V., and Streng, A. G. (1965). *Inorg. Chem.* **4**, 1370.
Sturgeon, G. D., Penneman, R. A., Kruse, F. H., and Asprey, L. B. (1965). *Inorg. Chem.* **4**, 748.
Sturm, B. J. (1960). *In* "Preparation of Inorganic Fluorides," React. Chem. Div. Annu. Prog. Rep., ORNL-2931, p. 186. Oak Ridge Natl. Lab., Oak Ridge, Tennessee.
Syono, Y., Akimoto, S., and Kohn, K. (1969). *J. Phys. Soc. Jpn.* **26**(4), 903.
Tedenac, J. C., Granier, W., Norbert, A., and Cot, L. (1969). *C. R. Hebd. Seances Acad, Sci.* **268**, 1368.
Tolberg, W. E., Rewick, R. T., Stringham, R. S., and Hill, M. E. (1967). *Inorg. Chem.* **6**, 1156.
Tressaud, A. (1969). Thesis, University of Bordeaux, France.
Tressaud, A., and Dance, J. M. (1977). *Adv. Inorg. Chem. Radiochem.* **20**, 133.
Tressaud, A., Galy, J., and Portier, J. (1969). *Bull. Soc. Fr. Mineral. Cristallogr.* **92**, 335.
Tressaud, A., de Pape, R., Portier, J., and Hagenmuller, P. (1970). *Bull. Soc. Chim. Fr.* p. 3411.
Tressaud, A., Pintchovski, F., Lozano, L., Wold, A., and Hagenmuller, P. (1976a). *Mater. Res. Bull.* **11**, 689.
Tressaud, A., Grannec, J., and Lozano, L. (1976b). *Commun. Int. Symp. Fluorine Chem. 8th, 1976, Kyoto.*
Tressaud, A., Wintenberger, M., Bartlett, N., and Hagenmuller, P. (1976c). *C. R. Hebd. Seances Acad. Sci.* **282**, 1069.
Tressaud, A., Dance, J. M., Parenteau, J. M., Launay, J. C., Portier, J., and Hagenmuller, P. (1976d). *J. Cryst. Growth* **32**, 211.
Tressaud, A., Soubeyroux, J. L., Touhara, H., Demazeau, G., and Langlais, F. (1981). *Mater. Res. Bull.* **14**, 1147.
Tressaud, A., Khaïroun, S., Dance, J. M., Grannec, J., Demazeau, G., and Hagenmuller, P. (1982). *C. R. Hebd. Seances Acad. Sci.* **295**, 183.
Trevorrow, L. E. (1958). *J. Chem. Soc.* p. 362.
Trevorrow, L. E., Fischer, J., and Steunenberg, R. K. (1957). *J. Am. Chem. Soc.* **79**, 5167.
Turner, J. J., and Pimentel, G. C. (1963). *Science* **140**, 974.
Vasile, M. J., Richardson, T. J., Stevie, F. A., and Falconer, W. E. (1976). *J. Chem. Soc., Dalton Trans.* p. 351.
Vedrine, A., Baraduc, L., and Cousseins, J. C. (1973). *Mater. Res. Bull.* **8**, 581.
Vlasse, M., Matejka, G. Tressaud, A., and Wanklyn, B. M. (1977). *Acta Crystallogr., Sect. B* **B33**, 3377.
Volavšek, B., Dobčnik, D., and Slivnik, J. (1972). *J. Inorg. Nucl. Chem.* **34**, 2483.

Von der Mühll, R., and Ravez, J. (1974). *Rev. Chim. Miner.* **11,** 652.

Walker, P. J. (1979). *J. Cryst. Growth* **46,** 709.

Walker, P. J. (1980). *J. Cryst. Growth* **49,** 77.

Wanklyn, B. M. (1969). *J. Cryst. Growth* **5,** 279.

Wanklyn, B. M. (1975). *J. Mater. Sci.* **10,** 1487.

Wanklyn, B. M., Garrard, B. J., Wondre, F., and Davison, W. (1976). *J. Cryst. Growth* **33,** 165.

Wapenaar, K. E. D., Van Koesveld, J. L., and Schoonman, J. (1981). *Solid State Ionics* **2,** 145.

Wartenberg, H. (1940). *Z. Anorg. Allg. Chem.* **244,** 337.

Wartenberg, H. (1941). *Z. Anorg. Allg. Chem.* **247,** 135.

Weaver, E. E., Weinstock, B., and Knop, C. P. (1963). *J. Am. Chem. Soc.* **85,** 111.

Weinstock, B. (1962). *Rec. Chem. Prog.* **23,** 23.

Weise, E. (1956). *Z. Anorg. Allg. Chem.* **283,** 377.

White, E. A. (1964). *GEC J. Sci. Technol.* **31,** 43.

Whitney, E. D., MacLaren, R. D., Fogle, C. E., and Hurley, T. J. (1964). *J. Am. Chem. Soc.* **86,** 2583.

Wilhelm, V., and Hoppe. R. (1976). *J. Inorg. Nucl. Chem., Suppl.* p. 113.

Wilson, W. W., and Christe, K. O. (1980). *J. Fluorine Chem.* **15,** 83.

Wilson, W. W., and Christe, K. O. (1982). *J. Fluorine Chem.* **19,** 253; see the other references given by the authors.

Woolf, A. A. (1964). *J. Chem. Soc.* p. 619.

Zalkin, A., and Templeton, D. H. (1953). *J. Am. Chem. Soc.* **75,** 2453.

Zalkin, A., Lee, K., and Templeton, D. H. (1962). *J. Chem. Phys.* **37,** 697.

Žemva, B., and Slivnik, J. (1971). *J. Inorg. Nucl. Chem.* **33,** 3952.

Žemva, B., Zupan, J., and Slivnik, J. (1973). *J. Inorg. Nucl. Chem.* **35,** 3941.

Žemva, B., Slivnik, J., and Šmalc, A. (1975). *J. Fluorine Chem.* **6,** 191.

Zupan, M., and Pollak, A, (1976). *J. Fluorine Chem.* **7,** 443, 445.

3

Crystal Chemistry of Fluorides

DIETRICH BABEL

Fachbereich Chemie
Philipps-Universität
Marburg, Federal Republic of Germany

ALAIN TRESSAUD

Laboratoire de Chimie du Solide du CNRS
Université de Bordeaux
Talence, France

INORGANIC
SOLID FLUORIDES

I. Introduction

Since the publication of the last reviews on the structural chemistry of transition metal fluorides (Bartlett, 1965; Babel, 1967; Canterford and Colton, 1968), not only have a large number of isomorphs of known structures been prepared, but also many new structures have been determined and older ones refined. There has been considerable interest in the connections between structure and properties (Portier, 1976), particularly magnetism of transition metal fluorides, which has been reviewed (Goodenough and Longo, 1970; Nomura, 1978; Tressaud and Dance, 1977, 1982) and is developed further in this volume. The present chapter is focused on well-established structures found in this class of compounds and generally does not discuss properties. The structures have been arranged according to the dimensional character-istics of their framework, however, which may be of great importance to their physical properties. Many compounds are listed in order to provide the solid-state chemist with a large body of materials for eventual selection.

The structures of the fluorides are in general less complex than those of materials such as oxides or sulfides as a consequence of the small polarizability of fluoride ion and of the ionic character of fluorine bonds, which make more difficult structural distortions. Binary fluorides have been reviewed (Edwards, 1982).

Structures of d transition metal fluorides, most of which exhibit octahedral coordination, are considered in more detail. Isostructural main-group-element fluorides, however, are generally included in this chapter. Emphasis is given to structures in which direct linking of octahedra occurs along one, two, or three dimensions (1D, 2D, or 3D). Mention of compounds containing isolated octahedra is made only as far as it seemed useful for understanding the structural relations. We have omitted all typically "molecular compounds" and those containing isolated ions such as BeF_4^{2-}, BF_4^-, SiF_6^{2-}, or PF_6^-. Nor is the structural diversity of f transition metal fluorides (Greis and Haschke, 1982; Penneman *et al.*, 1973) considered in this chapter.

In the tables an asterisk (*) denotes that full single-crystal structure determination has been performed; (*P) means refinement of powder data only. Many compounds listed in the tables have already been quoted by Pies and Weiss in the Landolt–Börnstein (LB) Tables, New Series, Group III, Volume 7, Parts a and g (1973), which give exhaustive references to the literature up to 1971. Therefore, in order to reduce our reference list, only the LB serial number (a-) is given as reference for such compounds. However, relevant new results have been added with full reference. Compounds with unknown structures have been omitted.

The following abbreviations are used: A, alkali or alkaline-earth ion (or other, if structurally isomorphic); CN, coordination number; fcc, face-centered cubic; hcp, hexagonal close-packed; M, d transition metal ion (or other, if structurally isomorphic).

II. Three-Dimensional Framework Structures

A. DIFLUORIDES AND RELATED STRUCTURES

A large variety of coordination types have been found in difluorides. They range from a trigonal pyramid in GeF_2 to a cubic environment in the fluorite-type structure (Table I). Structures with a low CN of 3 or 4 are briefly considered here, and those deriving from the fluorite type are examined in more detail in Chapter 11, devoted to fast fluorine ion conductors.

1. *Trigonal pyramid coordination.* Due to the presence of a lone pair in both Ge(II) and Sn(II), structures of GeF_2 and SnF_2 have similar features. The divalent cation is located at the apex of a trigonal pyramid with three fluorine atoms at the base. In GeF_2, GeF_3 pyramids are connected to form infinite chains, whereas the structure of SnF_2 is characterized by Sn_4F_8 tetramers.

TABLE I

Types of Environment in Difluorides

CN of Cation	Environment	Example
3	Trigonal pyramid	GeF_2, SnF_2
4	Tetrahedron	BeF_2
6	Octahedron	Rutile, trirutile, and α-PbO_2 types
6 + 2	Rhombohedron	High-pressure PdF_2, $Cd_{0.5}Pd_{0.5}F_2$
8	Cube	MF_2 fluorite and related structures
7 + 2	Distorted tricapped trigonal prism	α-PbF_2, high-pressure MnF_2 ($PbCl_2$ type)

2. *Tetrahedral coordination.* BeF_2 crystallizes in silica-like structures (cristobalite and quartz structures). Tetrahedral BeF_4 units are 3D linked by corners. A randomly disordered arrangement of the silica type has even been found from x-ray scattering curves in glassy BeF_2.

1. MF_2: Rutile and Related Structures

a. RUTILE STRUCTURE. The rutile structure corresponds to the tetragonal symmetry with space group $P4_2/mnm$ (no. 136), $Z = 2$. Most difluorides of d transition elements crystallize in this structure. Infinite chains of edge-shared MF_6 octahedra are directed along the c axis and are 3D connected to each other by common corners. The MF_6 octahedra are slightly distorted, with four identical equatorial $M-F_1$ distances larger than the two apical ones (axially compressed octahedra) (Baur, 1976). The geometry of the octahedron can be correlated with the x parameter of fluorine atoms and with the c/a ratio (Wells, 1973). Each fluorine atom has three M nearest neighbors in a same plane with two $M-F-M$ angles of $\sim 135°$ between two chains and one of $\sim 90°$ within a chain. Crystallographic data of rutile phases are given in Table II.

b. TRIRUTILE STRUCTURE. A cationic ordering of the trirutile type ($ZnSb_2O_6$ type; Byström *et al.*, 1941) occurs in $ABCF_6$ compounds when the involved cations have different charges (i.e., in the $A^IM^{II}M'^{III}F_6$ and $A_2^IM^{IV}F_6$ series). The space group is identical with that of the rutile structure, but the c constant is multiplied by 3 due to an additional ordering of one type of cation into every third layer (Fig. 1). Lattice constants of trirutile fluorides are given in Table III. A single-crystal study on a trirutile phase of the $LiZnCrF_6/ZnF_2$ system (composition: $Li_{0.75}Zn_{0.25}[Zn_{1.25}Cr_{0.75}]F_6$) has shown that the octahedra corresponding to the $(2a)$ sites ($Li_{0.75}Zn_{0.25}$) and $(4e)$ sites ($Zn_{1.25}Cr_{0.75}$) are of different size, but without any significant distortion (average $M-F = 2.045$ and 1.980 Å, respectively; Viebahn and Epple, 1976).

Concerning the cationic distribution in the $LiMM'F_6$ phases, when both M and M' are d transition metals, x-ray and neutron diffraction experiments have shown that the divalent and trivalent cations are randomly distributed in the $(4e)$ sites (Wintenberger *et al.*, 1972). In $LiFe_2F_6$, however, the narrowness of the linewidths of the Mössbauer spectra would indicate a complete ordering of the form $Li-Fe^{II}-Fe^{III}-Li\cdots$(Greenwood *et al.*, 1971). In contrast, a different ordering is observed in $LiMgFeF_6$ and in $LiZnCoF_6$ (see Table VI) (Portier *et al.*, 1970b; Hoppe, 1981). In $Li_2M^{IV}F_6$ trirutiles, octahedra corresponding to the tetravalent cation are isolated from each other (Portier *et al.*, 1969).

When Jahn–Teller ions such as Cu^{2+} are involved in an $LiM^{II}M'^{III}F_6$ phase, a dynamic effect occurs, resulting in a simple rutile structure with a random cationic distribution (see Table II) (Viebahn and Epple, 1976; Hoppe, 1981).

TABLE II

Crystallographic Data of Rutile Compounds

Compound[a]	a (Å)	c (Å)	x_F	4 ($M-F_1$) (Å)	2 ($M-F_2$) (Å)	Reference
MgF_2 (*)	4.621	3.052	0.3029	1.997	1.980	a-32; Baur (1976)
MgF_2 (*) (neutron diffraction, 52 K)	4.615	3.043	0.303	1.984	1.979	Vidal-Valat et al. (1979)
$Mg_{0.964}Sc_{0.036}F_{2.036}$	4.602	3.057	—	—	—	a-70
$Mg_{0.67}Mn_{0.33}F_2$	4.67	3.110	—	—	—	a-1741
$Mg_{0.67}Fe_{0.33}F_2$	4.65	3.130	—	—	—	a-1838
$Mg_{0.33}Mn_{0.67}F_2$	4.77	3.215	—	—	—	a-1742
$Mg_{0.33}Fe_{0.67}F_2$	4.69	3.120	—	—	—	a-1839
VF_2 (*)	4.804	3.237	0.306	2.083	2.090	Stout and Boo (1979); Cros (1978)
$Li_{0.33-x}Zn_{0.33+2x}Cr_{0.33-x}F_2$ ($x = 0.116$)	4.666	3.083	—	—	—	a-1636
MnF_2 (*)	4.873_4	3.309_9	0.305	2.13_2	2.10_2	a-279; Baur (1976)
$Mn_{1-x}Co_xF_2$ ($x = 0.06$)	4.82	3.27	—	—	—	a-301
FeF_2 (*)	4.696_6	3.309_1	0.301	2.11_8	1.99_8	a-293; Baur (1976)
$Fe_{0.33}Co_{0.33}Ni_{0.33}F_2$	4.68	3.175	—	—	—	a-1956
CoF_2 (*)	4.695_1	3.179_6	0.305	2.04_9	2.02_7	a-298; Baur (1976)
NiF_2 (*)	4.650_6	3.083_6	0.301	2.02_2	1.98_2	a-302; Baur (1976)
PdF_2 (*)	4.956	3.389	0.310	2.16	2.17	a-313; Baur (1976)
$Pd_{1-x}Zn_xF_2$ ($x = 0.50$)	4.832	3.257	—	—	—	Paus and Hoppe (1977)
ZnF_2 (*)	4.703_4	3.133_5	0.302	2.04_6	2.01_2	a-44; Baur (1976)
$\frac{1}{3}(LiCu^{II}CrF_6)$	4.616	3.057	—	—	—	Viebahn et al. (1969)
$\frac{1}{3}(LiCu^{II}FeF_6)$	4.651	3.082	—	—	—	Viebahn and Epple (1976)
$\frac{1}{3}(LiCu^IGaF_6)$	4.616	3.055	—	—	—	Viebahn and Epple (1976)
$\frac{1}{3}(LiCu^{II}Co^{III}F_6)$	4.609	3.105	—	—	—	Fleischer and Hoppe (1982b)

[a] For an explanation of asterisk (*) in tables, see Section I.

82   D. Babel and A. Tressaud

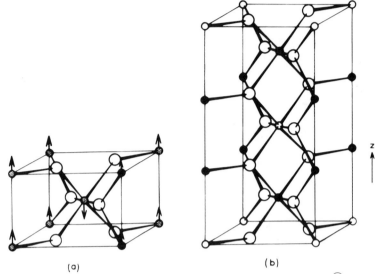

(a)   (b)

Fig. 1. Rutile (a) and trirutile (b) structures [○, (2*a*) sites; ●, (4*e*) sites; ○, F⁻].

c. RUTILE-RELATED STRUCTURE. Hydroxy fluorides Cd(OH)F and Hg(OH)F crystallize in the orthorhombic system with a doubled *c* axis (Table IV). Within distorted $M(OH)_3F_3$ octahedra, Cd–(O,F) and Hg–(O,F) distances range from 2.26 to 2.28 Å and from 2.10 to 2.69 Å, respectively (Grdenic and Sikirica, 1973; Volkova *et al.*, 1978; Nozik *et al.*, 1979). The corresponding zinc compound Zn(OH)F has the diaspore AlO(OH) structure, which can be derived from the rutile type; the edge-shared octahedra are linked into double columns by further edge sharing (Volkova *et al.*, 1978).

A rutile-derived structure has been found for tetrafluorides of the IrF_4 type (Bartlett and Tressaud, 1974). In this structure each octahedral IrF_6 unit shares four corners with other IrF_6 units, a pair of cis vertices being unshared. This structure is related to the rutile structure, and the fluorine-centered orthorhombic cell of MF_4 can be derived from a $2 \times 2 \times 2$ block of rutile cells by omitting in an ordered way half of the metal atoms (Rao *et al.*, 1976; and Fig. 2). Cell constants of isostructural phases are given in Table IV.

A monoclinic distortion is observed in CrF_2 and CuF_2, in which the transition element has a Jahn–Teller configuration. The distorted octahedral coordination of M(II) is characterized by three types of M–F distances (Table IV). The two axes of shorter bonds correspond to empty [Cr(II)] or half-filled [Cu(II)] $d_{x^2-y^2}$ orbitals, and the longer one to half-filled (or filled) d_{z^2} (Taylor and Wilson, 1974). Although the crystallographic features are similar, the magnetic structures are different. Whereas CrF_2 has an ordering identical with those of the other rutile phases (magnetic moments located on the corners of the pseudotetragonal cell and antiparallel to that of the central metal), CuF_2 shows ferromagnetic "layers" of CuF_4 squares linked by corners [ferromagnetic planes parallel to (102) in space group $P2_1/c$]. These "layers" are shifted with respect to each other in order to form distorted CuF_6 octahedra and are coupled antiferromagnetically (Fischer et al., 1974).

TABLE III
Lattice Constants of Trirutile Compounds

Compound[a]	a (Å)[b]	c (Å)[b]	Reference
Li_2TiF_6 (*P)	4.630	8.93_5	a-1297
$LiMgTiF_6$	4.66_5	9.21_1	Gaile et al. (1977)
High-temperature β-$LiMnTiF_6$	4.76_8	9.47_5	Gaile et al. (1977)
$LiMgVF_6$ (*P)	4.62_5	9.11_0	a-1512 (*P)
High-temperature β-$LiMnVF_6$	4.72_5	9.35_5	Gaile et al. (1977)
$LiNiVF_6$	4.63_1	9.14_2	a-1533
$LiZnVF_6$ (*P)	4.65_5	9.20_9	a-1521
$LiMgCrF_6$ (*P)	4.60_3	9.01_7	a-1614
$LiCoCrF_6$ (*P)	4.63_5	9.13_3	a-1658; Viebahn and Epple (1976)
$LiNiCrF_6$	4.61_2	9.02_7	a-1662
$LiZnCrF_6$ (*P)	4.63_2	9.10_0	a-1635; Viebahn and Epple (1976)
$LiMgFeF_6$ (*P)	4.64_0	9.07	a-1840
$LiFeFeF_6$ (*P)	4.67_3	9.29	a-1783
$LiCoFeF_6$	4.66_5	9.15_9	a-1876
$LiNiFeF_6$ (*P)	4.64_8	9.12_8	a-1879; Viebahn and Epple (1976)
$LiZnFeF_6$	4.67_1	9.15_4	a-1860
$LiMgGaF_6$ (*P)	4.59_5	8.96	a-731
High-temperature β-$LiFeGaF_6$	4.64_1	9.25_6	Gaile et al. (1977)
$LiCoGaF_6$	4.62_9	9.118	a-751
$LiNiGaF_6$	4.60_9	9.02_5	a-752
Low-temperature α-Li_2GeF_6 (*P)	4.58_0	8.80_5	a-1231 (*P)
$LiNiRhF_6$ (*P)	4.641	9.188	Viebahn and Epple (1976)
$LiZnRhF_6$ (*P)	4.663	9.235	Viebahn and Epple (1976)
Li_2PdF_6	4.61	8.98	Portier et al. (1970a)
$Li_{0.75}Zn_{1.50}Cr_{0.75}F_6$ (*)	4.632(5)	9.10(1)	Viebahn and Epple (1976)
Li_2PtF_6	4.65	9.14	Portier et al. (1970a)
Li_2MoF_6 (space group, $P4_22_12$, rutile related) (*)	4.686	9.19_1	a-1669 (*)
$LiMgCoF_6$ (*P)	4.60_2	9.05_4	Fleischer and Hoppe (1982b)
$LiNiCoF_6$ (*P)	4.61_1	9.07_1	Fleischer and Hoppe (1982b)
$LiZnCoF_6$ (*P)	4.62_1	9.16_5	Fleischer and Hoppe (1982b)
Na_2NbF_6 (*P)	5.04_2	10.27_7	Bournonville et al. (1984)

[a] For an explanation of (*P) in tables, see Section I.

[b] Numbers in parentheses and in subscripts indicate the standard deviation of the last figure given.

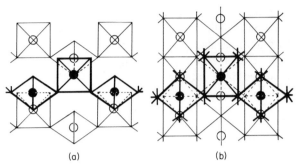

Fig. 2. Network of MF_6 octahedra in (a) IrF_4 and (b) rutile structures.

2. AgF₂ and High-Pressure PdF₂ Structures

AgF_2 and the high-pressure form of PdF_2 can also be described in terms of puckered "layers" of MF_4 square planes, connected in such a way that MF_6 distorted octahedra only share corners. In AgF_2, the AgF_4 "layers" are shifted relative to one another halfway between close-packed and primitive hexagonal arrangements; CN of Ag(II) is 4 + 2 (a-24).

High-pressure PdF_2 crystallizes with a cubic symmetry. The cations form an fcc sublattice and have six fluorine atoms as nearest neighbors. In this PdF_6 unit, the Pd–F distances are identical, but F–Pd–F angles are different from 90°, leading to different lengths of F–F edges. The coordination polyhedron is therefore a flattened octahedron, with two larger opposite faces. Two additional fluorine atoms belonging to other distorted octahedra cap these larger faces, forming thus around the central palladium atom a regular rhombohedron of 2.84-Å edge and 71° angle (Tressaud *et al.*, 1981a). It can be noted that this structure has strong relationships with both fluorite and pyrite types (Fig. 3). In can also be correlated to CuF_2 and rutile structures in which the staggered MF_4 "layers" are deduced by a simple translation along the c axis instead of a translation by a glide plane (Tressaud and Demazeau, 1984; Wells, 1975; a-24). Several fluorides crystallize in the high-pressure PdF_2 type: $CdPdF_4$, $HgPdF_4$, and high-pressure form of AgF_2 (Müller, 1982b; and Table IV).

3. Influence of High Pressure on Rutile-Type Fluorides

High-pressure phase transitions in rutile compounds have been followed *in situ* up to 300 kbars (for references, see Ming *et al.*, 1980).

In addition to distorted rutile phases, which are found at lower pressures for MnF_2 and NiF_2, a common feature of these systems is the presence of an intermediate "distorted fluorite" form (Table V). These phases, which generally crystallize in a simple cubic symmetry, have been proposed to be

TABLE IV

Crystallographic Data of Some Rutile-Related Compounds

Compound	Symmetry	Space Group	Z	a (Å)	b (Å)	c (Å)	β (deg)	M–F Distances (Å)	Reference
Cd(OH)F (*)	Orthorhombic	$P2_12_12_1$	4	4.832	5.516	6.856	—	2.26–2.28	Stålhandske (1979a,b)
Hg(OH)F (*)	Orthorhombic	$P2_12_12_1$	4	4.9568	5.9042	6.863	—	2.10–2.69	Stålhandske (1979a,b)
CrF_2 (*)	Monoclinic	$P2_1/n$	2	4.732	4.718	3.505	96.52	$2 \times 1.98; 2 \times 2.01; 2 \times 2.43$	a-258
CuF_2 (*)	Monoclinic	$P2_1/n$	2	4.599	4.546	3.307	96.57	$2 \times 1.90; 2 \times 1.93; 2 \times 2.30_5$	Taylor and Wilson (1974)
AgF_2 (*P)	Orthorhombic	$Pbca$	4	5.073	5.529	5.813	—	$4 \times 2.07; 2 \times 2.58$	a-24
High-pressure PdF_2 (*P)	Cubic	$Pa3$ (or $P2_13$)	4	5.322	—	—	—	$6 \times 2.18; 2 \times 3.17$	Tressaud et al. (1981a)
$Cd_{0.5}Pd_{0.5}F_2$ (*)	Cubic	$Pa3$	4	5.40_3	—	—	—	$6 \times 2.20_5; 2 \times 3.17$	Müller (1982b)
$Hg_{0.5}Pd_{0.5}F_2$ (*)	Cubic	$Pa3$	4	5.43	—	—	—	$6 \times 2.21_7; 2 \times 3.17_6$	Müller (1982b)
RhF_4 (*P)	Orthorhombic	$Fdd2$	8	9.71(2)	9.05(2)	5.63(1)	—	—	Rao et al. (1976)
IrF_4 (*P)	Orthorhombic	$Fdd2$	8	9.64(2)	9.25(2)	5.67(1)	—	—	Rao et al. (1976)
PdF_4 (*P)	Orthorhombic	$Fdd2$	8	9.37(2)	9.24(2)	5.84(1)	—	—	Rao et al. (1976)
PtF_4	Orthorhombic	$Fdd2$	8	9.61(2)	9.28(2)	5.71(1)	—	—	Rao et al. (1976)

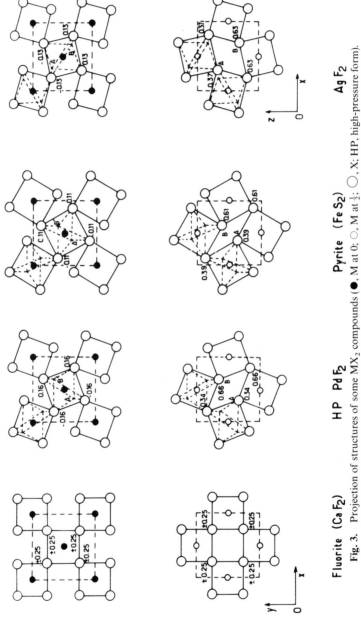

Fig. 3. Projection of structures of some MX_2 compounds (● M at 0; ○, M at $\frac{1}{2}$; ○, X; HP, high-pressure form).

Fluorite (CaF$_2$)   HP PdF$_2$ Pyrite (FeS$_2$)   AgF$_2$

TABLE V

High-Pressure Phase Transformations in Rutile Fluorides at $25°C^a$

$$MgF_2 \underset{}{\overset{\sim 250\,kbars}{\rightleftharpoons}} MgF_2\,(II')$$

$$MnF_2 \underset{}{\overset{\sim 15\,kbars}{\rightleftharpoons}} \underset{(orthorhombic)}{MnF_2\,(I'')} \underset{}{\overset{\sim 20\,kbars}{\rightleftharpoons}} \underset{\substack{(tetragonal \\ symmetry)}}{MnF_2\,(II')} \underset{}{\overset{\sim 100\,kbars}{\rightleftharpoons}} \underset{(PbCl_2\ type)}{MnF_2\,(III)}$$

$$\underset{(\alpha\text{-}PbO_2\ structure)}{MnF_2\,(I'')} \underset{(unloading\ to\ 1\ bar)}{\longleftarrow}$$

$$FeF_2 \underset{}{\overset{\sim 45\,kbars}{\rightleftharpoons}} FeF_2\,(II') \overset{\sim 220\,kbars}{\longrightarrow} \underset{\substack{[hexagonal\ symmetry; \\ ZnBr_2\,(III)\ type?]}}{FeF_2\,(III)}$$

$$CoF_2 \underset{}{\overset{\sim 65\,kbars}{\rightleftharpoons}} CoF_2\,(II') \underset{}{\overset{\sim 300\,kbars}{\rightleftharpoons}} \underset{\substack{(hexagonal\ symmetry; \\ modified\ NiAs\ type)}}{CoF_2\,(III)}$$

$$NiF_2 \overset{\sim 18\,kbars}{\longrightarrow} \underset{(orthorhombic)}{NiF_2\,(I')} \underset{}{\overset{\sim 85\,kbars}{\rightleftharpoons}} NiF_2\,(II')$$

$$ZnF_2 \overset{\sim 70\,kbars}{\longrightarrow} ZnF_2\,(II')$$

$$\underset{(\alpha\text{-}PbO_2\ structure)}{ZnF_2\,(I'')} \underset{(unloading\ to\ 1\ bar)}{\longleftarrow}$$

$$PdF_2 \overset{\sim 50\,kbars}{\longrightarrow} high\text{-}pressure\ PdF_2$$

$$high\text{-}pressure\ PdF_2 \underset{(unloading\ to\ 1\ bar)}{\longleftarrow}$$

^{a} Key: I', I'', rutile-related structures; II', "distorted fluorite" structures (cubic unless specified); III, post-"distorted fluorite" structures. From Ming *et al.* (1980).

isostructural with high-pressure PdF_2 (Tressaud and Demazeau, 1984). Post-high-pressure cubic phases are obtained at hyperpressures of several hundred kilobars and derive from the $PbCl_2$, $ZnBr_2$, or NiAs type (Ming *et al.*, 1980). These transformations are reversible except for PdF_2, MnF_2, and ZnF_2, in which high-pressure PdF_2 and α-PbO_2 types are respectively recovered at room pressure (see Table V).

B. STRUCTURAL RELATIONSHIPS IN $A_2M^{IV}F_6$ AND $AM^{II}M'^{III}F_6$ COMPOUNDS

Among the large variety of $AMM'F_6$ and A_2MF_6 compounds (A = Li or Na), only those in which all cations are in octahedral sites will be considered because of their structural relationships with the rutile structure. For further information and references on other A_2MF_6 phases (e.g., α-K_2GeF_6, α-K_2MnF_6, and K_2PtCl_6 types), see Babel (1967).

Most of those phases crystallize in either trirutile, Na_2SiF_6, Li_2ZrF_6 types or related structures (Table VI). Common features are quasi-hcp of anions and an occupancy of half the octahedral sites by cations. The relationships between their crystallographic cells are given in Fig. 4. A redetermination of the structural features of Na_2SnF_6 has shown that this compound could be, in fact, of the trirutile type with $a = 5.06$ Å and $c = 10.11$ Å (Bournonville et al., 1984).

Differences arise from the type of sharing between MF_6 octahedra; in Na_2SiF_6, for instance, SiF_6 units share three edges with NaF_6 octahedra, whereas in the trirutile structure only two edges are in common. In Li_2ZrF_6 and related structures (e.g., $LiCaAlF_6$ type) only corners are shared between A^IF_6 and $M^{IV}F_6$ octahedra.

In the Li_2MF_6 series the stability range of the different phases at room pressure and under high pressure has been correlated with the ionic radius of

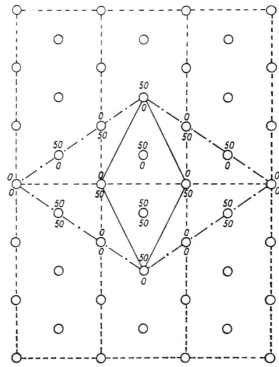

Fig. 4. Relationships between crystallographic cells of trirutile [projection on ($h01$) plane], Na_2SiF_6, and Li_2ZrF_6 types [projection on ($hk0$) plane] (---, trirutile; ·—·, Na_2SiF_6; ——, Li_2ZrF_6). From Gaile et al. (1977).

TABLE VI

Lattice Constants of Some A_2MF_6 Fluorides

I. Na_2SiF_6 Type

Compound	a_{hex} (Å)	c_{hex} (Å)	Reference
Li_2SiF_6	8.22	4.56	a-1209
Li_2MnF_6	8.42	4.59	a-1706
β-Li_2GeF_6	8.41	4.62	a-1230
$LiMgAlF_6$	8.315	4.525	a-684
Low-temperature α-$LiMnTiF_6$	8.75_3	4.71_5	Gaile *et al.* (1977)
Low-temperature α-$LiMnVF_6$	8.70_6	4.65_3	Gaile *et al.* (1977)
$LiMnCrF_6$	8.654	4.731	a-1655
$LiMnGaF_6$	8.638	4.738	Viebahn (1975)
Low-temperature α-$LiFeGaF_6$	8.57_2	4.66_5	Gaile *et al.* (1977)
$LiMgInF_6$	8.67_4	4.65_9	Gaile *et al.* (1977)
$LiMnInF_6$	8.84_7	4.78_0	Gaile *et al.* (1977)
$LiCoInF_6$	8.77_6	4.69_0	Gaile *et al.* (1977)
$LiNiInF_6$	8.72_6	4.62_8	Gaile *et al.* (1977)
$LiZnInF_6$	8.74_0	4.69_5	Gaile *et al.* (1977)
$LiCdInF_6$	8.96_5	4.92_9	Gaile *et al.* (1977)
$LiCaInF_6$	8.97_2	5.04_6	Gaile *et al.* (1977)
Na_2SiF_6	8.859	5.038	a-1210
Na_2TiF_6	9.20	5.13	a-1301
Na_2CrF_6	9.14	5.15	a-1566
Na_2MnF_6	9.03	5.13	a-1709
Na_2GeF_6	8.99	5.12	a-1232
Na_2RuF_6	9.32	5.15	a-1959
Na_2RhF_6	9.32	5.22	a-1971
Na_2PdF_6	9.23	5.25	a-1979
Na_2OsF_6 (I)	9.36	5.11	a-1997
Na_2IrF_6	9.34	5.14	a-2009
Na_2PtF_6	9.41	5.16	a-2023
$NaMnCrF_6$	8.993	5.003	Courbion *et al.* (1977)

II. Na_2SnF_6 Type (trirutile related)

Compound	a (Å)	b (Å)	c (Å)	β (deg)	Reference
Li_2CrF_6	4.587	4.584	9.993	117.27	Siebert and Hoppe (1972a)
Li_2RhF_6	4.63_7	4.63_4	10.18_5	117.10	Wilhelm and Hoppe (1974a)
β-Li_2SnF_6	4.74	4.68	10.31	116.9	a-1242
Na_2MoF_6	5.07	5.07	11.52	117.9	a-1671
Na_2ReF_6	5.05	5.05	11.62	117.2	a-1766
Na_2OsF_6 (II)	5.07	5.07	11.60	117.4	a-1998
Na_2SnF_6	5.06	5.06	11.31	116.6	a-1245
Na_2PbF_6	5.10	5.11	11.48	116.4	a-1273

(*cont.*)

TABLE VI (*cont.*)

III. Li_2ZrF_6 Type

Compound	a_{hex} (Å)	c_{hex} (Å)	Reference
HP-Li_2TiF_6	4.880	4.550	Demazeau *et al.* (1971)
Li_2ZrF_6	4.937	4.664	a-1335
Li_2HfF_6	4.969	4.637	a-1383
Low-temperature α-Li_2SnF_6	4.95	4.56	a-1243
High-pressure α-Li_2SnF_6	4.88	4.55	Demazeau *et al.* (1971)
Li_2PbF_6	5.01	4.66	a-1272
Li_2PrF_6	5.04_7	4.84_6	Feldner and Hoppe (1983)
Li_2NbOF_5	4.966	4.572	Galy *et al.* (1969)

IV. $LiCaAlF_6$ Type (Li_2ZrF_6 related)

Compound	a_{hex} (Å)	c_{hex} (Å)	Reference
$LiCaTiF_6$	5.13_3	9.91_1	Gaile *et al.* (1977)
$LiSrTiF_6$	5.23_3	10.04_4	Gaile *et al.* (1977)
$LiCaVF_6$	5.131	9.78	a-1515
$LiSrVF_6$	5.202	10.39	a-1518
$LiCdVF_6$	5.131	9.520	a-1522
$LiPbVF_6$	5.23_9	10.48_1	Gaile and Rüdorff (1976)
$LiCaCrF_6$	5.095	9.720	a-1622
$LiSrCrF_6$	5.170	10.340	a-1626
$LiCdCrF_6$	5.086	9.495	a-1642
$LiPbCrF_6$	5.20_8	10.44_2	Gaile and Rüdorff (1976)
$LiCaFeF_6$	5.128	9.767	a-1843
$LiCdFeF_6$	5.118	9.543	a-1861
$LiPbFeF_6$	5.23_6	10.45_2	Gaile and Rüdorff (1976)
$LiCaAlF_6$	4.996	9.636	a-690
$LiSrAlF_6$	5.084	10.21	a-693
$LiCaGaF_6$	5.10	9.55	a-735
$LiSrGaF_6$	5.164	10.33	a-738
$LiPbGaF_6$	5.19_8	10.42_3	Gaile and Rüdorff (1976)
$LiCaCoF_6$	5.10_2	9.78_3	Fleischer and Hoppe (1982b)
$LiCdCoF_6$	5.08_6	9.51_9	Fleischer and Hoppe (1982b)
$LiCaNiF_6$	5.060	9.74_5	Fleischer and Hoppe (1982b)
$LiSrNiF_6$	5.124	10.341	Fleischer and Hoppe (1982b)

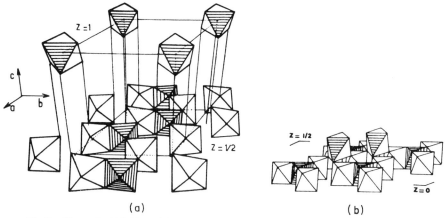

Fig. 5. Network of MnF_6 and CrF_6 (shaded) octahedra in structures deriving from Na_2SiF_6: (a) $NaMnCrF_6$ and (b) $LiMnCrF_6$. From Courbion *et al.* (1978).

the involved tetravalent cation. The results are in good agreement with Pauling's rules (Portier *et al.*, 1970a; Demazeau *et al.*, 1971; Menil *et al.*, 1972):

$$r_{M^{4+}} (\text{Å}): \quad r \leq 0.54 \qquad 0.54 < r < 0.71 \qquad 0.71 \leq r$$

$$\text{Structure:} \quad Na_2SiF_6 \qquad \text{trirutile } Li_2TiF_6 \qquad Li_2ZrF_6$$

The number of common edges in the structure decreases with increasing size of the M cation.

Different cationic distributions observed in A_2MF_6 structures are given in Table VII. It can be noted that, among the six cationic distributions theoretically possible in the Na_2SiF_6 structure, many of them have already been reported. Sharing of octahedra containing the paramagnetic species is given in Fig. 5 for two of those distributions: $NaMnCrF_6$ and $LiMnCrF_6$ (Courbion *et al.*, 1978).

C. TRIFLUORIDES AND RELATED STRUCTURES

In trifluorides, an important variation of the CN is observed with increasing size of the trivalent cation. For MF_3 with $0.60 \text{ Å} \leq r_{M^{3+}} \leq 0.80 \text{ Å}$, the CN of the cation is 6, and the structural arrangement can be described in terms of octahedra 3D connected to each other by corners. For larger M^{3+} cations (M = Y, rare earth, actinide, Tl, or Bi), CN is generally 9. A similar behavior occurs for the $M^{II}M'^{IV}F_6$ series. A CN of 4, found for AuF_3, is described in Section VII.

TABLE VII
Cationic Distributions in Some A_2MF_6 Structures

Trirutile: tetragonal, space group $P4_2/mnm$ (no. 136), $Z = 2$

Compound	Crystallographic Site				Reference
	$(2a)$ $(0,0,0; \frac{1}{2},\frac{1}{2},\frac{1}{2})$	$(4e)$ $\pm(0,0,z;\ldots)$ $z \simeq 0.33$	$(4f)$ $\pm(x,x,0;\ldots)$ $x \simeq 0.30$	$(8j)$ $(x,x,z;\ldots)$ $x \simeq 0.30, z \simeq 0.33$	
$LiM^{II}M^{III}F_6$ (M and M' = d transition metal)	Li	M^{II}, M^{III}	F_1	F_2	Wintenberger et al. (1972) (*P); Viebahn and Epple (1976)
$LiMg^{II}Fe^{III}F_6$	Fe^{III}	Li, Mg^{II}	F_1	F_2	Portier et al. (1970b) (*P)
$LiZn^{II}Co^{III}F_6$	Co^{III}	Li, Zn^{II}	F_1	F_2	Hoppe (1981) (*P)
$Li_2M^{IV}F_6$	M^{IV}	Li	F_1	F_2	Portier et al. (1969) (*P)

Na_2SiF_6 structure: trigonal, space group $P321$ (no. 150), $a_{hex} = 8.859$ Å, $c = 5.038$ Å, $Z = 3$ Zalkin et al. (1964)

Compound	Crystallographic Site				Reference
	$(1a)$ $(0,0,0)$	$(2d)$ $\pm(\frac{1}{3},\frac{2}{3},z)$ $z \simeq 0.50$	$(3e)$ $(x,0,0;\ldots)$	$(3f)$ $(x,0,\frac{1}{2};\ldots)$	
Na_2SiF_6	Si	Si	Na	Na	Zalkin et al. (1964) (*)
$NaMn^{II}Cr^{III}F_6$	Cr	Cr	Na	Mn	Courbion et al. (1977) (*)
$LiMn^{II}M^{III}F_6$ (M = Ga, Cr)	M^{III}	M^{III}	Mn	Li	Viebahn (1975); Courbion et al. (1978) (*)
$LiMn^{II}In^{III}F_6$	Li	Li	In	Mn	Gaile et al. (1977) (*P)
α-$LiMn^{II}M^{III}F_6$ (M = Ti, V, Fe), with F_1, F_2, and F_3 in general ($6g$) positions	Li	Li	Mn	M^{III}	Courbion et al. (1983a) (*)

92

Li₂ZrF₆ and related structures

Compound	Space Group	Cell Constants (Å)	Atomic Positions	Reference
Li$_2$ZrF$_6$ (*)	$P\bar{3}1m$ (no. 162)	$a = 4.98$ $c = 4.66$ $Z = 1$	Li: $(2d)$ site, $\pm(\frac{1}{3}, \frac{2}{3}, \frac{1}{2})$ Zr: $(1a)$ site, $(0, 0, 0)$ F: $(6k)$ site, $\pm(x, 0, z; \ldots)$	Hoppe and Dähne (1960)
LiCaAlF$_6$ (*)	$P\bar{3}1c$ (no. 163)	$a = 4.996$ $c = 9.636$ $Z = 2$	Li: $(2c)$ site, $\pm(\frac{1}{3}, \frac{2}{3}, \frac{1}{4})$ Ca: $(2b)$ site, $(0, 0, 0; 0, 0, \frac{1}{2})$ Al: $(2d)$ site, $\pm(\frac{2}{3}, \frac{1}{3}, \frac{1}{4})$ F: $(12i)$ site, $(x, y, z; \ldots)$	Viebahn (1971)
Li$_2$RhF$_6$ (*P) (Na$_2$SnF$_6$ type)	$P2_1/c$ (no. 14)	$a = 4.63_7$ $b = 4.63_4$ $c = 10.18_5$ $\beta = 117.1^0$ $Z = 2$	Li: $(4e)$ site, $\pm(x, y, z; \ldots)$ with $x = 0.34$, $y = 0.028$, $z = 0.33$ Rh: $(2a)$ site, $(0, 0, 0; 0, \frac{1}{2}, \frac{1}{2})$ F: $(12i)$ site, $(x, y, z; \ldots)$	Wilhelm and Hoppe (1974a)

1. ReO₃ Structure

The symmetry is cubic with space group $Pm3m$ (no. 221), $Z = 1$. A simple cubic ReO_3 structure had been previously claimed for niobium, molybdenum, and tantalum trifluorides. More recent investigations have shown, however, that these phases were in fact oxide–fluorides $MO_{3x}F_{3-3x}$, with x up to 0.66 (see Babel, 1967). These compounds are discussed in Chapter 4, devoted to oxide–fluorides. Genuine NbF_3 has been obtained by high-pressure synthesis. The compound crystallizes in the ReO_3 type. The cell constant ($a = 3.93_9$ Å) is significantly larger than those of the corresponding oxide–fluorides $[a\,(NbO_2F) = 3.89_9$ Å; $a\,(NbO_{0.40}F_{2.60}) = 3.90_3$ Å$]$ (Pouchard et al., 1971).

The high-temperature form of trifluorides crystallizing in the VF_3 type appears to be of the ReO_3 type. For instance, FeF_3 is cubic above $T_{tr} = 410°C$ (Tressaud et al., 1973b).

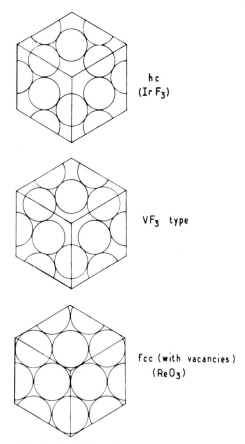

h c
(Ir F₃)

VF₃ type

fcc (with vacancies)
(ReO₃)

Fig. 6.   Anionic stackings in MF_3 compounds.

2. VF₃-Type Structure

The symmetry is rhombohedral with space group $R3c$ (no. 167), $a = 5.373$ Å, $\alpha = 57.52°$, $Z = 2$. The structure, which concerns d transition elements and scandium and indium trifluorides can be derived from ReO_3 by a rhombohedral distortion (at 25°C). Regular MF_6 octahedra are slightly rotated around their threefold axis, and F^- ions are therefore shifted from the edges of the pseudocell.

The anionic arrangement can be described as a compact stacking per-perpendicular to the ternary axis. The cations are located at the corners and at the center of the rhombohedron.

With increasing radius $r_{M^{3+}}$, the packing varies from the fcc type with 1-over-4 vacancies (ReO_3 type) to the hcp type (IrF_3). Simultaneously, the x parameter of the fluorine atoms varies from $-\frac{1}{4}$ to $-\frac{1}{12}$, the M–F–M angle decreases from 180 to 132°, and α decreases from 60 to 54° (Fig. 6; Table VIII).

The monoclinic distortion (space group $C2/c$) observed in MnF_3 is due to the Jahn–Teller configuration ($t_{2g}^3 e_g^1$) of Mn(III). The MnF_6 octahedra are characterized by three Mn–F distances (1.79, 1.91, and 2.09 Å), and the most elongated axis is alternatively directed along the a and b axes of the cell (Hepworth and Jack, 1957). However, a single-crystal determination is still needed. Mössbauer experiments performed on ^{57}Fe-doped MnF_3 have

TABLE VIII
VF₃- and ReO₃-Type Trifluorides

Compound	a_{rh} (Å)	α (deg)	x_F	a_{hex} (Å)	c_{hex} (Å)	Reference
AlF₃	5.03	58.62	—	4.925	12.448	a-56; Hoppe and Kissel (1984)
ScF₃ (*)	5.708	59.53	—	5.667	14.03	a-69; Lösch et al. (1982)
InF₃ (*P)	5.738	56.35	−0.166	5.419	14.43	a-59
TiF₃ (*P)	5.519	59.07	−0.183	5.44	13.61	a-218
VF₃ (*P)	5.373	57.52	−0.145	5.170	13.40	a-241
CrF₃ (*P)	5.264	56.56	−0.136	4.989	13.219	a-260
FeF₃ (*P)	5.362	57.99	−0.164	5.196	13.33	a-295
CoF₃ (*P)	5.279	56.97	−0.15	5.035	13.22	a-300
GaF₃ (*P)	5.20	57.5	−0.136	5.00	12.97	a-58
MoF₃ (*P)	5.666	54.72	−0.12	5.208	14.409	a-263
RuF₃ (*P)	5.408	54.67	−0.100	4.966	13.76	a-305
RhF₃ (*P)	5.330	54.42	−0.083	4.875	13.58	a-309
IrF₃ (*P)	5.418	54.13	−0.083	4.930	13.83	a-319
FeF₃	$a_{cubic} = 3.85_7$ Å (420°C)					Tressaud et al. (1973b)
NbF₃	$a_{cubic} = 3.93_9$ Å (25°C)					Pouchard et al. (1971)

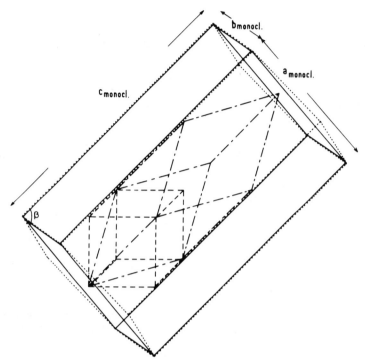

Fig. 7. Relationships between cell constants of MF_3 types [–––, cubic cell; –·–, rhombo-hedral cell; ----, hexagonal cell; ——, monoclinic cell (pseudo-orthorhombic)].

confirmed the presence at 4.2 K of two nonequivalent crystallographic sites (Lorin, 1980).

Relationships between the cell constants of the cubic (ReO_3), rhombohedral (VF_3), and monoclinic pseudo-orthorhombic MnF_3 types are given in Fig. 7.

3. Rare-Earth Trifluorides

Two main structural types are found for larger trivalent cations ($4f$ and $5f$ series) and are briefly considered: the orthorhombic YF_3 type adopted by smaller rare earths and the LaF_3 type occurring for larger $4f$ elements and for $5f$ trifluorides (Thoma and Brunton, 1966; Asprey *et al.*, 1965).

a. YF_3 STRUCTURE (LOW-TEMPERATURE FORM). In this struc-ture, which is orthorhombic (space group *Pnma*, $a = 6.353$ Å, $b = 6.850$ Å, $c = 4.393$ Å, $Z = 4$), Y has a tricapped trigonal prismatic coordination ($CN = 9$) with eight Y–F distances within 2.25 to 2.32 Å and a further one at ~ 2.6 Å (a-74; Cheetham and Norman, 1974). This structure is adopted by smaller rare-earth trifluorides (SmF_3 to LuF_3: Okamura and Yajima, 1974; Bukvetskii and Garashina, 1977; Piotrowski et al., 1979; TlF_3 and $\beta\text{-}BiF_3$: a-65; Greis and Martinez-Ripoll, 1978). In $\beta\text{-}BiF_3$, the ninth Bi–F distance is larger; the coordination can better be described as $CN = 8$, suggesting a sterically active lone electron pair (Cheetham and Norman, 1974).

b. LaF_3 STRUCTURE (TYSONITE). The tysonite structure, which crystallizes in the trigonal space group $P\bar{3}c1$ (no. 165), $a = 7.190$ Å, $c = 7.367$ Å, $Z = 6$ (a-78), is adopted by $4f$ elements from lanthanum to holmium; within this series and from SmF_3 to HoF_3 it constitutes a high-temperature variety. The metal ion is 11 coordinated: 7 M–F at 2.42 to 2.48 Å; 2 M–F at 2.63 Å, and 2 M–F at 3.04 Å. The coordination polyhedra can be described either with a $9:3$ coordination—if only the first 9 neighbors are taken into account—or as a distorted trigonal prism capped on all faces (11 neighbors) (a-78). Although space group $P\bar{3}c1$ has been confirmed by neutron diffraction experiments, $P6_3cm$ fits equally well (Cheetham et al., 1976).

In addition to $M^{II}M'^{IV}F_6$ phases, this structure is adopted by compounds with varied formulas, such as $\beta\text{-}SrTl_2F_8$ (a-792); $Ca_2M_5F_{19}$, with M = rare earth from samarium to thulium (Garashina and Sobolev, 1971a,b); solid solution $M^{II}_{1-x}La_xF_{2+x}$, with $0.84 \leq x \leq 1$ and $M^{II} = $ Ca, Sr, or Ba (a-80, a-82, a-84); solid solutions $La_{1-x}M^{IV}_xF_{3+x}$, with $M^{IV} = $ Th, U, or Zr and $0 \leq x \leq 0.35$ for U (a-183).

D. MM'F$_6$ COMPOUNDS

For $M^{II}M'^{IV}F_6$ compounds in which 0.7 Å $< r_{M^{II}} < 1.1$ Å, a sequence is found which is similar to that of the MF_3 series: for $r_{M'^{IV}} < 0.8$ Å the structures derive from VF_3, $LiSbF_6$, and ordered ReO_3 types, whereas LaF_3 structure is found for $r_{M'^{IV}} \geq 0.8$ Å. A summary of the distribution of the different structural types of $MM'F_6$ compounds is given in Fig. 8 (partly from Reinen and Steffens, 1978), and the lattice constants are compiled in Table IX.

1. LiSbF$_6$-Type Structure

The $LiSbF_6$ structure derives from VF_3 by a cationic ordering (rhombohedral, space group $R\bar{3}$, $a = 5.43$ Å, $\alpha = 56.97°$, $Z = 1$). Both cell constants are identical, but in $LiSbF_6$ the anionic arrangement is close to hcp. Two different types of regular octahedra (LiF_6 and SbF_6) arise from the occupancy of a more general site by fluorine atoms (Fig. 9). The structural arrangement of $LiSbF_6$ can also be described as a rock-salt lattice of Li^+ and $(SbF_6)^-$. Most of the LiM^VF_6 phases crystallize in this structural type (Babel, 1967) due to the high charge difference between the cations.

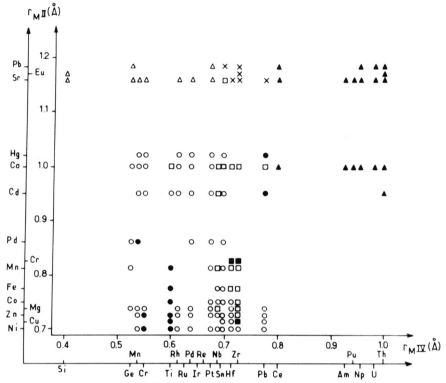

Fig. 8. Structures of $M^{II}M'^{IV}F_6$ compounds at room temperature ($\bigcirc$, LiSbF$_6$; $\bullet$, VF$_3$; $\square$, ordered ReO$_3$; $\blacksquare$, distorted ReO$_3$; $\blacktriangle$, LaF$_3$; $\triangle$, BaSiF$_6$; x, various). Partly from Reinen and Steffens (1978).

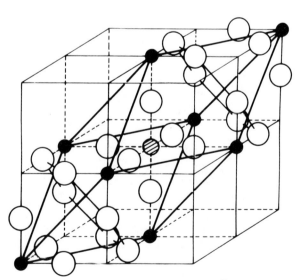

Fig. 9. Rhombohedral cell of the LiSbF$_6$-type structure [$\bigcirc$, F; ⌀, M(II) (Li); $\bullet$, M(IV) (Sb)].

TABLE IX

Lattice Constants of $M^{II}M'^{IV}F_6$ Conpounds[a]

$M^{II}M'^{IV}F_6$	a (Å)	c (Å)	Type[b]	Reference
$MgGeF_6$	5.09	13.04	L	Reinen and Steffens (1978)
$MnGeF_6$	5.14_6	13.61_5	L	Reinen and Steffens (1978)
$PdGeF_6$	5.02	14.13	L	a-1241
$CaGeF_6$	5.44_1	13.98	L	Reinen and Steffens (1978)
$CaGeF_6$	7.96	—	R for $T \geq 390$ K	Reinen and Steffens (1978)
$NiMnF_6$	4.91	13.17	L	Lorin et al. (1981)
$ZnMnF_6$	4.96_6	13.29	L	Lorin et al. (1981)
$MgMnF_6$	5.01	13.17	L	a-1740
$PdMnF_6$	5.00	13.93	V	Lorin et al. (1981)
$CdMnF_6$	5.08_7	14.01	L	a-1750
$CaMnF_6$	5.21	14.13	L	a-1743
$HgMnF_6$	5.08_4	14.12_5	L	a-1752
$NiCrF_6$	4.97_5	13.26	V	Siebert and Hoppe (1972c)
$ZnCrF_6$	5.02_6	13.33_7	V	Siebert and Hoppe (1972c)
$MgCrF_6$	5.09_1	13.14_3	L	Siebert and Hoppe (1972c)
$CdCrF_6$	5.14_0	14.07_5	L	a-1641
$CaCrF_6$	5.33_6	14.15_3	L	a-1620
$HgCrF_6$	5.12_8	14.26_5	L	a-1643
$NiTiF_6$	5.15	13.39	V	a-1334
$CuTiF_6$	5.2_7	13.3_1	L	Steffens and Reinen (1976)
$ZnTiF_6$	5.23	13.42	V	a-1327
$CoTiF_6$	5.24	13.48	V	a-1333
$FeTiF_6$	5.31	13.60	L	Reinen and Steffens (1978)
$MnTiF_6$	5.42	13.70	V	a-1332
$MnTiF_6$	7.80_8	—	R for $T \geq 398$ K	Reinen and Steffens (1978)
$CdTiF_6$	5.36	14.19	L	a-1326
$CaTiF_6$	8.16_1	—	R	Reinen and Steffens (1978)
$CaTiF_6$	5.77_1	14.13_6	L (space group $R\bar{3}m$)	a-1320
$NiRhF_6$	4.96_0	13.51_4	L	Wilhelm and Hoppe (1974b)
$ZnRhF_6$	4.99_6	13.68_3	L	Wilhelm and Hoppe (1974b)

(cont.)

TABLE IX (*cont.*)

$M^{II}M'^{IV}F_6$	a (Å)	c (Å)	Type[b]	Reference
$MgRhF_6$	5.02_7	13.51_1	L	Wilhelm and Hoppe (1974b)
$CdRhF_6$	5.12_8	14.44_7	V	Wilhelm and Hoppe (1974b)
$CaRhF_6$	5.26_7	14.61_2	L	Wilhelm and Hoppe (1974b)
$HgRhF_6$	5.13_3	14.67_6	L	Wilhelm and Hoppe (1974b)
$NiPdF_6$	4.93	13.70	L	Lorin et al. (1981)
$ZnPdF_6$	4.95	13.6_9	L	a-1991
$MgPdF_6$	4.98	13.48	L	a-1985
$PdPdF_6$	5.01	14.12	L	a-314
$CdPdF_6$	5.08	14.3_9	L	a-1993
$CaPdF_6$	5.19	14.5_9	L	a-1987
$HgPdF_6$	5.0	14.2	L	a-1994
$NiPtF_6$	4.93_7	13.68_7	L	Wilhem and Hoppe (1975b)
$ZnPtF_6$	4.98_0	13.82_8	L	Wilhem and Hoppe (1975b)
$MgPtF_6$	5.00	13.59	L	a-2029
$CoPtF_6$	5.00_2	13.81_5	L	Wilhem and Hoppe (1975b)
$MnPtF_6$	5.09_4	14.23_1	L	Wilhem and Hoppe (1975b)
$PdPtF_6$	5.04	14.18	L	a-2040
$PtPtF_6$	5.04	14.23	L	Tressaud et al. (1976a)
$CdPtF_6$	5.11_8	14.62_3	L	Wilhem and Hoppe (1975b)
$CaPtF_6$	5.24_5	14.78_4	L	Wilhem and Hoppe (1975b)
$HgPtF_6$	5.13_2	14.81_4	L	Wilhem and Hoppe (1975b)
$NiSnF_6$	5.15_2	13.76_3	L	Hoppe et al. (1972)
$CuSnF_6$	5.1_3	13.82	L	a-1257
$ZnSnF_6$	5.21_6	13.80_1	L	Hoppe et al. (1972)
$MgSnF_6$	5.31	13.58	L	a-1258
$CoSnF_6$	5.21_6	13.93_3	L	Hoppe et al. (1972)
$FeSnF_6$	5.27_0	14.06_3	L	Hoppe et al. (1972)
$MnSnF_6$	5.41_0	14.13_6	L	Hoppe et al. (1972)
$MnSnF_6$	7.96_8	—	R for $T \geq 530$ K	Reinen and Steffens (1978)
$PdSnF_6$	5.09_8	14.64	L	a-1271
$CdSnF_6$	5.3_7	14.4_7	L	a-1267
$CaSnF_6$	8.35_3	—	R	a-1259

TABLE IX (*cont.*)

$M^{II}M'^{IV}F_6$	a (Å)	c (Å)	Type[b]	Reference
$HgSnF_6$	5.32_1	15.02_7	L	Hoppe *et al.* (1972)
$SrSnF_6$	8.594	—	R for $T \leq 953$ K	a-1260
$NiHfF_6$	5.42_8	13.84_2	L	Steffens and Reinen (1976)
$NiHfF_6$	7.91_4	—	R for $T \geq 455$ K	Reinen and Steffens (1978)
$ZnHfF_6$	5.51_0	13.89_1	L	Steffens and Reinen (1976)
$CoHfF_6$	5.54_9	13.94_3	L	Steffens and Reinen (1976)
$CoHfF_6$	7.99_3	—	R for $T \geq 338$ K	Reinen and Steffens (1978)
$FeHfF_6$	5.61_4	14.03	L for $T \leq 258$ K	Reinen and Steffens (1978)
$FeHfF_6$	8.06_7	—	R	Steffens and Reinen (1976)
$MnHfF_6$	5.72_9	14.11	L for $T \leq 185$ K	Reinen and Steffens (1978)
$MnHfF_6$	8.15_5	—	R	Steffens and Reinen (1976)
$CaHfF_6$	8.462	—	R	a-1394
$NiZrF_6$	5.48_3	13.84_9	L	Reinen and Steffens (1978)
$NiZrF_6$	7.94_5	—	R for $T \geq 390$ K	a-1382
$ZnZrF_6$	7.99_5	—	R	a-1367
$MgZrF_6$	7.93_8	—	R	a-1365
$CoZrF_6$	5.59_2	13.96	L for $T \leq 273$ K	Reinen and Steffens (1978)
$CoZrF_6$	8.021	—	R	a-1381
$FeZrF_6$	5.59_4	14.04	L for $T \leq 219$ K	Pebler *et al.* (1978)
$FeZrF_6$	8.086	—	R	a-13
$MnZrF_6$	5.75_3	14.12	L for $T \leq 143$ K	Reinen and Steffens (1978)
$MnZrF_6$	8.162	—	R	a-1379
$CaZrF_6$	8.468	—	R	a-1366
$Ca_{2-x}Zr_xF_{4+2x}$	8.47–8.11	—	R for $1 \leq x \leq 1.5$	l'Helgouach *et al.* (1971)
$NiPbF_6$	5.21	14.00	L	a-1295
$CuPbF_6$	5.1_4	14.2_5	L	Steffens and Reinen (1976)
$ZnPbF_6$	5.21	14.15	L	a-1288
$MgPbF_6$	5.25	13.96	L	a-1284
$CdPbF_6$	5.36	15.09	V	a-1290
$CaPbF_6$	8.47_6	—	R	a-1285
$HgPbF_6$	5.27	15.90	V	a-1291

(*cont.*)

TABLE IX (cont.)

$M^{II}M'^{IV}F_6$	a (Å)	c (Å)	Type[b]	Reference
$MgNbF_6$	7.85_9	—	R	Chassaing et al. (1982)
$CaNbF_6$	8.40_9	—	R	Chassaing et al. (1982)
$MnNbF_6$	8.03_2	—	R	Chassaing et al. (1982)
$FeNbF_6$	5.42_2	14.09_3	L	Chassaing et al. (1982)
$CoNbF_6$	5.35_2	13.99_7	L	Chassaing et al. (1982)
$NiNbF_6$	5.40_0	13.59_3	L	Chassaing et al. (1982)
$ZnNbF_6$	5.32_5	13.67_8	L	Chassaing et al. (1982)
$CdNbF_6$	8.17_4	—	R (metastable at 300 K)	Chassaing et al. (1982)
$CdNbF_6$	5.55_4	14.59_5	L	Chassaing et al. (1982)

[a] Data measured at room temperature or at the transition temperature.
[b] Abbreviations: R, ordered ReO_3 type; V, VF_3 type; L, $LiSbF_6$ type with hexagonal cell constants.

For $M^{II}M'^{IV}F_6$ compounds, such an ordering is obtained as long as the cationic radii are sufficiently different. Neutron diffraction measurements have often been required to distinguish between VF_3 and $LiSbF_6$ arrangements, for instance, in Pd_2F_6 and in $M^{II}MnF_6$ (M^{II} = Ni or Zn) compounds (Lorin et al., 1981). In Pd_2F_6, two different PdF_6 octahedra are characterized by Pd^{II}–F = 2.17 Å and Pd^{IV}–F = 1.90 Å, respectively. The anionic stacking differs slightly from hc packing: $x_{theor} = -0.083$; $x_{Pd_2F_6} = -0.11$ (Tressaud et al., 1976b).

Several $M^{II}M'^{IV}F_6$ compounds have been shown to crystallize with the disordered VF_3 type (see Table IX). It can be noted that most of these phases have been characterized in powder samples, using only x-ray techniques. No cationic ordering is observed in $PdMnF_6$, however, even using neutron diffraction techniques (Lorin et al., 1981).

2. Ordered ReO_3 Structure

For larger tetravalent cations (M^{IV} = Zr, Hf, Nb, or Sn), several $M^{II}M^{IV}F_6$ compounds crystallize in an ordered ReO_3 structure ($NaSbF_6$ type, cubic, space group $Fm3m$, $a = 8.184$ Å, $Z = 4$). Transitions from a low-temperature form of the $LiSbF_6$ type to a high-temperature form of this ordered ReO_3 type generally occur in the temperature range 150–500 K (see Table IX) (Reinen and Steffens, 1978; Köhl et al., 1980).

We have already given the crystallographic data of $NaSbF_6$, which displays this type of ordering. However, the exact space group of cubic $MM'F_6$ compounds has not been clearly determined [$Fm3m$ in $NaSbF_6$ type; $Pa3$ in KPF_6 type; and $Pn3m$ versus $Pn3$ in $M^{II}Sn^{IV}(OH)_6$ types].

Ordered ReO_3 phases possessing fluorine atoms in excess are found in the CaF_2/ZrF_4 system, between compositions $CaZrF_6$ and $Ca_{0.50}Zr_{1.50}F_7$. The additional anions are located in the empty central site with $CN = 12$ (l'Helgouach et al., 1971; Poulain et al., 1975). Identical results have been obtained in $MNbF_6/NbF_4$ systems (Chassaing et al., 1982).

3. $CuM^{IV}F_6$ and $CrM^{IV}F_6$ Compounds

When M^{II} is a Jahn–Teller ion (Cr^{II} or Cu^{II}), $M^{II}M'^{IV}F_6$ phases often show additional distortions (Table X). $CuMF_6$ ($M^{IV} = Mn$, Pd, or Pt) can be indexed on the basis of a monoclinic cell deriving from that of MnF_3 (Lorin, 1980). Complex powder patterns have also been found in $Ag^{II}M^{IV}F_6$ phases (Müller and Hoppe, 1972a).

Phase transitions occur in $Cu(Cr)ZrF_6$ and in $Cu(Cr)HfF_6$. They are induced by crystal packing effects and vary according to static or dynamic Jahn–Teller character of $Cu(Cr)F_6$ octahedra distortions. Four structural types are observed (Reinen and Friebel, 1979; Friebel et al., 1983; Mayer et al., 1983):

$$CrZrF_6: II' \underset{415\,K^a}{\overset{150\,K}{\rightleftharpoons}} I' \overset{393\,K}{\rightleftharpoons} I \qquad CrHfF_6: II' \overset{200\,K}{\rightleftharpoons} I' \overset{393\,K}{\rightleftharpoons} I$$

$$CuZrF_6: II' \underset{353\,K}{\rightleftharpoons} II \underset{383\,K}{\rightleftharpoons} I \qquad CuHfF_6: II' \underset{298\,K}{\rightleftharpoons} II \underset{435\,K}{\rightleftharpoons} I$$

where I = ordered ReO_3 type, II = $LiSbF_6$ type, I' = tetragonal phase, II' = monoclinic (triclinic?) phase, and the superscript a means from neutron diffraction data.

In the distorted phases, the c/a ratio of the pseudo–cubic cell ($a_{tetrag} = a^*_{cubic}/\sqrt{2}$; $c_{tetrag} = a^*_{cubic}$) is lower than 1 (Table X). The low symmetry of phases II' results from the superposition of two types of distortions: one occurring along the threefold axis of the $LiSbF_6$-type unit cell and the other one along the fourfold axes of the $Cr(Cu)F_6$ octahedra (Reinen and Friebel, 1979).

4. $M^{II}M'^{IV}F_6$ Compounds with $r_{M^{IV}} \geq 0.8$ Å

As shown in Fig. 8, for $r_{M^{IV}} \geq 0.8$ Å, $MM'F_6$ compounds ($M'^{IV} = Ce$, Th, U, Np, Pu, or Am) crystallize in the LaF_3-type structure (Keller and Salzer, 1967). Metal atoms are statistically distributed in the $(6f)$ site. Solid solutions

TABLE X

MnF_3 and $M^{II}M'^{IV}F_6$ Compounds of Lower Symmetry

Compound	Symmetry	a (Å)	b (Å)	c (Å)	β (deg)	Z	Reference
MnF_3 (*P)	Monoclinic	8.904	5.037	13.448	92.74	12	a-283
$CuMnF_6$	Monoclinic	8.57	4.85	13.46	92.1	6	Lorin et al. (1981)
$CuPdF_6$	Monoclinic	8.59	4.87	13.61	91.34	6	Lorin (1980)
$CuPtF_6$	Monoclinic	8.61	4.91	13.79	91.6	6	Lorin (1980)
$CrHfF_6$	Tetragonal (I')	5.76_9	—	7.99_7 $(c/a\sqrt{2} = 0.98)$	—	4	Steffens and Reinen (1976); Propach and Steffens (1978);
$CrZrF_6$	Tetragonal (I')	5.80_9	—	8.01_6 $(c/a\sqrt{2} = 0.97)$	—	4	Friebel et al. (1983); Reinen and Steffens (1978);
$CuHfF_6$	Pseudotetragonal cell (II')	5.5_7	—	7.7_6 $(c/a\sqrt{2} = 0.98_5)$	—	4	Mayer et al. (1983)
$CuZrF_6$	Pseudotetragonal cell (II')	5.61_1	—	7.79_4 $(c/a\sqrt{2} = 0.98)$	—	4	

showing anionic vacancies are obtained in the $MF_2/MM'F_6$ systems up to concentrations of 25% MF_2 (in moles).

5. $M^IM'^VF_6$ Compounds with CsCl-Related Structures

The relationships between $MM'F_6$ structures deriving from CsCl packing $[M^+(M'F_6)^-$ ordering] have been developed by Babel (1967). Several structural types showing cubic, tetragonal, or rhombohedral symmetry are listed in Table XI. The results of Kruger et al. (1976) and Heyns and Pistorius (1976) have shown discrepancies with previous data, and some older structural determinations should be reconsidered.

In addition to $MM'F_6$ compounds, hexahydrates $M^{II}M'^{IV}F_6 \cdot 6H_2O$, namely, $[Ni(H_2O)_6]SnF_6$ (space group $R\bar{3}$) versus $[Fe(H_2O)_6]SiF_6$ (space group $R\bar{3}m$) and the hydroxy fluoride $Cs[SbF_5(OH)]$ derive from the $KOsF_6$ structure (a-2104 and following; Nolte and De Beer, 1979).

E. A_xMF_3 ($x \leq 1$) COMPOUNDS AND RELATED STRUCTURES

Framework structures in ternary and quaternary fluorides of composition $(MF_3)^{n-}$ are generally formed by 3D corner sharing of octahedra. The smallest structural units may comprise eight, six, or four octahedra, the centers of which form a cube, a trigonal prism, or a tetrahedron (Fig. 10) (Hartung et al., 1979). The first case, which has already been discussed in the class of trifluorides, is also realized in perovskites, the second in tetragonal (TTB) and hexagonal tungsten bronze (HTB) structures, and the last in pyrochlores.

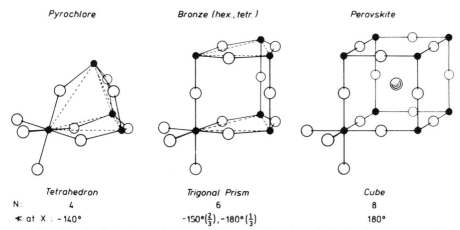

Fig. 10. Smallest structural units of 3D corner-shared octahedra in the structures of perovskite, bronzes, and pyrochlore.

TABLE XI

$M^IM^VF_6$ Compounds with CsCl-Related Structures

Structural Type	Symmetry, Space Group	Lattice Constants	Atomic Positions	Structural Data	Reference
KSbF₆ (low-temperature form)	Cubic, $I23$ or $I2_13$	$a = 10.176$ Å, $Z = 8$	—	Powder data; CsCl-type ordering	Heyns and Pistorius (1976) a-1537
KNbF₆ (*)	Tetragonal, $P\bar{4}c2$ (no. 116)	$a = 5.18$ Å, $c = 10.05$ Å, $Z = 2$	Nb: $(2a)$, $\pm(0,0,\frac{1}{4})$; K: $(2d)$ $(\frac{1}{2},\frac{1}{2},0; \frac{1}{2},\frac{1}{2},\frac{1}{2})$, F₁: $(4e)$ $(x,x,\frac{1}{4};\ldots)$, $x = 0.29$, F₂: $(8j)$ $(x,y,z;\ldots)$, $x = 0.28$, $y = 0.13$, $z = 0.11$	Tetragonally compressed CsCl lattice; doubling of c due to different orientations of $(NbF_6)^-$ octahedra; irregular CN = 8 + 4 for K (combination of two tetrahedra.)	
KSbF₆ (*) (room temperature form)	Tetragonal, $P\bar{4}2m$ (no. 111)	$a = 5.16$ Å, $c = 10.07$ Å, $Z = 2$	Sb: $(2g)$, $\pm(0,0,z)$, $z = 0.250$, K₁: $(1d)$ $(\frac{1}{2},\frac{1}{2},0)$; K₂: $(1b)$, $(\frac{1}{2},\frac{1}{2},\frac{1}{2})$, 3 F: $(4n)$, $(x,x,z;\ldots)$	KNbF₆-Related structure; dodecahedral environment (CN = 8) of K⁺	Kruger *et al.* (1976)
KOsF₆ (*)	Rhombohedral, $R\bar{3}$ (no. 148)	$a = 4.991$ Å, $\alpha = 97.18°$, $Z = 1$	Os: $(1a)$, $(0,0,0)$; K: $(1b)$ $(\frac{1}{2},\frac{1}{2},\frac{1}{2})$, F: $(6f)$ $\pm(x,y,z; z,x,y; y,z,x)$, $x = 0.72$, $y = 0.79$, $z = 0.103$	$(OsF_6)^-$ at the vertices of the rhombohedral cell; rhombohedral distortion of the octahedra; CN = 6 + 6 for K⁺	a-1999; Gafner and Kruger (1974)
BaSiF₆ (*)	Rhombohedral, $R\bar{3}m$ (no. 166)	$a = 4.75$ Å, $\alpha = 97.58°$, $Z = 1$	Si: $(1a)$, $(0,0,0)$; Ba: $(1b)$, $(\frac{1}{2},\frac{1}{2},\frac{1}{2})$, F: $(6h)$, $\pm(x,x,z; z,z,x; x,z,x)$, $x = 0.75$, $z = 0.08_5$	Particular case of the KOsF₆ type	a-1228; Nolte and De Beer (1979)

Which of these structures occurs depends largely on the quantity and size of the A cations. Including the structures derived from these prototypes, a large number of fluorides have been found since the early 1960s in each of three mentioned groups.

1. Perovskite, Elpasolite, and Related Structures ($A^IM^{II}F_3$ and $A_2^IB^IM^{III}F_6$ Compounds)

A perovskite or related structure is expected to occur for a ternary fluoride $A^IM^{II}F_3$ if the tolerance factor $t = (r_A + r_F)/\sqrt{2}(r_M + r_F)$ of the ions involved yields a value within the range $0.76 \leq t \leq 1.13$ (Babel, 1967, 1969). The same rule may be applied to the so-called elpasolites $A_2^IB^IM^{III}F_6$, in which a $2\,M^{II} = B^I + M^{III}$ substitution has been realized. How the different variants of perovskite and elpasolite structures can be rationalized using this radius ratio and ion-contact criterion, illustrated in Fig. 11, has been discussed elsewhere (Arndt et al., 1975; Babel, 1967, 1969; Babel et al., 1973a).

The possibility of varying the tolerance factor, not only by changes in cationic composition including partial substitution (Dance et al., 1979; Dance and Tressaud, 1979), but also by high-pressure experiments yielding metastable high-pressure polymorphs, has contributed considerably to confirming

Tolerance factor $t = (r_A + r_F)/\sqrt{2}(r_M + r_F)$

Plane (110) of cubic perovskites AMF_3

A   M   F

$t = 1$ 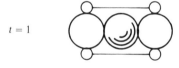 Both cations, A (CN = 12) and M (CN = 6), are in contact with the anions

$t < 1$ 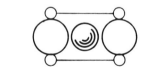 Cation A is too small to touch the anions, but the structure distorts only if $t < 0.88$ by bending the M–F–M bridges, which were linear before

$t > 1$  Cation A is too large; to avoid energetically unfavorable loosening of M–F contacts, transformation to hexagonal (chain) structures occurs; retransformation under *high pressure* (face → corner-sharing octahedra)

Fig. 11. Geometric meaning of the tolerance factor t.

these arguments (Longo and Kafalas, 1969; Goodenough *et al.*, 1972; Arndt *et al.*, 1975).

Because the large A cations are more compressible than the smaller M cations, the tolerance factor of a given compound appears to be smaller in its high-pressure modification. This will be obvious from examples shown in the tables that follow.

The tolerance factor values quoted in this chapter were calculated using Shannon's radii of hexacoordinated M ions with single alterations (Babel and Binder, 1983; Shannon, 1976), but for better consistency with former specifications (Arndt *et al.*, 1975; Babel, 1967; Babel *et al.*, 1973a), the Ahrens radii (Ahrens, 1952) have been chosen for the A cations, multiplied by 1.06 to correct for CN = 12.

Because the A_2BMF_6 compounds (generally obeying $r_A > r_B > r_M$) crystallize in perovskite-related superstructures due to B(I)/M(III) ordering in the octahedrally coordinated positions, they are treated here along with the perovskite group. These elpasolites contain isolated MF_6^{3-} octahedra, however, and, as far as M–F–M bridging is concerned, they do not belong to the 3D skeleton structures. In the following subsections we show how the value of the tolerance factor plays a decisive role in structural classification.

a. ORTHORHOMBIC PEROVSKITES (GdFeO$_3$ TYPE) AND CRYOLITES A_3MF_6. For the lowest range of tolerance factors observed in perovskite-like fluorides ($0.76 \leq t \leq 0.88$), the orthorhombically distorted GdFeO$_3$-type structure is found [space group *Pbnm* (no. 62), $Z = 4$ (Geller, 1956)] (Fig. 12). Not only sodium compounds but also some potassium compounds with larger M(II) ions adopt this structure (Alter and Hoppe, 1974e; Hidaka and Hosogi, 1982; Odenthal and Hoppe, 1971; and Table XII).

Fig. 12. Octahedral network in orthorhombically distorted perovskites. From Geller and Wood (1956).

TABLE XII

Lattice Constants of Orthorhombic Perovskites AMF_3 ($GdFeO_3$ Type) and Monoclinic
Cryolites Na_3MF_6 (Na_3AlF_6 Type)

Compound	t	a (Å)	b (Å)	c (Å)	β (deg)	Reference
$NaMgF_3$	0.83	5.350	5.474	7.652	—	a-564
$NaNiF_3$ (*)	0.83	5.365	5.528	7.689	—	Hidaka and Ono (1977)
$NaZnF_3$	0.81	5.415	5.587	7.775	—	a-588
$NaCoF_3$	0.81	5.420	5.603	7.793	—	a-1885
$NaFeF_3$	0.79	5.495	5.672	7.890	—	a-1785
$NaVF_3$	0.79	5.490	5.684	7.887	—	Williamson and Boo (1977a)
$NaMnF_3$	0.77	5.551	5.749	8.005	—	a-1707
$KPdF_3$	0.88	5.986	6.001	8.503	—	Alter and Hoppe (1974e)
$KAgF_3$	0.85	6.270	6.489	8.30	—	a-427A
$KCdF_3$ (*)	0.85	6.107	6.127	8.671	—	E. Herdtweck (unpublished work, 1981); Hidaka and Hosogi (1982)
$KCaF_3$	0.83	6.164	6.209	8.757	—	a-580; Wall (1971)
$KHgF_3$	0.82	6.20	6.28	8.81	—	a-621
Na_2LiNiF_6	0.83	5.309	5.415	7.515	90.4	Sorbe (1977)
Na_2LiCuF_6	0.83	5.314	5.424	7.528	90.3	Sorbe (1977)
Na_3AlF_6 (*)	0.82	5.402	5.596	7.756	90.28	Hawthorne and Ferguson (1975)
Na_3CuF_6	0.81	5.454	5.666	7.850	90.1	Grannec et al. (1976)
Na_3NiF_6	0.81	5.441	5.671	7.849	90.12	a-1926
Na_3GaF_6	0.80	5.47	5.68	7.88	~90	a-713
Na_3CoF_6	0.80	5.485	5.714	7.899	90.14	a-1886
Na_3CrF_6 (*)	0.80	5.491	5.702	7.913	90.39	a-1567
Na_3VF_6	0.80	5.513	5.721	7.963	90.47	Alter and Hoppe (1975)
Na_3FeF_6 (*)	0.80	5.52	5.74	7.97	90.40	a-1788; Matvienko et al. (1981)
Na_3TiF_6	0.79	5.53	5.83	7.99	~90	a-1302
Na_3ScF_6	0.78	5.60	5.81	8.12	90.7	a-799
Na_3InF_6	0.77	5.643	5.844	8.190	90.9	a-755
Na_3TlF_6	0.76	5.723	5.897	8.287	90.7	a-778
Ag_3ScF_6	0.83	5.854	6.037	8.415	90.8	Lösch and Hebecker (1979b)
Ag_3InF_6	0.82	5.934	6.094	8.514	91.6	Lösch and Hebecker (1979b)
Ag_3TlF_6	0.81	5.955	6.147	8.549	91.3	Lösch and Hebecker (1979b)

The existence of tetragonal fluoroperovskites, intermediate between the cubic and the orthorhombic structures, has been corroborated by complete single-crystal structure analysis of low-temperature $KMnF_3$ ($t = 0.90$), but only at $T < 200$ K (Okazaki and Ono, 1978; Hidaka, 1975). In contrast, orthorhombic $KCdF_3$ ($t = 0.85$) is tetragonal in a small high-temperature range before it becomes cubic at $T \geq 485$ K (Hidaka et al., 1977). No fluoroperovskite appears to be tetragonal at room temperature. Therefore, except for Jahn–Teller distorted structures, tetragonal AMF_3 phases are not further considered here, although they were well established in the course of phase transition studies (Arakawa and Ebisu, 1982; Helmholdt et al., 1980; Sakashita et al., 1981).

The main difference between cubic and orthorhombic perovskites is that the M–F–M bridge angles in the latter significantly deviate from linearity. The results for $KCdF_3$ ($t = 0.85$) yield Cd–F–Cd $= 157.1$ and $159.1°$; E. Herdtweck, unpublished work, 1981; Hidaka and Hosogi, 1982) compared with Ni–F–Ni $= 147.8$ and $148.0°$ in $NaNiF_3$ ($t = 0.83$; Hidaka and Ono, 1977; O'Keeffe et al., 1979). It may be thus assumed that within the group of orthorhombic perovskites the bridge angle itself depends on the tolerance factor. In fact, the tolerance factor is a measure of misfit between the size of the MF_6 octahedra and that of the A cations. The smaller its value, the stronger the distortion due to tilting of octahedra. This bending of bridges is required to bring some anions in contact with the A cations (see Fig. 11).

Most A_3MF_6 compounds, which are actually A_2BMF_6 elpasolites with B $=$ A, yield tolerance factors similar to those found for the orthorhombic perovskites. However, only the structures of sodium and of some silver compounds, which are isostructural with the mineral cryolite Na_3AlF_6 [monoclinic symmetry, space group $P2_1/n$ (no. 14), $Z = 2$ (Naray-Szabo and Sasvari, 1938; Hawthorne and Ferguson, 1975)], are closely related to the orthorhombic perovskite structure. This is also obvious from the cell dimensions given in Table XII. Furthermore, cations exhibit quite similar coordinations, which are octahedral for M^{3+} and one-third of sodium (or silver) ions and irregular (6 + 6) for the other two-thirds.

The orthorhombic low-temperature form of Li_3MF_6 compounds is cryolite related and contains all cations in CN $= 6$ [space group $Pna2_1$ (no. 33), $Z = 4$ (Burns et al., 1968)] (Table XIII). In the high-temperature form, lithium atoms exhibit tetrahedral coordination (see Section VII,D).

For A_3MF_6 compounds with larger alkali metals, different structural types have been assigned (Bode and Voss, 1957; Greis, 1982; Babel, 1967): tetragonal perovskite-related structures of the Rb_3TlF_6 and β-$(NH_4)_3ScF_6$ types (Bode and Voss, 1957), cubic $(NH_4)_3FeF_6$ type, and pseudotetragonal K_3FeF_6 type (Bode and Voss, 1957; Pauling, 1924). With the exception of cubic K_3MoF_6 (Toth et al., 1969) and orthorhombic $(N_2H_5)_3CrF_6$ (Kojic-

TABLE XIII

Lithium Cryolites α-Li_3MF_6

Compound	a (Å)	b (Å)	c (Å)	Reference
α-Li_3AlF_6 (*)	9.510	8.230	4.876	a-651; Burns et al. (1968)
α-Li_3GaF_6	9.60	8.38	4.97	a-712; Massa and Rüdorff (1971)
α-Li_3TiF_6	9.62	8.55	5.09	a-1300
α-Li_3VF_6	9.59	8.49	5.04	a-1477
α-Li_3CrF_6	9.60	8.35	5.02	a-1562
α-Li_3FeF_6	9.62	8.49	5.03	a-1782
Li_3NiF_6	9.60	8.37	4.90	Grannec et al. (1975a)

Prodic et al., 1972), however, no single-crystal structure analyses have been carried out recently. Therefore, these structures are not treated further here. We mention only some studies concerning the size of the unit cells that seem to indicate that many compounds such as K_3MF_6 and Rb_3MF_6 (M = Cr, Fe, or In), formerly assigned to the tetragonal β-$(NH_4)_3ScF_6$ type, $Z = 2$ (Babel, 1967; Bode and Voss, 1957), in fact crystallize in supercells with fivefold axes and $Z = 250$ (Alter and Hoppe, 1974a; Haegele, 1974; Champarnaud, 1978; Peradeau, 1978; de Kozak, 1971). Such complex structural features are possibly present in most A_3MF_6 compounds, but the details are still unknown.

b. CUBIC PEROVSKITES AMF_3 AND ELPASOLITES A_2BMF_6 OF CUBIC K_2NaAlF_6 TYPE. The tolerance factor range $0.88 \le t \le 1.00$ corresponds to the existence field of cubic perovskite and elpasolite structures, which are shown in Fig. 13 (Babel, 1967; Pausewang and Rüdorff, 1969). As can be seen from Tables XIV and XV, these structures [space group $Pm3m$ (no. 221), $Z = 1$, and $Fm3m$ (no. 225), $Z = 4$ (Morss, 1974), respectively], are adopted mainly by AMF_3 and A_2BMF_6 fluorides in which the alkali cations (A^I = K, Rb, NH_4, or Tl) are not very different in size from that of F^- and which therefore better fit the requirements of cubic close packing $(AF_3)^{2-}$. However, several rubidium and cesium compounds with $t > 1.0$ and hexagonal structures under normal conditions form metastable cubic high-pressure phases, too (Arndt et al., 1975; Longo and Kafalas, 1969). The compounds $CsMgF_3$ and $CsZnF_3$, although quoted in the earlier literature (Ludekens and Welch, 1952; Schmitz-DuMont and Bornefeld, 1956), probably exist only as hexagonal high-pressure phases, because normal-pressure conditions lead only to $Cs_4M_3F_{10}$ compositions (Babel, 1965a). Examples such as $CsNiF_3$ (Haegele et al., 1976) and Cs_2NaFeF_6 (Arndt et al., 1975), however, show that low-temperature preparation may lead to phases identical to those obtained under high pressure.

Single-crystal data on elpasolites (Haegele et al., 1975; Massa, 1982b) have confirmed the $Fm3m$ space group, which was ambiguous in early studies on the

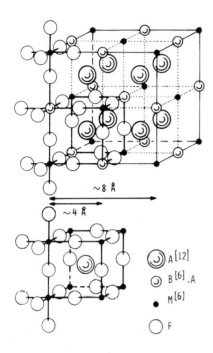

~8 Å

~4 Å

⊙ A [12]

⊚ B [6] , A

● M [6]

○ F

Fig. 13. The elpasolite structure A_2BMF_6 (above) as a superstructure of the perovskite $A_2MMF_6 = 2\,AMF_3$ (below) [CN of atoms in brackets].

cubic prototype (Frondel, 1948). Although this space group has also been assigned to K_3MoF_6 (Toth et al., 1969), A_3MF_6 "cryolites" generally adopt the cubic elpasolite structure only in their high-temperature form. Finally, it is worth noting that many oxide–fluorides of each composition between $A_2BM^{IV}OF_5$ and $A_2BM^{VI}O_3F_3$ (Pausewang and Rüdorff, 1969; Ravez et al., 1980) are known to crystallize with the cubic or related elpasolite structure, in contrast to the limited number of oxide–fluorides exhibiting the perovskite structure (Rüdorff and Krug, 1964). One reason for this may be that only the "pure" ionic bonds in the $B^{I}X_6$ octahedra (X = O, F), which separate the MX_6 octahedra in elpasolites, allow the latter to orient randomly along each of the three cell axes. The well-defined lower symmetry found for the MO_nF_{6-n} octahedra by vibrational spectroscopy (Pausewang and Dehnicke, 1969; Dehnicke et al., 1969) therefore is not in contradiction to the overall O_h symmetry observed by the averaging diffraction methods.

c. JAHN–TELLER DISTORTED PEROVSKITES AND ELPA-SOLITES. The examples of some Cr(II), Cu(II), and Mn(III) compounds (Table XVI) show that the Jahn–Teller effect associated with d^4 and d^9 configurations generally leads to distortions of the basic structures, which are imposed by simple geometric size relations only. Thus, $NaCuF_3$ [space group $P\bar{1}$ (no. 2), $Z = 4$] has a triclinically distorted $GdFeO_3$ structure (average

TABLE XIV

Lattice Constants of Cubic Perovskites AMF_3, Including Some Phases Prepared at High Pressure or Low Temperature[a]

$M\ (r_{M^{2+}})$	Na (1.03)	Ag (1.33)	K (1.41)	Rb (1.56)	NH₄ (1.57)	Tl (1.58)	Cs (1.77)	Reference
Mg (0.68)	O	3.918	3.989	LT: 4.006	4.06	H	HP: H	a-565; a-567; a-570; a-577
	0.83	*0.94*	*0.96*	*1.02*	*1.02*	*1.02*	*1.09*	
Ni (0.69)	O	3.936	4.015	HP: 4.077	2 × 4.08	HP: 4.110	H	a-1927; a-1932; a-1935; a-1942; a-1951
	0.83	*0.93*	*0.96*	*1.01*	*1.02*	*1.02*	*1.09*	
Zn (0.72)	O	3.972	4.055	LT: 4.110	4.116[b]	H	H	a-589; a-592; a-601; Babel (1969)
	0.81	*0.92*	*0.95*	*1.00*	*1.00*	*1.00*	*1.07*	
Co (0.74)	O	3.983	4.071	4.127	4.132[b]	4.138	H	a-1888; a-1894; a-1895; a-1903; a-1914
	0.81	*0.88*	*0.94*	*0.99*	*0.99*	*0.99*	*1.06*	
Fe (0.78)	O	—	4.121	4.172	4.177	4.188	HP: 4.283	a-1795; a-1806; a-1812; a-1827; a-1865
	0.79	—	*0.92*	*0.97*	*0.97*	*0.98*	*1.04*	
V (0.79)	O	—	4.131	4.182		—	—	Williamson and Boo (1977a)
	0.79	—	*0.91*	*0.96*				
Mn (0.83)	O	4.03	4.189	4.242	4.242[b]	4.251	HP: 4.331	a-1710; a-1721; a-1724; a-1732; a-1739; a-1753
	0.77	*0.87*	*0.90*	*0.95*	*0.95*	*0.95*	*1.01*	
Pd (0.86)			O	4.298		4.301	HP: 4.13	Alter and Hoppe (1974e)
			0.88	*0.93*		*0.94*	*1.00*	
Cd (0.95)			O	4.399		4.395	4.465	a-613; a-616; a-620; Fischer (1982)
			0.85	*0.90*		*0.90*	*0.96*	
Ca (1.00)			O	4.455			4.524	a-581; a-582: Fischer (1982)
			0.83	*0.88*			*0.94*	
Hg (1.02)			O	4.47			4.57	a-622; a-623
			0.82	*0.87*			*0.93*	

[a] The values in *italics* are the tolerance factors, as calculated from the ionic radii and $r_{F^-} = 1.33$ Å. The occurrence of orthorhombic (O) and hexagonal (H) compounds is mentioned; HP, high pressure; LT, low-temperature phase.

[b] Phase transitions studied by Helmholdt et al. (1980).

TABLE XV

Cubic Elpasolites A_2BMF_6

Elpasolite[a]	a (Å)	n^b	Reference
K_2NaAlF_6	8.122	+	a-688
K_2AgAlF_6	8.360	+10	Setter and Hoppe (1976b)
$Cs_2NH_4AlF_6$	9.056	—	Massa (1976)
$(NH_4)_2NaAlF_6$	8.337	—	Massa (1976)
$(NH_4)_2KAlF_6$	8.731	—	Massa (1976)
$Cs_2NH_4GaF_6$	9.206	+2	Massa (1976)
Cs_2TlGaF_6	9.182	+3	Babel et al. (1973a)
Rb_2NaInF_6	8.674	+3	Grannec and Ravez (1970)
Rb_2AgInF_6	8.897	+3	Setter and Hoppe (1976a)
Rb_2NaTlF_6	8.793	+4	Grannec et al. (1970)
Rb_2NaBiF_6	9.003	+5	Cousson et al. (1972)
$(NH_4)_2NaScF_6$	8.599	+	a-806
Cs_2NaYF_6	9.056	+	a-831
Cs_2KLaF_6	9.65	+	a-853
$Cs_2A^ILnF_6$ (A = Na, K, Rb)		+ Many	Feldner and Hoppe (1980)
K_2LiTiF_6	8.094	+5	Alter and Hoppe (1974a); Weissenhorn (1972)
Cs_2TlVF_6	9.233	+7	Alter and Hoppe (1975); Babel et al. (1973a)
$(NH_4)_2NaVF_6$	8.485	+2	Massa (1976)
K_2NaCrF_6 (*)	8.275	—	Massa (1982b)
Rb_2NaCrF_6	8.418	+7	Babel et al. (1973a); Siebert and Hoppe (1972b)
$(NH_4)_2KCrF_6$	8.898	+2	Massa (1976)
K_2NaMoF_6	8.501	—	Hoppe and Lehr (1975)
Rb_2NaMoF_6	8.632	—	Hoppe and Lehr (1975)
Tl_2KMoF_6	8.977	+4	Hoppe and Lehr (1975)
$K_{2.2}Li_{0.8}MnF_6$ (*)	8.110	—	Massa (1982a)
Rb_2NaFeF_6 (*)	8.465	—	Massa (1982b); Haegele et al. (1975)
Rb_2KFeF_6 (*)	8.867	—	Massa (1982b); Haegele et al. (1975)
Cs_2TlFeF_6	9.222	+5	Alter and Hoppe (1974c); Babel et al. (1973a)
$(NH_4)_2NaFeF_6$ (*)	8.484	+2	Massa (1976; 1982b)
Cs_2KCoF_6	8.998	+3	Alter and Hoppe (1974d); Babel et al. (1973a)
K_2LiCoF_6	7.995	—	J. Grannec and A. Tressaud (unpublished work, 1982)
K_2NaRhF_6	8.362	—	Wilhelm and Hoppe (1975a)

TABLE XV *(cont.)*

Elpasolite[a]	a (Å)	n^b	Reference
Rb_2KRhF_6	8.876		Wilhelm and Hoppe (1975a)
K_2NaNiF_6	8.211	$+3$	Alter and Hoppe (1974b)
K_2LiPdF_6	8.154	$+5$	Tressaud et al. (1984b)
K_2NaCuF_6	8.203	$+$	a-404
Rb_2NaCuF_6	8.368	$+$	Grannec et al. (1975b)
K_2LiCuF_6	7.935	$-$	J. Grannec and A. Tressaud (unpublished work, 1982)
HP K_2LiAlF_6	7.865	$-$	Arndt et al. (1975)
HP Cs_2NaAlF_6	8.628	$-$	Arndt et al. (1975)
HP Cs_2AgAlF_6	8.729	$-$	Setter and Hoppe (1976b)
HP Rb_2LiVF_6	8.248	$-$	Arndt et al. (1975)
HP Cs_2NaVF_6	8.752	$-$	Arndt et al. (1975)
HP Cs_2NaCrF_6	8.706	$-$	Arndt et al. (1975)
HP Cs_2NaMnF_6 (*P)	8.762	$-$	Massa (1982a)
HP Rb_2LiFeF_6 (*)	8.244	$-$	Arndt et al. (1975); Massa (1982b)
HP Cs_2NaFeF_6 (*)	8.739	$-$	Arndt et al. (1975); Massa (1982b)
HP Tl_2LiFeF_6	8.329	$-$	Arndt et al. (1975)

[a] HP, High-pressure form.
[b] The number of additional cubic isomorphs published for the same M cation in the same paper(s).

bridge angle Cu–F–Cu $= 145°$; Babel et al., 1974b; Binder, 1973; Haegele and Babel, 1974). $KCuF_3$ [space groupe $I4/mcm$ (no. 140), $Z = 4$ (Okazaki and Suemune, 1961)] has a tetragonal structure with linear bridges characteristic of the cubic perovskites. Nearly isodimensional orthorhombically distorted $(CuF_6)^{4-}$ octahedra appear in both copper compounds. The antiferrodistortive order of the long axes is schematically shown in Fig. 14, which also makes obvious how polytypism may occur in $KCuF_3$ (Okazaki, 1969; Tanaka et al.,

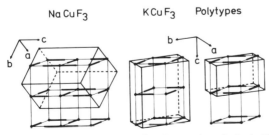

NaCuF₃   KCuF₃   Polytypes

Fig. 14. Antiferrodistortive order of Jahn–Teller elongated octahedra in $NaCuF_3$ (idealized) and polytypes of $KCuF_3$ (Okazaki and Suemune, 1961; Okazaki, 1969).

TABLE XVI

Jahn–Teller Distorted Perovskites and Elpasolites[a]

Compound	Space Group	a (Å)	b (Å)	c (Å)	Reference
$NaCrF_3$	$P\bar{1}$	5.52 $\alpha = 90.3°$	5.67 $\beta = 92.5°$	8.19 $\gamma = 86.0°$	Binder (1973); Vollmer (1966)
$KCrF_3$	$I4/mcm$, $Z = 4$	6.036	(8.536)	8.010	a-1573
$RbCrF_3$		6.149	(8.696)	8.088	a-1587
NH_4CrF_3		6.232	(8.813)	7.954	a-1585
$TlCrF_3$		6.194	(8.760)	8.064	a-1644
$NaCuF_3$ (*)	$P\bar{1}$	5.383 $\alpha = 90.67°$	5.547 $\beta = 92.05°$	7.912 $\gamma = 86.96°$	Babel et al. (1974b); Binder (1973)
$KCuF_3$ (*)	$I4/mcm$, $Z = 4$	5.855	(8.280)	7.852	Okazaki and Suemune (1961); a-401
$RbCuF_3$		6.001	(8.487)	7.894	a-406
NH_4CuF_3		6.09	(8.613)	7.78	a-405
$TlCuF_3$		6.083	(8.603)	7.866	a-424
$AgCuF_3$	Monoclinic	4.06	3.83 $\beta = 91.2°$	4.06	a-413
$RbAgF_3$	$I4/mcm$, $Z = 4$	6.335	(8.959)	8.44	a-428A
$CsAgF_3$		6.489	(9.177)	8.52	a-428B
K_2NaMnF_6 (*)	$I4/mmm$, $Z = 2$	5.778	(8.171)	8.577	Knox (1963); a-1720
Rb_2NaMnF_6		6.106	(8.365)	8.660	Siebert and Hoppe (1972b)
Rb_2KMnF_6		6.101	(8.628)	9.160	Siebert and Hoppe (1972b)
Cs_2KMnF_6		6.317	(8.933)	9.265	a-1736
$(NH_4)_2NaMnF_6$		5.915	(8.365)	8.740	Massa (1975)
$(NH_4)_2KMnF_6$	Monoclinic	6.153	6.151 $\beta = 91.37°$	9.346	Massa (1975)
K_2NaPdF_6 (*P)	$I4/mmm$, $Z = 2$	5.87	(8.30)	8.72	Tressaud et al. (1984b)
Rb_2NaPdF_6 (*P)		5.99	(8.47)	8.76	Tressaud et al. (1984b)
Rb_2KPdF_6 (*P)		6.18	(8.74)	9.23	Tressaud et al. (1984b)
Cs_2KPdF_6 (*P)		6.39	(9.04)	9.32	Tressaud et al. (1984b)
Cs_2RbPdF_6 (*P)		6.41	(9.06)	9.57	Tressaud et al. (1984b)

[a] The lattice constants b in parentheses are $a\sqrt{2}$ of the tetragonal I cells and refer to the corresponding F cells, which show $c/a < 1$ for the perovskites and $c/a > 1$ for the elpasolites.

1979, 1980). In contrast, ferrodistortive ordering is observed for tetragonal Mn(III) elpasolites of the K_2NaMnF_6 type [space group $I4/mmm$ (no. 139), $Z = 2$ (Knox, 1963)]. All long axes are oriented here along [001], and the four shorter ones have the same length by symmetry.

d. HEXAGONAL PEROVSKITES AMF_3 AND ELPASOLITES A_2BMF_6.

AMF_3 and A_2BMF_6 fluorides with a tolerance factor $t > 1$ adopt a variety of hexagonal structures. To simplify, we may call them "hexagonal perovskites" (or elpasolites) because they are also built up from close-packed $(AF_3)^{2-}$ layers. The stacking sequence of the layers is different, however, due to the change from cubic ABC to hexagonal AB packings and to mixed forms of both. The possibilities are shown in Table XVII.

As can be seen from Fig. 15, in all hexagonal structures, face sharing of octahedra occurs, and in the simplest hexagonal 2L structures of $CsNiF_3$ (Babel, 1969) and Cs_2LiGaF_6 (Babel and Haegele, 1976) this is the only kind of linking. These are therefore chain structures, and the other variants, exhibiting additional corner linkings, are mixed types of chain and cubic framework structures. Their hexagonal framework is widened by units of two,

TABLE XVII

Stacking Sequence and Structural Prototypes of Hexagonal Perovskites and Elpasolites.

N^a	Perovskites	N^a	Elpasolites
6L	ABC ACB, hexagonal $BaTiO_3$ type Space group $P6_3/mmc$ (no. 194), $Z = 6$ (Brown and Evans, 1948)	6L	ABC ACB, trigonal low-temperature K_2LiAlF_6 type Space group $R\bar{3}m$ (no. 166), $Z = 3$ (Winkler, 1954; Grjotheim et al., 1971)
12L	ABAB CACA BCBC, rhombohedral $CsMnNiF_6$ type Space group $R\bar{3}m$ (no. 166), $Z = 6$ (Dance et al., 1977)	12L	ABAB CACA BCBC, rhombohedral Cs_2NaCrF_6 type Space group $R\bar{3}m$ (no. 166), $Z = 6$ (hexagonal cell) (Babel and Haegele, 1976)
9L	ABA BCB CAC, rhombohedral $BaRuO_3$ type Space group $R\bar{3}m$ (no. 166), $Z = 9$ (hexagonal cell) (Donohue et al., 1965)		
10L	ABABA CBCBC, hexagonal $Cs_5Ni_4CdF_{15}$ type Space group $P6_3/mmc$ (no. 194), $Z = 2$ (Dance et al., (1984)		
2L	AB, hexagonal $BaNiO_3$ type Space group $P6_3/mmc$ (no. 194), $Z = 2$ (Lander, 1951)	2L	AB, trigonal Cs_2LiGaF_6 type Space group $P3m1$ (no. 156), $Z = 1$ (Babel and Haegele, 1976)

^{a} Number of close-packed layers (designates "type" in the following tables).

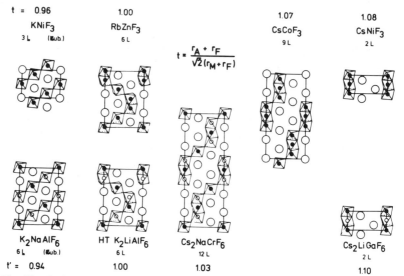

Fig. 15. Cations in the (110) planes of the structures of hexagonal (and cubic) perovskites and elpasolites and surrounding octahedra [○, A(I); ○, B(I); ●, M(II), M(III); HT, high-temperature form].

three, or four face-sharing octahedra, 3D corner linked at their ends only. The transition from the cubic structures ($t \leq 1.0$) to the final $2L$ chain structure ($t \geq 1.08$) goes stepwise with increasing tolerance factor in the sequence $6L \rightarrow 9L$ (or $12L$) $\rightarrow 10L \rightarrow 2L$ (Dance et al., 1979; Dance and Tressaud, 1979). It can be reversed by use of high pressure, as discussed elsewhere (Arndt et al., 1975, 1977; Babel, 1967). The paradoxical (but predicted) expansion of the M—F bonds (see Fig. 11) in the metastable high-pressure phases has been verified by single-crystal structure determinations of high-pressure forms of $CsNiF_3$ ($9L$) (Schmidt, 1982; Schmidt and Babel, 1983) and Rb_2LiFeF_6 and Cs_2NaFeF_6 (both cubic) (E. Herdtweck, W. Massa, and D. Babel, unpublished work, 1981; Massa, 1982b). An increase of about 1% compared with the M–F distances in the corresponding normal-pressure forms was observed. This high-pressure M—F bond lengthening, which is accompanied by shortening of the A—F bonds, is also compensated for by reduced M–M cation repulsion as corner sharing of the octahedra is enhanced. These compensations allow for higher densities in the high-pressure varieties despite longer M–F distances. Hexagonal perovskites are listed in Table XVIII and hexagonal elpasolites are given in Table XIX along with some metastable high-pressure polymorphs. For the elpasolite compounds it is noteworthy that a $12L$ structure is also found in some compounds $Cs_2M^{II}M'^{II}F_6$ (Dance et al., 1975, 1977; Dance and Tressaud, 1979) in which only the size, and not the charge, makes the difference

TABLE XVIII

Hexagonal Perovskites AMF_3 Listed by Increasing t within $6L$, $9L$, $10L$, and $2L$ Types

Perovskite[a]	t	a (Å)	c (Å)	Type	Reference
HT RbZnF$_3$	1.00	5.896	14.44	6L	Babel (1969)
TlZnF$_3$	1.00	5.934	14.52	6L	Vollmer (1966)
CsMnF$_3$ (*)	1.01	6.213	15.074	6L	a-1731
RbNiF$_3$ (*)	1.01	5.843	14.31	6L	Babel (1969); Weidenborner and Bednowitz (1970)
TlNiF$_3$	1.01	5.878	14.364	6L	Kohn et al. (1967)
RbMgF$_3$ (*)	1.02	5.857	14.259	6L	R. Schmidt, J. M. Dance, A. Tressaud, and D. Babel (unpublished work, 1983)
Cs$_{0.5}$Rb$_{0.5}$CoF$_3$	1.02	5.987	14.512	6L	Dance et al. (1979)
CsFeF$_3$	1.04	6.158	14.855	6L	a-1826
HP CsCoF$_3$	1.06	6.09	14.67	6L	Longo and Kafalas (1969)
HP CsZnF$_3$	1.07	6.09	14.67	6L	Longo and Kafalas (1969)
HP CsNiF$_3$ (II)	1.09	6.05	14.55	6L	Longo and Kafalas (1969)
HP CsMgF$_3$ (II)	1.09	6.04	14.45	6L	Longo and Kafalas (1969)
Cs$_{0.57}$Rb$_{0.43}$NiF$_3$	1.05	6.04	22.14	9L	Dance et al. (1979)
CsCoF$_3$ (*P)	1.06	6.194	22.61	9L	Babel (1969)
CsNi$_{0.75}$Mn$_{0.25}$F$_3$	1.07	6.187	22.46	9L	Dance and Tressaud (1979)
HP CsNiF$_3$ (I) (*)	1.09	6.147	22.35	9L	Haegele et al. (1975); Longo and Kafalas (1969); Schmidt and Babel (1983)
HP CsMgF$_3$ (I)	1.09	6.16	22.13	9L	Longo and Kafalas (1969)
HT CsNi$_{0.75}$Cd$_{0.25}$F$_3$ (*)	1.05	6.238	25.362	10L	Dance et al. (1984)
CsNi$_{0.8}$Cd$_{0.2}$F$_3$ (*P)	1.05	6.214$_5$	25.215	10L	Dance et al. (1984)
Cs$_{0.95}$Rb$_{0.05}$NiF$_3$	1.08	6.22	5.20	2L	Dance et al. (1979)
CsNiF$_3$ (*)	1.09	6.236	5.225	2L	Babel (1969)

[a] HP, High-pressure form; HT, high-temperature form.

TABLE XIX

Hexagonal $6L$, $12L$, and $2L$ Elpasolites A_2BMF_6 Listed by Increasing t

Elapsolite[a]	t	a (Å)	c (Å)	Type	Reference
LT K_2LiAlF_6	1.00	5.574	13.65	$6L$	Winkler (1954); Grjotheim et al. (1971) (*)
Cs_2NaTiF_6	1.02	6.281	15.31	$6L$	Weissenhorn (1972)
HP Rb_2LiFeF_6 (I)	1.03	5.881	14.36	$6L$	
HP Cs_2NaFeF_6 (I)	1.03	6.241	15.22	$6L$	Arndt et al. (1975)
HP Cs_2NaCrF_6 (I)	1.03	6.213	15.11	$6L$	
HP Cs_2NaAlF_6 (I)	1.05	6.149	14.93	$6L$	
HT K_2LiAlF_6 (*)	1.00	5.61	27.50	$12L$	Tressaud et al. (1984a)
Cs_2AgAlF_6	1.01	6.267	30.75	$12L$	Setter and Hoppe (1976b)
Rb_2LiTiF_6	1.02	5.930	29.11	$12L$	Weissenhorn (1972)
Cs_2NaTiF_6	1.02	6.272	30.91	$12L$	Alter and Hoppe (1974a)
Rb_2LiFeF_6 (*)	1.03	5.880	28.790	$12L$	Massa and Babel (1980)
Cs_2NaFeF_6 (*)	1.03	6.267	30.48	$12L$	Alter and Hoppe (1974c); Babel and Haegele (1976)
Rb_2LiVF_6	1.03	5.891	28.83	$12L$	Babel and Haegele (1976); Babel et al. (1973a)
Cs_2NaVF_6	1.03	6.267	30.40	$12L$	Alter and Hoppe (1975); Babel et al. (1973a)
Rb_2LiGaF_6	1.03	5.86	28.59	$12L$	
Cs_2NaGaF_6	1.03	6.22	30.19	$12L$	
Rb_2LiCrF_6	1.03	5.865	28.61	$12L$	Babel et al. (1973a); Babel and Haegele (1976)
Cs_2NaCrF_6 (*)	1.03	6.231	30.24	$12L$	
Rb_2LiCoF_6	1.03	5.856	28.55	$12L$	
Cs_2NaCoF_6	1.03	6.240	30.30	$12L$	
Rb_2LiCuF_6	1.04	5.844	28.30	$12L$	Grannec et al. (1975b)
Cs_2NaCuF_6	1.04	6.214	30.03	$12L$	
Cs_2NaNiF_6	1.04	6.20	30.03	$12L$	Alter and Hoppe (1974b)
Rb_2LiAlF_6	1.05	5.802	28.02	$12L$	Babel et al. (1973a)
Cs_2NaAlF_6 (*)	1.05	6.176	29.82	$12L$	Babel et al. (1973a); Golovastikov and Belov (1978)
HT Cs_2NaMnF_6 (*)	1.03	6.265	30.54	$12L$	Massa (1982a)

HP Tl$_2$LiFeF$_6$	1.03	5.925	29.05	12L	
HP Tl$_2$LiVF$_6$	1.03	5.940	29.02	12L	
HP Tl$_2$LiGaF$_6$	1.04	5.921	28.77	12L	
HP Tl$_2$LiCrF$_6$	1.04	5.915	28.78	12L	Arndt et $al.$ (1975)
HP Tl$_2$LiAlF$_6$	1.06	5.867	28.31	12L	
HP Cs$_2$LiFeF$_6$	1.10	6.187	29.67	12L	
HP Cs$_2$LiGaF$_6$	1.11	6.203	29.26	12L	
HP Cs$_2$LiCrF$_6$	1.11	6.162	29.38	12L	
Cs$_2$CdCoF$_6$	1.01	6.288	39.76	12L	
Cs$_2$CdZnF$_6$	1.01	6.295	30.30	12L	
Cs$_2$CdNiF$_6$	1.02	6.291	30.50	12L	Dance et $al.$ (1975); Dance et $al.$ (1977)
Cs$_2$MnCoF$_6$	1.04	6.213	30.22	12L	
Cs$_2$MnNiF$_6$	1.05	6.201	29.99	12L	
Cs$_2$MnMgF$_6$	1.05	6.219	29.995	12L	
Cs$_2$LiGaF$_6$ (*)	1.11	6.249	5.086	2L	Babel and Haegele (1976)
Cs$_2$LiCrF$_6$	1.11	6.248	5.106	2L	
Cs$_2$LiAlF$_6$	1.13	6.024	4.990	2L	Babel and Haegele (1976); Setter and Hoppe (1976a)

[a] LT, Low-temperature form; HP, high-pressure form; HT, high-temperature form.

between the two cations in octahedral sites. The existence of many oxide–fluorides with hexagonal elpasolite structures should also be mentioned (Arndt et al., 1977); it will be considered in Chapter 4.

2. Weberites and Pyrochlores ($A_2^I M^{II} M^{III} F_7$, $NaA^{II} M_2 F_7$, and $RbNiCrF_6$-Type Compounds)

Formally, the $Na_2 M^{II} M^{III} F_7$ and $NaA^{II} M_2^{II} F_7$ compounds differ from those with $(MF_3)^{n-}$ stoichiometry treated in this chapter and belong to the "odd"-composition compounds considered in Section VI. However, because the seventh anion in the pyrochlore-type mineral $CaNaNb_2 O_6 F$ (Gaertner, 1930) and in the isostructural $NaA^{II} M_2 F_7$ compounds does not belong to the octahedral coordination sphere and is "independent" (see Section VII,A), these compounds are structurally governed by the $M_2 X_6$ framework. Because of the related $M_2 X_7$ framework of the weberites, this class of materials is treated with the pyrochlores. The relationship between both structures becomes obvious in the oxide–fluoride β-$Na_2 Ta_2 O_5 F_2$, which combines both component frameworks (Chaminade and Pouchard, 1975; Vlasse et al., 1975, 1977a).

a. ORTHORHOMBIC AND TRIGONAL WEBERITES $A_2^I M^{II} M^{III} F_7$. The structure of the mineral weberite $Na_2 MgAlF_7$, discovered a long time ago [space group $Imm2$ (no. 44), $Z = 4$ (Byström, 1944)], has been refined and the question of the real space group thoroughly discussed (Giuseppetti and Tadini, 1978; Knop et al., 1982). The structure of $Na_2 NiFeF_7$ has also been investigated (Haegele et al., 1978). Many sodium and silver compounds, listed in Table XX, crystallize with the orthorhombic weberite structure. As shown in Fig. 16, the framework of octahedra contains M(II) and

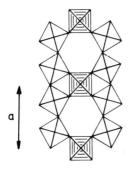

a

$\frac{1}{2} \sqrt{b^2 + c^2}$

Fig. 16. Layer from the $A_2^I M^{II} M^{III} F_7$ weberite structure; interconnection of chains containing the M(II) ions (open octahedra) by the M(III)-centered octahedra (shaded).

Orthorhombic and Trigonal Weberites $A_2^I M^{II} M^{III} F_7$

Weberite	a (Å)	b (Å)	c (Å)	n^a	Reference
Na$_2$MgAlF$_7$ (*)	7.051	9.968	7.285		Byström (1944); Knop et al. (1982)
Ag$_2$Mn	7.360	10.32	7.601	+4	Koch et al. (1982)
Na$_2$Ni	7.07	10.04	7.31	+1	Tressaud et al. (1974)
Ag$_2$Ni	7.210	10.139	7.564		Dance et al. (1974)
Na$_2$MgGaF$_7$	7.16	10.16	7.42		Chassaing (1969)
Ag$_2$Cu	7.200	10.34	7.755	+5	Koch et al. (1982)
Na$_2$MgInF$_7$	7.37	10.44	7.57		Chassaing (1969)
Ag$_2$Co	7.544	10.72	7.851	+2	Koch et al. (1982)
Ag$_2$Ni	7.499	10.622	7.822		Dance et al. (1974)
Na$_2$MgScF$_7$	7.34	10.43	7.55		Chassaing (1969)
Ag$_2$Mg	7.425	10.52	7.782	+4	Koch et al. (1982)
Na$_2$MgVF$_7$	7.24	10.30	7.45		Chassaing (1969)
Na$_2$MgCrF$_7$	7.15	10.20	7.39		Chassaing (1969)
Na$_2$Ni	7.204	10.223	7.408	+2	Haensler and Rüdorff (1970)
Ag$_2$Ni	7.305	10.285	7.673		Dance et al. (1974)
Na$_2$MgFeF$_7$	7.25	10.26	7.49		Chassaing (1969)
Na$_2$Co	7.329	10.432	7.384	+1	Cosier et al. (1970); Haensler and Rüdorff (1970)
Na$_2$Ni (*)	7.245	10.320	7.458		Haegele et al. (1978)
Ag$_2$Ni	7.345	10.345	7.692		Dance et al. (1974)
Na$_2$NiCoF$_7$	7.20	10.24	7.40		Cosier et al. (1970)
Na$_2$MnAlF$_7$	7.26	(Trigonal)	17.96		G. Courbion and G. Ferey (unpublished work, 1977)
Na$_2$MnGaF$_7$ (*)	7.401	—	18.091		Holler (1983)
Na$_2$MnTiF$_7$	7.535	—	18.423		G. Courbion and G. Ferey (unpublished work, 1977)
Na$_2$MnVF$_7$ (*)	7.492	—	18.261		Verscharen and Babel (1978)
Na$_2$MnCrF$_7$ (*)	7.481	—	18.166		G. Courbion and G. Ferey (unpublished work, 1977)
Na$_2$MnFeF$_7$ (*)	7.488	—	18.257		Verscharen and Babel (1978)
Na$_2$FeVF$_7$	7.436	—	18.262		Verscharen and Babel (1978)
Na$_2$FeFeF$_7$	7.431	—	18.258		Verscharen and Babel (1978)

[a] Number of additional isomorphs published for same cation combination A(I)/M(III) in same paper.

M(III) ions in different positions. The $(M^{II}F_{6/2})^-$ octahedra share trans corners to form chains along [100], which are linked together by isolated $(M^{III}F_2F_{4/2})^-$ octahedra possessing two terminal anions in a trans position. The 3D framework results from the intersection of layers shown in Fig. 16, which lie parallel to the (011) and (0$\bar{1}$1) planes of the lattice. The same layers occur in the trigonal weberites and, neglecting some different tilting of octahedra, in the structures of pyrochlores and hexagonal tungsten bronzes, to be discussed later (Section II,E,2,b,c and 3,b).

The trigonal weberite structure of Na_2MnFeF_7 type [space group $P3_121$ (no. 152), $Z = 6$ (Verscharen and Babel, 1978)] is obviously favored by larger M(II) ions, because most of the isostructural fluorides listed in Table XX are Mn(II) compounds. The cell dimensions are related by $a\sqrt{2} \simeq c/\sqrt{3} \simeq a_c$ to the lattice constant a_c of a cubic pyrochlore. The base of the relationships between trigonal and orthorhombic weberites and the pyrochlore structure has been discussed in detail elsewhere (Verscharen and Babel, 1978). The main difference between the trigonal and orthorhombic types is that neighboring layers, which extend parallel to the trigonal plane, are rotated by 60° relative to each other. The $(M^{III}F_2F_{4/2})^-$ octahedron, which makes the interlayer connection, consequently has its terminal ligands in a cis position instead of a trans one, as shown in Fig. 17.

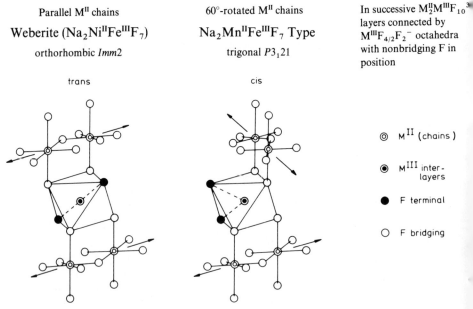

Fig. 17. Terminal ligands of the interlayer octahedra around M(III); turning from trans position in orthorhombic to cis position in trigonal weberites by rotation of layers containing the M(II) chains.

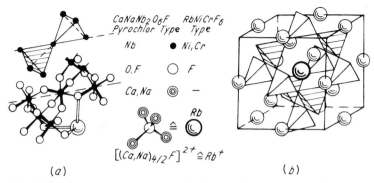

Fig. 18. RbNiCrF$_6$-type structure compared with pyrochlore structure of CaNaNb$_2$O$_6$F.

b. CUBIC PYROCHLORES OF CaNaNb$_2$O$_6$F TYPE. Pyrochlores belong to space group $Fd3m$ (no. 227), Z = 8. Although the structure of the mineral pyrochlore (CaNaNb$_2$O$_6$F) (Gaertner, 1930) is quite common among oxides, in fluorides it is found only in compounds like NaA^{II}M$_2^{II}$F$_7$ (A = Ca or Cd) (Haensler and Rüdorff, 1970) or A$_2^{II}$M$_2^{II}$F$_6$X (A = Cd or Hg; X = O or S) (Bernard *et al.*, 1975; Pannetier *et al.*, 1972), which are listed in Table XXI. The latter composition illustrates especially the peculiar role of the seventh anion. It is located outside the octahedral (M$_2$F$_6$)$^{2-}$ network, which is shown in Fig. 18 forming units of four corner-shared octahedra, as already mentioned (see Fig. 10). Together with the remaining cations, the seventh anion makes a separate anti-cristobalite arrangement A$_2$X, which interpenetrates the rather spacious M$_2$F$_6$ lattice.

c. MODIFIED PYROCHLORES OF CUBIC RbNiCrF$_6$ TYPE. Inside the M$_2$F$_6$ framework of octahedra of the pyrochlore structure, the A$_2$X sublattice may be replaced by large monovalent cations. These ions become located then at the former site of the seventh anion. For fluorides the prototype is RbNiCrF$_6$ [space group $Fd3m$ (no. 227), Z = 8 (Babel, 1972; Babel and Binder, 1983; Babel *et al.*, 1967)], which is characterized by Ni(II) and Cr(III) ions randomly distributed on crystallographically identical positions. The structure is illustrated in Fig. 18; it is adopted not only by many fluorides, some of which are listed in Table XXII, but also by oxides such as CsM$_2$O$_6$ and oxide–fluorides CsM$_2$O$_n$F$_{6-n}$, of all intermediate compositions (Babel *et al.*, 1967).

d. DISTORTED PYROCHLORES. Whereas the presence of Jahn–Teller active ions such as Cr(II), Cu(II), and Mn(III) in modified pyrochlores seems to have no influence on the cubic symmetry, distortions have been observed in compounds with ions in mixed-valency states or in obviously less fitting size relations (Tressaud *et al.*, 1970; Fleischer and Hoppe, 1982c).

TABLE XXI

Fluorides, Oxyfluorides, and Thiofluorides with the Pyrochlore Structure ($NaCaNb_2O_6F$ Type)

Compound	a (Å)	Reference	Compound	a (Å)	Reference
$NaCdMg_2F_7$	10.20	Haensler and Rüdorff (1970)	$Hg_2Mg_2F_6O$	10.277	Bernard et al. (1975)
$NaCaCo_2F_7$	10.37		$Hg_2Mg_2F_6S$	10.697	
$NaCaNi_2F_7$	10.29		$Hg_2Mn_2F_6S$ (*P)	10.955	
$NaCdNi_2F_7$	10.25		$Hg_2Co_2F_6O$	10.419	
$NaCdCu_2F_7$	10.35		$Hg_2Co_2F_6S$	10.806	
$NaCdZn_2F_7$	10.31		$Hg_2Ni_2F_6O$ (*P)	10.326	
			$Hg_2Ni_2F_6S$ (*P)	10.708	
$Cd_2Mn_2F_6S$	11.04	Pannetier et al. (1972)	$Hg_2Cu_2F_6O$	10.358	
$Cd_2Fe_2F_6S$	10.895			(c = 10.440)	
$Cd_2Co_2F_6S$	10.686		$Hg_2Cu_2F_6S$ (*P)	10.735	
$Cd_2Ni_2F_6S$	10.523		$Hg_2Zn_2F_6O$ (*P)	10.333	
$Cd_2Cu_2F_6S$	10.713		$Hg_2Zn_2F_6S$	10.780	
	(α = 90.65°)				
$Cd_2Zn_2F_6S$	10.670				

TABLE XXII

Lattice Constants of Selected Modified Pyrochlores (RbNiCrF$_6$ Type)[a]

Compound	a (Å)	Reference
NH$_4$NiAlF$_6$	9.96	Babel (1972)
CsMnGaF$_6$	10.42	Babel and Binder (1983); Jesse and Hoppe (1977c)
CsPdInF$_6$	10.89	Jesse and Hoppe (1977b)
CsAgInF$_6$	10.84	Müller and Hoppe (1973)
CsCuTlF$_6$	10.64	Hoppe and Jesse, (1973)
CsAgTlF$_6$	10.89	Müller and Hoppe (1973)
CsVScF$_6$	10.627	Hartung and Babel (1982)
CsPdScF$_6$	10.82	Jesse and Hoppe (1977b)
CsAgScF$_6$	10.79	Müller and Hoppe (1973)
CsCuTiF$_6$	10.39	Jesse and Hoppe (1974)
CsZnTiF$_6$	10.498	Jesse and Hoppe (1977a); Griebler and Babel (1980)
CsVVF$_6$	10.481	Cros *et al.* (1975); Hartung and Babel (1982); Hong *et al.* (1982)
RbFeVF$_6$	10.321	Banks *et al.* (1973)
CsVCrF$_6$	10.436	Hartung and Babel (1982)
RbFeCrF$_6$	10.274	Banks *et al.* (1973)
TlFeCrF$_6$	10.300	Banks *et al.* (1973)
RbCoCrF$_6$	10.277	Fourquet *et al.* (1973)
RbNiCrF$_6$ (*)	10.21	Babel (1972)
RbZnMnF$_6$	10.14	Massa (1982b)
CsZnMnF$_6$	10.40	Jesse and Hoppe (1977a); Massa (1982b)
TlMgFeF$_6$	10.306	Banks *et al.* (1973)
RbNiFeF$_6$	10.266	Banks *et al.* (1973)
CsPdFeF$_6$	10.64	Jesse and Hoppe (1977b)
RbCuFeF$_6$	10.216	Fleischer and Hoppe (1982c)
RbMgCoF$_6$	10.185	Fleischer and Hoppe (1982c)
RbNiCoF$_6$	10.183	Fleischer and Hoppe (1982c)
CsNiCoF$_6$	10.271	Fleischer and Hoppe (1982c)
RbCuCoF$_6$	10.17	Hoppe and Jesse (1973)
RbZnCoF$_6$	10.207	Fleischer and Hoppe (1982c)
CsPdRhF$_6$	10.65	Jesse and Hoppe (1977b)
RbMgNiF$_6$	9.978	Fleischer and Hoppe (1982c)
CsMgNiF$_6$	10.121	Fleischer and Hoppe (1982c)
CsMgCuF$_6$	10.169	Hartung and Babel (1982)
CsZnCuF$_6$	10.252	Hartung and Babel (1982); Jesse and Hoppe (1977a)
CsLi$_{0.5}$Sc$_{1.5}$F$_6$	10.543	Griebler (1978)
CsNa$_{0.5}$Sc$_{1.5}$F$_6$	10.743	Griebler (1978)
CsK$_{0.5}$Ti$_{1.5}$F$_6$	10.746	Griebler and Babel (1980)
CsRb$_{0.5}$Ti$_{1.5}$F$_6$	10.806	Griebler and Babel (1980)
RbLi$_{0.5}$Fe$_{1.5}$F$_6$	10.180	Courbion *et al.* (1974)
CsLi$_{0.5}$Co$_{1.5}$F$_6$	10.235	Hartung and Babel (1982)

[a] More than 70 cesium compounds are compiled in Babel and Binder (1983).

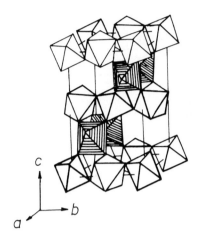

c

b

a

Fig. 19. Ordering of M(II) (shaded) and M(III) octahedra (empty) in the modified pyrochlore structure of orthorhombic $CsAgFeF_6$ (or $NH_4Fe_2F_6$) type.

Structure determinations have been performed for orthorhombic $CsAgFeF_6$ [space group *Pnma* (no. 62), $Z = 4$ (Müller, 1981a)], isostructural with $NH_4Fe_2F_6$ (Ferey *et al.*, 1981), and for the related body-centered compounds $CsNi_2F_6$ (Fleischer and Hoppe, 1982c) and $CsM^{II}PdF_5$ (M^{II} = Pd, Mg, Ni, Co, or Zn; Müller, 1982), both of space group *Imma* (no. 74), $Z = 4$. In contrast to cubic pyrochlore, ordering is observed in these structures between M(II) and M(III) [or Pd(II)] cations, which are arranged in chains along [100] and [010], respectively (Fig. 19). A quite different $A^I M^{II} M^{III} F_6$ structure with ordered M(II)/M(III) distribution has been found in hydrothermally synthesized NH_4MnFeF_6 [orthorhombic, space group *Pb2n* (Leblanc *et al.*, 1983a) or *Pbcn* (Marsh, 1984)]. Figure 20 shows that edge-shared $(MnFeF_{10})^{5-}$ units, which are linked together by their corners, build up the

x = 0.25   x = 0.75

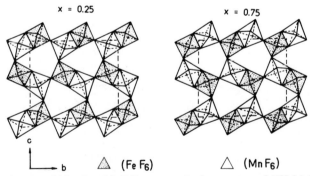

c

b   △ (Fe F₆)   △ (Mn F₆)

Fig. 20. Edge-shared binuclear $MnFeF_{10}$ groups in the structure of NH_4MnFeF_6. From Leblanc *et al.* (1983a).

framework of the structure, which is actually related to that of high-temperature $BaTa_2O_6$ (Galasso *et al.*, 1968). In Table XXIII are listed orthorhombic compounds with pyrochlore-related structures.

3. Tungsten Bronze-Related Structures A_xMF_3 ($x < 1$)

Tetragonal and hexagonal tungsten bronze structures A_xWO_3 (Magnéli, 1949, 1953) are characterized by tunnels of different cross sections. The smaller tunnels consist of trigonal prisms already shown in Fig. 10; in fluorides they always remain empty. The larger ones in TTB lattices have square and pentagonal cross sections; they are hexagonal in HTB lattices (Fig. 21). The full occupancy of the tunnels by the larger alkali ions A corresponds to $x = 0.60$ and 0.33, respectively. The presence of mixed-valency states, as found in tungsten oxide compounds, which contain mostly delocalized electrons in a conduction band, is not essential to stabilize the structure. In fluorides, distinct $M(II)$ and $M(III)$ belonging to different elements may be combined to build up tetragonal or hexagonal $(MF_3)^{x-}$ frameworks (de Pape, 1965).

a. FLUORIDES WITH STRUCTURES RELATED TO TETRAG-ONAL TUNGSTEN BRONZE STRUCTURE. Tetragonal structures related to TTB have been found preferentially in potassium compounds $KM^{II}M^{III}F_6$, which generally do not form the $RbNiCrF_6$-type pyrochlore structure (Hardy *et al.*, 1973). If they do so, however, they may transform into a tighter TTB phase under high pressure, as has been shown in the case of $KNiCrF_6$ (Babel, 1972). There is a certain range of composition in which such TTB phases can be obtained. For $K_yFe_2F_6$ the limits $0.80 \leq y \leq 1.20$ have been reported (Tressaud *et al.*, 1970). By single-crystal investigation on $K_{1.2}Fe_2F_6$, an orthorhombic deformation of the TTB space group $P4/mbm$

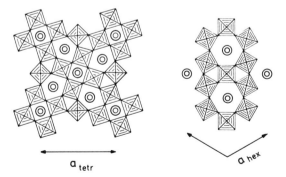

Fig. 21.   The [001] projections of tetragonal and hexagonal tungsten bronze structures.

TABLE XXIII

Orthorhombic Pyrochlores ($CsAgFeF_6$, $NH_4Fe_2F_6$, and $CsNi_2F_6$ Type)[a] and Compounds of NH_4MnFeF_6 and $CsPd_2F_5$ Structures

Compound	a (Å)	b (Å)	c (Å)	Type[a]	Reference
$RbCuAlF_6$	6.860	7.077	9.982	I	Fleischer and Hoppe (1982c)
$CsCuAlF_6$	7.094	7.134	10.244	I	Fleischer and Hoppe (1982c)
$RbAgAlF_6$	7.152	7.182	10.135	I	Müller (1981a)
$CsAgAlF_6$ (*)	7.380	7.241	10.352	I	Müller (1981a)
$CsZnAlF_6$	7.212	7.078	10.135	I	Fleischer and Hoppe (1982c)
$RbAgGaF_6$	7.153	7.419	10.250	I	Müller (1981a)
$CsAgGaF_6$	7.318	7.469	10.456	I	Müller (1981a)
$RbVVF_6$	7.252	7.464	10.186	—	Cros et al. (1975); Hong et al. (1982)
$CsVVF_6$	7.445	7.474	10.442	—	Cros et al. (1975); Hong et al. (1982)
$RbCuVF_6$	6.883	7.459	10.157	I	Fleischer and Hoppe (1982c)
$Rb_{1.06}Cr_2F_6$	7.042	7.959	10.182	—	Dumora et al. (1972)
$CsCrCrF_6$	7.243	7.416	10.486	—	Dumora et al. (1972)
NH_4MnCrF_6	7.808	12.755	10.501	III	Leblanc et al. (1983a)
$CsMnMnF_6$ (*P)	7.301	7.711	10.204	I	Massa (1982b)

130

			$10.03 = c' \sin \beta$		
High-temperature $RbMnFeF_6$	7.153	7.434		I?	Leblanc et al. (1983a)
$RbMnFeF_6$	7.913	12.858	10.619	III	Leblanc et al. (1983a)
NH_4MnFeF_6	7.844	12.819	10.582	III	Leblanc et al. (1983a)
$RbFeFeF_6$	7.02	7.44	10.14	—	Tressaud et al. (1970); Babel (1972)
$CsFeFeF_6$	7.22	7.46	10.38	—	Tressaud et al. (1970); Babel (1972)
NH_4FeFeF_6 (*)	7.045	7.454	10.116	I	Tressaud et al. (1970); Ferey et al. (1981)
$TlFeFeF_6$	6.97	7.40	10.18	—	Tressaud et al. (1970)
$RbAgAl_{0.45}Fe_{0.55}F_6$ (*)	7.19	7.39	10.32	I	Müller (1981a)
$RbAgFeF_6$	7.175	7.521	10.365	I	Müller (1981a)
$CsAgFeF_6$ (*)	7.338	7.564	10.554	I	Müller (1981a)
$RbNiNiF_6$	6.946	7.333	9.768	II	Fleischer and Hoppe (1982c)
$CsNiNiF_6$ (*P)	7.122	7.350	10.025	II	Fleischer and Hoppe (1982c)
$CsMgPdF_5$	6.603	7.415	10.548	IV	Müller (1982a)
$CsCoPdF_5$	6.527	7.553	10.659	IV	Müller (1982a)
$CsNiPdF_5$	6.499	7.504	10.575	IV	Müller (1982a)
$CsZnPdF_5$	6.576	7.483	10.645	IV	Müller (1982a)
$CsPdPdF_5$ (*)	6.534	7.811	10.774	IV	Müller (1982a)

[a] Type I, $CsAgFeF_6$ and $NH_4Fe_2F_6$; II, $CsNi_2F_6$; III, NH_4MnFeF_6; IV, $CsPd_2F_5$.

(no. 127), $Z = 5$ (for $A_yM_2X_6$) (Magnéli, 1949) has been found, which can be described in the space group $Pba2$ (no. 32), $Z = 5$ (Hardy et al., 1973). A tetragonal cell with doubled c axis appears for $K_{1.08}(Mn, Fe)_2F_6$ (Banks et al., 1979) [space group $P4_2bc$ (no. 106), $Z = 10$]. Table XXIV gives the lattice constants of these phases and similar ones.

b. FLUORIDES WITH STRUCTURES RELATED TO HEXAGONAL TUNGSTEN BRONZE STRUCTURE. Since the discovery of the first mixed-valency fluorides $A_yM_2F_6$ by de Pape (1965), a large number of structurally related compounds containing different M elements have been recognized (de Pape et al., 1968; Tressaud et al., 1972; Dumora et al., 1972; Cros et al., 1975; Hartung and Babel, 1982), including many oxide–fluorides (Babel et al., 1973b). They contain large A cations due to the large cross section of the tunnels. For cesium compounds $Cs_yM_y^{II}M_{2-y}^{III}F_6$, which are the most stable, the composition range is about $0.35 \leq y \leq 0.55$. For higher cesium content, pyrochlore phases ($0.80 \lesssim y \leq 1.05$) occur in which more space is available in the absence of the seventh anion. The intermediate range is a two-phase domain (Babel et al., 1973b).

A single-crystal structure determination of $Cs_{0.4}Zn_{0.4}Fe_{1.6}F_6$ revealed a (pseudohexagonal) monoclinic symmetry [space group $P2_1$ (no. 4), $Z = 3$ (Hartung et al., 1979)]. The main difference with the HTB structure of space group $P6_3/mcm$ (no. 193), $Z = 3$ (Magnéli, 1953), is the appreciable deviation from linearity of the bridge angles M–F–M $= 159°$ along the pseudohexagonal c axis.

In the orthorhombic structure found for $(H_2O)_{0.33}FeF_3$, [space group $Cmcm$ (no. 63), $Z = 12$], the Fe–F–Fe bridge angles are even smaller: $150° //c$ and $143° \perp c$ (Leblanc et al., 1983b); the latter approaches the typical value for pyrochlores, $\sim 140°$ (Babel and Binder, 1983). Similar orthorhombic lattice constants that are related to the hexagonal ones by $a \simeq a_h$, $b \simeq a_h\sqrt{3}$, $c = c_h$ were reported for some of the HTB-related fluorides listed in Table XXIV. So far there seems to be no proof that any of these compounds actually has the hexagonal symmetry (Eyring et al., 1978; Rieck et al., 1982).

III. Layer Structures

In a structure, if only four of the six corners of the octahedra are shared with other ones in the same way, the resulting $(MF_4)^{n-}$ composition generally corresponds to layer-type networks. Differences in the layer structures are due to the cis or trans position of the terminal ligands and to the angles found at the bridging ligands. In addition, there are different possible stackings of these layers, which are influenced by the coordination requirements of the larger

TABLE XXIV

Fluorides with Structures Related to Tetragonal or Hexagonal Tungsten Bronzes

Compound	a (Å)	b (Å)	c (Å)	Type[a]	n^b	Reference
$KVVF_6$	12.61	—	3.930	TTB		Cros et al. (1975, 1976); Hong et al. (1980); Williamson and Boo (1977b)
$KMnCrF_6$	12.649	—	7.935	TTB		Banks et al. (1971, 1982)
$KFeCrF_6$	12.455	—	3.900	TTB	+2	Hardy et al. (1973)
High-pressure $KNiCrF_6$	12.41	—	3.894	TTB		Babel (1972)
$KMn_{1.08}Fe_{0.92}F_6$ (*)	12.765	—	8.002	TTB		Banks et al. (1979)
$K_{1.2}Fe_2F_6$ (*)	12.750	12.637	3.986	TTB, o		Hardy et al. (1973)
$KFeFeF_6$	12.60	—	3.936	TTB		de Pape (1965); Hardy et al. (1973)
$KCoFeF_6$	12.505	—	3.920	TTB	+4	Banks et al. (1977); Hardy et al. (1973)
$K_{0.5}V_2F_6$	7.41	—	7.51	HTB		Cros et al. (1975); Williamson and Boo (1977b)
$Rb_{0.5}V_2F_6$	7.40	—	7.55	HTB		Hong et al. (1979); Eyring et al. (1978)
$Cs_{0.5}V_2F_6$	7.45	—	7.61	HTB		Hong et al. (1981); Rieck et al. (1982)
$Tl_{0.5}V_2F_6$	7.479	12.841	7.572	HTB, o		Cros et al. (1977b)
$K_{0.6}Nb_2F_6$	7.540(3)	13.06(2)	7.750(3)	HTB, o		Masse et al. (1984)
$K_{0.44}Cr_2F_6$	7.287	—	7.365	HTB		Dumora et al. (1970, 1972)
$Rb_{0.48}Cr_2F_6$	7.291	—	7.421	HTB		Dumora et al. (1970, 1972)
$Cs_{0.4}Cr_2F_6$	7.410	—	7.500	HTB		Dumora et al. (1970, 1972)
$K_{0.5}Fe_2F_6$	7.38	—	7.51	HTB		Tressaud et al. (1970, 1972)
$Rb_{0.56}Fe_2F_6$	7.36	—	7.53	HTB		Tressaud et al. (1970, 1972)
$Cs_{0.5}Fe_2F_6$	7.47	—	7.63	HTB		Tressaud et al. (1970, 1972)
$(NH_4)_{0.5}Fe_2F_6$	7.42	—	7.54	HTB		Tressaud et al. (1970, 1972)
$Tl_{0.6}Fe_2F_6$	7.35	—	7.52	HTB		Tressaud et al. (1970, 1972)

(cont.)

TABLE XXIV (*cont.*)

Compound	a (Å)	b (Å)	c (Å)	Type[a]	n^b	Reference
$(H_2O)_{0.66}Fe_2F_6$ (*)	7.423	12.730	7.526	HTB, o		Leblanc *et al.* (1983b)
$Cs_{0.4}Mn_{0.4}Ga_{1.6}F_6$	7.384	—	7.549	HTB	$+5$	Babel *et al.* (1973b); Binder (1973)
$Cs_{0.4}V_{0.4}Sc_{1.6}F_6$	7.336	—	8.061	HTB		Hartung and Babel (1982)
$K_{0.4}Mg_{0.4}Ti_{1.6}F_6$	7.498	12.984	7.654	HTB, o	$+5$	Hartung and Babel (1982)
$Cs_{0.4}V_{0.4}$	7.557	—	7.752	HTB	$+3$	Hartung and Babel (1982)
$Rb_{0.4}Zn_{0.4}V_{1.6}F_6$	7.420	12.817	7.564	HTB, o	$+5$	Hartung (1978)
$Cs_{0.4}Ni_{0.4}$	7.458	—	7.614	HTB	$+7$	Babel *et al.* (1973b); Bolte *et al.* (1972); Hartung (1978)
$Tl_{0.4}Mg_{0.4}Cr_{1.6}F_6$	7.277	12.555	7.416	HTB, o	$+3$	Hartung (1978)
$Cs_{0.4}V_{0.4}$	7.395	—	7.523	HTB	$+6$	Babel *et al.* (1973); Hartung and Babel (1982)
$Tl_{0.4}Mg_{0.4}Fe_{1.6}F_6$	7.401	12.797	7.557	HTB, o	$+5$	Hartung (1978)
$Cs_{0.4}Cu_{0.4}$	7.458	—	7.589	HTB	$+6$	Babel *et al.* (1973b); Binder (1973)
$Cs_{0.4}Zn_{0.4}$ (*)	7.474	7.461	7.636	$P2_1$ $\gamma = 120.0°$		Hartung *et al.* (1979)
$Rb_{0.4}Zn_{0.4}Co_{1.6}F_6$	7.292	12.597	7.425	HTB, o	$+1$	Hartung and Babel (1982)
$Cs_{0.4}Mg_{0.4}$	7.331	—	7.521	HTB	$+2$	Hartung and Babel (1982)
$Cs_{0.4}Mg_{0.4}Cu_{1.6}F_6$	7.231	—	7.419	HTB		Hartung and Babel (1982)
$Cs_{0.4}Zn_{0.4}$	7.249	—	7.449	HTB		Hartung and Babel (1982)

[a] TTB, Tetragonal tungsten bronze; HTB, hexagonal tungsten bronze; o, orthorhombic deformation.
[b] Number of isomorphs published in same paper(s) for same M(III) cation.

cations located between them. The SnF_4 structure (Hoppe, 1962), also known from NbF_4 (Gortsema and Didchenko, 1965; Schäfer *et al.*, 1965), is an example of a planar layer structure for a binary tetrafluoride. Only ternary representatives are treated in the following.

A.   AMF₄ COMPOUNDS WITH TRANS,TRANS-CONNECTED OCTAHEDRA

The structure of the simple tetragonal prototype $TlAlF_4$ described by Brosset (1937) [space group $P4/mmm$ (no. 123), $Z = 1$] (Fig. 22) is less common than formerly thought (Brosset, 1938b; Gladney and Street, 1968). In the meantime, many variants and polymorphs have been found and phase transitions studied (Bulou and Nouet, 1982; Hidaka *et al.*, 1979, 1982). All single-crystal structure refinements show, contrary to the findings for some weberites (Haegele *et al.*, 1978; Verscharen and Babel, 1978), the usual shortening of the terminal bonds (Boca, 1981). These structures can be derived from the aristotype as illustrated in Fig. 23 and discussed as follows:

1. *The M–F–M bridges are bent.* This induces enlargement of the *a* axes by a factor of $\sqrt{2}$ or 2; often, a small orthorhombic distortion occurs. Within the pseudocell ($Z = 1$) the A cations may also have off-centered positions.

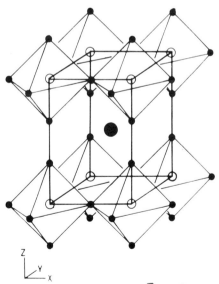

Fig. 22.   Unimolecular cell of $TlAlF_4$-type structure (●, Tl; ○, Al; ●, F). From Hidaka *et al.* (1981).

$KFeF_4$: $a = 7.596$ Å, $b = 3.884$ Å, $c = 12.27$ Å, $V/Z = 90.5$ Å^3

Amma, $Z = 4$ $\downarrow$ 1 hr, 40 kbars, 500°C

                $a = 7.228$ Å, $c = 6.409$ Å, $V/Z = 83.7$ Å^3

                $CsFeF_4$, $P4/nmm$, $Z = 4$:

                $a = 7.794$ Å, $c = 6.553$ Å, $V/Z = 99.5$ Å^3

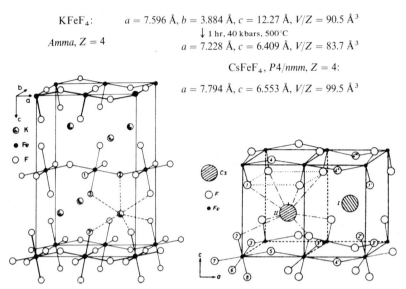

Fig. 23. Superstructures of $TlAlF_4$: $CsFeF_4$ type (only front half of the unit cell is illustrated) and $KFeF_4$ type (pseudocell front half of the $KTiF_4$ type).

Depending on that shift and on the different tilting possibilities of the octahedra, several superstructure types with bimolecular (e.g., $RbAlF_4$) and tetramolecular cells (e.g., $CsFeF_4$) have been found. They are listed in Table XXV along with other compounds for which only powder data are available. Space-group assignments should be made cautiously in such cases, owing in particular to possible phase transitions (Hidaka *et al.*, 1981, 1982; Kleemann *et al.*, 1982; Leble *et al.*, 1982).

2. *The rotation or tilting of octahedra is opposite in neighboring layers* (NH_4AlF_4) *or the layers are shifted relative to each other* $(KFeF_4/KTiF_4)$. In either case, the c axis becomes doubled, too (see Fig. 23). Data for the structures mentioned are given in Table XXVI, where isostructural or related compounds are also listed.

A special case is the structure of α-$NaTiF_4$ (Omaly *et al.*, 1976). Although $TlAlF_4$ related, it shows essentially hcp arrangement of anions, as found in α-PbO_2 (Zaslavskii and Tolkacev, 1952). However, the distribution of the cations differs from α-PbO_2 and from the related $NaNbO_2F_2$ (Andersson and Galy, 1969; Omaly *et al.*, 1976). The Ti–F–Ti bridging angles are 140°, compared with 132° in an ideal hcp structure.

3. *Layers containing triangular meshes.* The layers that derive directly from the $TlAlF_4$ structure consist of more or less square meshes and all varieties may be viewed as perovskites disconnected in one dimension. If similar

TABLE XXV

$A^I MF_4$ Compounds Related to the $TlAlF_4$ Structure (c Axis Unchanged)

Compound	Space Group	Z	a (Å)	b (Å)	c (Å)	Reference
$TlAlF_4$ (*P)	$P4/mmm$	1	3.616	Tetragonal	6.366	a-702; Fourquet et al. (1979)
NH_4GaF_4	—	1	3.71	—	6.39	a-718
NH_4ScF_4	—	1	4.06	—	6.67	a-803
$KAlF_4$	$P4/mbm$	2	5.043	—	6.164	a-662; Fourquet et al. (1979)
$RbAlF_4$ (*P)	$P4/mbm$	2	5.125	—	6.283	Bulou and Nouet (1982); Fourquet et al. (1979)
$CsGaF_4$	—	4	7.690	—	6.479	Wall (1971); Babel et al. (1974a)
$CsScF_4$	—	4	7.984	—	6.796	Hartung (1978)
$CsTiF_4$ (*P)	$P4/nmm$	4	7.897	—	6.505	Sabatier et al. (1982); Weissenhorn (1972)
$CsVF_4$	—	4	7.796	—	6.574	Hidaka et al. (1981)
Low-temperature $RbCrF_4$	—	4	7.438	—	6.442	Knoke (1977); a-1589
NH_4CrF_4	—	4	7.432	—	6.504	Knoke (1977); Knoke et al. (1979b)
$CsMnF_4$ (*)	$P4/nmm$	4	7.944	—	6.338	Massa and Steiner (1980)
$CsFeF_4$ (*)	$P4/nmm$	4	7.794	—	6.553	Eibschütz et al. (1972); a-1828; Babel et al. (1974a)
$RbTlF_4$ (*)	$Pb2_1a$	4	8.252	8.359	6.244	Hebecker (1975)
$CsTlF_4$	$Pb2_1a$	4	8.405	8.484	6.566	Hebecker (1975)
$TlTlF_4$	$Pb2_1a$	4	8.228	8.348	6.262	Grannec et al. (1971b); Hebecker (1975)
$RbMnF_4$	—	4	7.773	7.814	6.176	Köhler et al. (1978)
$RbFeF_4$ (*)	$Pbma$	4	7.615	7.620	6.245	Tressaud et al. (1969); Abrahams and Bernstein (1972)
NH_4FeF_4	$P2_12_12$	4	7.58	7.58	6.36	Ferey et al. (1975)
$KCoF_4$	—	4	7.526	7.584	5.792	Fleischer and Hoppe (1982d); Hartung (1978)
$KMnF_4$ (*)	$P2_1/a$	4	7.685	7.653 ($\beta = 90.74°$)	5.784	Köhler et al. (1978); Massa (1982b)
NH_4MnF_4	Triclinic	4	7.731	7.752	6.113	Massa (1977a)

$\alpha = 90.02°,\ \beta = 90.65°,\ \gamma = 90.46°$

disconnection is carried out in the frameworks of TTBs and HTBs new layer structures are obtained in which triangular meshes appear (see Fig. 21). In the β-RbAlF$_4$ structure deriving from TTB, there are additional square and pentagonal meshes (Fourquet et al., 1980). In $Cs_{0.67}Na_{0.33}AlF_4$ deriving from HTB, only triangles and hexagons appear (Courbion et al., 1976). The structure of $Cs_{0.67}Na_{0.33}AlF_4$ (i.e., $Cs_2NaAl_3F_{12}$) is best described as an ordered pyrochlore in which the upper apex of the cationic tetrahedron shown in Fig. 10 is occupied by sodium ions. There also exist cubic and obviously disordered compounds of formulation $CsA^I_{0.5}M^{III}_{1.5}F_6$, which have already been listed as pyrochlores in Table XXII. Some ordered variants of this type and similar bronze-related layer structures are given in Table XXVI.

B. AMF$_4$ COMPOUNDS WITH CIS,TRANS-CONNECTED OCTAHEDRA

Two structural types with cis,trans-connected octahedra and therefore characteristically puckered layers have been identified in fluorides, one for $A^I M^{III} F_4$, the other for $A^{II} M^{II} F_4$ compounds. Significant shortening of the terminal bonds, which here are in the cis position, was found only in the $A^I M^{III} F_4$ compounds, whereas in $A^{II} M^{II} F_4$ one of the terminal bonds is more within the intermediate range.

1. Fluorides $A^I M^{III} F_4$ of Monoclinic NaNbO$_2$F$_2$ Type

The refined structures of NaCrF$_4$ (Knoke, 1977; Knoke et al., 1979a) and NaFeF$_4$ (Dance et al., 1981) show a less pronounced off-center position of the octahedrally coordinated M cations than the oxide–fluoride prototype [space group $P2_1/c$ (no. 14), $Z = 4$] (Andersson and Galy, 1969). As illustrated in Fig. 24 the M–F–M bridging angle is about 150° in both directions of cis and trans connection. In this structural type, which is related to α-PbO$_2$, sodium cations have CN = 6 + 2 (Table XXVII).

2. Fluorides $A^{II} M^{II} F_4$ of Orthorhombic BaZnF$_4$ Type

Fluorides of this kind were first detected in BaFeF$_4$ by de Pape and Ravez (1966), but the structure, illustrated in Fig. 25, was first determined for BaZnF$_4$ [space group $A2_1am$ (standard: $Cm2_1$, no. 36), $Z = 4$ (von Schnering and Bleckmann, 1968)] and more accurately for BaMnF$_4$ (Keve et al., 1969) and BaCoF$_4$ (Keve et al., 1970). This class of compounds, listed in Table XXVII, is of great interest because of the pyro- or ferroelectric and related elastic properties (Eibschütz and Guggenheim, 1968) connected with the

TABLE XXVI

Layer Structures: $NaTiF_4$ Type and $TlAlF_4$ Structure Variants with Doubled c Axis, Tetragonal Tungsten Bronze–Related β-$RbAlF_4$, and Hexagonal Tungsten Bronze–Related $Cs_2NaAl_3F_{12}$ Layer Structures

Compound	Space Group	Z	a (Å)	b (Å)	c (Å)	Reference
$NaTiF_4$ (*)	$Pbcn$	4	4.976	5.755	11.070	Omaly et al. (1976)
$NaCoF_4$	$Pbcn$	4	4.965	5.516	10.800	Fleischer and Hoppe (1982d); Hartung (1978)
NH_4AlF_4 (*)	$I4/mcm$	4	5.078	(Tetragonal)	12.715	Bulou et al. (1982); Fourquet et al. (1979)
$TlGaF_4$	Orthorhombic	4	5.271	5.235	12.71	Wall (1971)
$TlTiF_4$	—	4	5.478	5.384	12.85	Weissenhorn (1972)
$TlVF_4$	—	4	5.353	5.305	12.88	Babel et al. (1974a)
$TlMnF_4$	—	4	5.445	5.401	12.491	Köhler et al. (1978)
$TlFeF_4$	—	4	5.364	5.308	12.88	Wall (1971)
$KInF_4$	—	8	7.930	7.760	12.57	a-757
$KTiF_4$ (*)	$Pcmn$	8	7.944	7.750	12.195	Hidaka et al. (1979); Sabatier et al. (1979)
$RbTiF_4$		8	7.868	7.843	13.22	Belmont et al. (1975); Weissenhorn (1972)
KVF_4		8	7.738	7.596	12.28	Hidaka et al. (1979); a-1482
$RbVF_4$		8	7.710	7.682	12.638	Hidaka et al. (1982); a-1491
$TlCrF_4$		8	7.38	(Tetragonal)	12.87	a-1645
$KFeF_4$ (*)	$Pcmn$	8	7.768	7.596	12.27	Hidaka et al. (1979); a-1797
Pseudocell	$Bmmb$	4	3.884	7.596	12.27	Heger et al. (1971)
β-$RbAlF_4$ (*)	$I\bar{4}c2$	20	11.66	(Tetragonal)	12.551	Fourquet et al. (1980)
β-$CsFeF_4$	$I\bar{4}c2$	20	12.491	—	13.272	Fleischer and Hoppe (1982d)
$RbCoF_4$	?	20	12.222	—	12.356	Fleischer and Hoppe (1982d)
$CsCoF_4$	$I\bar{4}c2$	20	12.478	—	12.971	Fleischer and Hoppe (1982d)
$Cs_2LiAl_3F_{12}$	$R\bar{3}m$	3	6.975	(Hexagonal)	17.563	Courbion et al. (1974)
$Cs_2NaAl_3F_{12}$ (*)	$R\bar{3}m$	3	7.026	—	18.244	Courbion et al. (1974)
$Cs_2KAl_3F_{12}$	$R\bar{3}m$	3	7.071	—	18.778	Courbion et al. (1974)
$Cs_2RbAl_3F_{12}$	?	12	14.156	—	19.085	Courbion et al. (1974)
$Cs_2LiGa_3F_{12}$	$R\bar{3}m$	3	7.23	—	17.63	Courbion et al. (1971)
$Cs_2KFe_3F_{12}$	?	12	14.935	—	18.630	Griebler (1978)

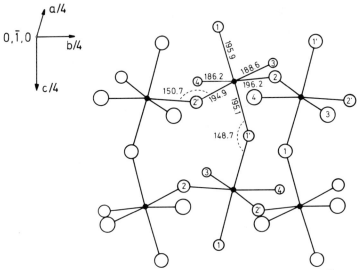

Fig. 24. The cis and trans connection of octahedra in NaFeF₄ (●, Fe; ○, F).

polarity of the space group. The puckered layers are similar to those observed in NaMF₄ compounds only concerning the cis bridging angle, which is somewhat below 150°. In the direction of the trans connection, however, the bridge is not far from linearity. The barium cations between the layers are 11 coordinated.

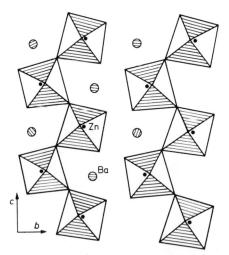

Fig. 25. Projection along the puckered layers of the BaZnF₄ structure. From von Schnering and Bleckmann (1968).

TABLE XXVII

Layer Structures of the $NaNbO_2F_2$ and $BaZnF_4$ Types with Cis, Trans-Connected Octahedra

Compound	Space Group	a (Å)	b (Å)	c (Å)	β (deg)	Reference
$NaVF_4$	$P2_1/c$,	7.896	5.355	7.542	101.7	Knoke et al. (1979a)
$NaCrF_4$ (*)	$Z = 4$	7.862	5.328	7.406	101.65	Knoke et al. (1979a)
$NaFeF_4$ (*)		7.908	5.351	7.531	101.92	Dance et al. (1981); a-1786
$BaMgF_4$	$A2_1am$,	5.810	14.509	4.125	—	Keve et al. (1969); a-578
$BaMnF_4$ (*)	$Z = 4$	5.985	15.098	4.222	—	Keve et al. (1969); a-1746
$BaFeF_4$		5.829	14.837	4.238	—	de Pape and Ravez (1966); a-1847
$BaCoF_4$ (*)		5.852	14.628	4.210	—	Keve et al. (1970); a-1910
$SrNiF_4$		5.653	14.435	3.935	—	a-1944; von Schnering and Bleckmann (1968)
$BaNiF_4$		5.799	14.458	4.153	—	a-1945; von Schnering and Bleckmann (1968)
$BaCuF_4$		5.551	13.972	4.476	—	a-421; von Schnering et al. (1971)
$BaZnF_4$ (*)		5.841	14.563	4.206	—	a-604; von Schnering and Bleckmann (1968)

142 D. Babel and A. Tressaud

C. K₂NiF₄-LIKE COMPOUNDS AND RELATED STRUCTURES

1. A_2MF_4 and $A_3M_2F_7$ Fluorides of Tetragonal K_2NiF_4 or $Sr_3Ti_2O_7$ Type

The relation between these two structural types [both in space group $I4/mmm$ (no. 139), $Z = 2$ (Balz and Plieth, 1955; Ruddlesden and Popper, 1958)], which may be called 1D-disconnected perovskites, is obvious from Fig. 26. Structural refinements of several K_2MF_4 compounds do not show (Babel and Herdtweck, 1982; Herdtweck and Babel, 1980) significant shortening of the trans terminal bonds, as claimed in the earlier literature (Winkler and Brehler, 1954). This can be explained electrostatically from the observed geometry of the KF_9 polyhedra, which involves one very short K—F bond just to the "terminal" F^- of the nearest MF_6 octahedron. Similar conditions are found for the double-layer compounds $K_3M_2F_7$, for which refinements have also been performed (Babel and Herdtweck, 1982). Only one of the octahedron ligands is terminal in this structure. The M cations are attracted by this terminal ligand and therefore are not strictly in the same plane as the equatorial anions. Neglecting these small differences, effectively undistorted

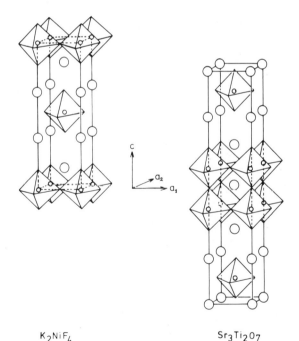

K₂NiF₄   Sr₃Ti₂O₇

Fig. 26. Single- and double-layer structures of K_2NiF_4 and $Sr_3Ti_2O_7$ (○, K, Sr; ○, Ni, Ti).

octahedra and linear bridges are present in the K_2MF_4 and $K_3M_2F_7$ compounds listed in Table XXVIII. The known isomorphs, including Jahn–Teller distorted Cu^{2+} compounds, are added (Hirakawa *et al.*, 1982).

2. $A_2^{II}MF_6$ Fluorides of Tetragonal Bi_2NbO_5F Type

Structures of this type belong to space group $I422$ (no. 97), $Z = 2$ (Aurivillius, 1952). The structure, in which two independent anions are not coordinated to the M cations, must be written $(A^{II}F)_2MF_4$ if one wishes to show its relation to the K_2NiF_4 type. It is illustrated in Fig. 27, which includes other layer structures. Examples of shortened (M = Ni or Zn) as well as of lengthened (M = Co) terminal bonds were found in refined $(BaF)_2MF_4$ compounds (von Schnering, 1966). There is also evidence, however, that some of the compounds listed in Table XXIX have superstructures induced by deviation from bridge linearity in the tetragonal plane of octahedral linkings (Samouel, 1971; E. Herdtweck, 1981, unpublished; Renaudin *et al.*, 1983). Bridge angles bent like Cu–F–Cu = 154° have been reported from neutron diffraction studies (Reinen and Weitzel, 1977) on the orthorhombic Jahn–Teller distorted fluoride $(BaF)_2CuF_4$ [space group *Bbam*; standard: *Cmca*, (no. 64), $Z = 4$ (von Schnering, 1973)].

D.   CHIOLITES $Na_5M_3F_{14}$ AND OTHER LAYER STRUCTURES

1. $Na_5M_3F_{14}$ Fluorides Related to Chiolite $Na_5Al_3F_{14}$

If in the tetragonal arrangement of four square meshes (as found within a layer of $TlAlF_4$ or K_2NiF_4 structures) the cation is removed from the center, a "diluted" layer is achieved. It is realized in the structure of mineral chiolite [tetragonal, space group $P4/mnc$ (no. 128), $Z = 2$ Brosset, 1938a; Jacoboni

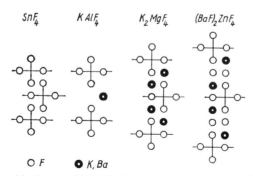

Fig. 27.   Relationships between $(BaF)_2ZnF_4$ and other layer structures. From von Schnering (1967).

TABLE XXVIII

Tetragonal Fluorides of K_2NiF_4- and $Sr_3Ti_2O_7$-Type Structures

Compound	a (Å)	c (Å)	Reference
K_2MgF_4 (*)	3.980	13.179	Babel and Herdtweck (1982)
Rb_2MgF_4	4.055	13.79	a-571
$(NH_4)_2MgF_4$	4.07	13.88	a-568
Tl_2MgF_4	4.007	14.22	a-579
K_2MnF_4 (*)	4.174	13.272	Babel and Herdtweck (1982)
Rb_2MnF_4	4.23	13.92	a-1725
Cs_2MnF_4	4.31	14.63	a-1733
K_2FeF_4	4.140	12.98	a-1798
Rb_2FeF_4	4.20	13.38	a-1815
Tl_2FeF_4	4.194	13.91	a-1867
K_2CoF_4 (*)	4.073	13.087	Babel and Herdtweck (1982)
Rb_2CoF_4	4.135	13.67	a-1897
Tl_2CoF_4	4.114	14.05	a-1915
K_2NiF_4 (*)	4.012	13.076	Babel and Herdtweck (1982)
Rb_2NiF_4	4.087	13.71	a-1936
$(NH_4)_2NiF_4$	4.084	13.79	a-1933
Tl_2NiF_4	4.051	14.22	a-1952
K_2CuF_4 (*)	4.147	12.734	Haegele and Babel (1974); Herdtweck and Babel (1981); Hidaka and Walker (1979)
Rb_2CuF_4	4.238	13.28	a-407
Tl_2CuF_4	4.199	13.66	a-425
Cs_2CuF_4	4.408	14.03	Dance et al. (1976)
Cs_2AgF_4	4.581	14.192	Odenthal et al. (1974)
K_2ZnF_4 (*)	4.058	13.109	Herdtweck and Babel (1980)
Rb_2ZnF_4	4.104	13.28	a-596
$(NH_4)_2ZnF_4$	4.14	13.97	a-593
Tl_2ZnF_4	4.105	14.10	a-609
Rb_2CdF_4	4.414	13.98	a-614
$K_3Mn_2F_7$ (*)	4.187	21.586	Babel and Herdtweck (1982)
$Rb_3Mn_2F_7$	4.222	22.26	Navarro et al. (1976)
$K_3Fe_2F_7$	4.130	21.15	a-1802
$K_3Co_2F_7$ (*)	4.074	21.163	Babel and Herdtweck (1982)
$K_3Ni_2F_7$ (*)	4.015	21.073	Babel and Herdtweck (1982)
$K_3Cu_2F_7$ (*)	4.156	20.520	Herdtweck and Babel (1981)
$K_3Zn_2F_7$ (*)	4.060	21.171	Herdtweck and Babel (1980)
$Rb_3Cd_2F_7$	4.403	22.71	a-615

et al., 1981)] with meshes of doubled edge length, as shown in Fig. 28. The transition metal compounds listed in Table XXIX were shown to possess several polymorphs of lower symmetry (Dance and Ravez, 1979; Knox and Geller, 1958; Vlasse et al., 1976), but in any case the high-temperature

TABLE XXIX

$A_2^{II}M^{IV}F_6$ and $A_5^{I}M_3^{III}F_{14}$ Fluorides with Layer Structures Related to Bi_2NbO_5F and Chiolite Types, Respectively

Compound	Space Group	a (Å)	b (Å)	c (Å)	Reference
Ba_2MgF_6		4.043	(Tetragonal)	16.515	Chenavas et al. (1974)
Ba_2MnF_6		4.148	—	16.791	Chenavas et al. (1974)
Ba_2FeF_6 (*)	$I422$,	4.154	—	16.084	Renaudin et al. (1983); a-1849
Ba_2CoF_6 (*)	$Z = 2$	4.101	—	16.284	a-1911
Ba_2NiF_6 (*)		4.054	—	16.341	Renaudin et al. (1983); a-1946
Ba_2AgF_6		4.321	—	17.61	a-433
Ba_2ZnF_6 (*)		4.101	—	16.263	a-605
Ba_2CrF_6	$Bham$,	6.117	5.887	16.20	a-1631
Ba_2CuF_6 (*)	$Z = 4$	5.937	5.837	15.852	von Schnering (1973)
Pb_2CuF_6		5.78	5.65	15.80	a-426
Pb_2MnF_6	$P4_2/nbc$,	7.98	(Tetragonal)	16.92	a-1292
Pb_2FeF_6	$Z = 8$	7.97	—	16.66	a-1293
Pb_2CoF_6		7.92	—	16.42	a-1294
Pb_2NiF_6		7.88	—	16.22	a-1296
Pb_2ZnF_6		7.92	—	16.33	a-1289
$Na_5Al_3F_{14}$ (*)		7.014	—	10.402	Brosset (1938a); Jacoboni et al. (1981)
$K_5In_3F_{14}$		7.933	—	11.882	a-759
$(NH_4)_5In_3F_{14}$		8.046	—	12.658	Champarnaud et al. (1974)
$K_5Tl_3F_{14}$	$P4/mnc$,	8.112	—	11.802	a-782
$Na_5Ti_3F_{14}$	$Z = 2$	7.497	—	10.287	Avignant et al. (1976)
$K_5Ti_3F_{14}$		7.771	—	11.630	a-1309A
$Na_5V_3F_{14}$		7.33	—	10.36	McKinzie et al. (1972)
$K_5V_3F_{14}$		7.71	—	11.57	Wanklyn et al. (1976); Cros et al. (1977a)
$Na_5Cr_3F_{14}$		7.32	—	10.24	Miranday et al. (1975)
γ-$Na_5Fe_3F_{14}$ (*)	$P4_21_2$	7.345	—	10.400	Vlasse et al. (1976)
$Na_5Fe_2CoF_{14}$		7.33	—	10.34	McKinzie et al. (1972)
$Na_5Co_3F_{14}$		7.30	—	10.21	McKinzie et al. (1972)

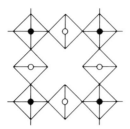

Fig. 28. Structural unit of the chiolite layer structure. The M(III) cations possess either two ($\bullet$) or four terminal ligands ($\circ$).

modification is tetragonal. As determined for γ-$Na_5Fe_3F_{14}$ (Vlasse *et al.*, 1976), however, it crystallizes in the noncentrosymmetric space group $P4_22_12$ (no. 94). Not only are the bridge angles (Fe–F–Fe = 147°) considerably bent, but also the octahedra themselves—one having two, the other four terminal ligands—exhibit strong distortions in angles and distances. There are notable ferrimagnetic properties in connection with this structure (Knox and Geller, 1958; Wintenberger *et al.*, 1975) and nonlinear properties for some oxide–fluorides (Elaatmani *et al.*, 1981; Ravez *et al.*, 1979a,b, 1981).

2. Other Layer Structures

There are some other layer structures known among fluorides that generally correspond to more complex formulas. Some of them contain units of face-shared octahedra ($Cs_7Ni_4F_{15}$, $Cs_4Ni_3F_{10}$, and $Cs_6Ni_5F_{16}$), which are linked via terminal corners. Others show a pyrochlore-related triple layer ($Cs_4CoCr_4F_{18}$) or include cations of higher coordination number ($BaMnGaF_7$). These examples are briefly treated as odd-composition compounds in Section VI.

IV. Chain Structures

Most fluorides showing 1D crystallographic features correspond to the general formula A_xMF_5. These phases are generally characterized by isolated $(MF_5)_n$ chains of octahedra sharing two vertices.

A. Chain Structures in Pentafluorides

Two types of pentafluorides, the VF_5 type and the low-temperature form of UF_5 (α-UF_5), show neutral $(MF_5)_n$ strings. Differences in structures arise from the various sharings within the chains. Lattice constants are given in Table XXX.

TABLE XXX

Chain Structures in Pentafluorides[a]

Pentafluoride	a (Å)	b (Å)	c (Å)	Reference
VF_5	5.40	<u>16.72</u>	7.53	a-243 (*)
CrF_5	5.5	<u>16.3</u>	7.4	a-261
TcF_5	5.76	<u>17.01</u>	7.75	a-285
ReF_5	5.70	<u>17.23</u>	7.67	a-289
$\alpha\text{-}UF_5$	6.518	(Tetragonal)	<u>4.470</u>	Eller et al. (1979) (*)
BiF_5	6.581	(Tetragonal)	<u>4.229</u>	a-237

[a] The chain direction is underlined.

1. VF₅ Structure: Orthorhombic, Space Group Pmcn (No. 62), Z = 8

In the $(VF_5)_n$ chains, VF_6 octahedra are linked by cis fluorine atoms. Significant differences in V–F distances are found between $V-F_{term} = 1.69$ Å and $V-F_{bridg} = 1.97$ Å (averaged). The V–F–V angles are about 150°, and the anionic stacking is somewhat related to the hexagonal type.

2. α-UF₅ Structure: Tetragonal, Space Group I4/m (No. 87), Z = 2

Trans fluorine chains are observed in $\alpha\text{-}UF_5$ (a-177; Eller et al., 1979). The UF_6 octahedra are elongated along the chain direction (c axis): $U-F_{term} = 1.99_5$ Å; $U-F_{bridg} = 2.235$ Å; U–F–U angle = 180°. It can be noticed that, above 130°C, $\alpha\text{-}UF_5$ undergoes a transition to a high-temperature β form. In this variety uranium exhibits CN = 8, intermediate between a square antiprism and a dodecahedron (Ryan et al., 1976; Taylor and Waugh, 1980).

B. STRUCTURES OF A^II M^III F₅ Compounds

The arrangement of $(MF_5)_n^{2n-}$ infinite chains can be correlated with both $r_{A^{2+}}$ ionic radius and $r_{A^{2+}}/r_{M^{3+}}$ ratio (Dance and Tressaud, 1973; von der Mühll and Ravez, 1974). The different structural types of the AMF_5 phases are given in Fig. 29. $M^{II}M^{III}F_5$ phases containing a divalent transition element (i.e., Cr_2F_5, $MnAlF_5$, and $MnCrF_5$ types) are discussed in Section VI,A.

In $CaCrF_5$, the framework is composed of zigzag chains of $(CrF_6)^3$ octahedra sharing trans fluorine atoms. Within those chains the Cr–F–Cr angle is bent down to 152.2°. Calcium atoms exhibit a pentagonal bipyramid

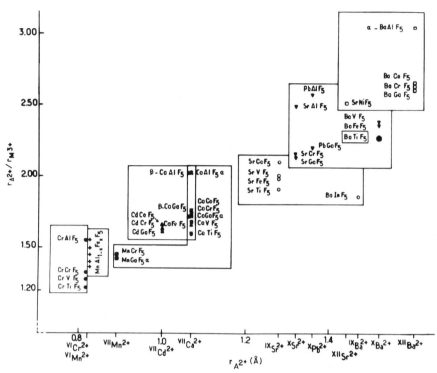

Fig. 29. $A^{II}M^{III}F_5$ structural types ($r_{A^{2+}}$ and $r_{M^{3+}}$ from Shannon, 1976) ($\bullet$, Cr_2F_5 type; $+$, $MnAlF_5$ type; $\blacksquare$ $MnCrF_5$ type; $\blacktriangle$, $CaFeF_5$ type; $\blacksquare$, $CaCrF_5$ type; $\bigcirc$, $SrFeF_5$ type; $\blacktriangledown$, $BaFeF_5$ type; $\square$, $BaGaF_5$ type).

coordination between four neighboring chains (Dumora *et al.*, 1971; Wu and Brown, 1973). The relationships between $CaCrF_5$ and $CaFeF_5$ types can be explained on the basis of a rotation of MF_6 octahedra with respect to the chain direction (Fig. 30). In addition, half of the calcium positions are shifted by $c/2$ (Dumora *et al.*, 1971). Cell constants of the fluorides crystallizing with the $CaCrF_5$ and $CaFeF_5$ structures are compiled in Table XXXI together with those of other AMF_5 compounds.

For larger divalent cations, the $SrFeF_5$-type structure is composed of helicoidal $(FeF_5)_n^{2n-}$ chains with octahedra sharing corners in the cis position (von der Mühll *et al.*, 1973); the bridging angle is therefore canted (Fe–F–Fe $= 138°$). Two opposite directions of rotation are present, and strontium atoms are located between three neighboring chains (Fig. 31). Cell constants of isostructural fluorides are given in Table XXXI.

The $BaFeF_5$ structure shows two different types of chains: a single one with octahedra sharing trans fluorine atoms and a branched one (von der Mühll

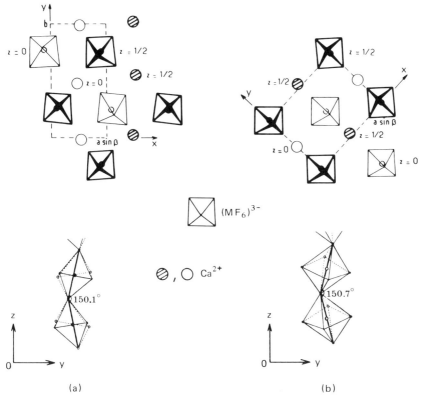

Fig. 30. Relationships between (a) CaFeF$_5$ and (b) CaCrF$_5$. From von der Mühll and Ravez (1974).

et al., 1971; and Fig. 32). The structures of Sr$_3$Fe$_2$F$_{12}$ and Pb$_5$Fe$_3$F$_{19}$, which have been solved, are closely related to that of BaFeF$_5$ (von der Mühll, 1973; Jacoboni *et al.*, 1983; also see Section VII,A). It has been shown that dimeric units (Ti$_2$F$_{10}$)—instead of single-chain (MF$_5$)$_n$—are associated with branched chains in BaTiF$_5$ (space group $I4/m$) (Eicher and Greedan, 1984).

For higher values of $r_{A^{2+}}/r_{M^{3+}}$, the BaGaF$_5$ structure is observed: GaF$_6$ octahedra share cis corners to form chains directed along the c axis (Domesle and Hoppe, 1978; and Fig. 33a). Cell constants of isostructural fluorides BaM^{III}F$_5$ and SrNiF$_5$ are given in Table XXXI.

Structural changes can also be correlated with the environment of the divalent cation: CN ranges from 7 in CaFeF$_5$ and CaCrF$_5$ types to 9 in SrFeF$_5$, whereas it is 9 and 11 in the BaFeF$_5$-type structure. Barium atoms are at the center of a distorted cubooctahedron of 12F in BaGaF$_5$.

TABLE XXXI

$A^{II}M^{III}F_5$ Structures Listed by Increasing Cell Constants Corresponding to the Chain Direction[a]

Structural Type	Compound	a (Å)	b (Å)	c (Å)	β (deg)	Reference
CaCrF$_5$ type; monoclinic, space group C2/c (no. 15), Z = 4 (Cr–F–Cr angle = 150.7°)	α-CaAlF$_5$	8.70(1)	6.32(1)	7.30(1)	115.3(2)	a-689
	α-CaGaF$_5$	8.88(1)	6.42(1)	7.47(1)	115.2(2)	a-734
	CdCoF$_5$	8.78(1)	6.410(1)	7.495(2)	115.27(2)	Fleischer and Hoppe (1982a)
	CaCrF$_5$ (*)	9.005(5)	6.472(5)	7.533(5)	115.8(5)	Dumora et al. (1971); Wu and Brown (1973)
	CaCoF$_5$	8.920(4)	6.444(3)	7.522(4)	115.6(2)	a-1905
	CaVF$_5$	8.99(1)	6.47(1)	7.67(1)	115.5(2)	a-1514
	CaTiF$_5$ (*)	9.10(1)	6.577(7)	7.70(1)	115.7(5)	a-1319; Eicher and Greedan (1984)
CaFeF$_5$ type; monoclinic, space group P2$_1$/c (no. 14), Z = 4 (Fe–F–Fe angle = 150.1°)	β-CaAlF$_5$	5.35[b]	9.81(2)	7.31(1)	110.5	a-689
	β-MnGaF$_5$	5.25[b]	9.67	7.35	110.5	Chassaing and Julien (1972)
	CdCrF$_5$	5.42	9.98	7.47	110.4	Samouël and de Kozak (1983); a-1640
	CdGaF$_5$	5.41[b]	9.82	7.49	110.5	Chassaing and Julien (1972)
	β-CaGaF$_5$	5.44[b]	9.981(5)	7.541(4)	110.5	a-733
	CaFeF$_5$ (*)	5.500(5)	10.05(1)	7.58(1)	110.49	von der Mühll and Ravez (1974)
SrFeF$_5$ type; monoclinic, space group P2$_1$/c (no. 14), Z = 8 (Fe–F–Fe angle = 137.9°)	SrVF$_5$	7.01	7.19	14.517	95.3	a-1516
	SrCoF$_5$	6.993(4)	7.232(4)	14.602(7)	94.8(2)	a-1907
	SrTiF$_5$	7.099(6)	7.280(6)	14.787(8)	96.6(2)	a-1321
	SrFeF$_5$ (*)	7.062(1)	7.289(1)	14.704(1)	95.40(5)	von der Mühll et al. (1973)
	BaInF$_5$	7.593(5)	7.718(4)	15.575(8)	95.6(5)	a-772

150

BaFeF$_5$ type: tetragonal
space group P4 (no. 75),
Z = 32 (Fe–F–Fe angles = 180°)

SrAlF$_5$ (*)	14.089(8)	—	14.33_4	von der Mühll et al. (1971)
EuAlF$_5$	14.12	—	14.37	a-704
PbAlF$_5$	14.25(2)	—	14.46	a-705
SrGaF$_5$	14.23(1)	—	14.54	a-736
SrCrF$_5$	14.34(1)	—	14.59	a-1624
PbGaF$_5$	14.489(6)	—	14.65	a-746
BaVF$_5$	14.97	—	15.06	a-1519
BaFeF$_5$ (*)	14.919(8)	—	15.21_8	von der Mühll et al. (1971)

BaTiF$_5$ type, tetragonal,
space group I4/m (no. 87)

BaTiF$_5$ (*)	15.091(5)	—	7.670(3)	Eicher and Greedan (1984)

BaGaF$_5$ type; orthorhombic,
space group P2$_1$2$_1$2$_1$ (no. 19),
Z = 4 (Ga–F–Ga angle = 141°)

BaMnF$_5$ (*)	14.115(2)	5.811(1)	4.881(1)	Bukovec and Hoppe (1984)
SrNiF$_5$	13.265(2)	5.4404(6)	4.9270	Fleischer and Hoppe (1982a)
α-BaAlF$_5$	13.71_0	5.60_4	4.93_0	Domesle and Hoppe (1982a)
BaCrF$_5$ (*)	13.93_8	5.71_1	4.94_7	Holler et al. (1982)
BaGaF$_5$ (*)	13.93_0	5.66_8	4.97_8	Domesle and Hoppe (1978)

[a] c constant, except for SrFeF$_5$ type (b constant).
[b] a constant, calculated for β = 110.5°.

151

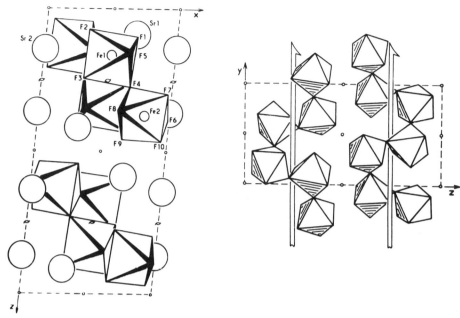

Fig. 31. SrFeF$_5$ structure.

It can be noted that for a higher $r_{A^{2+}}/r_{M^{3+}}$ ratio (BaBF$_5$) (Ravez and Hagenmuller, 1964) or for larger M(III) cations (BaTlF$_5$) (Grannec *et al.*, 1971a) the structural data are still not determined. It should also be mentioned that no decisive information has been published on the structural arrangements of A^{I}M^{IV}F$_5$ phases since Babel (1967).

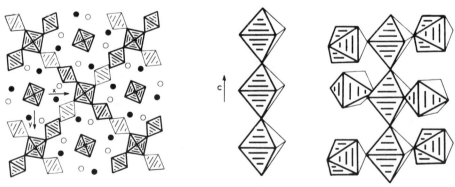

Fig. 32. BaFeF$_5$ structure [●, Sr (Ba), $z = 0$; ○, Sr (Ba), $z = \frac{1}{2}$].

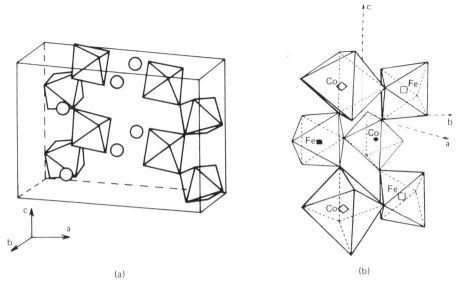

(a)   (b)

Fig. 33. Octahedra sharings in (a) $BaCrF_5$ ($\bigcirc$, Ba) ($BaGaF_5$-type) and (b) Ba_2CoFeF_9. From de Kozak *et al.* (1981).

C. STRUCTURES OF $A_2^I M^{III} F_5$ COMPOUNDS

A_2MF_5 fluorides have long been assigned to the Tl_2AlF_5 type (a-703). In fact, three main structures are presently known for these compounds: K_2FeF_5 (Vlasse *et al.*, 1977b), Rb_2CrF_5 (Jacoboni *et al.*, 1974), which is isostructural with $K_2VO_2F_3$ (Ryan *et al.*, 1971), and Tl_2AlF_5.

It can be pointed out that compounds such as Li_2MF_5 and Na_2MF_5 are unusual and have been reported only for M(III) = Cr or Mn without detailed structural information (Babel, 1967).

For larger trivalent cations [M(III) = rare earth], another chain structure has been proposed: the K_2SmF_5 type (Bochkova *et al.*, 1974; Pistorius, 1975). The coordination polyhedra of Sm(III) are pentagonal bipyramids connected by edges to form infinite chains. These chains are separated by potassium atoms with CN = 8.

Structures of several A_2MF_5 phases are shown in Fig. 34, and their crystallographic data are grouped in Table XXXII.

1. Cis-Connected $(MF_5)_n^{2n-}$ Chains

K_2MF_5 compounds crystallize with the K_2FeF_5 type. MF_6 octahedra are connected by cis corners to form infinite zigzag chains parallel to the *a* axis. In

TABLE XXXII

Cristallographic Data of A_2MF_5 Fluorides[a]

A	M				
	Al	Cr	Mn	Fe	Ga
K	$Pna2_1$: $a = 19.6$ Å $b = 12.6$ Å $c = 7.1$ Å $Z = 16$ (a-663)	$Pna2_1$: $a = 19.60$ Å $b = 12.84$ Å $c = 7.37$ Å $Z = 16$ (a-1577)	—	$Pna2_1$ (*): $a = 20.39$ Å $b = 12.84$ Å $c = 7.40$ Å $Z = 16$ (Vlasse et al., 1977b)	$Pna2_1$: $a = 20.29$ Å $b = 12.83$ Å $c = 7.34$ Å $Z = 16$ (Chassaing, 1968; Pistorius, 1975)
Rb	—	$Pnma$ (*): $a = 7.515$ Å $b = 5.724$ Å $c = 11.985$ Å $Z = 4$ (Jacoboni et al., 1974)	$P4/m, P4/mmm$: $a = 6.10$ Å $c = 4.14$ Å $Z = 1$ (Günther et al., 1978)	$Pnma$ (*): $a = 7.53$ Å $b = 5.78$ Å $c = 11.985$ Å $Z = 4$ (a-1816; Tressaud et al., 1981b)	$Pnma$: $a = .7.49$ Å $b = 5.77$ Å $c = 11.96$ Å $Z = 4$ (Chassaing, 1968; Pistorius, 1975)

Cs   —

$C2_1 2 2$ (∗P):
$a =$ 7.46 Å
$b =$ 10.06 Å
$c =$ 8.24 Å
$Z =$ 4
(a-703)

Tl   —

—

—

—

Pnma:
$a =$ 7.84 Å
$b =$ 5.95 Å
$c =$ 12.57 Å
$Z =$ 4
(a-1829; Dance et al., 1980)

NH₄   —

Pnma (∗):
$a =$ 6.20 Å
$b =$ 7.94 Å
$c =$ 10.72 Å
$Z =$ 4
(a-1722)

—

—

[a] Chain direction underlined.

155

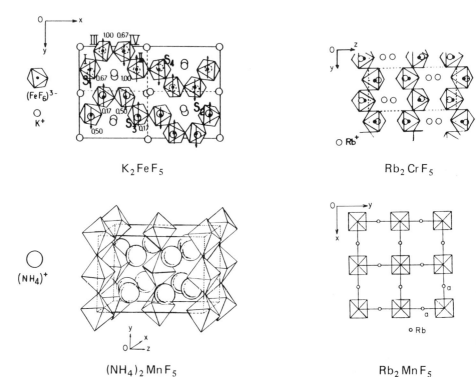

K_2FeF_5

Rb_2CrF_5

$(NH_4)_2MnF_5$

Rb_2MnF_5

Fig. 34. A_2MF_5 chain structures.

K_2FeF_5, the octahedra are characterized by two different bond lengths: four $Fe-F_{term} = 1.88$ Å and two $Fe-F_{bridg} = 2.02$ Å. The bridging $Fe-F-Fe$ angles range from 162 to 173°. Potassium atoms connect the chains to one another with $CN = 9$ and 10 (Vlasse et al., 1977b).

The Rb_2MF_5 fluorides are isostructural with Rb_2CrF_5 (Table XXXII). Their structure is formed by linear chains of MF_6 octahedra linked as in K_2MF_5 by cis corners. Here also, $M-F_{term}$ distances are shorter than $M-F_{bridg}$ distances [1.867 (average) and 1.99 Å, respectively]. The bridging $M-F-M$ angles are close to 180°, and Rb^+ ions have $CN = 9$ and 10.

2. Trans-Connected $(MF_5)_n^{2n-}$ Chains

Although Rb_2MnF_5 (Günther et al., 1978), $(NH_4)_2MnF_5$ (a-1722), and Tl_2AlF_5 (a-703) crystallize in different space groups, they have a common feature: MF_6 octahedra are bridged by trans fluorine atoms. The $M-F-M$ angles vary from 143° in $(NH_4)_2MnF_5$ to 180° in Rb_2MnF_5.

Concerning the A_2MnF_5 compounds, the octahedra are elongated along the chain direction due to a ferrodistortive ordering of Mn^{3+} d_{z^2} orbitals. A reversible hydration $\rightleftharpoons$ dehydration reaction,

$$A_2MF_5 \xrightleftharpoons[\text{heating}]{H_2O} A_2MF_5 \cdot H_2O$$

$$(A = Rb, Cs; M = Mn, Fe)$$

occurs via a chain-controlled topotactic mechanism (Günther *et al.*, 1978; Fourquet *et al.*, 1981a).

D. CONDENSED-CHAIN STRUCTURES

Different chains of the kind just discussed may be connected or octahedra may be linked in a different way by sharing of edges or faces (instead of vertices only) to form condensed-chain structures. In all these cases the A_xMF_5 formula is no longer valid. In Table XXXIII are compiled structural data of several phases showing infinite condensed chains.

Double chains of MF_6 octahedra sharing only corners have been characterized in Ba_2CoFeF_9, for instance. The structure is described in Section VI,C with other $A_2MM'F_9$ phase, but we can report here the similarities between M–F–M angles in the $BaGaF_5$ type (Ga–F–Ga $= 140.8°$) and the $BaCoFeF_9$ type (average M–F–M $= 143°$). Octahedra sharings in both structures are compared in Fig. 33.

In the $CsCrF_4$ structure, three strings of CrF_6 octahedra connected by trans fluorine atoms additionally share cis equatorial ligands, thus forming infinite triple chains of $(Cr_3F_{12})_n^{3n-}$ formula (Babel and Knoke, 1978; Fourquet *et al.*, 1979; and Fig. 35).

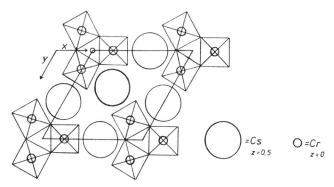

Fig. 35. Projection of the structure of $CsCrF_4$ on the (001) plane.

TABLE XXXIII
Condensed-Chain Structures

Structural Type	Compound	a (Å)	b (Å)	c (Å)	β (deg)	Reference
$CsCrF_4$ type; hexagonal, space group $P\bar{6}2m$, $Z = 3$	$CsAlF_4$ (*)	9.500	—	3.713	—	Lösch and Hebecker (1979a)
	$CsCrF_4$ (*)	9.650(5)	—	3.857(3)	—	Babel and Knoke (1978)
Na_2CuF_4 type; monoclinic, space group $P2_1/c$, $Z = 2$	Na_2CrF_4	3.34_4	9.53_3	5.65_7	87.2	a-1565
	Na_2CuF_4 (*)	3.26_1	9.35_4	5.60_1	87.5	a-400
$CsNiF_3$ type; hexagonal, space group $P6_3/mmc$, $Z = 2$	$CsNiF_3$ (*)	6.236	—	5.225	—	a-1938

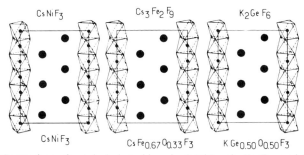

Fig. 36. Comparison of octahedral stackings in $CsNiF_3$-, $Cs_3Fe_2F_9$-, and K_2GeF_6-type structures.

No mixed sharings of corners and edges have hitherto been found in fluoride chain systems. However, the hydrated oxide–fluoride $CsVOF_3 \cdot \frac{1}{2}H_2O$ shows infinite double chains; two VOF_5 octahedra are linked by an edge, forming $V_2O_2F_8$ groups. These units are connected in a complex way via corner sharing (Waltersson, 1979).

In Na_2CuF_4 and Na_2CrF_4, chains are formed of octahedra sharing opposite edges; they are isolated by sodium atoms with $CN = 7$. The MF_6 units exhibit a tetragonal elongation due to the Jahn–Teller effect (Babel, 1965b).

In the series of hexagonal perovskites described earlier, $CsNiF_3$ consists of infinite chains of NiF_6 octahedra sharing two opposite faces (Fig. 36). Whereas the six Ni–F distances are equivalent, two different F–F spacings lead to a trigonal distortion of the octahedra, which are elongated in the c direction (Babel, 1969).

V. Structures Containing Polynuclear Units

In many structural types showing a 3D framework it is possible, using suitable cationic substitutions, to obtain phases in which polyhedra containing transition metals are separated by diamagnetic elements. However, the basic structure of these phases (hexagonal polytypes of perovskite, for instance) has a 3D character.

We shall deal only with compounds in which the presence of polynuclear units is a specific character of the structure—for example, (1) polynuclear units isolated within a 3D lattice, i.e., $(Fe_2F_9)^{3-}$ dimers in $Cs_3Fe_2F_9$ or $(Al_4F_{20})^{8-}$ tetramers in $Ba_3Al_2F_{12}$; and (2) neutral M_4F_{20} tetramers linked together by van der Waals interactions (RhF_5 and MoF_5 types). It can be noted that M_3X_{15} trimers consisting of octahedra sharing cis fluorine atoms have been found in oxide–tetrafluorides $MoOF_4$ and $TcOF_4$ (Edwards et al., 1970) and also in gaseous TaF_5 (Brunvoll et al., 1979).

A. $Cs_3Fe_2F_9$ TYPE

The symmetry is hexagonal [space group $P6_3/mmc$ (no. 194), $Z = 2$, according to unpublished work by W. Massa et al. (1984), slightly modifying former results (Wall et al., 1971)]. The structural arrangement of $Cs_3Fe_2F_9$ corresponds to a $CsNiF_3$ type (see Section II) in which 1-over-3 nickel atoms would be missing within the infinite $(MF_3)_n{}^{n-}$ chain. The structure is therefore composed of isolated $(Fe_2F_9)^{3-}$ groups of two octahedra sharing a face. These dimers are connected to each other by the coordination polyhedra of cesium atoms. The structure of $Cs_3Fe_2F_9$ is compared in Fig. 36 with those of $CsNiF_3$ and K_2GeF_6. Lattice constants are given in Table XXXIV.

B. $Ba_3Al_2F_{12}$ TYPE

The symmetry is orthorhombic with $Pnnm$ space group (no. 58), $Z = 4$, Domesle and Hoppe, 1982b). The structure of $Ba_3Al_2F_{12}$ contains $(Al_4F_{20})^{8-}$ tetramers. Within a tetramer, the bridging fluorine atoms are arranged in such a way as to suggest a "chair" conformation. However, it should be pointed out that, owing to the alternative formula $Ba_6F_4(Al_4F_{20})$, the tetrameric units are part of a 3D framework. They are connected to each other by the coordination polyhedra of barium atoms including "independent" fluorine atoms.

Due to the 3D character of the structure, two groups of Al–F distances arise: two bridging and one "terminal" are longer (1.83 Å) than the three other "terminal" distances (1.77 Å) (Domesle and Hoppe, 1980b, 1982b).

C. TRANSITION METAL PENTAFLUORIDES

The structures of most $4d$ and $5d$ transition metal pentafluorides are characterized by neutral M_4F_{20} tetrametric units. Within a tetramer MF_6 octahedra share two corners in a cis position.

1. RhF_5 Type: Monoclinic, Space Group $P2_1/a$ (No. 14), $Z = 8$

In MF_5 pentafluorides of the platinum-related metals (M = Ru, Os, Rh, Ir, or Pt), tetramers show hcp stacking of fluorine atoms. As a consequence, M–F–M angles are about 135° (Fig. 37). As found in other fluoride series, the bridging fluorine atoms are more distant from the metal

TABLE XXXIV

Lattice Constants of Structures Containing Polynuclear Units

Structural Type	Compound	a (Å)	b (Å)	c (Å)	β (deg)	Reference
$Cs_3Fe_2F_9$ type; hexagonal, space group $P6_3/mmc$, $Z=2$	$Cs_3Fe_2F_9$ (*)	6.345	—	14.82	—	a-1831A; W. Massa, S. Kummer, J. M. Dance, and D. Babel (1984, unpublished) a-725 A
	$Cs_3Ga_2F_9$	6.319	—	14.762	—	Waltersson (1978)
	$Cs_3V_2O_2F_7$ (*)	6.335	—	14.871	—	Wall et al. (1971)
	$Cs_3V_2O_4F_5$	6.31_3	—	14.76_6	—	Wall et al. (1971)
	$Cs_3Mo_2O_6F_3$ (*)	6.389	—	15.10	—	Mattes et al. (1980)
$Ba_3Al_2F_{12}$ type; orthorhombic, space group $Pnnm$, $Z=4$	$Ba_3Al_2F_{12}$ (*)	10.18_7	9.86_9	9.50_2	—	Domesle and Hoppe (1982b)
	$\alpha\text{-}Ba_3Ti_2O_2F_{10}$	10.32_0	10.07_2	9.70_5	—	Domesle and Hoppe (1982b)
RhF_5 type; monoclinic, space group $P2_1/a$, $Z=8$	RuF_5 (*)	12.47(1)	10.01(1)	5.42(1)	99.8	a-306
	OsF_5 (*)	12.59(10)	9.91(10)	5.53(3)	99.5	a-316
	RhF_5 (*)	12.33_8	9.91_7	5.51_7	100.4	Morrell et al. (1973)
	IrF_5	12.25	10.0	5.40	99.8	a-320
MoF_5 type; monoclinic, space group $C2/m$, $Z=8$	NbF_5 (*)	9.62(1)	14.43(2)	5.12(1)	96.1	a-249
	TaF_5 (*)	9.64(1)	14.45(2)	5.12(1)	96.3	a-251
	MoF_5 (*)	9.61(1)	14.22(2)	5.16(1)	94.4	a-264
	WF_5 (*)	9.61	14.26	5.32	94.6	a-270

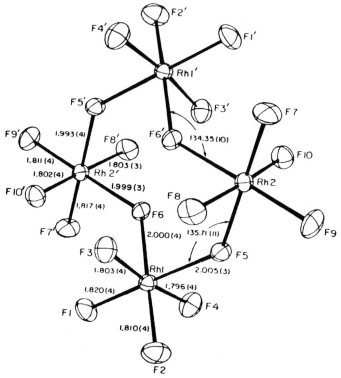

Fig. 37. Rh_4F_{20} tetramer [angles (deg), distances (Å)].

atom ($M-F_{bridg} \simeq 2.00$ Å) than the terminal one ($M-F_{term} = 1.81$ Å; Morrell *et al.*, 1973).

2. MoF_5 Type: Monoclinic, Space Group $C2/m$ (No. 12), $Z = 8$

Although pentafluorides of niobium, tantalum, molybdenum, and tungsten also exhibit M_4F_{20} tetramers, linear $M-F-M$ angles arise from a different anionic packing. The lattice constants of transition metal pentafluorides are given in Table XXXIV. Complete solid solution exists in the NbF_5/TaF_5 system (*a*-252).

VI. Structures Corresponding to Odd Compositions

A number of compounds that have a noninteger F/M ratio are treated under this heading, but only structures that are fully characterized by single-crystal work are discussed.

A. STRUCTURES OF $M^{II}M^{III}F_5$ COMPOUNDS

Three different $M^{II}M^{III}F_5$ structural types are so far known. They are closely related and contain trans-connected $(MF_5)_n^{2n-}$ chains like those already discussed: Cr_2F_5, $MnAlF_5$, and $MnCrF_5$. Cell sizes and space groups are given in Table XXXV as are $M^{III}-F-M^{III}$ bridge angles within the trans chains. In all three structural types, the larger M(II) ions are incorporated between four parallel running chains, as shown in Fig. 38. In $MnAlF_5$, where the chains are nearly linear, the coordination of Mn(II) is also octahedral. In $MnCrF_5$, which is in fact isostructural with $CaCrF_5$ (Wu and Brown, 1973), an exceptional CN = 7 (distorted pentagonal bipyramid) results for Mn(II) due to considerably bent trans bridging angles in the corner-linked chain. The angle is intermediate in Cr_2F_5, where the Cr(II) ion favors strongly distorted octahedra because of the Jahn–Teller effect. Some isomorphic $Cr^{II}MF_5$ compounds are known (Tressaud et al., 1973a). It should be pointed out that the structure of Cr_2F_5 is correlated to those of $CaCrF_5$ and $MnCrF_5$ by $z/4$ translation of chromium(II) ions, thus showing the distorted CN = 6 instead of CN = 7 (Dumora et al., 1971). As can be seen from Fig. 38, the coordination polyhedra of M(II) ions are connected by two opposite edges, thus forming infinite chains, too. Therefore, in such $M^{II}M^{III}F_5$ structures, essential features of the two component structures MF_2 (rutile) and MF_3 are preserved in the common framework. The difference in cationic radii required to fit edge and corner sharings of this kind explains why no structurally related compounds $M^{II}M^{III}F_5$ can be prepared in which the cation sizes are relatively similar (Ferey et al., 1971). Except rutile-related lattice constants, nothing is known about the structures of the mixed-valency compounds V_2F_5 and Fe_2F_5, also listed in Table XXXV; only Mn_2F_5 is reported to be $MnCrF_5$ related.

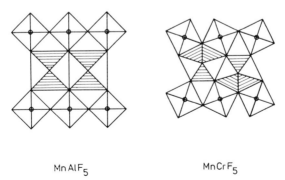

MnAlF$_5$ MnCrF$_5$

Fig. 38. Chains of corner- and edge-sharing polyhedra in the $MnAlF_5$ and $MnCrF_5$ lattices [●, M(III)].

TABLE XXXV

Crystal Data of $M^{II}M^{III}F_5$ Compounds

Compound	Symmetry, Space Group	Z	a (Å)	b (Å)	c (Å)	β (deg)	$M^{III}-F-M^{III}$ Angle (deg)	Reference
V_2F_5	Tetragonal	2	4.77	—	3.20	—	—	a-240; Seifert et al. (1968)
Cr_2F_5 (*)	Monoclinic, $C2/c$ (no. 15)	4	7.773	7.540	7.440	124.25	158	a-259; Steinfink and Burns (1964)
$CrAlF_5$	Monoclinic	4	7.58	7.46	7.25	123.7	—	Tressaud et al. (1973a)
$CrTiF_5$	Monoclinic	4	7.98	7.65	7.70	125.2	—	Tressaud et al. (1973a)
$CrVF_5$	Monoclinic	4	7.91	7.60	7.63	125.0	—	Tressaud et al. (1973a)
Mn_2F_5	Orthorhombic	8	15.44	7.27	6.17	—	—	Tressaud and Dance (1974)
$MnAlF_5$ (*)	Orthorhombic, $Ama2$ (no. 40)	4	9.54	9.85	3.58	—	175	Rimsky et al. (1970)
$MnCrF_5$ (*)	Monoclinic, $C2/c$ (no. 15)	4	8.586	6.291	7.381	115.46	150	Ferey et al. (1977, 1978)
α-$MnGaF_5$	Monoclinic	4	8.56	6.26	7.41	115.7	—	Chassaing and Julien (1972)
Fe_2F_5	Tetragonal	6	8.05	—	9.56	—	—	a-294; Brauer and Eichner (1958)

B. STRUCTURES OF BaM^{II}M^{III}F$_7$ COMPOUNDS

Barium compounds BaM^{II}M^{III}F$_7$ are observed with essentially the same
M(II) and M(III) ions that combine with NaF or AgF to form the weberites
A$_2^I$M^{II}M^{III}F$_7$ already discussed.

Three structural types are so far known. Those of BaMnFeF$_7$ and high-
temperature BaZnFeF$_7$ show 3D-connected frameworks containing edge-
sharing binuclear M$_2$F$_{10}$ units, whereas BaMnGaF$_7$ is a layer structure,
characterized by an exceptional CN = 8 for one-half of the Mn(II) ions. A
fourth type, found for BaCaCrF$_7$ and BaCaGaF$_7$, is a layer structure as
well, in which all calcium ions exhibit CN = 8 in a fluorite-related arrangement (Holler and Babel, 1985). As no replacement of calcium by transition
metal M(II) ions seems possible in the latter structure, it is not treated here
further.

1. Monoclinic BaMnFeF$_7$ Type

Figure 39 shows how binuclear $(Mn_2F_{10})^{6-}$ groups are connected in
BaMnFeF$_7$ [space group $P2_1/c$ (no. 14), $Z = 4$ (Holler et $al.$, 1981)] via four
corners of isolated $(FeF_6)^{3-}$ octahedra. The cis position of the two terminal
ligands of these octahedra is alike that observed for half of the $(FeF_6)^{3-}$

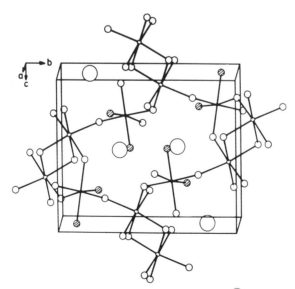

Fig. 39. Octahedral network in the BaMnFeF$_7$ structure [○, Ba; ○, shared and ◍,
terminal, ligands: ○, Mn(II); ●, Fe(III)].

octahedra in trigonal Na_2MnFeF_7. But the corner-linked trans chain of composition $(MnF_5)_n^{3n-}$ in the weberite structure is replaced here by the quite different binuclear unit. The cell dimensions of the prototype and some isostructural compounds, of which the mixed-valency iron fluoride $BaFe_2F_7$ was the first example (Ravez et al., 1967), are listed in Table XXXVI.

2. High-Temperature BaZnFeF₇ Type

The low-temperature form of $BaZnFeF_7$ is isostructural with $BaMnFeF_7$. A high-temperature modification crystallizes with the same space group $[P2_1/c$ (no. 14), $Z = 4$, (Holler and Babel, 1982)] (see Table XXXVI). Although it is the only representative so far known, its structure is worth consideration because it contains both cations ordered in an edge-sharing unit $(ZnFeF_{10})^{5-}$. In contrast to the analogous groups observed in the NH_4MnFeF_6 structure mentioned earlier, the $(ZnFeF_{10})^{5-}$ units have two terminal ligands, one at each cation and in a trans position with respect to the plane of the common edge. The 3D framework is built up from these units only, which are linked together via the remaining six corners.

3. Monoclinic BaMnGaF₇/BaCdGaF₇ Type

In this structural type [space group $C2/c$ (no. 15), $Z = 8$ (Holler et al., 1984)], well established by several full determinations on isostructural compounds (see Table XXXVI), M(II) ions are located in chains of corner-linked polyhedra, which are alternately octahedra and distorted square antiprisms. Octahedrally coordinated M(III) ions connect these chains, giving layers as illustrated in Fig. 40. The unusual octacoordinated Mn(II) could also be achieved in high-pressure polymorphs of $BaMnFeF_7$ and $BaMnVF_7$, which normally crystallize in the structural type discussed earlier. Another way to realize this structure is by size differentiation between divalent ions [M(II) > M'(II)], which leads to isostructural compounds $Ba_2M^{II}M'^{II}M_2^{III}F_{14}$, showing additional M(II)/M'(II) ordering within the chains on the octa- and hexacoordinated sites, respectively (Holler, 1983).

C. STRUCTURES OF Ba₂MⁱⁱMⁱⁱⁱF₉ AND RELATED (M₂F₉)ⁿ⁻ COMPOUNDS

Contrary to the $Cs_3Fe_2F_9$ structure, in which the $(M_2F_9)^{n-}$ composition is realized by separated binuclear units of two face-shared octahedra, extended linking is observed in the structures of Ba_2ZnAlF_9, Ba_2CoFeF_9, $KPbCr_2F_9$, and the unique dioxygenyl compound $(O_2)Mn_2F_9$. All these structures contain double chains of cis-connected octahedra, as illustrated in Fig. 41. The

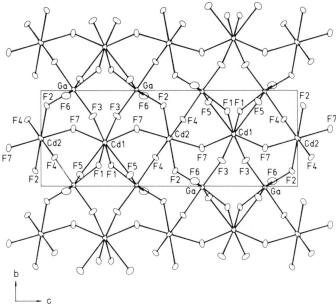

Fig. 40. Layer structure of BaCdGaF$_7$ showing M(II) chains of polyhedra with CN = 8 (Cd1) and CN = 6 (Cd2); x = 0.0.

resulting $(M_2F_9)^{n-}$ string has the form of a stair in the first three cases, which therefore are intermediate between the chain structure of BaGaF$_5$ and the layer structure of BaZnF$_4$. The crenelled strings of $(O_2)Mn_2F_9$ make the structure of this $(O_2)^+$ compound unique, too. The structure of Ba$_2$MnFeF$_9$

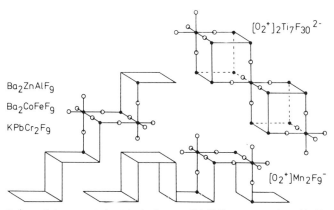

Fig. 41. Strings of cis-connected octahedra in some $(M_2F_9)^{n-}$ compounds and in $(O_2)_2Ti_7F_{30}$.

TABLE XXXVI

Monoclinic Compounds of $BaMnFeF_7$ (I), High-Temperature $BaZnFeF_7$ (II), and $BaMnGaF_7$ (III) Types

Compound	a (Å)	b (Å)	c (Å)	β (deg)	Type	Reference
$BaMnVF_7$	5.544	11.017	9.181	94.67	I	de Kozak and Samouel (1978)
$BaMnCrF_7$	5.514	10.978	9.192	94.58	I	de Kozak and Samouel (1978)
$BaMnFeF_7$ (*)	5.532	10.980	9.183	94.67	I	Holler et al. (1981); Samouel and de Kozak (1978)
$BaFeFeF_7$	5.47	10.93	9.10	94.83	I	Ravez et al. (1967)
Low-temperature $BaZnFeF_7$	5.439	10.841	9.082	94.88	I	de Kozak and Samouel (1978)
High-temperature $BaZnFeF_7$ (*)	5.603	9.971	9.584	92.80	II	Holler and Babel (1982)
$BaMnAlF_7$	13.653	5.206	14.461	91.52	III	de Kozak and Samouel (1981); A. de Kozak (personal communication, 1982); Holler (1983); Holler et al. (1984, 1985)
$BaCdAlF_7$	13.883	5.266	14.777	91.84	III	
$BaMnGaF_7$ (*)	13.808	5.308	14.688	91.13	III	
$BaCdGaF_7$ (*)	13.896	5.354	14.950	91.55	III	
High-pressure $BaMnVF_7$	13.799	5.342	14.776	91.03	III	
$BaCdVF_7$	13.853	5.400	15.043	91.50	III	
$BaCdCrF_7$	13.908	5.362	14.982	91.36	III	
High-pressure $BaMnFeF_7$	13.809	5.337	14.754	90.81	III	
$BaCdFeF_7$	13.852	5.390	15.023	91.27	III	
$Ba_2CaNiAl_2F_{14}$	13.623	5.216	14.654	91.97	III	Holler (1983); Holler et al. (1985)
$Ba_2CdMgAl_2F_{14}$	13.607	5.207	14.482	91.46	III	
$Ba_2CdMnAl_2F_{14}$	13.741	5.250	14.642	91.70	III	
$Ba_2CdFeAl_2F_{14}$	13.692	5.250	14.594	91.73	III	
$Ba_2CdCoAl_2F_{14}$	13.668	5.230	14.562	91.70	III	
$Ba_2CdNiAl_2F_{14}$	13.614	5.209	14.504	91.53	III	
$Ba_2CaCuGa_2F_{14}$ (*)	13.760	5.340	14.743	91.76	III	
$Ba_2CdCoGa_2F_{14}$	13.736	5.316	14.690	91.31	III	
$Ba_2CaCoV_2F_{14}$ (*)	13.697	5.384	13.916	91.49	III	
$Ba_2CdMnFe_2F_{14}$ (*)	13.812	5.372	14.895	91.11	III	

TABLE XXXVII

Tetragonal Compounds of $Ba_2M^{II}M^{III}F_9$ and $Ba_5Mn^{II}M_2^{III}F_{18}$ Types

Compound	a (Å)	c (Å)	Z	Reference
Ba_2MgAlF_9	20.521	15.082	32	Fleischer and Hoppe (1982f);
Ba_2MnGaF_9	20.631	15.283	32	de Kozak and Samouel (1978, 1981);
Ba_2MnVF_9	20.649	15.498	32	Samouel and de Kozak (1978)
Ba_2MnCrF_9	20.606	15.507	32	
Ba_2MgFeF_9	20.796	15.322	32	
Ba_2MnFeF_9	20.700	15.511	32	
Ba_2MgCoF_9	20.631	15.316	32	
Ba_2NiCoF_9	20.553	15.355	32	
Ba_2ZnCoF_9	20.517	15.398	32	
$Ba_5MnGa_2F_{18}$	20.515	15.403	16	de Kozak and Samouel
$Ba_5MnV_2F_{18}$	20.577	15.388	16	(1978, 1981)
$Ba_5MnCr_2F_{18}$	20.558	15.422	16	
$Ba_5MnFe_2F_{18}$	20.564	15.380	16	

and of some isomorphous tetragonal $Ba_2M^{II}M'^{III}F_9$ and $Ba_5Mn^{II}M_2'^{III}F_{18}$ compounds listed in Table XXXVII is not yet known but supposed to be related to those of β-$BaFeF_5$ (von der Mühll et al., 1971) and $Sr_3(FeF_6)_2$ (von der Mühll, 1974).

1. Orthorhombic Ba₂ZnAlF₉ Type

Zn(II) and Al(III) cations located in the octahedra forming the strings shown in Fig. 41 are statistically distributed in this structure [space group $Pnma$ (no. 62), $Z = 4$ (Fleischer and Hoppe, 1982e, 1982f)]. The bond lengths of the three terminal ligands at each cation are significantly shorter than those of the bridging ones. The bridge angles are 141 and 149°. The barium ions show 10 and 12 coordination. Isostructural compounds are listed in Table XXXVIII.

2. Monoclinic Ba₂CoFeF₉ Type

This structure [space group $P2_1/n$, $Z = 4$ (de Kozak et al., 1981; Fleischer and Hoppe, 1982e)] is a slightly distorted and more frequently observed variant of the Ba_2ZnAlF_9 structure. The main difference is the obviously fairly ordered distribution of cations. However, the differentiation between terminal and bridging bond lengths seems less significant in this structure. Anion bridging angles ($\simeq 143°$) are quite the same as in Ba_2ZnAlF_9. It is interesting that the difference in the coordination numbers 10 and 12 of the barium

TABLE XXXVIII

Compounds of Orthorhombic Ba$_2$ZnAlF$_9$ (I), Monoclinic Ba$_2$CoFeF$_9$ (II), and Orthorhombic KPbCr$_2$F$_9$ (III) Types

Compound	a (Å)	b (Å)	c (Å)	β (deg)	Type	Reference
Ba$_2$CoAlF$_9$	17.678	5.591	7.382	—	I	
Ba$_2$NiAlF$_9$	17.625	5.556	7.360	—	I	Fleischer and Hoppe (1982e,f)
Ba$_2$ZnAlF$_9$ (*)	17.841	5.542	7.362	—	I	
Ba$_2$NiGaF$_9$	17.713	5.616	7.417	—	I	
Ba$_2$ZnGaF$_9$	7.427	17.884	5.618	90.34	II	
Ba$_2$FeVF$_9$	7.604	17.661	5.745	90.56	II	
Ba$_2$CoVF$_9$	7.534	17.795	5.728	90.95	II	Fleischer and Hoppe (1982e,f)
Ba$_2$NiVF$_9$	7.488	17.791	5.685	90.63	II	
Ba$_2$ZnVF$_9$	7.504	17.920	5.692	90.72	II	
Ba$_2$FeCrF$_9$	7.503	17.700	5.676	90.71	II	
Ba$_2$CoCrF$_9$	7.467	17.756	5.667	90.73	II	
Ba$_2$NiCrF$_9$	7.423	17.721	5.633	90.55	II	de Kozak and Samouel (1977)
Ba$_2$ZnCrF$_9$	7.418	17.885	5.617	90.46	II	
Ba$_2$FeFeF$_9$	7.602	17.608	5.762	90.86	II	
Ba$_2$CoFeF$_9$ (*)	7.486	17.757	5.687	90.87	II	de Kozak et al. (1981)
Ba$_2$NiFeF$_9$	7.445	17.731	5.654	90.62	II	de Kozak and Samouel (1977)
Ba$_2$ZnFeF$_9$	7.508	17.882	5.687	90.78	II	
Ba$_2$NiNiF$_9$	7.396	17.590	5.631	90.13	II	Fleischer and Hoppe (1982f)
Ba$_2$ZnNiF$_9$	7.393	17.860	5.594	90.15	II	
NaBaGa$_2$F$_9$	7.315	17.265	5.364	91.13	II	
NaBaV$_2$F$_9$	7.372	17.555	5.491	91.60	II	
NaBaCr$_2$F$_9$	7.318	17.328	5.406	91.09	II	
NaBaFe$_2$F$_9$	7.371	17.533	5.475	91.66	II	de Kozak et al. (1982)
KBaGa$_2$F$_9$	7.460	17.468	5.517	91.08	II	
KBaV$_2$F$_9$	7.520	17.695	5.601	91.27	II	
KBaCr$_2$F$_9$	7.481	17.523	5.536	91.21	II	
KBaFe$_2$F$_9$	7.515	17.687	5.615	91.09	II	
KPbCr$_2$F$_9$ (*)	9.812	5.412	13.93	—	III	Vlasse et al. (1982)

ions could be successfully applied to prepare some obviously isostructural compounds, $NaBaM_2F_9$ and $KBaM_2F_9$ (de Kozak et al., 1982), which are also listed in Table XXXVIII.

3. KPbCr₂F₉ Type

$KPbCr_2F_9$ [orthorhombic, space group $Pnma$ (no. 62), $Z = 4$ (Vlasse et al., 1982)], which is isostructural with $K_2Ta_2O_3F_6$ (Chaminade et al., 1974), deviates in cell symmetry and dimensions from the $A^IBaM_2^{III}F_9$ compounds just mentioned (see Table XXXVIII). Although it crystallizes in the same space group as Ba_2ZnAlF_9 and its $(M_2F_9)^{n-}$ strings run in the same direction of the shortest b axis, the relative orientation of the strings and the distance between them is different. The difference is also reflected by a change in CN for the larger cations, which becomes 9 for both potassium and lead. In addition, the bridging angles Cr–F–Cr (152 and 159°) are less bent in $KPbCr_2F_9$ than in the other structural types, showing such stair-related strings (see Fig. 41).

4. Structures of the Dioxygenyl Compounds $(O_2)Mn_2F_9$ and $(O_2)_2Ti_7F_{30}$

The cell data of these two dioxygenyl compounds (Müller, 1981b,c) are given in Table XXXIX. The linking type of octahedra, as already mentioned for $(O_2)Mn_2F_9$, results in a crenelled string (see Fig. 41). The angles at the bridging anions lie between 140 and 145°. The difference between bridging and terminal bond lengths is more pronounced here and reaches more than 0.1 Å, as would be expected for a high-valency cation such as Mn(IV).

In spite of its strange structure and composition ($F/Ti = 4.286 = 1 \times 3 + 6 \times 4.5$), $(O_2)_2Ti_7F_{30}$ has some related features (Müller, 1981c). As can be seen from Fig. 41 the string present in $(O_2)Mn_2F_9$ requires only a shift of some connecting squares to form a row of separated cubes. Such cubical units are found in $(O_2)_2Ti_7F_{30}$. In the structure the cubes are rotated and are connected by diagonally opposite corners, as shown in Fig. 41. The resulting columns of composition $(Ti_7F_{30})^{2-}$ are directed along the c axis. The $(Ti_7F_{30})^{2-}$ stoichiometry results from the addition of $2/2$ $TiF_{6/2}$ linking octahedra (i.e., TiF_3 per cube), to six $TiF_3F_{3/2}$ (i.e., six $TiF_{4.5}$) being at the remaining corners. The cube distortion of the eight titanium atoms is less than 4° from right angles, but Ti–F–Ti bridges deviate up to 20° from linearity. Each of the six $TiF_3F_{3/2}$ octahedra surrounding the string direction has three cis terminal ligands, which constitute the peripheral faces. The shortening of the terminal bonds of about 0.2 Å compared with the bridging ones is even more pronounced than in $(O_2)Mn_2F_9$.

TABLE XXXIX

Crystal Data of the Dioxygenyl Compounds $[O_2^+]Mn_2F_9^-$ and $[O_2^+]_2Ti_7F_{30}^{2-}$, of $A_nM_3^{II}F_{10}$, and of Some Additional Fluorides of Odd Composition

Compound	Space Group	Z	a (Å)	b (Å)	c (Å)	β (deg)	Reference
$[O_2^+]Mn_2F_9$ (*)	$C2/c$	8	17.552	8.373	9.101	102.3	Müller (1981b)
$[O_2^+]_2Ti_7F_{30}^{2-}$ (*)	$P\bar{3}$	1	10.192	(Trigonal)	6.500	—	Müller (1981c)
$Ba_2Co_3F_{10}$	$C2/m$	4	18.54	6.02	7.75	111.8	a-1913
$Ba_2Ni_3F_{10}$ (*)			18.542	5.958	7.821	111.92	a-1947; Leblanc et al. (1980)
$Ba_2Zn_3F_{10}$			18.78	6.07	7.91	112.3	a-606
$Cs_4Mg_3F_{10}$ (*)			6.133	14.561	13.653	—	a-575; Steinfink and Brunton (1969)
$Cs_4Fe_3F_{10}$			6.26	14.4	14.0	(90.3?)	a-1831B
$Cs_4Co_3F_{10}$ (*)	$Cmca$	4	6.214	14.513	13.892	—	R. Schmidt and D. Babel (unpublished work, 1983)
$Cs_4Ni_3F_{10}$ (*)			6.158	14.514	13.736	—	Schmidt and Babel (1983)
$Cs_4Zn_3F_{10}$			6.23	14.65	13.90	—	a-599
$Cs_6Ni_5F_{16}$ (*)	$Cmca$	4	6.184	14.555	21.451	—	Schmidt (1982); Schmidt and Babel (1984)
$Cs_7Ni_4F_{15}$ (*)	$P2_1/c$	2	7.872	10.897	11.495	92.74	Schmidt and Babel (1983, 1985)
$Cs_7Co_4F_{15}$ (*)	$P2_1/c$	2	7.883	10.966	11.649	92.59	Schmidt and Babel (1985)
$Cs_4MgCr_4F_{18}$	$P\bar{3}m1$	1	7.169	(Trigonal)	10.744	—	Courbion et al. (1983b)
$Cs_4Cr_5F_{18.24}$ (*)			7.200		10.679	—	Courbion et al. (1983b)
$Cs_4CoCr_4F_{18}$ (*)			7.203		10.761	—	Courbion et al. (1983b)
$Cs_4NiCr_4F_{18}$			7.181		10.725	—	Courbion et al. (1983b)

D. STRUCTURES OF $Ba_2Ni_3F_{10}$, $Cs_4Ni_3F_{10}$, AND OTHER COMPOUNDS

Cell constants of the structures of $Ba_2Ni_3F_{10}$, $Cs_4Ni_3F_{10}$, $Cs_6Ni_5F_{16}$, $Cs_7Ni_4F_{15}$, and $Cs_4CoCr_4F_{18}$ and some isostructural compounds are given in Table XXXIX.

1. $Ba_2Ni_3F_{10}$

Many $A^IM^{III}_3F_{10}$ fluorides are known, M(III) being a rare-earth ion (Aléonard et al., 1976; Podberezskaya et al., 1976). Their structures contain M(III) ions with CN > 6. In contrast, the $(M_3F_{10})^{n-}$ composition in $Ba_2Ni_3F_{10}$ (Leblanc et al., 1980) leads to a 3D framework, which is built up from octahedra only. Parallel rutile-like chains of edge-sharing octahedra are connected in a complex manner via corners of "satellite" groups: an edge-sharing double group and a single octahedron, as shown in Fig. 42.

2. $Cs_4Ni_3F_{10}$ and $Cs_4Mg_3F_{10}$

The layer structure independently determined for $Cs_4Ni_3F_{10}$ (Babel, 1968; Schmidt and Babel, 1983) and $Cs_4Mg_3F_{10}$ (Steinfink and Brunton, 1969) was first detected in an oxide, $Ba_4(Ti, Pt)_3O_{10}$ (Blattner et al., 1948; Fischer and Tillmanns, 1981). Quite different from the $Ba_2Ni_3F_{10}$ structure, there are units of three face-shared octahedra, just like those found in the 9L perovskites. These units, however, are linked here only by two corners at each end in order to form the puckered layer shown in Fig. 43.

3. $Cs_6Ni_5F_{16}$, $Cs_7Ni_4F_{15}$, and $Cs_4CoCr_4F_{18}$

The structure of a compound $Cs_6Ni_5F_{16}$ has been determined (Schmidt and Babel, 1983, 1984) in which units of five face-shared octahedra form quite similar layers as those of three do in $Cs_4Ni_3F_{10}$ (Fig. 44). A layer structure made up of double groups of face-shared octahedra was found in $Cs_7Ni_4F_{15}$ (Fig. 45) (Schmidt and Babel, 1983). As well as in $Cs_4Ni_3F_{10}$, there are nonbridging anions in these compounds, which show considerably shortened Ni–F distances. The difference of about 0.1 Å compared to the bridging anions seems more pronounced than those sometimes found for higher-valency cations. The triple layer structure of $Cs_4CoCr_4F_{18}$ (Courbion et al., 1983b), which derives from the pyrochlore structure, is shown in Fig. 46 (p. 176). The central layer contains mixed cations—Co(II) and Cr(III)—but in the neighboring octahedra at the upper and lower side, only Cr(III) ions are present, which show three terminal bonds, about 0.07 Å shorter than the bridging ones. In this context it should be noted that an isolated unit deriving

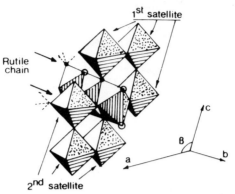

Fig. 42. Part of octahedral network in the $Ba_2Ni_3F_{10}$ structure. From Leblanc *et al.* (1980).

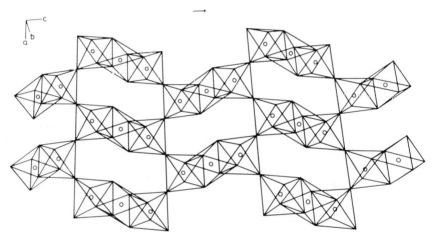

Fig. 43. Layer structure of $Cs_4Ni_3F_{10}$.

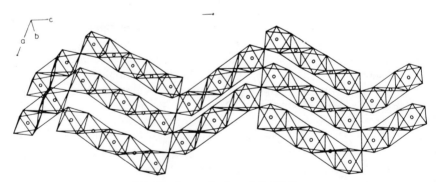

Fig. 44. Layer structure of $Cs_6Ni_5F_{16}$.

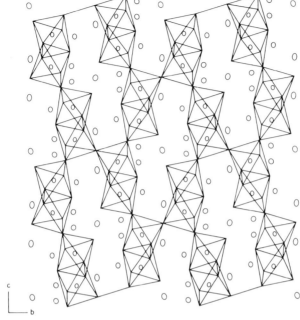

Fig. 45. Layer structure of $Cs_7Ni_4F_{15}$.

from the pyrochlore structure (see Figs. 10 and 18) has been found in the oxide–fluoride anion $(Ta_4O_6F_{12})^{4-}$ with oxygen bridging the tetrahedrally arranged tantalum atoms (Sala-Pala *et al.*, 1982).

VII. Special Structures and Behavior

A. STRUCTURES CONTAINING "INDEPENDENT" FLUORIDE IONS

In most of the structures that have been previously described, F^- ions are directly coordinated to the M^{n+} cation to form the MF_6 octahedra constituting the 3D, layer, chain, or isolated unit network. In this section we deal with compounds in which some F^- ions do not belong to the octahedral coordination sphere and may be thus called "independent" (or "lone" or "single") fluoride ions. Only those compounds corresponding to a definite stoichiometry are considered; excluded here are intercalation compounds and structures in which F^- can be progressively introduced either in interstitial positions or in vacant sites [e.g., ReO_3-related phases: MF_{3+x},

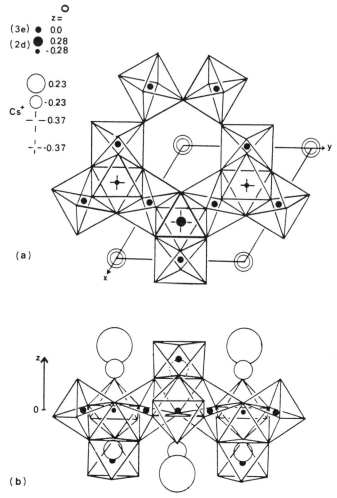

Fig. 46. Pyrochlore-related triple layer structure of $Cs_4CoCr_4F_{18}$: (a) [001] projection; (b) [210] projection.

$Ca_{2-x}Zr_xF_{4+2x}$; fluorite-related phases: $M^{II}_{1-x}M'^{IV}_xF_{2+2x}$, $M^I_{1-x}M'^{III}_xF_{1+2x}$, $M^{II}_{1-x}M'^{III}_xF_{2+x}$; tysonite-related series: $La_{1-x}M^{IV}_xF_{3+x}$; $(NH_4)_3FeF_6$-related phases: $(NH_4)_3ZrF_{6+x}$].

1. Anhydrous Fluorides

In anhydrous fluorides that exhibit a network of MF_6 octahedra, the presence of independent F^- is still an unusual phenomenon. It has already been mentioned in Section III,C that compounds of Ba_2ZnF_6 type and related structures may be written $(BaF)_2(ZnF_4)$ in order to point out the presence of layers of barium and fluorine atoms between 2D layers of composition $(ZnF_4)_n^{2n-}$ (see Fig. 27; Table XXIX). A few series of fluorides, oxide–fluorides, and thiofluorides $NaA^{II}M_2^{II}F_7$ and $A_2^{II}M_2^{II}F_6X$ (X = O or S) have been shown (Table XXI) to crystallize in the pyrochlore structure ($NaCaNb_2O_6F$ type). This structure may be described as being formed of a 3D network of octahedra linked by corner sharing and therefore showing an M_2F_6 formulation (see Fig. 18). The seventh anion is not directly connected to the M(II) elements, but it is tetrahedrally surrounded by the remaining A cations (sodium and calcium)

a. K_3SiF_7 TYPE. Table XL lists the cell parameters of phases crystalliz-ing in the K_3SiF_7 [tetragonal, space group $P4/mbm$ (no. 127), $Z = 2$)], or alternatively $K_2SiF_6 \cdot KF$ type (Deadmore and Bradley, 1962). In this structure SiF_6 octahedra are isolated by a network of A^+ and F^-. Independent F^- ions are surrounded by six A^+ at a distance of about 2.9 Å. The structure consists of a tetragonal framework of composition A_2SiF_6 with additional strings of alternating A^+ and F^-.

b. $Sr_3Fe_2F_{12}$ TYPE. The structure of $Sr_3Fe_2F_{12}$ [tetragonal, space group $I4_1$ (no. 80), $Z = 24$ (Ravez et al., 1975)], which was obtained after several annealings, derives from that of $BaFeF_5$ (Section IV,B). Single chains $(FeF_5)_n$ are maintained, but the central chain of the branched strings of formulation $(Fe_3F_{15})_n$ are no longer present in $Sr_3Fe_2F_{12}$. They are sub-stituted, giving rise to single octahedra. Half of the remaining empty sites are filled with one strontium atom and two independent fluorine atoms (von der Mühll, 1974; Ravez et al., 1975; and Fig. 47). The strontium atoms are coordinated to eight fluorine atoms: four independent fluorine atoms and four others belonging to the isolated FeF_6 octahedra. The occurrence of a related structure has been proposed for $A_5^{II}M_3^{III}F_{19}$ phases, with $(M_3F_{15})_n$ branched chains totally substituted by chains of composition $(AM_2F_{14})_n$ (von der Mühll, 1973). The determination of the structure of $Pb_5Fe_3F_{19}$ (tetragonal, space group $I4cm$) has confirmed this proposal (Jacoboni et al., 1983) (Table XL).

c. Ca_2AlF_7 TYPE. In this structure [orthorhombic, space group $Pnma$ (no. 62), $Z = 4$], isolated AlF_6 octahedra are present, but here independent fluorine atoms are coordinated to three calcium atoms in a plane with Ca–F distances ranging from 2.22 to 2.30 Å (Domesle and Hoppe, 1980a) (Fig. 48). The structure may be viewed alternatively as a 3D network with Ca_2F_7 general formula. Two types of CaF_7 polyhedra are present: distorted pentagonal bipyramids and strongly distorted monocapped trigonal prisms, aluminum atoms occupying sites with CN = 6 between them.

TABLE XL

Lattice Constants of Some Structural Types Exhibiting "Independent" Fluoride Ions

Compound	a (Å)	b (Å)	c (Å)	Reference
K_3SiF_7 (*)	7.740(4)	—	5.564(3)	a-1215
$(NH_4)_3SiF_7$ (*)	8.04(1)	—	5.845(10)	a-1219
Cs_3SiF_7	8.306	—	6.170 ⎫	
Rb_3SiF_7	7.959	—	5.823 ⎪	
Cs_2RbSiF_7	8.198	—	6.019 ⎪	
Cs_2KSiF_7	8.115	—	5.972 ⎬	Hoppe (1981)
Rb_2CsSiF_7	8.099	—	5.899 ⎪	
Rb_2KSiF_7	7.883	—	5.724 ⎭	
$(NH_4)_3TiF_7$	8.33(1)	—	5.96(1)	a-1312
Cs_3TiF_7	8.473	—	6.313 ⎫	
Rb_3TiF_7	8.202	—	5.979 ⎪	
Cs_3CrF_7	8.390	—	6.247 ⎪	
Rb_3CrF_7	8.084	—	5.902 ⎪	
Cs_3MnF_7	8.369	—	6.233 ⎬	Hoppe (1981)
Rb_3MnF_7	8.050	—	5.890 ⎪	
Cs_3NiF_7	8.307	—	6.192 ⎪	
Rb_3NiF_7	7.978	—	5.857 ⎪	
K_3MnF_7	11.146	11.005	5.631 ⎭	
$Sr_3Fe_2F_{12}$ (*)	20.338	—	14.668	von der Mühll (1974)
$Sr_5Ga_3F_{19}$	14.22	—	7.272	a-737
$Sr_5Cr_3F_{19}$	14.24	—	7.290	Ravez et al. (1975)
$Sr_5Co_3F_{19}$	14.241	—	7.296	a-1909
$Pb_5Fe_3F_{19}$ (*)	14.48	—	7.46	Jacoboni et al. (1983)
Ca_2AlF_7 (*)	7.685	6.998	9.549	Domesle and Hoppe (1978)
Ca_2CrF_7	7.81	7.09	9.70	Dumora and Ravez (1969)
Ca_2RhF_7	7.871	7.147	9.780	Domesle and Hoppe (1978)
Sr_2RhF_7 (*)	5.56_9	11.85_4	8.83_2 ($\beta = 91.00°$)	Domesle and Hoppe (1983)
Pb_2RhF_7	5.51_9	11.66_8	8.63_5 ($\beta = 90.94°$)	Domesle and Hoppe (1983)
Sr_2InF_7 (*)	5.466(1)	12.243(1)	8.255(1) ($\beta = 90.54°$)	Scheffler and Hoppe (1984)

d. Pb_2RhF_7 TYPE. The CN of the d cation in Pb_2RhF_7 [monoclinic, space group $P2_1/c$ (no. 14), $Z = 4$] is different from that of K_2NbF_7, although both phases have similar cell constants (Domesle and Hoppe, 1983; and Table XL). Whereas in K_2NbF_7 niobium atoms form CN = 7 monocapped trigonal prisms, in Pb_2RhF_7 rhodium atoms occupy the center of distorted octahedra separated by independent fluorine atoms. These are coordinated to three lead atoms, as in Ca_2AlF_7. Such phases can be alternatively written A_2F (MF_6).

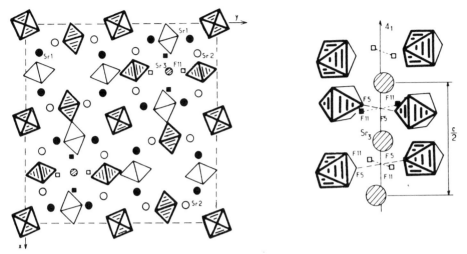

Fig. 47. Structure of $Sr_3Fe_2F_{12}$ ($\bigcirc$, $\bullet$, $\oslash$, Sr; $\square$, $\blacksquare$, F). From von der Mühll (1974).

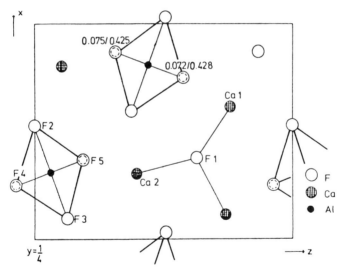

Fig. 48. Projection of the structure of Ca_2AlF_7 on the (101) plane. From Domesle and Hoppe (1978).

2. Hydrated Metal Fluorides

In hydrated fluorides, the presence of H—O—$H \cdots F$ hydrogen bonds may stabilize units in which F^- ions are disconnected from the network of $M(H_2O,F)_6$ octahedra (Table XLI). Short distances, $O \cdots F$ with an average of 2.56 Å, are observed in this case (Massa, 1982b).

TABLE XLI

Hydrated Fluorides with "Independent" Fluoride Ions

Compound	Space Group	Z	a (Å)	b (Å)	c (Å)	β (deg)	Reference
$CrF_3 \cdot 9H_2O$ (*)	$R3$	3	10.837(9)	—	8.157(3)	—	Epple and Massa (1978)
$(NH_4)_2CrF_5 \cdot 6H_2O$ (*)	$C2/c$	4	11.997(1)	6.928(1)	13.574(2)	90.0	Massa (1977b)
$NH_4CuSiF_7 \cdot 4H_2O$ (*)	$P4/n$	2	7.560(5)	—	8.065(5)	—	De Cian et al. (1967)
$NH_4CuTiF_7 \cdot 4H_2O$			7.671(4)	—	8.271(5)	—	De Cian et al. (1967)
$NH_4CuSnF_7 \cdot 4H_2O$			7.743(5)	—	8.422(7)	—	De Cian et al. (1967)
$Ba_4Fe_3F_{17} \cdot 3H_2O$ (*)	$Imma$	4	10.252	12.295	13.547	—	Massa and Pebler (1983)

Figure 49 shows the unique structure of $CrF_3 \cdot 9H_2O$, that is, $[Cr(H_2O)_6][F_3(H_2O)_3]$, in which puckered $[F_3(H_2O)_3]^{3-}$ rings connect the (red) hexa-aquo cations (O—H$\cdots$F distances = 2.61–2.65 Å; Epple and Massa, 1978). In the $(NH_4)_2[Cr(H_2O)_6]F_5$ structure, $[Cr(H_2O)_6]^{3+}$ octahedra are linked to each other by a 3D network of NH_4^+ and F^- with tetrahedral coordinations (Massa, 1977b). Three different tetrahedral groups have been found: $NH_4(F_4)$, $F(NH_4(H_2O)_3)$, and $F((H_2O)_2(NH_4)_2)$. In this case again, the network of short —O—H$\cdots$F$\cdots$H—O— bonds (2.53–2.57 Å) stabilizes the $Cr(H_2O)_6$ complex despite the presence of "independent" fluoride ions.

In the $(NH_4)CuMF_7 \cdot H_2O$ series (M = Si, Ti, Sn), chains built up of elongated $Cu(H_2O)_4F_2$ octahedra and nearly regular TiF_6 octahedra connected by trans corners are separated by NH_4^+ and F^- (De Cian et al., 1967). Independent F^- ions are tetrahedrally coordinated to four H_2O molecules with short O—H$\cdots$F distances (2.52 Å) (Massa, 1977b).

The structure of $Ba_4Fe_3F_{17} \cdot 3H_2O$ shows isolated FeF_6 octahedra and chains of FeF_6 units connected in turn by cis and trans corners (Massa and Pebler, 1983). These groups are isolated from each other by $[F(H_2O)_3]^-$ anions, which form disordered four-membered rings.

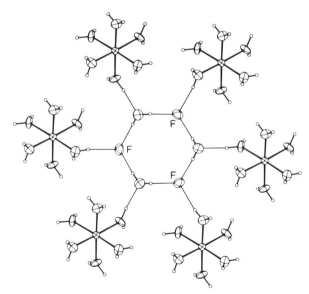

Fig. 49. The $(F \cdots HOH)_3^{3-}$ ring linking six $[Cr(H_2O)_6]^{3+}$ octahedra in $[Cr(H_2O)_6]F_3 \cdot 3H_2O$.

B.   HYDRATED METAL FLUORIDES

Numerous hydrated metal fluorides are known. Their structures are of interest not only because of the HO—H $\cdots$ F hydrogen bridges (Simonov and Bukvetskii, 1978), but also because water molecules may compete with fluoride ions to act as ligands (Avkhutskii et al., 1979). Some cases, just mentioned, lead to "independent" fluoride ions. In the context of structures possessing linked octahedra, the most important $A_nMF_x \cdot y (H_2O)$ hydrates are those in which $x + y$ is less than 6 or in which ligand-sharing occurs in spite of $x + y \geq 6$. It is beyond the scope of this chapter to give more than some selected examples. Most are chosen from the work of Massa (1982b).

1. $(MF_6)^{n-}$ Coordination and Noncoordinating "Crystal Water"

Hydrates such as $Li_2MF_6 \cdot 2H_2O$ (M = Ti or Sn) (Marseglia and Brown, 1973) are less interesting in the context of this chapter because of the presence of isolated $(MF_6)^{n-}$ octahedra. Table XLII lists several compounds in which the water molecules do not enter the M coordination sphere, in spite of a ratio F/M < 6. Therefore, in contrast to isolated $[FeF_5(H_2O)]^{2-}$ octahedra in $K_2FeF_5 \cdot H_2O$ (Edwards, 1972), $(MF_5)_n^{2n-}$ chains are observed, as discussed earlier for nonhydrated compounds. Long hydrogen bonds, with average $O \cdots F$ distances of 2.79 Å, occur in this class of hydrates (Massa, 1982b). We may quote also the extremely long $O \cdots F$ distance (3.09 Å) in the HTB structure of $(H_2O)_{0.33}FeF_3$ (Leblanc et al., 1983b) already mentioned. One should actually speak of "zeolitic water" in this case. Reversible dehydration is possible in the pyrochlore $KNiCrF_6 \cdot H_2O$, in which $O \cdots F > 3.3$ Å and where only short K—O bonds fix the water molecules (Babel, 1972).

2.   Hydrates Exhibiting Mixed F^-/H_2O Coordination $[MF_x(H_2O)_{6-x}]^{n-}$

In most hydrates, F^- ions as well as water molecules are coordinated to the M cations. In this class of materials the hydrogen bond distance is intermediate, with an average $O \cdots F$ distance of 2.69 Å (Massa, 1982b). The trans-aquo coordination found in compounds such as $RbVF_4 \cdot 2H_2O$ (Bukvetskii et al., 1976) and $CsMnF_4 \cdot 2H_2O$ (Bukovec and Kaucic, 1977; Dubler et al., 1977) is also observed in the monohydrate $RbMnF_4 \cdot H_2O$ (Kaucic and Bukovec, 1979). The latter compound contains in turn $[MnF_6]^{3-}$ and $[MnF_4(H_2O)_2]^-$ octahedra sharing fluorine corners. This hydrate belongs, therefore, to the class of chain compounds listed in Table XLII. Some hydrates exhibiting a 3D connection of octahedra are listed in Table XLIII. The mixed-valency iron compound $Fe_2F_5 \cdot 2H_2O$ (Hall et al., 1977) has

TABLE XLII

Hydrates Containing Chains of Fluoro-Bridged Octahedra

Hydrate	Space Group	Z	a (Å)	b (Å)	c (Å)	β (deg)	Reference
$K_2AlF_5 \cdot H_2O$ (*)	$Cmcm$	4	9.19	8.11	7.45	—	Brosset (1942)
$Rb_2AlF_5 \cdot H_2O$ (*)	$Cmcm$	4	9.604	8.378	7.542	—	Fourquet et al. (1981a)
$Tl_2AlF_5 \cdot H_2O$	$Cmcm$	4	10.06	8.24	7.47	—	Fourquet et al. (1981a)
$Hg_2AlF_5 \cdot 2H_2O$ (*)	$I4cm$	4	9.353	—	7.241	—	Fourquet et al. (1981b)
$HTiF_5 \cdot H_2O$ (*)	$C2/c$	8	14.528	4.839	13.798	115.59	Cohen et al. (1982)
$K_2MnF_5 \cdot H_2O$ (*)	$P2_1/m$	2	6.04	8.20	5.94	96.5	Edwards (1971)
$Rb_2MnF_5 \cdot H_2O$ (*)	$Cmcm$	4	9.383	8.214	8.348	—	Bukovec and Kaucic (1978)
$Cs_2MnF_5 \cdot H_2O$ (*)	$Cmmm$	2	9.727	8.668	4.254	—	Kaucic and Bukovec (1978)
$RbMnF_4 \cdot H_2O$ (*)	$C2/c$	8	13.932	6.471	10.635	104.54	Kaucic and Bukovec (1979)
$SrMnF_5 \cdot H_2O$ (*)	$P2_1/m$	2	5.108	7.920	6.106	110.24	Massa (1983)
$BaMnF_5 \cdot H_2O$ (*)	$P2_1/m$	2	5.370	8.172	6.280	111.17	Massa (1983)
β-$Rb_2FeF_5 \cdot H_2O$	$Cmcm$	4	9.691	8.446	7.947	—	Fourquet et al. (1981a, 1982)

TABLE XLIII

Hydrates with Framework Structures of $Fe_2F_5 \cdot 2H_2O$ and $Fe_3F_8 \cdot 2H_2O$ Types

Hydrate	Space Group	Z	a (Å)	b (Å)	c (Å)	β (deg)	Reference
$Fe_2F_5 \cdot 2H_2O$ (*)	$Imma$	4	7.489	10.897	6.671	—	Hall et al. (1977)
$MnScF_5 \cdot 2H_2O$	$Immm$	4	7.745	11.231	6.795	—	E. Herdtweck (unpublished work, 1981)
$Fe_3F_8 \cdot 2H_2O$ (*)	$C2/m$	2	7.612	7.500	7.469	118.38	Herdtweck (1983a)
$MnFe_2F_8 \cdot 2H_2O$ (*)	$C2/m$	2	7.602	7.519	7.465	118.09	Herdtweck (1983b)
$CoAl_2F_8 \cdot 2H_2O$ (*)	$C2/m$	2	7.211	7.157	7.146	117.92	Herdtweck (1983b)

a structure closely related to weberite structure but with an interchange of M(II) and M(III) positions. $(FeF_6)^{3-}$ octahedra are trans connected into $(FeF_5)_n^{2n-}$ chains. Those chains are further cross-linked by four fluorine atoms to trans-dihydrated $[Fe(H_2O)_2]^{2+}$ cations (cf. Fig. 16). In the structure of $Fe_3F_8 \cdot 2\,H_2O$, there are two $(FeF_4)_n^{n-}$ layers instead of $(FeF_5)_n^{2n-1}$ chains, which are connected by $[Fe(H_2O)_2]^{2+}$ cations in a similar way (Herdtweck, 1983a,b).

C. JAHN–TELLER DISTORTED STRUCTURES

The Jahn–Teller effect (Jahn and Teller, 1937; Reinen and Friebel, 1979) occurring, for instance, in octahedrally coordinated d^4 and d^9 configuration cations, generally leads to tetragonal (or orthorhombic) elongation of the $(MF_6)^{n-}$ octahedra. Examples have been discussed or are given in previous tables, especially for compounds of Cr(II), Mn(III), and Cu(II). Space groups with lower symmetry compared with "normal'" compounds are often observed as a consequence of the Jahn–Teller effect (Reinen, 1979).

It has been already mentioned that the modified pyrochlore structure of the cubic $RbNiCrF_6$ type seems insensitive to the Jahn–Teller effect, at least for the cell symmetry observable by current diffraction methods. However, detailed structure analysis in similar cases has shown that the Jahn–Teller distortion can be obscured by orientation disordering (Mullen et al., 1975). This is reflected by anomalies in the anisotropic temperature factors of the ligands. Concerning the Mn(III) compounds $K_2(K_{0.2}Li_{0.8})MnF_6$ (cubic elpasolite) (Massa, 1978, 1982b), $\gamma\text{-}Cs_2NaMnF_6$ (12L elpasolite; Massa, 1982a), and $CsMnF_4$ ($CsFeF_4$ type; Massa and Steiner, 1980), the thermal ellipsoids either of all or of some ligands elongate along the bond direction, compared with the flattening observed in isostructural Fe(III) compounds. This can be explained by a disorder of long and short axes in Jahn–Teller systems. Spectroscopic methods are more powerful for detecting local Jahn–Teller distortions, if such distortions do not lead to long-range order and cell symmetry reduction (Reinen and Friebel, 1979). Examples are given in Chapter 18, including an esr study of $Ba_2CaCuGa_2F_{14}$ (Friebel et al., 1985), which crystallizes in the undistorted $BaMnGaF_7$ structure but nevertheless exhibits significant shifts of anion positions (Holler, 1983).

Another question concerns the orientation of long and short octahedral axes with respect to the neighboring octahedra (i.e., the Jahn–Teller ordering, which may be either ferro- or antiferrodistortive). Following proposals of Massa (1982b, 1983), which are derived from bond strength calculations (Allman, 1975), in chain and layer structures the long axes of Jahn–Teller distorted octahedra preferentially orientate along the direction in which the linking of octahedra occurs. The preference of chain structures in Mn(III) systems that exhibit trans connection of octahedra via the long axes serve as an example of the important influence of the Jahn–Teller effect on the selection of an appropriate structural type (Massa, 1982b, 1983).

D. STRUCTURES WITH UNUSUAL COORDINATION NUMBERS

Most previously described structures are characterized by 3D, 2D, or 1D networks of MF_6 octahedra. We have selected here some examples of compounds that are related to these phases either by the formula or by the structural arrangements and the cations of which have CN < 6 due to electronic or size conditions. Some Mn(II) compounds of CN > 6 have already been mentioned.

1. Square-Planar Coordination

Although the term *square-planar* coordination has sometimes been used for networks of axially elongated octahedra (e.g., Jahn–Teller distortions), it would be better reserved for structures that can be described only in terms of linked MF_4 groups.

a. AuF_3 STRUCTURE: HEXAGONAL, SPACE GROUP $P6_122$ (NO. 178), $Z = 6$, $a = 5.149$ Å, $c = 16.26$ Å (a-25). In this peculiar structure, each AuF_4 square plane is linked to two adjacent units by cis fluorine bridges, giving an infinite hexagonal helix ($Au–F_{bridg} = 2.04$ Å, $Au–F_{term} = 1.91$ Å). Another Au–F distance of 2.69 Å completes a tetragonally elongated octahedral configuration (Einstein *et al.*, 1967; and Fig. 50). This arrangement is due to the low-spin configuration (d_{xz}^2, d_{yz}^2, $d_{z^2}^2$, d_{xy}^2) of Au(III), which induces a very strong Jahn–Teller distortion.

b. $KBrF_4$ TYPE: TETRAGONAL, SPACE GROUP $I4/mcm$ (NO. 140), $Z = 4$. Many $A^I M^{III} F_4$ and $A^{II} M^{II} F_4$ compounds crystallize in the $KBrF_4$ type (Sly and March, 1957). This structure can be derived from the fluorite type by a cationic ordering, yielding a doubling of the c axis. Whereas cations show an fcc stacking, an important shift of the anionic sublattice defines two types of environment for the cations: layers of AF_8 square antiprisms connected by edges at $z = \frac{1}{4}$ and $\frac{3}{4}$ and MF_4 square planes at $z = 0$ and $\frac{1}{2}$, four more remote anions giving a complete CN = 4 + 4 around M cations (Fig. 51).

In d transition element series, this type of structure is consistent with Jahn–Teller configurations, but in $KClF_4$ and $KBrF_4$ it is due to the presence of unshared electronic pairs E occupying the apical positions of the $X_4 E_2$ coordination octahedron. Lattice constants of $KBrF_4$-type compounds are given in Table XLIV.

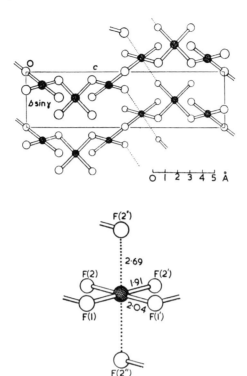

Fig. 50. Structure of AuF_3. From Einstein *et al.* (1967).

Sr_2CuF_6 also derives from the fluorite. Cations form an fcc lattice with one layer of CuF_4 square planes isolated by two layers of SrF_8 polyhedra (von Schnering *et al.*, 1971) (Fig. 51). The relations between Sr_2CuF_6 and $SrCuF_4$ may be viewed as homologous to those between K_2NiF_4 and perovskite structures.

c. STRUCTURES CONTAINING PALLADIUM(II). Connections between $CsPd_2F_5$ and pyrochlore structures have been delineated in Section II,C. An ordering of palladium atoms occurs in $CsPd_2F_5$, and half of Pd(II) species are arranged in chains of corner-shared PdF_6 octahedra. These chains are linked by PdF_4 square planes (Müller, 1982a).

In the Rb_3PdF_5-type structure [tetragonal, space group $P4/mbm$ (no. 127), $Z = 2$, $a = 7.467$ Å, $c = 6.497$ Å] PdF_4 square planes are isolated from each other by rubidium atoms with CN = 10. These layers are separated along the c axis by other rubidium atoms with CN = 8. Not all fluorine atoms are directly bound to palladium (Müller, 1982a). In Rb_2CsPdF_5, cesium atoms adopt the interlayer sites with the larger CN = 10.

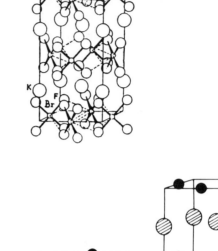

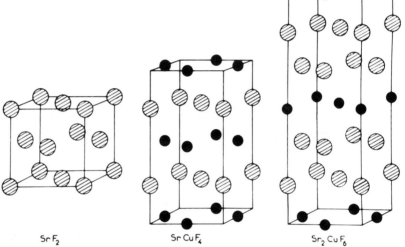

Fig. 51. KBrF$_4$ structure (above) and cationic orderings in SrF$_2$, SrCuF$_4$ (KBrF$_4$ type), and Sr$_2$CuF$_6$ (below) (●, Cu; ◍, Sr).

2. Tetrahedral Coordination

We have previously noted that in many fluorides of LiMF$_6$, LiMM'F$_6$, Li$_2$MF$_6$, and Li$_3$MF$_6$ formulations, lithium atoms have an octahedral coordination. In several structural types, however, the presence of lithium in tetrahedral sites induces peculiar structures. As indicated in the introduction of this chapter, we will not consider here compounds such as fluoroberyllates or fluoroborates that are basically characterized by a framework of MF$_4$ tetrahedra.

TABLE XLIV

$KBrF_4$- Type and Related Fluorides

Compound	a (Å)	c (Å)	Reference
$CaCrF_4$	5.45(2)	10.62(2)	a-1617
$SrCrF_4$ (*)	5.673(3)	10.920(6)	a-1623
$CaCuF_4$ (*)	5.377(2)	10.32(5)	a-416
$SrCuF_4$ (*)	5.643(2)	10.672(5)	a-417
$CaPdF_4$ (*)	5.521	10.57	a-1986
$SrPdF_4$	5.793	10.747	Müller and Hoppe (1972b)
$BaPdF_4$	6.120	10.981	Müller and Hoppe (1972b)
$PbPdF_4$	5.873	10.833	Müller and Hoppe (1972b)
$NaAgF_4$	5.54	10.56	a-427
$KAgF_4$	5.90	11.15	a-428
$CaAgF_4$	5.493	10.86	a-430
$SrAgF_4$	5.740	11.12	a-431
$BaAgF_4$	6.038	11.46	a-432
$CdAgF_4$	5.423	10.80	a-434
$HgAgF_4$	5.524	10.92	a-435
$LiAuF_4$ $(C2/c)$	5.41	11.16	a-436
	$(b = 5.39)$	$(\beta = 115.7°)$	
$NaAuF_4$	5.64	10.49	a-437
$KAuF_4$	5.99	11.38	a-438
$RbAuF_4$	6.18	11.85	a-439
$KClF_4$	6.18	10.94	a-1696
$KBrF_4$ (*)	6.174(2)	11.103(2)	a-1698
$(NO)^+BrF_4^-$	6.28	10.80	a-1703
Sr_2CuF_6 $(Bbam)$	5.70_8	16.44_6	von Schnering et al. (1971)
	$(b = 5.70_8)$		
$Ca_{2-2x}Sr_{2x}CuF_6$ $(Bbam)$,	5.54	16.10	Dumora (1971)
$1 \geq x \geq 0.40$	$(b = 5.54)$		
	for $x = 0.40$		

a. SCHEELITE STRUCTURE ($CaWO_4$ TYPE).   Many $LiMF_4$ phases crystallize with the inverse scheelite type. Here, M(III) has CN = 8, whereas lithium atoms are in a tetrahedral coordination. The 8 fluorine atoms surrounding each M(III) ion are shared with 12 other metal ions: 4 Li^+ at $z = 0$, 4 M(III) and 4 Li^+ at $z = \pm\frac{1}{4}$. The M–F–M bridging angles are about 110°. The tetrahedral coordination of lithium supposes a competing M—F bond that is relatively weak. Such a feature is indeed illustrated by the high CN of M (8).

Due to possible sp^3 hybridation of Zn^{2+} orbitals, $CaZnF_4$ and $SrZnF_4$ also exhibit this structure with Zn^{2+} in tetrahedral coordination. Lattice constants of scheelite-type fluorides are given in Table XLV. The prototype $CaWO_4$ has tetragonal symmetry and belongs to space group $I4_1/a$ (no. 88), $Z = 4$.

TABLE XLV

Scheelite-Type Fluorides

Compound	a (Å)	c (Å)	Reference
$LiYF_4$	5.175(5)	10.74(1)	a-817
$LiEuF_4$	5.228(5)	11.03(1)	a-927
$LiGdF_4$	5.219(5)	10.97(1)	a-934
$LiTbF_4$	5.200(5)	10.89(1)	a-943
$LiDyF_4$	5.188(5)	10.83(1)	a-957
$LiHoF_4$	5.175(5)	10.75(1)	a-965
$LiErF_4$	5.162(5)	10.70(1)	a-976
$LiTmF_4$	5.145(5)	10.64(1)	a-987
$LiYbF_4$ (*)	5.1335(2)	10.588(2)	a-997
$LiLuF_4$	5.124(5)	10.54(1)	a-1010
$LiTlF_4$	5.118(3)	10.63(1)	a-775
$LiBiF_4$	5.264(5)	11.26(1)	a-1467
$CaZnF_4$ (*)	5.279(1)	11.021(3)	a-602; von Schnering et al. (1983)
$SrZnF_4$ (*)	5.440(2)	11.645(3)	a-603; von Schnering et al. (1983)
Li_2CaUF_8 ($I\bar{4}m2$) (*)	5.229	11.013	Vedrine et al. (1979)
Li_2CaPrF_8 (*P)	5.196	10.88_6	Feldner and Hoppe (1983)
Li_2CdPrF_8 (*P)	5.17_7	10.82_2	Feldner and Hoppe (1983)

b. SPINEL STRUCTURE. Although the spinel structure [cubic, space group $Fd3m$ (no. 227), $Z = 8$] is widely adopted by oxide materials, only one fluoride has been shown to have this structure: Li_2NiF_4, $a = 8.313$ Å (a-1922). It is of the inverse spinel type $Li[LiNi]F_4$, due to the strong preference of nickel(II) for octahedral positions. Isolated LiF_4 tetrahedra are linked to edge-shared MF_6 octahedra via Li–F–M angles of about 125° ($M = Li_{0.5}Ni_{0.5}$).

c. $LiBaCrF_6$ STRUCTURE. This structure [monoclinic, space group $P2_1/c$ (no. 14) $Z = 4$ (Babel, 1974)] is built up of a framework of CrF_6 octahedra (Cr–F = 1.90_3 Å) and LiF_4 tetrahedra (Li–F = 1.86_8 Å) mutually linked by corners. Barium atoms are located in icosahedral BaF_{12} sites with Ba–F = 2.89 Å. The $LiBaCrF_6$ structure may be included in the structural sequence of $A^IM^{II}M'^{III}F_6$ phases with CN of monovalent, divalent, and trivalent cations equal to 6, 6, 6; 6, 6 + 2, 6; 4, 12, 6 in $LiCaAlF_6$-, $LiSrFeF_6$-, and $LiBaCrF_6$-type structures, respectively. Here again, the lower CN of lithium seems favored by larger cation of higher CN and weaker competing M—F bonds. Lattice constants are given in Table XLVI together with those of other lithium compounds.

d. β-Li_3VF_6 STRUCTURE. In the metastable high-temperature form of $Li_3M^{III}F_6$ compounds, M atoms and 7/9 of lithium atoms possess

TABLE XLVI

Examples of Structures Containing LiF_4 Tetrahedra

Compound	a (Å)	b (Å)	c (Å)	β (deg)	Reference
$LiBaAlF_6$	5.328(2)	10.142(5)	8.521(3)	90.22(4)	Babel (1974); Viebahn and Babel (1974)
$LiBaGaF_6$	5.404(4)	10.33(1)	8.619(10)	90.41(9)	
$LiBaTiF_6$	5.533(6)	10.51(2)	8.52(1)	91.3(1)	
$LiBaVF_6$	5.468(1)	10.400(5)	8.535(3)	91.0(1)	
$LiBaCrF_6$ (*)	5.397(3)	10.355(5)	8.638(5)	90.72(5)	
$LiBaFeF_6$	5.453(3)	10.380(8)	8.588(6)	90.9(1)	Fleischer and Hoppe (1982b)
$LiBaCoF_6$	5.506(1)	10.263(1)	8.438(1)	90.99(1)	
β-Li_3AlF_6	14.23	8.14	9.88	94.1	a-650
β-Li_3GaF_6	14.37	8.57	9.98	94.7	a-711
β-Li_3TiF_6	14.45	8.80	10.12	96.4	a-1299
β-Li_3VF_6 (*)	14.405	8.688	10.082	95.84	Massa (1980); a-1476
β-Li_3CrF_6	14.382	8.594	10.011	94.81	Massa (1980)
β-Li_3FeF_6	14.42	8.67	10.05	95.3	a-1781
$Na_3Li_3Al_2F_{12}$ (*)	12.121				Geller (1971); a-660
$Na_3Li_3Ga_2F_{12}$	12.305				a-714
$Na_3Li_3In_2F_{12}$	12.693				a-756
$Na_3Li_3Sc_2F_{12}$	12.607				a-800
$Na_3Li_3Ti_2F_{12}$	12.498				a-1304
$Na_3Li_3V_2F_{12}$	12.409				a-1480
$Na_3Li_3Cr_2F_{12}$	12.328				a-1569
$Na_3Li_3Mn_2F_{12}$	12.116				Langley and Sturgeon (1979)
$Na_3Li_3Fe_2F_{12}$ (*)	12.387				Massa et al. (1982); a-1792
$Na_3Li_3Co_2F_{12}$	12.326				a-1887
$Na_3Li_3Rh_2F_{12}$	12.415				a-1972
$Na_3Li_3Ni_2F_{12}$	12.165				Langley and Sturgeon (1979)

octahedral coordination. The remaining 2/9 are located in a distorted tetrahedral environment and form binuclear Li_2F_6 groups by edge sharing. The complete structure is built up of complex arrangements of MF_6, LiF_6, and Li_2F_6 groups. The prototype is monoclinic and belongs to space group $C2/c$ (no. 15), $Z = 12$ (Massa, 1980).

e. GARNET STRUCTURE. Tetrahedral coordination is also found in fluoride garnets $Na_3Li_3M_2^{III}F_{12}$ [cubic, space group $Ia3d$ (no. 230) $Z = 8$] (Geller, 1971; Menzer, 1930; de Pape et $al.$, 1967). The structure has been refined for $Na_3Li_3Fe_2F_{12}$ (Massa et $al.$, 1982). Three nonequivalent sites are present: MF_6 octahedra [(a) sites], LiF_4 tetrahedra [(d) sites], and NaF_8 dodecahedra [(c) sites]. The site ratio is $2:3:3$ respectively, as in usual oxide garnets (Table XLVI).

References

Abrahams, S. C., and Bernstein, J. L. (1972). $Mater. Res. Bull.$ **7**, 715.
Ahrens, L. H. (1952). $Geochim. Cosmochim. Acta$ **2**, 155.
Aléonard, S., Guitel, J. C., Lefur, Y., and Roux, M. T. (1976). $Acta Crystalloger., Sect. B$ **B32**, 3227.
Allman, R. (1975). $Monatsh. Chem.$ **106**, 779.
Alter, E., and Hoppe, R. (1974a). $Z. Anorg. Allg. Chem.$ **403**, 127.
Alter, E., and Hoppe, R. (1974b). $Z. Anorg. Allg. Chem.$ **405**, 167.
Alter, E., and Hoppe, R. (1974c). $Z. Anorg. Allg. Chem.$ **407**, 305.
Alter, E., and Hoppe, R. (1974d). $Z. Anorg. Allg. Chem.$ **407**, 313.
Alter, E., and Hoppe, R. (1974e). $Z. Anorg. Allg. Chem.$ **408**, 115.
Alter, E., and Hoppe, R. (1975). $Z. Anorg. Allg. Chem.$ **412**, 110.
Andersson, S., and Galy, J. (1969). $Acta Crystallogr., Sect. B$ **B25**, 847.
Arakawa, M., and Ebisu, H. (1982). $J. Phys. Soc. Jpn.$ **51**, 191.
Arndt, J., Babel, D., Haegele, R., and Rombach, N. (1975). $Z. Anorg. Allg. Chem.$ **418**, 193.
Arndt, J., Rombach, N., and Pausewang, G. (1977). $Mater Res. Bull.$ **12**, 803.
Asprey, L. B., Keenan, T. K., and Kruse, F. H. (1965). $Inorg. Chem.$ **4**, 985.
Aurivillius, B. (1952). $Ark. Kemi$ **5**, 39.
Avignant, D., Cousson, A., Cousseins, J.-C., and Védrine, A. (1976). $Bull. Soc. Chim. Fr.$ No. 11, p. 1662.
Avkhutskii, L. N., Oktiabrskii, G. A., and Sokolova, G. V. (1979). $Dokl. Akad. Nauk SSSR$ **245**, 388.
Babel, D. (1965a). $Z. Naturforsch., A$ **20A**, 165.
Babel, D. (1965b). $Z. Anorg. Allg. Chem.$ **336**, 200.
Babel, D. (1967). $Struct. Bonding (Berlin)$ **3**, 1.
Babel, D. (1968). $Conf. Rep., 2nd Eur. Symp.Fluorine Chem., August, 1968, Göttingen.$
Babel, D. (1969). $Z. Anorg. Allg. Chem.$ **369**, 117.
Babel, D. (1972). $Z. Anorg. Allg. Chem.$ **387**, 161.
Babel, D. (1974). $Z. Anorg. Allg. Chem.$ **406**, 23.
Babel, D., and Binder, F. (1983). $Z. Anorg. Allg. Chem.$ **505**, 153.
Babel, D., and Haegele, R. (1976). $J. Solid State Chem.$ **18**, 39.
Babel, D., and Herdtweck, E. (1982). $Z. Anorg. Allg. Chem.$ **487**, 75.
Babel, D., and Knoke, G. (1978). $Z. Anorg. Allg. Chem.$ **442**, 151.

Babel, D., Pausewang, G., and Viebahn, W. (1967). *Z. Naturforsch.* **22B**, 1219.

Babel, D., Haegele, R., Pausewang, G., and Wall, F. (1973a). *Mater. Res. Bull.* **8**, 1371.

Babel, D., Binder, F., and Pausewang, G. (1973b). *Z. Naturforsch. B: Anorg. Chem., Org. Chem.* **B28**, 213.

Babel, D., Wall, F., and Heger, G. (1974a). *Z. Naturforsch. B: Anorg. Chem. Org. Chem.* **B29**, 139.

Babel, D., Binder, F., and Haegele, R. (1974b). *Conf. Rep. 5th Eur. Symp. Fluorine Chem., September 1974, Aviemore, Scotland.*

Balz, D., and Plieth, K. (1955). *Z. Elektrochem.* **59**, 545.

Banks, E., Berkooz, O., and Deluca, J. A. (1971). *Mater. Res. Bull.* **6**, 659.

Banks, E., Deluca, J. A., and Berkooz, O. (1973). *J. Solid State Chem.* **6**, 569.

Banks, E., Torre, G., and Deluca, J. A. (1977), *J. Solid State Chem.* **22**, 95.

Banks, E., Nakajima, S., and Williams, G. J. B. (1979). *Acta Crystallogr., Sect. B* **B35**, 46.

Banks, E., Shone, M., Hong, Y. S., Williamson, R. F., and Boo, W. O. J. (1982). *Inorg. Chem.* **21**, 3894.

Bartlett, N. (1965). *Prep. Inorg. React.* **2**, 301.

Bartlett, N., and Tressaud, A. (1974). *C. R. Hebd. Seances Acad. Sci.* **278**, 1501.

Baur, W. H. (1976), *Acta Crystallogr., Sect B* **B32**, 2200.

Belmont, J. P., Bolte, M., and Charpin, P. (1975). *Rev. Chim. Miner.* **12**, 113.

Bernard, D., Pannetier, J., and Lucas, J. (1975). *J. Solid State Chem.* **14**, 328.

Binder, F. (1973). Thesis, Univ. of Tübingen.

Blattner, H., Graenicher, H., Kaenzig, W., and Merz, W. (1948). *Helv. Phys. Acta* **21**, 341.

Boca, R. (1981). *Chem. Zvesti* **35**, 769.

Bochkova, R. I., Sav'yanov, Y. N., Kuz'min, E. A., and Belov, N. V. (1974). *Sov. Phys.—Dokl. (Engl. Transl.)* **18**, 575.

Bode, H., and Voss, E. (1957). *Z. Anorg. Allg. Chem.* **290**, 1.

Bolte, M., Besse, J. P., and Capestan, M. (1972). *C. R. Hebd. Seances Acad. Sci.* **274**, 1051.

Brauer, G., and Eichner, M. (1958). *Z. Anorg. Allg. Chem.* **296**, 13.

Brosset, C. (1937). *Z. Anorg. Allg. Chem.* **235**, 139.

Brosset, C. (1938a). *Z. Anorg. Allg. Chem.* **238**, 201.

Brosset, C. (1938b). *Z. Anorg. Allg. Chem.* **239**, 301.

Brosset, C. (1942). Ph.D. Thesis, Univ. of Stockholm.

Brown, D. H., and Evans, H. T. (1948). *Acta Crystallogr., Sect. A* **A1**, 330.

Brunvoll J., Ischenko, A. A., Miakshin, I. N., Romanov, G. V., Sokolov, V. B., Spiridinov, V. P., and Strand, T. G. (1979). *Acta Chem. Scand. Ser. A* **A33**, 775.

Bukovec, P., and Hoppe, R. (1984). *Z. Anorg. Allg. Chem.* **509**, 138.

Bukovec, P., and Kaucic, V. (1977). *J. Chem. Soc., Dalton Trans.* **9**, 945.

Bukovec, P., and Kaucic, V. (1978). *Acta Crystallogr., Sect. B* **B34**, 3339.

Bukvetskii, B. V., and Garashina, L. S. (1977). *Koord. Khim.* **3**, 1024.

Bukvetskii, B. V., Muradyan, L. A., Davidovich, R. L., and Simonov, V. I. (1976). *Sov. J. Coord. Chem. (Engl. Transl.)* **2**, 869.

Bulou, A., and Nouet, J. (1982). *J. Phys. C* **15**, 183.

Bulou, A., Leble, A., Hewat, A. W., and Fourquet, J. L. (1982). *Mater. Res. Bull* **17**, 391.

Burns, J. H., Tennissen, A. C., and Brunton, G. D. (1968). *Acta Crystallogr., Sect. B* **B24**, 225.

Byström, A. (1944). *Ark. Kemi* **18**, 10.

Byström, A., Höl, B., and Mason, B. (1941). *Ark. Kemi* **15B**, 4.

Canterford, J. H., and Colton, R. (1968). "Halides of the Transition Elements." Wiley, New York.

Chaminade, J. P., and Pouchard, M. (1975). *Ann. Chim. (Paris)* **10**, 75.

Chaminade, J. P., Vlasse, M., and Pouchard, M. (1974). *Bull. Soc. Chim. Fr.* **9**, 1791.

Champarnaud, J. C. (1978). Ph.D. Thesis, Univ. of Limoges.

Champarnaud, J. C., Grannec, J., and Gaudreau, B. (1974). *C. R. Hebd. Seances Acad. Sci* **278**, 171.

Chassaing, J. (1968). *Rev. Chim. Miner.* **5**, 1115.

Chassaing, J. (1969). *C. R. Hebd. Seances Acad. Sci.* **268,** 2188.

Chassaing, J., and Julien, P. (1972). *C. R., Hebd. Seances Acad Sci.* **274,** 871.

Chassaing, J., Monteil, C., and Bizot, D. (1982). *J. Solid State Chem.* **43,** 327.

Cheetham, A. K., and Norman, N. (1974). *Acta Chem. Scand., Ser. A* **A28,** 55.

Cheetham, A. K., Fender, B., Fuess, H., and Wright, A. F. (1976). *Acta Crystallogr., Sect. B* **B32,** 94.

Chenavas, J., Capponi, J. J., Joubert, J. C., and Marezio, M. (1974). *Mater. Res. Bull.* **9,** 13.

Cohen, S., Selig, H., and Gut, R. (1982). *J. Fluorine Chem.* **20,** 349.

Cosier, R., Wise, A., Tressaud, A., Grannec, J., Olazcuaga, R., and Portier, J. (1970). *C. R. Hebd. Seances Acad. Sci.* **271,** 142.

Courbion, G., Jacoboni, C., and de Pape, R. (1971). *C. R. Hebd. Seances Acad. Sci.* **273,** 809.

Courbion, G., Jacoboni, C., and de Pape, R. (1974). *Mater. Res. Bull* **9,** 425.

Courbion, G., Jacoboni, C., and de Pape, R. (1976). *Acta Crystallogr., Sect. B* **B32,** 3190.

Courbion, G., Jacoboni, C., and de Pape, R. (1977). *Acta Crystallogr., Sect. B* **B33,** 1405.

Courbion, G., Ferey, G., and de Pape, R. (1978). *Mater Res. Bull.* **13,** 967.

Courbion, G., Jacoboni, C., and de Pape, R. (1983a). *Stud. Inorg. Chem.* **3,** 605.

Courbion, G., de Pape, R., Knoke, G., and Babel, D. (1983b), *J. Solid State Chem.* **49,** 353.

Cousson, A., Vedrine, A., and Cousseins, J.-C. (1972). *C. R. Hebd. Seances Acad Sci.* **274,** 864.

Cros, C. (1978). *Rev. Inorg. Chem.* **1,** 163.

Cros, C., Feurer, R. Pouchard, M., and Hagenmuller, P. (1975). *Mater. Res. Bull.* **10,** 383.

Cros, C., Feurer, R., Grenier, J. C., and Pouchard, M. (1976). *Mater. Res. Bull.* **11,** 539.

Cros, C., Dance, J. M., Grenier, J. C., Wanklyn, B. M., and Garrard, B. J. (1977a). *Mater. Res. Bull.* **12,** 415.

Cros, C., Feurer, R., Pouchard, M., and Hagenmuller, P. (1977b). *Mater. Res. Bull.* **12,** 745.

Dance, J. M., and Ravez, J. (1979). *C. R. Hebd. Seances Acad Sci., Ser. C* **289,** 247.

Dance, J. M., and Tressaud, A. (1973). *C. R. Hebd. Seances Acad. Sci.* **277,** 379.

Dance, J. M., and Tressaud, A. (1979). *Mater. Res. Bull.* **14,** 37.

Dance, J. M., Grannec, J., Jacoboni, C., and Tressaud, A. (1974). *C. R. Hebd Seances Acad. Sci., Ser. C* **279,** 601.

Dance, J. M., Grannec, J., and Tressaud, A. (1975). *C. R. Hebd. Seances Acad. Sci., Ser. C* **281,** 91.

Dance, J. M., Grannec, J., and Tressaud, A. (1976). *C. R. Hebd. Seances Acad. Sci., Ser. C* **283,** 115.

Dance, J. M., Grannec, J., Tressaud, A., and Perrin, M. (1977). *Mater. Res. Bull.* **12,** 989.

Dance, J. M., Kerkouri, N., and Tressaud, A. (1979). *Mater. Res. Bull.* **14,** 869.

Dance, J. M., Soubeyroux, J. L., Sabatier, R., Fournes, L., Tressaud, A., and Hagenmuller, P. (1980). *J. Magn. Magn. Mater.* **15-18,** 534.

Dance, J. M., Tressaud, A., Massa, W., and Babel, D. (1981). *J. Chem. Res. Synop.* p. 202.

Dance, J. M., Darriet, J., Tressaud, A., and Hagenmuller, P. (1984). *Z. Anorg. Allg. Chem.* **508,** 93.

Deadmore, D. L., and Bradley, W. F. (1962). *Acta Crystallogr.* **15,** 186.

de Bournonville, M. B., Bizot, D., Chassaing, J., and Quarton, M. (1984). *Conf. Rep. Solid State Symp., Bordeaux.*

De Cian, A., Fischer, J., and Weiss, R. (1967). *Acta Crystallogr.* **22,** 340.

Dehnicke, K., Pausewang, G., and Rüdorff, W. (1969). *Z. Anorg. Allg. Chem.* **366,** 64.

de Kozak, A. (1971). *Rev. Chim. Miner.* **8,** 301.

de Kozak, A., and Samouel, M. (1977). *Rev. Chim. Miner.* **14,** 553.

de Kozak, A., and Samouel, M. (1978). *Rev. Chim. Miner.* **15,** 406.

de Kozak, A., and Samouel, M. (1981). *Rev. Chim. Miner.* **18,** 255.

de Kozak, A., Leblanc, M., Samouel, M., Ferey, G., and de Pape, R. (1981). *Rev. Chim. Miner.* **18,** 659.

de Kozak, A., Samouel, M., Leblanc, M., and Ferey, G. (1982). *Rev. Chim. Miner.* **19,** 668.

Demazeau, G., Menil, F., Portier, J., and Hagenmuller, P. (1971). *C. R. Hebd. Seances Acad. Sci.* **273,** 1641.

de Pape, R. (1965). *C. R. Hebd. Seances Acad. Sci.* **260,** 4527.

de Pape, R., and Ravez, J. (1966). *Bull. Soc. Chim. Fr.* **33**, 3283.

de Pape, R., Portier, J., Gauthier, G., and Hagenmuller, P. (1967). *C. R. Hebd. Seances Acad. Sci.* **265**, 1244.

de Pape, R., Tressaud, A., and Portier, J. (1968). *Mater. Res. Bull.* **3**, 753.

Domesle, R., and Hoppe, R. (1978). *Rev. Chim. Miner.* **15**, 439.

Domesle, R., and Hoppe, R. (1980a). *Z. Kristallogr.* **153**, 317.

Domesle, R., and Hoppe, R. (1980b). *Angew. Chem., Int. Ed. Engl.* **19**, 489.

Domesle, R., and Hoppe, R. (1982a). *Z. Anorg. Allg. Chem.* **495**, 16.

Domesle, R., and Hoppe, R. (1982b). *Z. Anorg. Allg. Chem.* **495**, 27.

Domesle, R., and Hoppe, R. (1983). *Z. Anorg. Allg. Chem.* **501**, 102.

Donohue, P. C., Katz, L., and Ward, R. (1965). *Inorg. Chem.* **4**, 306.

Dubler, E., Linowsky, L., Matthieu, J. P., and Oswald, H. R. (1977). *Helv. Chim. Acta* **60**, 1589.

Dumora, D. (1971). Ph.D. Thesis, University of Bordeaux.

Dumora, D., and Ravez, J. (1969). *C. R. Hebd. Seances Acad Sci.*, Ser. C **268**, 1246.

Dumora, D., Ravez, J., and Hagenmuller, P. (1970). *Bull. Soc. Chim. Fr.* No. 5, p. 1751.

Dumora, D., von der Mühll, R., and Ravez, J. (1971). *Mater. Res. Bull.* **6**, 561.

Dumora, D., Ravez, J., and Hagenmuller, P. (1972). *J. Solid State Chem.* **5**, 35.

Edwards, A. J. (1971). *J. Chem. Soc. A* **16**, 2653.

Edwards, A. J. (1972). *J. Chem. Soc., Dalton Trans.* **7**, 816.

Edwards, A. J. (1982). *Adv. Inorg. Chem. Radiochem.* **27**, 83.

Edwards, A. J., Jones, G. R., and Sills, R. (1970). *J. Chem. Soc.* p. 2511.

Eibschütz, M., and Guggenheim, H. J. (1968). *Solid State Commun.* **6**, 737.

Eibschütz, M., Guggenheim, H. J., Holmes, L., and Bernstein, J. L. (1972). *Solid State Commun.* **11**, 457.

Eicher, S. M., and Greedan, J. E. (1984). *J. Solid State Chem.* **52**, 12.

Einstein, F., Rao, P. R., Trotter, J., and Bartlett, N. (1967). *J. Chem. Soc. A* p. 478.

Elaatmani, M., Ravez, J., Doumerc, J. P., and Hagenmuller, P. (1981). *Mater. Res. Bull.* **16**, 105.

Eller, P. G., Larson, A. C., Peterson, J. R., Ensor, D. D., and Yound, J. P. (1979). *Inorg. Chim. Acta* **37**, 129.

Epple, M., and Massa, W. (1978). *Z. Anorg. Allg. Chem.* **444**, 47.

Eyring, L., Langley, R., Rieck, D., Eick, H., Williamson, R. F., and Boo, W. O. J. (1978). *Mater. Res. Bull.* **13**, 1297.

Feldner, K., and Hoppe, R. (1980). *Z. Anorg. Allg. Chem.* **471**, 131.

Feldner, K., and Hoppe, R. (1983). *Rev. Chim. Miner.* **20**, 351.

Ferey, G., Leblanc, M., Jacoboni, C., and de Pape, R. (1971). *C. R. Hebd. Seances Acad Sci.*, Ser. C **273**, 700.

Ferey, G., Leblanc, M., de Pape, R., Passaret, M., and Bothorel-Razazi, M. (1975). *J. Cryst. Growth* **29**, 209.

Ferey, G., dePape, R., Poulain, M., Grandjean, D., and Hardy, A. (1977). *Acta Crystallogr., Sect. B* **B33**, 1409.

Ferey, G., de Pape, R., and Boucher, B. (1978). *Acta Crystallogr., Sect. B* **B34**, 1084.

Ferey, G., Leblanc, M., and de Pape, R. (1981). *J. Solid State Chem.* **40**, 1.

Fischer, M. (1982). *J. Phys. Chem. Solids* **43**, 673.

Fischer, P., Hälg, W., Schwarzenbach, D., and Gamsjäger, H. (1974). *J. Phys. Chem. Solids* **35**, 1683.

Fischer, R., and Tillmanns, E. (1981). *Z. Kristallogr.* **157**, 69.

Fleischer, T., and Hoppe, R. (1982a). *Z. Anorg. Allg. Chem.* **490**, 111.

Fleischer, T., and Hoppe, R. (1982b). *Z. Naturforsch., B: Anorg. Chem., Org. Chem.* **37B**, 988.

Fleischer, T., and Hoppe, R. (1982c). *J. Fluorine Chem.* **19**, 529.

Fleischer, T., and Hoppe, R. (1982d). *Z. Naturforsch., B: Anorg. Chem., Org. Chem.* **37B**, 1132.

Fleischer, T., and Hoppe, R. (1982e). *Z. Anorg. Allg. Chem.* **492**, 83.

196   D. Babel and A. Tressaud

Fleischer, T., and Hoppe, R. (1982f). *Z. Anorg. Allg. Chem.* **493**, 59.

Fourquet, J. L., Jacoboni, C., and de Pape, R. (1973). *Mater. Res. Bull.* **8**, 393.

Fourquet, J. L., Plet, F., Courbion, G., Bulou, A., and de Pape, R. (1979). *Rev. Chim. Miner.* **16**, 490.

Fourquet, J. L., Plet, F., and de Pape, R. (1980). *Acta Crystallogr. Sect. B* **B36**, 1997.

Fourquet, J. L., Plet, F., and de Pape, R. (1981a). *Rev. Chim. Miner.* **18**, 19.

Fourquet, J. L., Plet, F., and de Pape, R. (1981b). *Acta Crystallogr., Sect. B* **B37**, 2136.

Fourquet, J. L., de Pape, R., Teillet, J., Varret, F., and Papaefthymiou, G. C. (1982). *J. Magn. Mater.* **27**, 209.

Friebel, C., Pebler, J., Steffens, F., Weber, M., and Reinen, D. (1983). *J. Solid State Chem.* **46**, 253.

Friebel, C., Holler, H., and Babel, D. (1985). To be published.

Frondel, C. (1948). *Am. Mineral.* **32**, 84.

Gaertner, H. R. V. (1930). *Neues Jahrb. Mineral. Abh.* **61**, 1.

Gafner, G., and Kruger, G. J. (1974). *Acta Crystallogr., Sect. B* **B30**, 250.

Gaile, J., and Rüdorff, W. (1976). *Z. Naturforsch., B: Anorg. Chem., Org. Chem.* **31B**, 684.

Gaile, J., Rüdorff, W., and Viebahn, W. (1977). *Z. Anorg. Allg. Chem.* **430**, 161.

Galasso, F., Layden, G., and Ganung, G. (1968). *Mater. Res. Bull.* **3**, 397.

Galy, J., Andersson, S., and Portier, J. (1969). *Acta Chem. Scand.* **23**, 2949.

Garashina, L. S., and Sobolev, B. P. (1971a). *Kristallografiya* **16**, 307.

Garashina, L. S., and Sobolev, B. P. (1971b). *Sov. Phys.—Cryst. (Engl. Transl.)* **16**, 254.

Geller, S. (1956). *J. Chem. Phys.* **24**, 1236.

Geller, S. (1971). *Am. Mineral.* **56**, 18.

Geller, J., and Wood, E. A. (1956). *Acta Cryst.* **9**, 563.

Giuseppetti, G., and Tadini, C. (1978). *Tschermaks Mineral. Petrogr. Mitt.* **25**, 57.

Gladney, H. M., and Street, G. B. (1968). *J. Inorg. Nucl. Chem.* **30**, 2949.

Golovastikov, N. I., and Belov, N. V. (1978). *Kristallografiya* **23**, 42.

Goodenough, J. B., and Longo, J. M. (1970). *In* "Landolt–Börnstein Tables," Group III, Vol. 4a, p. 126. Springer-Verlag, Berlin and New York.

Goodenough, J. B., Kafalas, J. A., and Longo, J. M. (1972). *In* "Preparative Methods in Solid State Chemistry" (P. Hagenmuller, ed.), p. 1. Academic Press, New York.

Gortsema, F. P., and Didchenko, R. (1965). *Inorg. Chem.* **4**, 182.

Grannec, J., and Ravez, J. (1970). *C. R. Hebd. Seances Acad. Sci.* **271**, 1084.

Grannec, J., Tressaud, A., and Portier, J. (1970). *Bull. Soc. Chim. Fr.* **5**, 1719.

Grannec, J., Ravez, J., Portier, J., and Hagenmuller, P. (1971a). *Bull. Soc. Chim. Fr.* p. 804.

Grannec, J., Lozano, L., Portier, J., and Hagenmuller, P. (1971b). *Z. Anorg. Allg. Chem.* **385**, 26.

Grannec, J., Lozano, L., Sorbe, P., Portier, J., and Hagenmuller, P. (1975a) *J. Fluorine Chem.* **6**, 267.

Grannec, J., Sorbe, P., Portier, J., and Hagenmuller, P. (1975b). *C. R. Hebd. Seances Acad. Sci.* **280**, 45.

Grannec, J., Portier, J., Pouchard, M., and Hagenmuller, P. (1976). *J Inorg. Nucl. Chem., Suppl.* p. 119.

Grdenic, D., and Sikirica, M. (1973). *Inorg. Chem.* **12**, 544.

Greenwood, N. N., Howe, A. T., and Menil, F. (1971). *J. Chem. Soc. A* p. 2218.

Greis, O. (1982). *Rev. Inorg. Chem.* **4**, 1.

Greis, O., and Haschke, J. (1982). In Handbook on the Physics and Chemistry of Rare Earths" (K. A. Gschneidner, Jr. and L. Eyring, eds.), Vol. 5, p. 387. North-Holland Publ., Amsterdam.

Greis, O., and Martinez-Ripoll, M. (1978). *Z. Anorg. Allg. Chem.* **436**, 105.

Griebler, W. D. (1978). Ph.D. Thesis, Univ. of Marburg.

Griebler, W. D., and Babel, D. (1980). *Z. Anorg. Allg. Chem.* **467**, 187.

Grjotheim, K., Holm, J. L., Malinovsky, M., and Mikhaiel, S. A. (1971). *Acta Chem. Scand.* **25**, 1695.

Günther, J. R., Mathieu, J. P., and Oswald, H. R. (1978). *Helv. Chim. Acta* **61**, 328.

Haegele, R. (1974). Ph.D. Thesis, Univ. of Marburg.
Haegele, R., and Babel, D. (1974). Z. Anorg. Allg. Chem. **409**, 11.
Haegele, R., Verscharen, W., and Babel, D. (1975). Z. Naturforsch., B: Anorg. Chem., Org. Chem. **30B**, 462.
Haegele, R., Babel, D., and Reinen, D. (1976). Z. Naturforsch., B: Anorg. Chem., Org. Chem. **31B**, 60.
Haegele, R., Verscharen, W., Babel, D., Dance, J. M., and Tressaud, A. (1978). J. Solid State Chem. **24**, 77.
Haensler, R., and Rüdorff, W. (1970). Z. Naturforsch., B: Anorg. Chem., Org. Chem., Biochem., Biophys., Biol. **25B**, 1305.
Hall, W., Kim, S., Zubieta, J., Walton, E. G., and Brown, D. B. (1977). Inorg. Chem. **16**, 1884.
Hardy, A. M., Hardy, A., and Ferey, G. (1973). Acta Crystallogr., Sect. B **B29**, 1654.
Hartung, A. (1978). Ph.D. Thesis, Univ. of Marburg.
Hartung, A., and Babel, D. (1982). J. Fluorine Chem. **19**, 369.
Hartung, A., Verscharen, W., Binder, F., and Babel, D. (1979). Z. Anorg. Allg. Chem. **456**, 106.
Hawthorne, F. C., and Ferguson, R. B. (1975). Can. Mineral, **13**, 377.
Hebecker, C. (1975). Z. Anorg. Allg. Chem. **412**, 37.
Heger, G., Geller, R., and Babel, D. (1971). Solid State Commun. **9**, 355.
Helmholdt, R. B., Wiegers, G. A., and Bartolome, J. (1980). J. Phys. C **13**, 5081.
Hepworth, M. A., and Jack, K. H. (1957) Acta Crystallogr. **10**, 345.
Herdtweck, E. (1983a). Z. Anorg. Allg. Chem. **501**, 131.
Herdtweck, E. (1983b). Z. Kristallogr. **162**, 100.
Herdtweck, E., and Babel, D. (1980). Z. Kristallogr. **153**, 189.
Herdtweck, E., and Babel, D. (1981). Z. Anorg. Allg. Chem. **474**, 113.
Heyns, A. M., and Pistorius, C. (1976). Spectrochim. Acta, Part A **32A**, 535.
Hidaka, M. (1975). J. Phys. Soc. Jpn. **39**, 180.
Hidaka, M., and Hosogi, S. (1982). J. Phys. Orsay, Fr. **43**, 1227.
Hidaka, M., and Ono, M. (1977). J. Phys. Soc. Jpn. **43**, 258.
Hidaka, M., and Walker, P. J. (1979). Solid State Commun. **31**, 383.
Hidaka, M., Hosogi, S., Ono, M., and Horai, K. (1977). Solid State Commun. **23**, 503.
Hidaka, M., Garrard, B. J., and Wanklyn, B. M. (1979). J. Phys. C **12**, 2737.
Hidaka, M., Yamashita, S., Inoue, K., Tsukuda, N., Garrard, B. J., and Wanklyn, B. M. (1981). J. Phys. Soc. Jpn. **50**, 4022.
Hidaka, M., Inoue, K., Garrard, B. J., and Wanklyn, B. M. (1982). Phys Status Solidi A **72**, 809.
Hirakawa, K., Yoshizawa, H., and Ubukoshi, K. (1982). J. Phys. Soc. Jpn. **51**, 2151.
Holler, H. (1983). Ph.D. Thesis, Univ. of Marburg.
Holler, H., and Babel, D. (1982). Z. Anorg. Allg. Chem. **491**, 137.
Holler, H., and Babel, D. (1985). Z. Anorg. Allg. Chem., in press.
Holler, H., Babel, D., Samouel, M., and de Kozak, A. (1981). J. Solid State Chem. **39**, 345.
Holler, H., Babel, D. Samouel, M., and de Kozak, A. (1984). Rev. Chim. Minerale **21**, 358.
Holler, H., Kurtz, W., Babel, D., and Knop, W. (1982). Z. Naturforsch., B: Anorg. Chem., Org. Chem. **37B**, 54.
Holler, H., Pebler, J., and Babel, D. (1985). Z. Anorg. Allg. Chem., in press.
Hong, Y. S., Williamson, R. F., and Boo, W. O. J. (1979). Inorg. Chem. **18**, 2123.
Hong, Y. S., Williamson, R. F., and Boo, W. O. J. (1980). Inorg. Chem. **19**, 2229.
Hong, Y. S., Williamson, R. F., and Boo, W. O. J. (1981). Inorg. Chem. **20**, 403.
Hong, Y. S., Williamson, R. F., and Boo, W. O. J. (1982). Inorg. Chem. **21**, 3898.
Hoppe, R. (1962). Naturwissenschaften **49**, 254.
Hoppe, R. (1981). Angew. Chem., Int. Ed. Engl. **20**, 63.
Hoppe, R., and Dähne, W. (1960). Naturwissenschaften **47**, 397.

Hoppe, R., and Jesse, R. (1973). *Z. Anorg. Allg. Chem.* **402**, 29.

Hoppe, R., and Kissel, D. (1984). *J. Fluorine Chem.* **24**, 327.

Hoppe, R., and Lehr, K. (1975). *Z. Anorg. Allg. Chem.* **416**, 240.

Hoppe, R., Wilhelm, V., and Müller, B. (1972). *Z. Anorg. Allg. Chem.* **392**, 1.

Jacoboni, C., de Pape, R., Poulain, M., Le Marouille, J. Y., and Grandjean, D. (1974). *Acta Crystallogr., Sect. B* **B30**, 2688.

Jacoboni, C., Leble, A., and Rousseau, J. J. (1981). *J. Solid State Chem.* **36**, 297.

Jacoboni, C., Le Bail, A., de Pape, R., and Renard, J. P. (1983). *Stud. Inorg. Chem.* **3**, 687.

Jahn, H. A., and Teller, E. (1937). *Proc. R. Soc. London, Ser. A* **161**, 220.

Jesse, R. R., and Hoppe, R. (1974). *Z. Anorg. Allg. Chem.* **403**, 143.

Jesse, R. R., and Hoppe, R. (1977a). *Z. Anorg. Allg. Chem.* **428**, 83.

Jesse, R. R., and Hoppe, R. (1977b). *Z. Anorg. Allg. Chem.* **428**, 91.

Jesse, R. R., and Hoppe, R. (1977c). *Z. Anorg. Allg. Chem.* **428**, 97.

Kaucic, V., and Bukovec, P. (1978). *Acta Crystallogr., Sect. B* **B34**, 3337.

Kaucic, V., and Bukovec, P. (1979). *J. Chem. Soc., Dalton Trans.* **10**, 1512.

Keller, C., and Salzer, M. (1967). *J. Inorg. Nucl. Chem.* **29**, 2925.

Keve, E. T., Abrahams, S. C., and Bernstein, J. L. (1969). *J. Chem. Phys.* **51**, 4928.

Keve, E. T., Abrahams, S. C., and Bernstein, J. L. (1970). *J. Chem. Phys.* **53**, 3279.

Kleemann, W., Schaefer, F. J., and Nouet, J. (1982). *J. Phys. C* **15**, 197.

Knoke, G. (1977). Ph.D. Thesis, Univ. of Marburg.

Knoke, G., Verscharen, W., and Babel, D. (1979a). *J. Chem. Res., Synop.* p. 213.

Knoke, G., Babel, D., and Hinrichsen, T. (1979b). *Z. Naturforsch., B: Anorg. Chem., Org. Chem.* **34B**, 934.

Knop, O., Cameron, T. S., and Jochem, K. (1982). *J. Solid State Chem.* **43**, 213.

Knox, K. (1963). *Acta Crystallogr., Sect. A* **A16**, A45.

Knox, K., and Geller, S. (1958). *Phys. Rev.* **110**, 771.

Koch, J., Hebecker, C., and John, H. (1982). *Z. Naturforsch., B: Anorg. Chem., Org. Chem.* **37B**, 1659.

Köhl, P., Reinen, D., Decher, G., and Wanklyn, B. M. (1980). *Z. Kristallogr.* **153**, 211.

Köhler, P., Massa, W., Reinen, D., Hoffmann, B., and Hoppe, R. (1978). *Z. Anorg. Allg. Chem.* **446**, 131.

Kohn, K., Fukuda, R., and Iida, S. (1967). *J. Phys. Soc. Jpn.* **22**, 333.

Kojic-Prodic, B., Scavnicar, S., Liminga, R., and Sljukic, M. (1972). *Acta Crystallogr., Sect. B* **B28**, 2028.

Kruger, G. J., Pistorius, C., and Heyns, A. M. (1976). *Acta Cryst.* **B32**, 2916.

Lander, J. J. (1951). *Acta Crystallogr., Sect. A* **A4**, 148.

Langley, R. H., and Sturgeon, G. D. (1979). *J. Fluorine Chem.* **14**, 1.

Leblanc, M., Ferey, G., and de Pape, R. (1980). *J. Solid State Chem.* **33**, 317.

Leblanc, M. Ferey, G., Calage, Y., and de Pape, R. (1983a). *J. Solid State Chem.* **47**, 24.

Leblanc, M., Ferey, G., Chevallier, P., Calage, Y., and de Pape, R. (1983b). *J. Solid State Chem.* **47**, 53.

Leble, A., Rousseau, J. J., Fayet, J. C., Pannetier, J., Fourquet, J. L., and de Pape, R. (1982). *Phys. Status Solidi A* **69**, 249.

l'Helgouach, H., Poulain, M., Rannou, J. P., and Lucas, J. (1971). *C. R. Hebd. Seances Acad, Sci,* **272**, 1321.

Longo, J. M., and Kafalas, J. A. (1969). *J. Solid State Chem.* **1**, 103.

Lorin, D. (1980). Thesis, Univ. of Bordeaux.

Lorin, D., Dance, J. M., Soubeyroux, J. L., Tressaud, A., and Hagenmuller, P. (1981). *J. Magn. Magn. Mater.* **23**, 92.

Lösch, R., and Hebecker, C. (1979a). *Z. Naturforsch., B: Anorg. Chem., Org. Chem.* **34B**, 131.

Lösch, R., and Hebecker, C. (1979b). *Z. Naturforsch., B: Anorg. Chem., Org. Chem.* **34B,** 1765.

Lösch, R., Hebecker, C., and Ranft, Z. (1982). *Z. Anorg. Allg. Chem.* **491,** 199.

Ludekens, W. L. W., and Welch, A. J. E. (1952). *Acta Crystallogr.* **5,** 841.

McKinzie, H., Dance, J. M., Tressaud, A., Portier, J., and Hagenmuller, P. (1972). *Mater. Res. Bull.* **7,** 673.

Magnéli, A. (1949). *Ark. Kemi* **1,** 213.

Magnéli, A. (1953). *Acta Chem. Scand.* **7,** 315.

Marseglia, E. A., and Brown, I. D. (1973). *Acta Crystallogr., Sect. B* **B29,** 1352.

Marsh, R. E. (1984). *J. Solid State Chem.* **51,** 405.

Massa, W. (1975). *Z. Anorg. Allg. Chem.* **415,** 254.

Massa, W. (1976). *Z. Anorg. Allg. Chem.* **427,** 235.

Massa, W. (1977a). *Inorg. Nucl. Chem. Lett.* **13,** 253.

Massa, W. (1977b). *Z. Anorg. Allg. Chem.* **436,** 29.

Massa, W. (1978). Conference report, GDCh (Section Solid-State Chemistry) meeting, 1978, Giessen.

Massa, W. (1980) *Z. Kristallogr.* **153,** 201.

Massa, W. (1982a). *Z. Anorg. Allg. Chem.* **491,** 208.

Massa, W. (1982b). Habilitationsschrift, Univ. of Marburg.

Massa, W. (1983). *Z. Kristallogr.* **162,** 166.

Massa, W., and Babel, D. (1980). *Z. Anorg. Allg. Chem.* **469,** 75.

Massa, W., and Pebler, J. (1983). *Stud. Inorg. Chem.* **3,** 577.

Massa, W., and Rüdorff, W. (1971). *Z. Naturforsch., B: Anorg. Chem. Org. Chem., Biochem., Biophys., Biol.* **26B,** 1216.

Massa, W., and Steiner, M. (1980). *J. Solid State Chem.* **32,** 137.

Massa, W., Post, B., and Babel, D. (1982). *Z. Krystallogr.* **158,** 299.

Masse, R., Aleonard, J., and Averbuch-Pouchot, M. T. (1984). *J. Solid State Chem.* **53,** 136.

Mattes, R., Mennemann, K., Jäckel, N., Rieskamp, H., and Brosckmeyer, H. J. (1980). *J. Less-Common Met.* **76,** 199.

Matvienko, E. N., Iakubovich, O. V., Simonov, M. A., Ivashchenko, A. N., Melnikov, O. K., and Belov, N. V. (1981). *Dokl. Akad. Nauk SSSR* **257,** 105.

Mayer, H. W., Reinen, D., and Heger, G. (1983). *J. Solid State Chem.* **50,** 213.

Menil, F., Grannec, J., Demazeau, G., and Tressaud, A. (1972). *C. R. Hebd. Seances Acad. Sci.* **275,** 495.

Menzer, G. (1930). *Z. Kristallogr.* **75,** 265.

Ming, L. C., Manghnani, M. H., Matsui, T., and Jamieson, J. C. (1980). *Phys. Earth Planet. Inter.* **23,** 276.

Miranday, J. P., Ferey, G., Jacoboni, C., Dance, J. M., Tressaud, A., and de Pape, R. (1975). *Rev. Chim. Miner.* **12,** 187.

Morrell, B. K., Zalkin, A., Tressaud, A., and Bartlett, N. (1973). *Inorg. Chem.* **12,** 2640.

Morss, L. R. (1974). *J. Inorg. Nucl. Chem.* **36,** 3876.

Mullen, D., Heger, G., and Reinen, D. (1975). *Solid State Commun.* **17,** 1249.

Müller, B. G. (1981a). *J. Fluorine Chem.* **17,** 317.

Müller, B. G. (1981b). *J. Fluorine Chem.* **17,** 409.

Müller, B. G. (1981c). *J. Fluorine Chem.* **17,** 489.

Müller, B. G. (1982a). *Z. Anorg. Allg. Chem.* **491,** 245.

Müller, B. G. (1982b). *J. Fluorine Chem.* **20,** 291.

Müller, B. G., and Hoppe, R. (1972a). *Z. Anorg. Allg. Chem.* **392,** 37.

Müller, B. G., and Hoppe, R. (1972b). *Mater. Res. Bull.* **7,** 1297.

Müller, B. G., and Hoppe, R. (1973). *Z. Anorg. Allg. Chem.* **395,** 239.

Naray-Szabo, S. V., and Sasvari, K. (1938). Z. Kristallogr., Kristallgeom., Kristallphys., Kristallchem. 99, 27.

Navarro, R., Smit, J. J., de Jongh, L. J., Crama, W. J., and Ijdo, D. J. W. (1976). Physica B + C (Amsterdam) 83, 97.

Nolte, M. J., and De Beer, W. (1979). Acta Crystallogr., Sect. B B35, 1208.

Nomura, S. (1978). In "Landolt–Börnstein Tables," Group III, Vol. 12a, p. 368. Springer-Verlag, Berlin and New York.

Nozik, Ju. Z., Fykin, L. E., Bukin, V. I., and Laptaš, N. M. (1979). Koord. Khim. 5, 276.

Odenthal, R. H., and Hoppe, R. (1971). Monatsh. Chem. 102, 1340.

Odenthal, R. H., Paus, D., and Hoppe, R. (1974). Z. Anorg. Allg. Chem. 407, 144.

Okamura, K., and Yajima, S. (1974). Bull. Chem. Soc. Jpn. 47, 1531.

Okazaki, A. (1969). J. Phys. Soc. Jpn. 26, 870.

Okazaki, A., and Ono, M. (1978) J. Phys. Soc. Jpn. 45, 206.

Okazaki, A., and Suemune, Y. (1961). J. Phys. Soc. Jpn. 16, 176.

O'Keeffe, M., Hyde, B. G., and Bovin, J. D. (1979). Phys. Chem. Miner, 4, 299.

Omaly, J., Batail, P., Grandjean, D., Avignant, D., and Cousseins, J. C. (1976). Acta Crystallogr., Sect. B. B32, 2106.

Pannetier, J., Calage, Y., and Lucas, J. (1972). Mater. Res. Bull. 7, 57.

Pauling, L. (1924). J. Am. Chem. Soc. 46, 2738.

Paus, D., and Hoppe, R. (1977). Z. Anorg. Allg. Chem. 431, 207.

Pausewang, G., and Rüdorff, W. (1969). Anorg. Allg. Chem. 364, 69.

Pausewang, G., and Dehnicke, K. (1969). Z. Anorg. Allg. Chem. 369, 265.

Pebler, J., Reinen, D., Schmidt, K., and Steffens, F. (1978). J. Solid State Chem. 25, 107.

Penneman, R. A., Ryan, R., and Rosenzweig, A. (1973). Struct. Bonding (Berlin) 13, 1.

Peraudeau, G. (1978). Thesis, Univ. of Bordeaux, France.

Pies, W., and Weiss, A. (eds.) (1973). "Landolt–Börnstein Tables," New Series, Group III, Vol. 7, Parts a and g. Springer-Verlag, Berlin and New York.

Piotrowski, M., Ptasiewicz, M., and Murasik, A. (1979). Phys. Status Solidi A 55, K163.

Pistorius, C. (1975). Mater. Res. Bull. 10, 1079.

Podberezskaya, N. V., Potapova, D. G., Borisov, S. V., and Gatilov, Y. V. (1976). J. Struct. Chem. (Engl. Transl.) 17, 815.

Portier, J. (1976). Angew. Chem. 15, 475.

Portier, J., Tressaud, A., Menil, F., Claverie, J., de Pape, R., and Hagenmuller, P. (1969). J. Solid State Chem. 1, 100.

Portier, J., Menil, F., and Hagenmuller, P. (1970a). Bull. Soc. Chim. Fr. p. 3485.

Portier, J., Menil, F., and Tressaud, A. (1970b). Mater. Res. Bull. 5, 503.

Pouchard, M., Torki, M. R., Demazeau, G., and Hagenmuller, P. (1971) C.R. Hebd. Seances Acad. Sci. 273, 1093.

Poulain, M., Poulain, M., and Lucas, J. (1975). Rev. Chim. Miner. 12, 9.

Propach, V., and Steffens, F. (1978). Z. Naturforsch., B: Anorg. Chem. Org. Chem. 33B, 268.

Rao, P. R., Tressaud, A., and Bartlett, N. (1976). J. Inorg. Nucl. Chem., Suppl. p. 23.

Ravez, J., and Hagenmuller, P. (1964). Bull. Soc. Chim. Fr. p. 1811.

Ravez, J., de Pape, R., and Hagenmuller, P. (1967). Bull. Soc. Chim. Fr. p. 4375.

Ravez, J., von der Mühll, R., and Hagenmuller, P. (1975). J. Solid. State Chem. 14, 20.

Ravez, J., Elaatmani, M., von der Mühll, R., and Hagenmuller, P. (1979a). Mater. Res. Bull. 14, 1083.

Ravez, J., Elaatmani, M., and Chaminade, J. P. (1979b). Solid State Commun. 32, 749.

Ravez, J., Peraudeau, G., Arend, H., Abrahams, S. C., and Hagenmuller, P. (1980). Ferroelectrics 26, 767.

Ravez, J., Elaatmani, M., Cervera-Marzal, M., Chaminade, J. P. and Pouchard, M. (1981). *Mater. Res. Bull.* **16**, 1167.

Reinen, D. (1979). *J. Solid State Chem.* **27**, 71.

Reinen, D., and Friebel, C. (1979). *Struct. Bonding (Berlin)* **37**, 1.

Reinen, D., and Steffens, F. (1978). *Z. Anorg. Allg. Chem.* **441**, 63.

Reinen, D., and Weitzel, H. (1977). *Z. Naturforsch., B: Anorg. Chem., Org. Chem.* **32B**, 476.

Renaudin, J., Pannetier, J., Pelaud, S., Ducouret, A., Varret, F., and Ferey, G. (1983). *Solid State Commun.* **47**, 445.

Rieck, D., Langley, R., and Eyring, L. (1982). *J. Solid State Chem.* **45**, 259.

Rimsky, A., Thoret, J., and Freundlich, W. (1970). *C.R. Hebd. Seances Acad. Sci., Ser C* **270**, 407.

Ruddlesden, S. N., and Popper, P. (1958). *Acta Crystallogr.* **11**, 54.

Rüdorff, W., and Krug, D. (1964). *Z. Anorg. Allg. Chem.* **329**, 211.

Ryan, R. R., Martin, S. H., and Reisfeld, M. J. (1971). *Acta Crystallogr. Sect. B* **B27**, 1270.

Ryan, R. R., Penneman, R. A., Asprey, L. B., and Paine, R. T. (1976). *Acta Crystallogr. Sect. B* **B32**, 3311.

Sabatier, R., Charroin, G., Avignant, D., Cousseins, J. C., and Chevalier, R. (1979). *Acta Crystallogr., Sect. B.* **B35**, 1333.

Sabatier, R., Vasson, A. M., Vasson, A., Lethuiller, P., Soubeyroux, J. L., Chevalier, R., and Cousseins, J.-C. (1982). *Mater. Res. Bull.* **17**, 369.

Sakashita, H., Ohama, N., and Okazaki, A. (1981). *J. Phys. Soc. Jpn.* **50**, 4013.

Sala-Pala, J., Guerchais, J. E., and Edwards, A. J. (1982). *Angew. Chem., Int. Ed. Engl.* **21**, 870.

Samouel, M. (1971). *Rev. Chim. Miner.* **8**, 537.

Samouel, M., and de Kozak, A. (1978). *Rev. Chim. Miner.* **15**, 268.

Samouel, M., and de Kozak, A. (1983). *Rev. Chim. Miner.* **20**, 37.

Schäfer, H., von Schnering, H. G., Niehues, K. J., and Nieder-Vahrenholz, H. G. (1965). *J. Less-Common Met.* **9**, 95.

Scheffler, J., and Hoppe, R. (1984). *J. Fluorine Chem.* **25**, 27.

Schmidt, R. E. (1982). Diplomarbeit, Univ. of Marburg.

Schmidt, R. E., and Babel, D. (1983). *Z. Kristallogr.* **162**, 200.

Schmidt, R. E., and Babel, D. (1984). *Z. Anorg. Allg. Chem.* **516**, 187.

Schmidt, R. E., and Babel, D. (1985). *Z. Anorg. Allg. Chem.*, in press.

Schmitz-Dumont, O., and Bornefeld, H. (1956). *Z. Anorg. Allg. Chem.* **287**, 120.

Seifert, H. J., Loh, H. W., and Jungnickel, K. (1968). *Z. Anorg. Allg. Chem.* **360**, 62.

Setter, J., and Hoppe, R. (1976a). *Z. Anorg. Allg. Chem.* **423**, 125.

Setter, J., and Hoppe, R. (1976b). *Z. Anorg. Allg. Chem.* **423**, 133.

Shannon, R. D. (1976). *Acta Crystallogr., Sect. A* **A32**, 751.

Siebert, G., and Hoppe, R. (1972a). *Z. Anorg. Allg. Chem.* **391**, 113.

Siebert, G., and Hoppe, R. (1972b). *Z. Anorg. Allg. Chem.* **391**, 117.

Siebert, G., and Hoppe, R. (1972c). *Z. Anorg. Allg. Chem.* **391**, 126.

Simonov, V. I., and Bukvetskii, B. V. (1978). *Acta Crystallogr., Sect. B* **B34**, 355.

Sly, W. G., and March, R. E. (1957). *Acta Crystallogr.* **10**, 378.

Sorbe, P. (1977). Ph.D. Thesis, Univ. of Bordeaux.

Stålhandske, C. (1979a). *Acta Crystallogr., Sect. B* **B35**, 949.

Stålhandske, C. (1979b). *Acta Crystallogr., Sect. B* **B35**, 2184.

Steffens, F., and Reinen, D. (1976). *Z. Naturforsch., B: Anorg. Chem., Org. Chem.* **31B**, 894.

Steinfink, H., and Brunton, G. (1969). *Inorg. Chem.* **8**, 1665.

Steinfink, H., and Burns, J. H. (1964). *Acta Crystallogr.* **17**, 823.

Stout, J. W., and Boo, W. (1979). *J. Chem. Phys.* **71**, 1.

Tanaka, K., Konishi, M., and Marumo, F. (1979). *Acta Crystallogr., Sect. B* **B35**, 1303.

Tanaka, K., Konishi, M., and Marumo, F. (1980). *Acta Crystallogr., Sect. B* **B36**, 1264.

Taylor, J. C., and Waugh, A. B. (1980). *J. Solid State Chem.* **35**, 137.

Taylor, J. C., and Wilson, P. W. (1974). *J. Less-Common Met.* **34**, 257.

Thoma, R. E., and Brunton, G. D. (1966). *Inorg. Chem.* **5**, 1937.

Toth, L. M., Brunton, G. D., and Smith, G. P. (1969). *Inorg. Chem.* **8**, 2694.

Tressaud, A., and Dance, J. M. (1974). *C.R. Hebd, Seances Acad. Sci., Ser. C* **278**, 463.

Tressaud, A., and Dance, J. M. (1977). *Adv. Inorg. Chem. Radiochem.* **20**, 133.

Tressaud, A., and Dance, J. M. (1982). *Struct. Bonding (Berlin)* **52**, 87.

Tressaud, A., and Demazeau, G. (1984). *High Temp.—High Pressures* **16**, 303.

Tressaud, A., Galy, J., and Portier, J. (1969). *Bull. Soc. Fr. Mineral. Cristallogr.* **92**, 335.

Tressaud, A., de Pape, R., Portier, J., and Hagenmuller, P. (1970). *Bull. Soc. Chim. Fr.* No. 10, p. 3411.

Tressaud, A., Menil, F., Georges, R., Portier, J., and Hagenmuller, P. (1972). *Mater. Res. Bull.* **7**, 1339.

Tressaud, A., Dance, J. M., Ravez, J., Portier, J., Hagenmuller, P., and Goodenough, J. B. (1973a). *Mater. Res. Bull.* **8**, 1467.

Tressaud, A., Dance, J. M., Menil, F., Portier, J., and Hagenmuller, P. (1973b). *Z. Anorg. Allg. Chem.* **399**, 231.

Tressaud, A., Dance, J. M., Portier, J., and Hagenmuller, P. (1974). *Mater. Res. Bull.* **9**, 1219.

Tressaud, A., Pintchovski, F., Lozano, L., Wold, A., and Hagenmuller, P. (1976a). *Mater. Res. Bull.* **11**, 689.

Tressaud, A., Wintenberger, M., Bartlett, N., and Hagenmuller, P. (1976b). *C. R. Hebd. Seances Acad. Sci.* **282**, 1069.

Tressaud, A., Soubeyroux, J. L., Touhara, H., Demazeau, G., and Langlais, F. (1981a). *Mater. Res. Bull.* **16**, 207.

Tressaud, A., Soubeyroux, J. L., Dance, J. M., Sabatier, R., and Hagenmuller, P. (1981b). *Solid State Commun.* **37**, 479.

Tressaud, A., Darriet, J., Lagassie, P., Grannec, J., and Hagenmuller, P. (1984a). *Mater. Res. Bull.* **19**, 983.

Tressaud, A., Khaïroun, S., Dance, J. M., and Hagenmuller, P. (1984b). *Z. Anorg. Allg. Chem.* **517**, 43.

Vedrine, A., Trottier, D., Cousseins, J.-C., and Chevalier, R. (1979). *Mater. Res. Bull.* **14**, 583.

Verscharen, W., and Babel, D. (1978). *J. Solid State Chem.* **24**, 405.

Vidal-Valat, G., Vidal, J. P., Zeyen, C., and Kurki-Suonio, K. (1979). *Acta. Crystallogr., Sect. B* **B35**, 1584.

Viebahn, W. (1971). *Z. Anorg. Allg. Chem.* **386**, 335.

Viebahn, W. (1975). *Z. Anorg. Allg. Chem.* **413**, 77.

Viebahn, W., and Babel, D. (1974). *Z. Anorg. Allg. Chem.* **406**, 38.

Viebahn, W., and Epple, P. (1976). *Z. Anorg. Allg. Chem.* **427**, 45.

Viebahn, W., Rüdorff, W., and Hänsler, R. (1969). *Chimia* **23**, 503.

Vlasse, M., Chaminade, J. P., Massies, J. C., and Pouchard, M. (1975). *J. Solid State Chem.* **12**, 102.

Vlasse, M., Menil. F., Moriliere, C., Dance, J. M., Tressaud, A., and Portier, J. (1976). *J. Solid State Chem.* **17**, 291.

Vlasse, M., Chaminade, J. P., Saux, M., and Pouchard, M. (1977a). *Rev. Chim. Miner.* **14**, 429.

Vlasse, M., Matjeka, G., Tressaud, A., and Wanklyn, B. M. (1977b). *Acta Crystallogr., Sect. B* **B33**, 3377.

Vlasse, M., Chaminade, J. P., Dance, J. M., Saux, M., and Hagenmuller, P. (1982). *J. Solid State Chem.* **41**, 272.

Volkova, L. M., Samarec, L. V., Poliščuk, S. A., and Laptaš, N. M. (1978). *Kristallografiya* **23**, 951; *Sov. Phys.—Crystallogr. (Engl. Transl.)* **23**, 536.

Vollmer, G. (1966). Ph.D. Thesis, Univ. of Tübingen.

von der Mühll, R. (1973). Thesis, Univ. of Bordeaux.
von der Mühll, R. (1974). *C.R. Hebd. Seances Acad. Sci., Ser. C* **278**, 713.
von der Mühll, R., and Ravez, J. (1974). *Rev. Chim. Miner*, **11**, 652.
von der Mühll, R., Andersson, S., and Galy, J. (1971). *Acta Crystallogr., Sect. B* **B27**, 2345.
von der Mühll, R., Daut, F., and Ravez, J. (1973). *J. Solid State Chem.* **8**, 206.
von Schnering, H. G. (1967). *Z. Anorg. Allg. Chem.* **353**, 13.
von Schnering, H. G. (1973). *Z. Anorg. Allg. Chem.* **400**, 201.
von Schnering, H. G., and Bleckmann, P. (1968). *Naturwissenschaften* **55**, 342.
von Schnering, H. G., Kolloch, B., and Kolodziejczyk, A. (1971). *Angew. Chem.* **83**, 440.
von Schnering, H. G., Vu, D., and Peters, K. (1983). *Z. Kristallogr.* **165**, 305.
Wall, F. (1971) Ph.D. Thesis, Univ. of Tübingen.
Wall, F., Pausewang, G., and Babel, D. (1971). *J. Less-Common Met.* **25**, 257.
Waltersson, K. (1978). *Cryst. Struct. Commun.* **7**, 507.
Waltersson, K. (1979). *J. Solid State Chem.* **28**, 121.
Wanklyn, B. M., Garrard, B. J., Wondre, F., and Davidson, W. (1976). *J. Cryst. Growth* **33**, 165.
Weidenborner, J. E., and Bednowitz, A. L. (1970). *Acta Crystallogr., Sect. A* **A26**, 1464.
Weissenhorn, F. J. (1972). Ph.D. Thesis, Univ. of Tübingen.
Wells, A. F. (1973). *J. Solid State Chem.* **6**, 469.
Wells, A. F. (1975). "Structural Inorganic Chemistry." Oxford Univ. Press (Clarendon), London and New York.
Wilhelm, V., and Hoppe, R. (1974a). *Z. Anorg. Allg. Chem.* **405**, 193.
Wilhelm, V., and Hoppe, R. (1974b). *Z. Anorg. Allg. Chem.* **407**, 13.
Wilhelm, V., and Hoppe, R. (1975a). *Z. Anorg. Allg. Chem.* **414**, 91.
Wilhelm, V., and Hoppe, R. (1975b). *Z. Anorg. Allg. Chem.* **414**, 130.
Williamson, R. F., and Boo, W. O. (1977a). *Inorg. Chem.* **16**, 646.
Williamson, R. F., and Boo, W. O. (1977b). *Inorg. Chem.* **16**, 649.
Winkler, H. G. F. (1954). *Acta Crystallogr.* **7**, 33.
Winkler, H. G. F., and Brehler, B. (1954). *Beitr. Mineral. Petrogr.* **4**, 6.
Wintenberger, M., Tressaud, A., and Menil, F. (1972). *Solid State Commun.* **10**, 739.
Wintenberger, M., Dance, J. M., and Tressaud, A. (1975). *Solid State Commun.* **17**, 1335.
Wu, K. K., and Brown, I. D. (1973). *Mater. Res. Bull.* **8**, 593.
Zalkin, A., Forrester, J. D., and Templeton, H. (1964). *Acta Crystallogr.* **17**, 1408.
Zaslavskii, A. I., and Tolkacev, S. S. (1952). *Zh. Fiz. Khim.* **26**, 743.

4

The Crystal Chemistry of Transition Metal Oxyfluorides

B. L. CHAMBERLAND
Department of Chemistry
and Institute of Materials Science
University of Connecticut
Storrs, Connecticut

I. Introduction

I. Introduction

The preparation and characterization of metal oxides have developed into an extensive area in the field of solid-state inorganic chemistry. This increased activity has resulted mainly from a strong interest in new materials necessary for the sophisticated applications of solid-state components required in the electronic and magnetic industry. Furthermore, an increased effort has been made to screen new compounds as catalytic agents in processes based on heterogeneous catalysis. The major theme underlining all these materials, aside from their envisioned applications, is their crystal chemistry. Many of the chemical and physical properties are directly related to the transition metal ions in a particular anion environment predicated on the crystal structure.

Since the development of many useful oxide compounds, several researchers have focused their attention on the fluoride analogs of oxide crystal systems. These investigations have generated a variety of interesting fluoride derivatives for materials applications.

Solid-state inorganic chemistry has greatly developed in the United States, France, and Germany since the late 1950s. Several texts on the subject of metal oxides, their preparation and properties, have been and are continuing to be published. Several review articles concerning transition metal fluoride compounds have also appeared. No comprehensive overview has been written on transition metal oxyfluorides since the late 1960s, however, with the exception of a review article by Siegel (1968) on (non-transition) group III oxyhalides and a brief account of rare-earth oxyfluorides in a reference book (Brown, 1968). No review articles are available in which *transition metal* oxyfluoride systems are discussed in terms of their structure and physical properties.

The area of transition metal oxyfluorides bridges the large field of "partly covalent" metal oxides and the relatively new research on "mostly ionic" metal fluorides. It is expected that in this oxyfluoride area one would observe properties intermediate between the two limits of bonding character. This aspect would greatly influence the physical and chemical properties of these metal oxyfluorides. By the proper blending of ionic–covalent character in a given structure, one should be able to modify the electrical and magnetic properties of a chemical system.

In this chapter, attention is focussed on oxyfluorides of the transition metals, namely, the $3d$, $4d$, and $5d$ transition elements, the lanthanides, and the actinides. The elements scandium and yttrium are also included in this survey; however, the large family of niobium and molybdenum oxyfluorides are not. [A review on fluoride and oxyfluoride compounds of niobium, antimony, and tellurium has been prepared (Corbin, 1982).] This chapter excludes all nonmetal oxyfluorides, molecular metal oxyfluorides, fluoro acid salts such as fluorophosphates, and highly complex oxyfluorides such as fluoromicas, hydrated oxyfluorides, and complex ionic salts.

The transition metal oxyfluorides are listed according to structural and chemical composition (simple to complex). The known structural parameters and reported physical properties are summarized in tabular form.

The ionic radii of the elements are influenced by several factors, such as oxidation state, chemical environment, coordination, electronic spin state, "covalent" character, and polarizability. For the oxygen and fluorine ions (isoelectronic species), the values of ionic radii are very similar under normal conditions of temperature and pressure. The Shannon and Prewitt (1968) values of ionic radii for O^{2-} and F^- in different coordinations (C.N. = coordination number) are given in the following tabulation:

	Ionic radius (Å)			
Ion	C.N. = 2	C.N. = 3	C.N. = 4	C.N. = 6
O^{2-}	1.35	1.36	1.38	1.40
F^-	1.29	1.30	1.31	1.33

It would then seem reasonable that fluorine would be capable of substituting for oxygen in an oxide matrix (and, similarly, oxygen substituting for fluorine in a fluoride crystal structure) to generate oxyfluoride compounds. In such substitutions, however, there must be a simultaneous change in cationic charge (charge compensation) to maintain overall electrical neutrality. For this reason the metals studied are primarily those that exhibit variable valence or oxidation states, such as the transition metals, which include the lanthanide and actinide elements. We shall focus on the low oxidation states of the transition metals, for as the oxidation state of the central metal increases the system tends to become highly covalent and molecular.

In general, the preparative aspects include direct solid-state (ceramic-type) reactions between metal oxides and metal fluorides. In some instances, however, other techniques have been utilized, and these are mentioned in this chapter. The direct solid-state reactions can be carried out under a variety of conditions [i.e., in vacuo, under an inert atmosphere, at autogenous pressure, at moderate or high pressure, or under hydrothermal conditions (Munoz and Eugster, 1969) in sealed containers]. In certain cases, pressure was found to be advantageous in the isolation of stoichiometric products in good yields, whereas for other systems, high pressure always led to disproportionation of the desired phase.

The physical properties of transition metal oxyfluorides are of great technical importance and, as mentioned before, have been the major reason for studying these systems. Electrical property characterization has indicated that semiconductors, metallic conductors, insulators, ferroelectrics, and electronic switching materials exist in the family of oxyfluoride compounds. Several metal oxyfluorides have also been studied in magnetic experiments, and certain of these systems exhibit some of the different magnetic behaviors noted in three-dimensional (3D) solids (i.e., paramagnetism and diamagnetism, ferromagnetic, antiferromagnetic, metamagnetic, ferrimagnetic, and canted-spin systems). Some transition metal oxyfluorides have been investigated for magneto-optical application because of their Faraday or Kerr effect. Oxyfluoride compounds have also been found to be promising laser hosts and fluorescent materials for optical applications.

From a structural viewpoint it should be stated that, in most cases, ordering of the anions does not normally occur. In other words, the fluorine and oxygen atoms occupy random positions in the solid crystalline structure. In only a few cases is there any evidence for ordering of the anions, and examples include the lanthanide oxyfluorides, ScOF, and Sr_2FeO_3F. The atomic scattering factors for oxygen and fluorine are so similar that the normal technique for detecting positional order in an x-ray diffraction experiment cannot be used to discriminate easily between the exact location of these two similar ions. Fluorine nmr studies, however, have shown the existence of ordered phases in well-defined systems.

In addition to the problem of anion ordering, there is the additional question as to the exact stoichiometry and chemical composition of oxyfluoride compounds. In many instances, when transition elements are involved in oxyfluoride formation, there exists the possibility that variable-composition and nonstoichiometric products will be formed because of the various oxidation states exhibited by these metals. This problem is particularly prominent in the tetragonal, rare-earth oxyfluorides in which the lanthanide can yield several compositions of the type LnO_xF_{3-2x} ($0.7 \leq x \leq 1.0$). The study of solid solutions of two end members, a metal oxide and a metal fluoride, has been undertaken in several instances to obtain more information concerning this problem of nonstoichiometric or variable-composition phases.

Reliable analytical methods (Portier and Roux, 1968) have been developed for the accurate determination of the fluorine content of metal oxyfluoride compounds. The direct determination of oxygen content by the normal fusion technique (Smith and Krause, 1968) cannot reliably be used on these compounds. The use of ion-selective electrodes has been highly valuable in determining the fluorine content in systems that readily dissolve in appropriate solvents.

Aside from the abundant naturally occurring fluoromicas and fluoroapatites, very few metal oxyfluorides exist in nature. Essentially all transition metal oxyfluorides are synthetic, and many are structurally related to well-known oxide structures, such as spinel, cassiterite, and garnet. Some oxyfluorides are related to common fluoride structures such as the chiolite, K_2NiF_4, and fluorite structures. The fluoride mineral cryolithionite is structurally similar to its oxygen counterpart, the mineral garnet, both having the general composition $M_3M_2'(M''X_4)_3$ (X = F or O).

It should be cautioned that a small amount of fluorine in an oxide matrix can radically change the physical properties of a pure oxide material. The detection of low-level fluoride concentrations is difficult, and often a measure of change in physical property is more readily accomplished than the analytical determination of fluorine content. Many solid-state chemists and physicists use fluorides as low-melting and highly soluble fluxes [e.g., KF or PbF_2 (Weaver and Li, 1969)], but the possibility of introducing fluorine in the desired crystalline phase is quite probable. As an example, the dc resistivity in $BaTiO_3$ was greatly increased when the compound was heated (Richter and Cook, 1966) in the presence of BaF_2. The cubic-to-tetragonal transition was also shifted to lower temperatures by the fluorine substitution. Crystals of $BaTiO_3$ isolated from fused KF were found (Arend et al., 1967; Buessem et al., 1963) to contain 0.1 wt % of F^-. This foreign-ion incorporation led to defect structures and reduced forms of $BaTiO_3$. Sintering of $BaTiO_3$ with LiF has been studied by Haussonne et al. (1983), and they report the cubic phase as well as a second tetragonal form of $BaTiO_3$.

Different cyrstallographic forms of Ca_4PtO_6 were isolated when different fluxes were used (Shaplygin and Lazarev, 1975) in the same temperature region during crystal growth experiments. The use of $CaCl_2$ as a flux yielded the normal orthorhombic phase, whereas KF or PbF_2 formed hexagonal polymorphs of the "same" Ca_4PtO_6 compound.

Aqueous HF is often used to remove impurities, for example, to separate tantalum oxides from niobium oxides. In a sample of "Specpure" niobium pentoxide, obtained from a reliable chemical supply house, colorless, rectangular plates were handpicked, and single-crystal x-ray studies showed (Andersson, 1964) that these crystals had the general composition Nb_3X_8 and that the compound was not a polymorph of the oxide Nb_2O_5. Further studies showed this orthorhombic phase to be pure Nb_3O_7F, and its direct synthesis soon followed.

The oxidation rates of niobium, tantalum, and tungsten have been observed to be greatly retarded (Bhat and Khan, 1965, 1966) by a fluoride treatment. The metals, in powder or sheet form, were treated with KHF_2 solutions and were then found to have decreased reactivity to O_2 on heating. It was concluded that fluorine was substituted for oxygen in the protective oxide layer, which further assisted in the oxidation resistance.

210   B. L. Chamberland

II. Transition Metal Oxyfluoride Compositions

A. MOF COMPOUNDS

1. Rutile or Cassiterite Structure

Titanium dioxide has three different crystalline structures and as such is polymorphic. The most stable mineral form has the rutile structure. To circumvent the polymorphic problem in discussing the structure of TiO_2, it will be more advantageous to discuss the details of the rutile structure in terms of its isostructural analog, SnO_2, or the mineral cassiterite.

In cassiterite, $^{VI}Sn^{III}O_2$, the metal atom lies in an octahedral coordination of oxygen atoms. The anions are trigonally coordinated by the metal atoms. The octahedra in this structure are distorted by compressing along four of the anions (equatorial) to give four short metal–oxygen bonds and two long metal–oxygen links. The octahedra are interconnected by edge sharing along the c crystallographic axis and by vertex sharing with other chains of $SnO_6{}^{8-}$ octahedra (Fig. 1). The tetragonal crystal structure has the space group $P4_2/mnm$, and the unit cell contains two formula weights of SnO_2.

This structure is adopted by most of the tetravalent transition metals (excluding zirconium and hafnium) of ionic radii about 0.5–0.75 Å. The high-pressure form of SiO_2, stishovite, also possesses the cassiterite-type structure. Divalent transition metal fluorides also crystallize with this structure, CrF_2 and CuF_2 being monoclinic variants of the basic structure.

Stoichiometric oxyfluoride formation with transition metals requires a stable trivalent state as in $M^{3+}OF$. The trivalent metals capable of forming ternary MOF derivatives include titanium, vanadium, chromium, manganese,

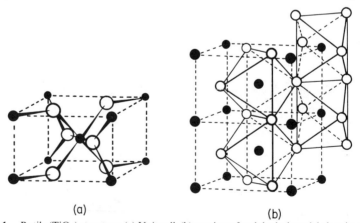

(a)   (b)

Fig. 1. Rutile (TiO_2) structure. (a) Unit cell; (b) portion of polyhedral model showing opposite edge and vertex bonding (●, Ti; ○, O). Adapted from Wells (1975) with permission of Oxford University Press.

iron, cobalt, and rhodium. Of these, only TiOF, VOF, and FeOF have been characterized as single-phase stoichiometric products. The exclusion of MnOF is believed to be caused by the Jahn–Teller distortion of Mn^{3+} and the high stability of the oxide (MnO_2) and fluoride (MnF_2), both possessing the cassiterite-type structure. The chromium system yields an oxyfluoride only at the oxygen-rich end of the compositional phase diagram, and stoichiometric CrOF is not obtained because of the high stability of the reactants or products: Cr_2O_3 and CrF_3.

Several mixed-metal or more complex derivatives and variable compositions in the solid solution $MO_{2-x}F_x$ have been reported with this crystal system. No evidence for an ordered anion arrangement has been presented for the cassiterite-type oxyfluoride products.

The first reported example of a cassiterite-type oxyfluoride was FeOF, prepared (Hagenmuller et al., 1965) by a solid-state reaction,

$$MF_3 + M_2O_3 \longrightarrow 3\,MOF$$

which was carried out at 950°C under an atmosphere of O_2 to ensure the trivalent state in the product. Several other transition metal oxyfluorides have since been prepared (Chamberland and Sleight, 1967; Chamberland et al., 1970) by solid-state reactions under high pressure. The high-pressure conditions were effective in increasing the yield, maintaining the stoichiometry of the desired product, and forming single crystals of the transition metal oxyfluorides.

The unit cell parameters of cassiterite-type oxyfluoride products prepared by different solid-state reactions are listed in Table I.

A 3D crystallographic study on single crystals of FeOF has been carried out, and the results indicate the cassiterite-type structure with space group $P4_2/mnm$ and a random arrangement of oxygen and fluorine in anion sites of the structure (Vlasse et al., 1973).

The magnetic structure in FeOF was derived from Mössbauer and neutron diffraction studies by Chappert and Portier (1966a,b). Antiferromagnetic spin alignment on iron sites was observed below the transition temperature of 315 K. The magnetic and electrical data obtained on oxyfluoride phases are presented in Table II. The resistivity data for TiOF, VOF, and FeOF are shown as a function of temperature in Fig. 2. The negative Seebeck coefficient ($\alpha = -290$ and -305 $\mu V/deg$) for TiOF measured (Chamberland et al., 1970) at 46°C indicates that electrons are the major charge carriers in this antiferromagnetic semiconductor.

A study on the cassiterite-type system $FeOF/FeF_2$ was carried out by Hagenmuller et al. (1965). They found a large immiscibility gap in the FeO_xF_{2-x} system at $0.2 \leq x \leq 0.9$ from crystallographic data. Solid-solution formation in the systems FeO/FeF_2 and MnO/MnF_2 was investigated by

TABLE I

Tetragonal Cell Parameters for Cassiterite Transition Metal Oxyfluorides

Composition	Cell Parameters		Density (g/cm³)		Reference[b]
	a (Å)[a]	c (Å)[a]	d_{obs}[a]	d_{calc}	
TiOF	4.651	3.013	—	4.22	1, 2
Ti_2O_3F	4.63	2.96	—	4.26	2
VOF	4.623	3.025	4.46	4.41	2
V_2O_3F	4.57	2.94	—	4.57	2
FeOF	4.662	3.043	—	4.56	1, 2
FeOF	4.654(3)	3.058(3)	4.53(1)	4.55	3
FeOF	4.647(5)	3.048(9)	4.53	4.58	4, 5
$FeVO_3F$	4.57	2.97	—	4.65	2
$TiVO_3F$	4.62	2.99	—	4.31	2
$FeTiO_3F$	~4.63	~3.01	—	4.39	2
$Ti_{0.25}Fe_{0.75}O_{0.5}F_{1.5}$	4.713	3.236	—	4.17	2
$Ti_{0.5}Fe_{0.5}OF$	4.644	3.010	—	4.44	2
$Ti_{0.25}Fe_{0.75}OF$	4.662	3.045	—	4.46	2
$Ti_{0.33}Fe_{0.67}OF$	4.656	3.026	—	4.47	2
$Cr_{0.33}Ti_{0.67}OF$	4.644	3.037	—	4.27	2
$V_{0.33}Ti_{0.67}OF$	4.643	3.013	—	4.29	2
$NbTi_2O_5F$	4 703(2)	3.008(2)	—	—	6
NbV_2O_5F	4.655(2)	3.032(2)	4.94	4.95	6
$NbCr_2O_5F$	4.700(5)	9.180(9)	—	-	6

[a] Numerals in parentheses indicate the standard deviation in the last figure given.

[b] 1, Chamberland and Sleight (1967); 2, Chamberland et al. (1970); 3, Vlasse et al. (1973); 4, Chappert and Portier (1966a); 5, Hagenmuller et al. (1965); 6, Senegas and Galy (1973).

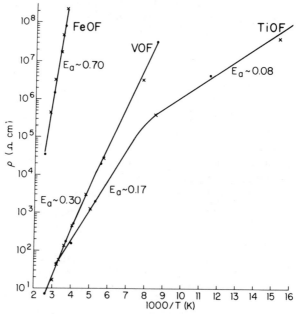

Fig. 2. Four-probe resistivity data versus reciprocal of temperature plot for single-crystal MOF compounds. From Chamberland et al. (1970).

TABLE II

Magnetic and Electrical Data for Rutile-Type Oxyfluorides

Composition	Magnetic Properties				Electrical Properties		
	Magnetic Type[a]	Néel Temp T_N (K)	Weiss Constant (K)	p_{eff}^2 $(\mu_B)^2$	Resistivity at 298 K $(\Omega \, cm)$[b]	E_a (eV)	Reference[c]
TiOF	AF	—	−77	—	4.5	0.17	1
VOF	AF	—	−10	~7	60	0.30	1
FeOF	AF	315	—	—	2×10^7	0.70	2, 3
FeO_xF_{2-x} ($0 \leq x \leq 0.18$)	—	—	—	—	$\sim 10^{10}$ (*)	—	4
$Ti_2O_{2.3}F_{1.7}$	AF	—	−77	2.4	17 (*)	0.14	1
Ti_2O_3F	—	—	—	—	~3.0 (*)	—	1
V_2O_3F	AF	—	−70	~11	0.9 (*)	~0.04	1
$VTiO_3F$	AF	19	18	7.2	9×10^3 (*)	0.25	1
$FeVO_3F$	—	—	—	—	1.6×10^2 (*)	0.22	1
$FeTiO_3F$	—	—	—	—	5×10^5 (*)	—	1
$Fe_{0.75}Ti_{0.25}OF$	AF	~100	—	23.4	—	—	1
$Fe_{0.67}Ti_{0.33}OF$	AF	~80	—	16	—	—	1
NbV_2O_5F	—	—	−17	—	—	—	5

[a] AF, Antiferromagnetic.

[b] (*) denotes powder data.

[c] 1, Chamberland et al. (1970); 2, Chappert and Portier (1966a); 3, Chappert and Portier (1966b); 4, Hagenmuller et al. (1965); 5, Senegas and Galy (1973).

Austin (1969) utilizing high-pressure conditions, but no evidence for reaction or single-phase products was found. Other studies, however, such as those on the NiO/NiF_2 and CoO/CoF_2 systems, at 50 kbars pressure and 800°C, yielded partial solubility in the MO_xF_{2-2x} series, where $0 \leq x \leq 0.37$ for nickel and an indefinite range for cobalt. The nickel oxyfluoride derivative possessed a powder diffraction pattern similar to that found for high-pressure NiF_2, indexed on an orthorhombic basis (see following tabulation).

	Unit Cell Parameters				
	a (Å)	b (Å)	c (Å)	b/a	Refractive Index
NiF_2 (I)	4.56(1)	4.77(1)	3.065(5)	1.02	1.600
$NiO_{0.37}F_{1.26}$	—	—	—	1.09	~1.654

The orthorhombic form may be anion deficient or might contain an excess of metal atoms.

In addition to these formulations, large families of oxyfluoride derivatives can be formed by changing the oxygen/fluorine ratio (yet maintaining the proper cation/anion ratio), as seen from the solid solutions $MO_{2-x}F_x$ obtained from the end members MO_2 and MF_2. In such a solid solution the metal ion must possess stable tetravalent, trivalent, and divalent oxidation states. For the compositions M_2O_3F and M_2OF_3, the cation/anion ratio is maintained at 1:2, but the metal atoms must exhibit mixed valence states of M^{3+}/M^{4+} and M^{2+}/M^{3+}, respectively. The latter type of oxyfluoride is found primarily for the early transition elements because these metals can exist in several different oxidation states. Studies on two first-row transition elements (vanadium and chromium) have been carried out and are discussed here in detail.

a. $V_2O_{4-x}F_x$ SYSTEM.   This system is of interest for two main reasons. The fluorine-rich end, where $x = 2.0$, is isoelectronic with CrO_2, and thus this oxyfluoride was predicted to possess some interesting ferromagnetic properties, as observed in the oxide analog. The oxide of tetravalent vanadium has an important semiconductor-to-metal transition near room temperature (68°C) and therefore has some important practical applications. The substitution of a small amount of fluorine for oxygen in this system could produce some interesting electronic oxyfluoride derivatives. Studies on this system have been reported by Chamberland (1970, 1971b) and by Bayard et al. (1971, 1975a,b).

The preparation of single-phase $V_2O_{4-x}F_x$ products was accomplished by the hydrothermal reaction of vanadium oxides and aqueous HF or NH_4HF_2 at 700°C and 3 kbars in sealed, collapsible gold capsules. The products exhibited semiconductor-to-metal transitions when the fluorine concentration

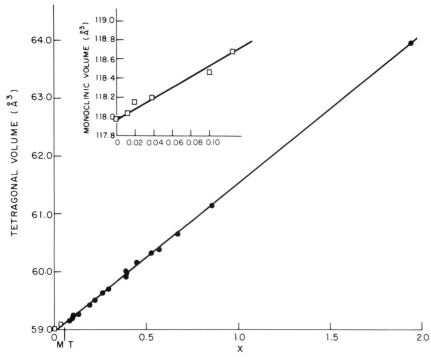

Fig. 3. Unit cell volume versus fluorine content (x) in $V_2O_{4-x}F_x$. M, Monoclinic; T, tetragonal. From Chamberland (1971b) with permission of Pergamon Press.

was in the region $0 \leq x \leq 0.20$. As x increased, the transition temperature T_t decreased. In addition to the decreased transition temperature, the magnitude of resistivity change diminished to a point that it was no longer detectable at x values greater than 0.17. The composition $V_2O_{3.92}F_{0.08}$ exhibited an electronic transition at room temperature, and at higher fluorine concentrations only semiconductors were formed, possessing the cassiterite-type structure at all temperatures.

The volume of the monoclinic cell (VO_2 structure at room temperature) in the cassiterite-type oxyfluoride products increased as the fluorine content increased (Fig. 3). This increased cell size is caused primarily by the larger V^{3+} ion in the charge-compensated products of composition $V_{2-x}^{4+}V_x^{3+}O_{4-x}^{2-}F_x^-$.

b. $CrO_{2-x}F_x$ SYSTEM. Attempts to prepare CrOF from Cr_2O_3 and CrF_3 under a variety of conditions and by several methods were unsuccessful, as described earlier. Several members of the $CrO_{2-x}F_x$ series with the tetragonal, cassiterite-type structure, however, could be isolated (Chamberland et al., 1973) under high-pressure conditions (60–65 kbars). Homogeneous,

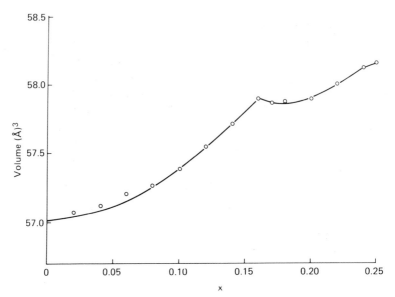

Fig. 4. Unit cell volume versus fluorine content (x) in the $CrO_{2-x}F_x$ system. From Chamberland *et al.* (1973).

single-phase products were formed in the region $0 \leq x \leq 0.28$. The tetragonal volume increased with increasing fluorine concentration, exhibiting an anomaly in the plot of volume versus composition at $x = 0.15$ (Fig. 4). The products were ferromagnetic, and the Curie temperature decreased from 398 K for pure CrO_2 to 213 K for the composition $CrO_{1.84}F_{0.16}$ (Fig. 5). Metallic conductivity was observed in the oxygen-rich phases, and semimetallic behavior was detected in fluorine-rich compositions, $0.10 \leq x \leq 0.20$. For the latter compositions low-temperature metamagnetic behavior was detected in the magnetic experiments. Charge compensation of the type $Cr_{1-x}^{4+}Cr_x^{3+}O_{2-x}^{2-}F_x^-$ has been used to explain the magnetic and electrical properties observed in this system.

A small amount of fluorine has also been introduced in the cassiterite-type CrO_2 ferromagnet by a hydrothermal process. Divalent transition metal fluorides were used as modifers, and the chromium oxyfluoride products were noted to possess decreased Curie temperature through the incorporation of fluorine and/or metal modifiers (Ingraham and Swoboda, 1962).

c. MIXED SYSTEMS.   A system intermediate between CrOF and VOF was studied by Bayard *et al.* (1975a). The general composition can be expressed by the formula $Cr_xV_{1-x}O_{2-x}F_x$, and a solid solution was obtained for the region $0 \leq x \leq 0.20$. x-Ray, magnetic, and electrical data for several of the intermediate compounds have been presented.

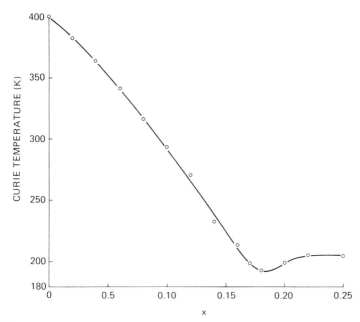

Fig. 5. Curie temperature (K) dependence as a function of the fluorine content (x) in the $CrO_{2-x}F_x$ system. From Chamberland *et al.* (1973).

The preparation of mixed transition metal oxyfluoride pigments possessing the cassiterite-type structure was discussed in general terms by Hund in 1962 (Hund, 1962a). The compositions $A^+M_2^{5+}O_5F$, $A^+M^{6+}O_3F$, $A_2^+M_3^{3+}OF_9$, $A^{2+}M^{3+}OF_3$, and $A_2^{3+}M^{3+}O_3F_3$ were disclosed. The overall cation/anion ratio is 1:2. The compound $Li_2Cr_3OF_9$ was claimed to be a purple pigment, but no crystallographic data were presented and the product was not further characterized. A broad patent covering these pigment phases has been issued (Hund, 1962b).

Senegas and Galy reported (1972) a study on the NiF_2/Nb_2O_5 system in which they attempted to prepare an ordered cassiterite phase. The stoichiometric phase $Ni_2NbO_3F_3$ was prepared, but no superstructure could be detected in the crystallographic data. The solid solution $Ni_{3-2x}Nb_{2x}O_{6x}F_{6-6x}$ was investigated, and single-phase products were obtained for $0 \le x \le 0.65$. These products possessed the cassiterite structure; ordering to the "trirutile" or the columbite structure was not observed. Ordering to a trirutile structure was observed, however, in the case $NbCr_2O_5F$ (Senegas and Galy, 1973). Several mixed-metal oxyfluorides of general composition $MM'M''O_3F_3$, $MM'M''O_2F_4$, and $MM'M''O_4F_2$ have been reported by Odenthal *et al.* (1974).

2. Fluorite Structure

a. OXYFLUORIDES OF THE LANTHANIDE AND ACTINIDE ELEMENTS. Lanthanide ions are larger than transition metal ions and therefore require a higher coordination than that found in the cassiterite structure. The common octacoordinated AX_2 structure is the fluorite structure, and several lanthanide oxyfluorides adopt this structure. In this cubic structure, however, the fluorine and oxygen atoms normally occupy random anion sites. When the anions order, as is sometimes observed in stoichiometric and nonstoichiometric derivatives, superstructural features are observed, and the simpler fluorite structure becomes less stable relative to these ordered structures. In this section, therefore, ordered structures formed for various compositions with formula MOF will be mentioned, as will fluorine-rich MO_xF_{3-2x} compounds.

Oxyfluorides of the lanthanide group adopt at least three different crystalline structures under normal preparative conditions. The crystalline phases with structural features and ordering information are listed in the following tabulation:

Structure Type	Anion Order	Crystal Class	Symbol	Chemical Composition
CaF_2, fluorite	Random	Cubic	α-MOF	Nearly stoichiometric
CaF_2, superstructure	Order along [111]	Rhombohedral	β-MOF	Stoichiometric
PbFCl	Order along [001]	Tetragonal	γ-MO_xF_{3-2x}	Excess fluorine $(0.7 < x < 1.0)$
$PbCl_2$	—	Orthorhombic	MOF (I)	Stoichiometric (high pressure)

The nonstoichiometric tetragonal phases have a wide homogeneity range and are stabilized by the presence of excess F^- over the ideal composition MOF. The ordered rhombohedral phases undergo a reversible crystallographic transition to the cubic fluorite structure at 500 to 625°C, which underlines the close structural relationship between the two stoichiometric systems.

b. CUBIC FLUORITE-TYPE OXYFLUORIDES. The mineral fluorite is chemically calcium fluoride, CaF_2. In the fluorite structure the cations form a face-centered cubic array, and the fluorine atoms occupy the eight tetrahedral interstices formed by the large cations. These anions can also be pictured as forming a simple cube interpenetrating the face-centered cubic array of cations. The geometry about the metal ion in the fluorite structure is

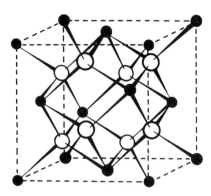

Fig. 6. Unit cell of the fluorite (CaF_2) structure (●, Ca; ○, F). Adapted from Wells (1975) with permission of Oxford University Press.

cubic or eightfold coordination, whereas the anion has fourfold coordination in a tetrahedral arrangement. The formula is expressed as $^{VIII}Ca^{IV}F_2$, and the unit cell representation is shown in Fig. 6. The crystal class is cubic with space group $Fm3m$, and Z equals four formula units.

The ionic radii of the trivalent, octacoordinated lanthanide ions vary between 1.18 Å for lanthanum and 0.97 Å for lutetium. The radii for the actinides are slightly smaller: 1.06 to ~0.92 Å at the end of the $5f$ series Shannon and Prewitt, 1968.

The preparation of these metal oxyfluorides can easily be accomplished by the pyrohydrolysis of the fluorides at 800°C or by the direct solid-state reaction of oxide with fluoride in an inert atmosphere.

The cubic fluorite structure is known for stoichiometric and nearly stoichiometric oxyfluorides of the lanthanide elements. There is evidence that slight oxygen deficiency or slight fluorine excess tends to stabilize this cubic structure. In all cases, however, the structure possesses a random distribution of the two kinds of anions. The cubic structure has also been observed when the rhombohedral (ordered) oxyfluorides are heated above 500 to 625°C. The cubic phase, however, cannot be quenched if the compound is stoichiometric. In contrast, the derivatives containing a slight excess of fluorine, when prepared at high temperature, always retain the cubic structure when quenched to room temperature.

The crystallographic unit cell dimensions for the fluorite-type oxyfluorides are given in Table III.

c. RHOMBOHEDRAL FLUORITE-TYPE (ORDERED) OXY-FLUORIDES. The cubic fluorite-type derivatives, containing a random arrangement of F^- and O^{2-} in the anion sites, can also be referred to a rhombohedral unit cell having $a = 7.02$ Å and $\alpha = 33.36°$ for the model compound CaF_2. The superstructural arrangement of anions in the rhombohedral form of LnOF requires an ordering of anions with threefold symmetry

TABLE III

Unit Cell Parameters for Lanthanide and Actinide Oxyfluorides Having the
Cubic Fluorite Structure

Composition	Color	Unit Cell Dimension ($\mathring{A}$)	Reference[a]
AcOF	Colorless	5.943(2)	1
$CeO_{1.14}F_{0.88}$	Black	5.703(1)	2
CeOF	—	5.66–5.73(1)	3
CeOF	—	5.697(3)	4
$CeO_{1.145}F$	—	5.66	4
$CeO_{1.1}F_{0.9}$	—	5.68	4
CfOF	—	5.561(4)	5
EuOF (at 800°C)	—	5.53(1)	6
HoOF	—	5.523(3)	7
LaOF	Colorless	5.756(3)–5.82	3
NdOF + (F)	Purple	5.595	8
PmOF	Pink	5.560	9
PrOF + (F)	Brown	5.644	8
PuOF	—	5.71(1)	1
SmOF	Gray-green	5.519	8
ThOF	Gray-white	5.685	10
YOF	Colorless	5.363	11

[a] 1, Zachariasen (1949); 2, Baenziger et al. (1954); 3, Finkelnberg and Stein (1950);
4, Pannetier and Lucas (1969); 5, Peterson and Burns (1968); 6, Tanguy et al. (1973);
7, Zalkin and Templeton (1953); 8, Mazza and Jandelli (1951); 9, Brown (1968); 10,
Lucas and Rannou (1968); 11, Zachariasen (1951).

about the body diagonal [111] of the original cube, as shown in Fig. 7. The rhombohedral phases are believed to be highly stoichiometric. The known compounds are listed in Table IV with their reported unit cell parameters. The rhombohedral structure has the space group $R\bar{3}m$ with $Z = 2$. The hexagonal representation has six formula weights per unit cell. The volume versus the cube of the radius for the lanthanide ion follows a linear plot. Colors for the different rare-earth oxyfluorides have been reported by Popov and Knudson (1954).

Several studies have been carried out on the high-temperature conversion of the rhombohedral LnOF to the disordered (cubic) fluorite structure. Accurate determination of the transition temperature T_t is difficult to obtain because of the sluggish transformation, which gives rise to broad peaks in the differential thermogram. The results of these studies are presented in Table V.

The fluorescence properties of LaOF and YOF (doped with 5% Th^{3+}) were investigated by Blasse and Bril (1967). These oxyfluoride compounds showed good quantum efficiency under ultraviolet excitation but relatively low radiant efficiency.

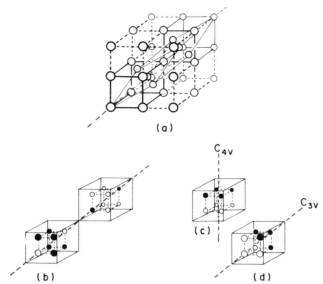

Fig. 7. Ordered fluorite structures for LnOF compounds. (a) Alternative rhombohedral unit cell wherein the Ca^{2+} ions lie at the centers of all faces in the smaller cubes; only a sufficient number of these ions are indicated to show the rhombohedral cell. (b) Ordered LaOF structure. (c) Tetragonal structure with C_{4v} symmetry. (d) Another ordered structure with C_{3v} symmetry. Adapted from Wells (1975) with permission of Oxford University Press.

d. TETRAGONAL LANTHANIDE OXYFLUORIDES. A third modification of the fluorite structure occurs in oxyfluorides when the ordering of anions is of another type. In this superstructural form the crystal class is tetragonal (C_{4v} symmetry). The model structure for this type of ordering is that of PbFCl, having the space group $P4/nmm$ and $Z = 2$. This tetragonal form is stable over a range of compositions expressed by the general formula LnO_xF_{3-2x}, where $0.7 \leq x \leq 1.0$. At the fluorine-rich limit there is an excess of anions over the amount necessary in the fluorite structure, and these "extra" fluorine atoms occupy interstitial sites along the edge of the tetragonal cell. The unit cell parameters for several of these oxyfluorides are presented in Table VI.

An interesting aspect of the structural data is that for the lanthanum and neodymium compounds the c/a ratio is greater than 1.414, the value for the ideal cubic fluorite structure referred to the tetragonal cell, whereas that for the gadolinium to erbium compounds is less than the theoretical value. This variation may be related to more efficient packing of anions in ordered layers and the size of the lanthanide ion in such an arrangement.

e. OXYFLUORIDES WITH THE PbCl$_2$ STRUCTURE. The rhombohedral LnOF compounds were treated (Pistorius, 1973) at high pressure to

TABLE IV

Crystallographic Data for Rhombohedral Oxyfluorides of the Lanthanide Elements

Composition	Color	Rhombohedral Parameters		Hexagonal Parameters		Reference[a]
		a (Å)	α (deg)	a (Å)	c (Å)	
CeOF	Black	7.075	33.00	4.010	20.05	1
DyOF	Colorless	6.719(3)	33.02(2)	3.819	19.04	2
ErOF	Pink	6.639(1)	33.10(1)	3.782	18.81	2
EuOF	Colorless	6.827(2)	33.05(2)	3.884	19.34	2–4
GdOF	Yellow	6.801(1)	33.05(1)	3.869	19.27	2
HoOF	Yellow	6.687(1)	33.02(2)	3.801	18.95	2
LaOF	Gray-lilac	7.132(1)	33.01(1)	4.051	20.21	5
LuOF	—	6.536(2)	33.05(2)	3.718	18.52	6
NdOF	Purple	6.953(1)	33.04(1)	3.953	19.70	3
PrOF	Brown	7.016(4)	33.03(3)	3.989	19.88	3
SmOF	Gray-green	6.863(2)	33.10(2)	3.907	19.45	2
TbOF	Colorless	6.751(5)	33.09(3)	3.844	19.13	7
TmOF	—	6.604(2)	33.06(2)	3.758	18.71	2
YOF	Colorless	6.697(5)	33.20(2)	3.826	18.97	8
YOF (stoichiometric)	—	6.666(2)	33.09(1)	3.797(1)	18.89(1)	9
YbOF	—	6.569(2)	33.04(2)	3.736	18.61	2

[a] 1, Pannetier and Lucas (1969); 2, Niihara and Yajima (1971); 3, Baenziger et al. (1954); 4, Tanguy et al. (1973); 5, Zachariasen (1951); 6, Niihara and Yajima (1972); 7, Templeton and Dauben (1954); 8, Templeton (1957); 9, Mann and Bevan (1970).

TABLE V

Rhombohedral-to-Cubic Transition Temperatures for Lanthanide Oxyfluorides

	Transition Temperature T_t (°C)[a]			
Composition	Niihara and Yajima (1972)	Shinn and Eick (1969)	Pistorius (1973)	Bedford and Catalano (1970)
DyOF	579	558	—	—
ErOF	595	592	592.5(1.0)	—
EuOF	495	513	—	503–505(10)
GdOF	613	606	601.5(2)	—
HoOF	590	588	—	—
LaOF	505	494	485.5(1.0)	—
NdOF	515	517, 524	—	—
SmOF	530	524	523	501–508(5)
TbOF	544	551	—	—
YOF	560	571	—	—

[a] Mean temperature underscored.

yield phases possessing the orthorhombic $PbCl_2$ structure. The reaction conditions and structural data are given in Table VII.

Little or no structural transformation was observed (Gondrand *et al.*, 1970) in TbOF, DyOF, and YOF at 80 to 90 kbars and 1000 to 1100°C after 1 hr. The high-pressure $PbCl_2$ phases are metastable and revert to their original STP forms on heating to 400°C. The reconversion of LaOF occurs at a lower temperature, 340°C.

3. Other Trivalent Metal Oxyfluorides

Most of the trivalent MOF compounds crystallize with the hexa-coordinated cassiterite or the octacoordinated fluorite structure, as previously noted. However, the trivalent metal ion scandium, which does not contain any *d* electrons, tends to adopt a different structure primarily because of its intermediate size (0.87 Å), which requires heptacoordination.

The compound ScOF, prepared from a solid-state reaction between ScF_3 and Sc_2O_3 in a sealed evacuated system, was first reported (Kutek, 1964) to crystallize with the cubic fluorite structure. The cubic unit cell parameter for the product obtained was reported to be 5.575 Å. A crude form of the same product was also obtained by hydrolysis of ScF_3 in moist nitrogen at 800°C after 5 hr.

More recently, Holmberg (1966) repeated the solid-state preparation and reported a different crystal structure. The product was observed to be isostructural with the monoclinic form of ZrO_2 (baddeleyite), space group

TABLE VI

Unit Cell Parameters for Tetragonal Oxyfluorides with the PbFCl Structure

Composition	Unit Cell Parameters			
	a (Å)	c (Å)	c/a Ratio	Reference[a]
$CeO_{0.75}F_{1.50}$	4.08	5.74	1.41	1
$DyO_{0.77}F_{1.46}$	3.933(1)	5.45(1)	1.386(1)	2
$ErO_{0.85}F_{1.3}$	3.893(1)	5.400(2)	1.387(1)	2
$ErO_{0.80}F_{1.4}$	3.907(1)	5.385(1)	1.387(1)	2
$GdO_{0.72}F_{1.56}$	3.977(1)	5.528(1)	1.390(1)	2
$LaO_{0.7}F_{1.6}$	4.106(2)	5.852(4)	1.425(2)	3
γ-LaOF	4.091(1)	5.837(1)	1.427(1)	3
$NdO_{0.85}F_{1.3}$	3.999(2)	5.704(3)	1.426(2)	2
$NdO_{0.73}F_{1.04}$	4.014(3)	5.720(4)	1.425(2)	2
PmOF	3.980	5.58	1.39	4
PuOF	4.05	5.72	1.41	3
$YO_{0.7}F_{1.6}$	3.930(5)	5.46(1)	1.389(1)	3
γ-YOF	3.910(5)	5.43(1)	1.389(4)	3

[a] 1, Pannetier and Lucas (1969); 2, Shinn and Eick (1969); 3, Zachariasen (1951); 4, Brown (1968).

TABLE VII

Lanthanide Oxyfluorides with the $PbCl_2$ Structure[a]

Composition	Preparative Conditions			Unit Cell Parameters			Volume Change (%)
	P (kbars)	T (°C)	t (hr)	a (Å)	b (Å)	c (Å)	
LaOF	40–87	800–1000	1	3.92	6.42	7.15	−6.1
PrOF	90	1000	$\frac{3}{4}$	3.84	6.32	7.03	−6.9
NdOF	5–90	1000–1180	1–2	3.84	6.28	6.93	−6.4
SmOF	90	1000	2	3.76	6.25	6.88	−6.0
EuOF	80	1000	1	3.76	6.16	6.81	−6.4
GdOF	90	1000	2	3.74	6.15	6.81	−5.5

[a] From Gondrand et al. (1970).

$P2_1/c$. Scandium, like zirconium in ZrO_2, is in sevenfold coordination. The monoclinic unit cell parameters for monoclinic ZrO_2 and ScOF are compared in the following tabulation:

Compound	a (Å)	b (Å)	c (Å)	β (deg)
ZrO_2	5.1454(5)	5.2075(5)	5.3107(5)	99°14′(5)
ScOF	5.1673(5)	5.1466(5)	5.2475(8)	99°42′(5)

Barker (1968) presented some theoretical energy calculations that suggested ordering of anions in the reported structure. ^{19}F-nmr studies support (Vlasse et al., 1979) the proposed anion ordering. Scandium is bonded to three fluorine and four oxygen atoms. The cationic coordination of fluorine is triangular, whereas that about oxygen is tetrahedral. The formula can therefore be written as [VII]Sc[IV]O[III]F.

Monoclinic ScOF has been treated (B. L. Chamberland, unpublished results, 1968) at high pressure and high temperature to yield a slightly distorted monoclinic phase. The expected cubic fluorite structure was not stabilized. The high-pressure form transforms to the ambient-pressure, monoclinic form at $\sim 418°C$.

B. MO_2F AND MOF_2 COMPOUNDS

For the stoichiometric compositions to be discussed, the M atom can possess either a pentavalent or a tetravalent oxidation state, respectively. When the M cation is small (ionic radius ~ 0.62 Å) the products may crystallize with the ReO_3 structure or one closely related to it. When the M metal ion is larger, as in the case of the lanthanides or actinides (ionic radius ~ 1.0 Å), a different structure arises.

1. ReO_3 Structure

The ReO_3 structure can be described as a continuous 3D array of vertex-shared octahedra (Fig. 8). The coordination of the ions is shown in the formula [VI]Re[II]O_3. Its crystal class is cubic, and the space group is $Pm3m$. Many trifluorides crystallize with a slightly distorted version of this structure in which the M–F–M angle is between 132 and 180°. All oxyfluoride compounds

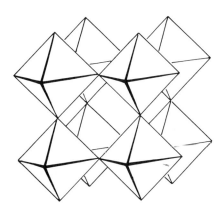

Fig. 8. Polyhedral model for the ReO_3 structure. From Wells (1975) with permission of Oxford University Press.

with this structure possess a random distribution of fluorine and oxygen in the anion sites. There is but one formula per unit cell.

The only reported stoichiometric oxyfluoride compounds with the cubic ReO_3 structure are $TiOF_2$, TaO_2F, and NbO_2F. The niobium system forms an entire solid solution between NbO_2 and NbF_3, all possessing a cubic unit cell (Dehnicke, 1965b; Ehrlich et al., 1955; Gortsema and Didchenko, 1965; Pouchard et al., 1971; Schäfer et al., 1964, 1965). Molybdenum, tungsten, and tantalum oxyfluorides have also been reported to generate variable compositions of general formula $MO_{3-x}F_x$ and MO_xF_{3-x}. Pure NbF_3 and MoF_3 are believed to contain traces of oxygen, which assist in stabilizing the cubic ReO_3 structure. Compounds reported to have the cubic ReO_3 structure are listed in Table VIII.

Metallic conductivity has been measured (Sleight, 1969b) in a series of molybdenum compounds $MoO_{3-x}F_x$ ($0.41 \leq x \leq 0.97$); however, semiconductivity behavior and a band gap of 0.13 eV were observed in (blue) $MoO_{2.75}F_{0.25}$. Magnetic susceptibility data in the cubic, metallic molybdenum oxyfluorides indicate temperature-independent behavior indicative of delocalized electrons.

Distorted forms of the ReO_3 structure are observed for some trivalent transition metal fluorides, such as VF_3 (in which the M–F–M interbond angle is closer to 105°) and PdF_3 (in which the M–F–M angle is 132°). Structural data on MOF_2 compounds, however, are not available, although several examples have been reported in the literature. These include the tetravalent oxyfluoride compounds VOF_2, $CrOF_2$, and an uncharacterized $Pt(O_{1-x}F_x)_3$ (Bartlett and Lohmann, 1964; Ruff and Lickfelt, 1911). The preparation of a rhenium analog, $ReOF_2$, has been claimed (Ruff and Kwasnik, 1934), but attempts by Aynsley et al. (1950) to isolate this compound were not successful.

The system FeF_3/WO_3 has been studied (Dance et al., 1977; Tressaud et al., 1973) by crystallographic, magnetic, and Mössbauer spectroscopy measurements. A solid solution exists up to 74% WO_3. All products were hexagonal with the FeF_3-type structure, and no superstructure was observed, even at the composition $FeWO_3F_3$. The antiferromagnetic ordering temperature T_N decreased from 363 K (for pure FeF_3) to 62 K (for $x = 0.74$ in $Fe_{1-x}W_xO_{3x}F_{3-3x}$).

2. Tysonite Structure

Transition metal ions with large ionic radii (i.e., > 0.80 Å) can no longer be accomodated by six oxygen or fluorine neighbors and must be satisfied with higher coordination numbers. One structure that has greater than sixfold coordination is the tysonite structure or LaF_3 structure. In this structure (Zalkin et al., 1966) there are seven fluorine atoms at normal bonding distances

TABLE VIII

Oxyfluoride Compositions with the Cubic ReO_3 Structure

Composition	Range	Unit Cell Parameter a (Å)	Density (g/cm³)		Reference[a]
			d_{obs}	d_{calc}	
$TiOF_2$	—	3.71	—	—	1, 2
$TiOF_2$ or $TiO(OH)F$	—	3.798(5)	2.92	3.09	3, 4
NbO_2F	—	3.902(1)	—	—	5
$NbO_{3-x}F_x$	$1.2 \leq x \leq 2.2$	3.889–3.917	3.98–4.10	—	6
$NbO_{3-x}F_x$	$1.05 \leq x \leq 3.0$	3.899–3.939	4.03–4.11(1)	—	7
$NbO_{x/2}F_{3-x}$	$x \simeq 0.1$	3.895	—	—	8
$NbO_{x/2}F_{3-x}$	$x \simeq 0.1$	3.890	—	—	9
$MoO_{3-x}F_x$	$0.74 \leq x \leq 0.97$	3.833–3.844	—	—	10
$MoO_{3-x}F_x$	$x = 0.6$	3.842(3)	4.11(1)	4.22	11, 12
MoO_xF_{3-x}	$x \simeq 0.1$	3.83	—	—	13
MoO_xF_{3-x}	$x \simeq 0.1$	3.8985(5)	—	—	14
MoO_xF_{3-x}	—	3.896(3)	5.99(13)	6.51	5
TaO_2F	—	3.90	—	—	9
TaO_xF_{3-x}	$x \simeq 0.1$	3.901	—	—	14
$WO_{3-x}F_x$	$0.17 \leq x \leq 0.66$	3.789–3.836	—	—	10, 15
$WO_{3-x}F_x$	$0.41 \leq x \leq 0.44$	3.810–3.816	—	—	16

[a] 1, Dehnicke (1965a); 2, Dehnicke (1965b); 3, Vorres and Donahue (1955); 4, Vorres and Dutton (1955); 5, Frevel and Rinn (1956); 6, Gortsema and Didchenko (1965); 7, Pouchard et al. (1971); 8, Ehrlich et al. (1955); 9, Muetterties and Castle (1961); 10, Sleight (1969b); 11, Pierce and Vlasse (1971); 12, Pierce et al. (1970); 13, La Valle et al. (1960); 14, Gutmann and Jack (1951); 15, Derrington et al. (1978); 16, Reynolds and Wold (1973).

($\sim$2.45 Å), two fluorine atoms slightly farther away (2.63 Å), and another two at a greater distance (3.04 Å). The three types of fluorine atoms are nonequivalent. If all these fluorine atoms are included as nearest bonded neighbors, the central metal atom would be 11 coordinated, the geometry of which is a distorted trigonal prism capped on all faces. The space group for tysonite is $P\bar{3}c1$ having hexagonal symmetry and six formula weights per unit cell.

Zachariasen (1949) first reported $ThOF_2$ to possess the tysonite structure in which oxygen and fluorine were randomly distributed over the anion positions. Another study (D'Eye, 1958) on the ThO_2/ThF_4 system yielded an orthorhombic product of composition $ThOF_2$ in which the oxygen and fluorine atoms were ordered in the structure.

Rannou and Lucas (1969) restudied the system and concluded that a superstructure did indeed exist in the $ThOF_2$ phase. Thorium atoms occupied centers of trigonal bipyramids, whereas oxygen atoms were located in the basal plane and the fluorine atoms occupied apical vertices of the polyhedron. This type of ordering lowers the symmetry from hexagonal to orthorhombic. A disordered phase could not be isolated by rapid quenching of the reactants from 1200°C.

The composition $ThO_{0.5}F_{2.5}$ also crystallized with the ordered tysonite structure, and the tri- and tetravalent metal ions are believed to be randomly distributed in the bipyramids. A complete solid solution exists between $ThOF_2$ and $ThO_{0.5}F_{2.5}$.

C. AMO_2F, $AMOF_2$, AND M_2O_2F COMPOUNDS

The compositions $AMOF_2$ and AMO_2F must contain one of the following possible oxidation-state combinations for the A and M cations, respectively:

Compound	A	M
$AMOF_2$	1+	3+
	2+	2+
AMO_2F	1+	4+
	2+	3+

The possible common structures for this cation/anion ratio are the perovskite structure (for a large, low-valent A cation) or the ilmenite and the $LiNbO_3$ structure (for two transition metal ions of similar size). The adoption of a definite structure type is determined principally by the radius of the cations and the possible ordering of metal ions in certain structures. Each structure type is discussed individually.

1. Perovskite Structure

The mineral perovskite is chemically $CaTiO_3$; it was first believed to possess a primitive cubic unit cell but was later found to be slightly distorted, possessing orthorhombic symmetry. Nevertheless, the structural name for many AMO_3 compounds having the cubic structure remains the perovskite structure. In this structure the A cation is rather large (0.8–1.2 Å) and is in a dodecahedral site of cuboctahedral geometry. The smaller transition metal ion M is in an octahedral environment of oxygen ions. All octahedral groups $(MO_6)^{n-}$ mutually share all vertices in a 3D array as in the ReO_3 structure.

A unit cell can be taken in which the M atoms occupy the corners of the cubic cell, with oxygen atoms at the center of the cell edges. The large A atom then occupies the central position of the cube and thus achieves 12 coordination with the oxygen atoms. The space group for the ideal cubic structure is $Pm3m$, and the coordination numbers are given in the formula $^{XII}A^{VI}M^{VI}O_3$ (Fig. 9).

The number of oxide perovskites is legion, and this structure is also adopted by a number of fluorides and chlorides of formula AMF_3 and $AMCl_3$, respectively. Even though many compounds exist with the perovskite or perovskite-related structure having oxygen or fluorine as the anion, only a few mixed-anion systems have been prepared.

A comprehensive discussion on multiple ion substitution in the perovskite structure has been given by Roy (1954), but his experimental attempts to prepare $KGeO_2F$, $KTiO_2F$, and $BaMgOF_2$ by solid-state reactions at atmospheric pressure were unsuccessful.

A few examples of metal oxyfluorides with the perovskite or perovskite-type structure corresponding to the general formula AMO_2F or $AMOF_2$ have been reported. Specifically, the compounds isolated are $KTiO_2F$, $KNbO_2F$, and Tl_2OF_2. The last compound is most interesting in that it contains thallium in two different oxidation states, namely, Tl^+ and Tl^{3+}, in 12 and 6 coordination, respectively. High-pressure conditions were utilized in the

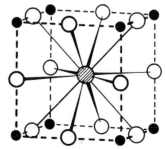

Fig. 9. Perovskite ($CaTiO_3$) structure, idealized ($\bullet$, Ti; ⊘, Ca; ○, O). Taken from Wells (1975) with permission of Oxford University Press.

formation of the compounds $KTiO_2F$ (Chamberland, 1971a) and Tl_2OF_2 (Demazeau et al., 1969). The reactions were carried out according to the following equations:

$$KF + TiO_2 \xrightarrow[1000°C]{65\,kbars} KTiO_2F$$

$$TlOF + TlF \xrightarrow[500°C]{5\,kbars} Tl_2OF_2$$

$$2\,TlF + \tfrac{1}{2}O_2 \xrightarrow[500°C]{4.5\,kbars} Tl_2OF_2$$

The compound $KNbO_2F$ was formed (Rüdorff and Krug, 1964) by the direct combination of KF and NbO_2 at 750°C and ambient pressure. All these oxyfluoride compounds are white and slightly to moderately hygroscopic.

The unit cell parameters and densities for the reported perovskite oxyfluorides are listed in Table IX.

Rüdorff and Krug (1964) also studied the reaction of other alkali fluorides with NbO_2 and obtained several $ANbO_2F$ compounds with related perovskite-type structures. Magnetic and electrical conductivity experiments were carried out, and the three compounds, $ANbO_2F$ (A = Li, Na, or K), showed temperature-independent paramagnetism when the susceptibility was extrapolated to infinite fields, and all were found to be semiconductors with conductivities in the range $10^{-6}–10^{-7}\ \Omega^{-1}\ cm^{-1}$.

In addition to these niobium oxyfluoride compounds, other metal oxyfluorides with related perovskite structures have been reported. These are listed in Table X with unit cell dimensions, crystal types, and other pertinent data. The overall cation/anion ratio varies between AMX_3 and $A_{0.25}MX_3$. Many of these compositions, in which the concentration of the large metal ion A is less than unity, are part of the well-known tungsten "bronze" family.

The tungsten oxyfluoride bronzes of general formula $A_xWO_{3-x}F_x$ (A = K, Rb, or Cs; $0.08 \le x \le 0.30$) were studied (Hubble et al., 1971) for superconductivity behavior. The dark blue and black hexagonal crystals were

TABLE IX

Unit Cell Parameters and Densities of Cubic Perovskite Oxyfluorides

Compound	Unit Cell Parameter, a (Å)	Density (g/cm³) d_{obs}	d_{calc}	Reference[a]
$KTiO_2F$	3.963(1)	—	3.68	1
$KNbO_2F$	4.00_6	4.60	4.72	2
$KNbO_2F$	4.023(1)	—	4.67	3
$TlTlOF_2$	4.59	—	7.94	4

[a] 1, Chamberland (1971a); 2, Rüdorff and Krug (1964); 3, B. L. Chamberland, unpublished results (1968); 4, Demazeau et al. (1969).

TABLE A

Unit Cell Parameters, Structure Type, and Densities of "Bronze"-Type Oxyfluorides

Compound	Structure Type	Crystal Class	Unit Cell Parameters			Density (g/cm³)		Reference
			a (Å)	b (Å)	c (Å)	d_{obs}	d_{calc}	
$NaNiO_2F$	$GdFeO_3$	Orthorhombic	5.50_7	5.52_0	7.81_9	4.47	4.66	1
$RbNbO_2F$	$BaTiO_3$	Tetragonal	3.70	—	6.05	—	—	1
$K_xNbO_{2+x}F_{1-x}$ $(x \sim 0.23)$	Bronze	Hexagonal	7.543	—	7.760	3.98	4.08	2
	Bronze $(x \sim 0.5)$	Tetragonal	12.64	—	3.950	4.24	4.35	2
$Na_xWO_{3-x}F_x$	Bronze $(x \sim 0.2)$	Orthorhombic	7.375	7.470	3.862	—	—	3
	Bronze $(x \sim 0.04)$	Tetragonal	5.256	—	3.890	—	—	3
	Bronze $(x \sim 0.15)$	Tetragonal	3.776	—	3.856	—	—	3
	Bronze $(x \sim 0.25)$	Cubic	3.811	—	—	—	—	3
$K_xWO_{3-x}F_x$	Bronze	Hexagonal	7.38–7.43	—	7.50–7.69	—	—	4
KNb_2O_5F	Bronze	Tetragonal	12.632	—	3.950	4.25	4.28	5
KTa_2O_5F	Bronze	Tetragonal	12.569	—	3.961	6.6	6.66	5
$NaNb_2O_5F$	Bronze	Tetragonal	12.29	—	3.93	—	—	6
$K_xVO_{3x}F_{3-3x}$ $(x \sim 0.25)$	Bronze	Hexagonal	7.347(3)	—	7.481(3)	3.27	3.29	7
$Na_{1-x}TaO_{3-x}F_x$ $(x = 0.05)$	Bronze	Tetragonal	3.897(2)	—	3.892(2)	—	—	8
$CaK_2Nb_5O_{14}F$	Bronze	Tetragonal	12.503(6)	—	3.912(3)	—	—	9
$BaK_2Nb_5O_{14}F$	Bronze	Tetragonal	12.569(8)	—	4.006(2)	—	—	9
$BaNa_2Nb_5O_{14}F$	Bronze	Tetragonal	12.403(6)	—	3.926(5)	—	—	9
$SrK_2Nb_5O_{14}F$	Bronze	Tetragonal	12.594(8)	—	3.945(2)	—	—	9, 11
LiW_3O_9F	Bronze	Orthorhombic	12.716(2)	15.230(2)	7.288(1)	6.76	6.79	11

[a] 1, Rüdorff and Krug (1964); 2, de Pape et al. (1969); 3, Doumerc and Pouchard (1970); 4, Gulick and Sienko (1970); 5, Magneli and Nord (1965); 6, Arend et al. (1967); 7, Cappy and Galy (1971); 8, Chaminade et al. (1972); 9, Ravez et al. (1973); 10, Ravez et al. (1972); 11, Moutou et al. (1984).

found to be superconductive in the temperature region 0.8–4.5 K, having an upper critical magnetic field in the range 4.0–9.0 Oe.

The crystal structure and dielectric properties of the $KNbO_{3-x}F_x$ compounds, possible ferroelectric materials, have been explored Malabry et al., 1973. The compounds $Bi_2TiO_4F_2$, Bi_2NbO_5F, and Bi_2TaO_5F have been discovered (Ismailzade and Ravez, 1978; Mirishli and Ismailzade, 1971) to be layer ferroelectrics that possess a 2D perovskite-like layer structure. The tetragonal $Bi_2MO_4F_2$ and Bi_2MO_5F compounds were found to have ferroelectric phase transitions in a low-temperature region as given in the following tabulation:

| Compound | Unit Cell Parameters | | $T_t\,(^\circ C)$ | Reference |
	$a\,(\text{Å})$	$c\,(\text{Å})$		
$Bi_2TiO_4F_2$	3.802	16.33	11.0(5)	Ismailzade and Ravez (1978)
Bi_2NbO_5F	3.805	16.35	30.0	Aurivillius (1952); Mirishli and Ismailzade (1971)
Bi_2TaO_5F	—	—	10.0	Aurivillius (1952); Mirishli and Ismailzade (1971)

A structural phase change from $I4/mmm$ to $Imm2$ at the Curie temperature has been proposed.

A large family of perovskite-related oxides are formed by the lanthanides, as represented by the general formula $LnMO_3$, where M is a trivalent transition metal ion and Ln is any lanthanide element. The reduced size of Ln^{3+} requires lower coordination than that found in the perovskite structure, where the A ion is an alkaline-earth species. This lower coordination site for Ln^{3+} decreases the symmetry from cubic to orthorhombic.

A series of fluorine-substituted orthochromites with general composition $LnCr_{1-x}M_xO_{3-x}F_x$ (Ln = Gd, Eu, or Y; M = Cr, Mn, Fe, Co, Ni, or Cu; x = 0.2) has been reported (Storm, 1970). These compounds were prepared for magnetic studies wherein the interaction between the divalent transition metal ions and the Cr^{3+} ions in the compounds could be investigated. A limit of substitution for single-phase materials appeared at $x \sim 0.3$ in the system $EuCr_{1-x}Cu_xO_{3-x}F_x$. All compounds in the three lanthanide series (gadolinium, europium, and yttrium) had orthorhombic unit cell parameters and showed no Ln_2O_3 impurities after regrinding and refiring of the samples. The magnetic properties of these phases were not reported. In a separate investigation (Robbins et al., 1971), fluorine substitution in $LnFeO_3$ was carried out to isolate $LnM_xFe_{1-x}O_{3-x}F_x$ compounds for magnetic measurements. The transition metals M^{2+} in this series were nickel, manganese, and copper. The lanthanide ions were yttrium, lanthanum through ytterbium (excluding cerium). The fluorine substitution level was kept at $x = 0.2$. Magnetic moments for the oxyfluoride materials when nickel was substituted for iron were of the order of three to four times those of the unsubstituted orthoferrites. The Curie temperatures decreased by about $20°C$. Copper and fluorine substitution in $LnFeO_3$ had little effect on the magnetic moments or the Curie temperature, whereas Mn^{2+} substitution resulted in an appreciable decrease in Curie temperature but did not produce a uniform trend in magnetic moment change.

a. OTHER STRUCTURES RELATED TO PEROVSKITE

i. CRYOLITE STRUCTURE. Cryolite is a perfluoro mineral of great importance in the manufacture of aluminum because it acts as the low-melting flux for refractory aluminum oxide. The chemical composition of cryolite is Na_3AlF_6, and its structure is related to that of perovskite $CaTiO_3$. The relationship can be seen when the formula is represented as $Na_2(NaAl)F_6$ with sodium in two different crystallographic sites. The presence of sodium in octahedral sites results in a distortion of the unit cell and lowers the symmetry.

The compound $Na_3VO_2F_4$ crystallizes (Pausewang and Dehnicke, 1969) in the monoclinic cryolite-type structure with space group $P2_1/c$. However, $K_3VO_2F_4$ has a pseudotetragonal unit cell. The ammonium salts also form compounds in this structural form, namely, $(NH_4)_3VO_2F_4$ and $(NH_4)_3MoO_3F_3$, and these compounds possess the ideal cubic symmetry.

ii. ELPASOLITE STRUCTURE. The elpasolite structure is closely related to the cryolite structure and therefore is also related to the perovskite structure in a similar manner. In the cryolite structure the anions normally lie along the edges of the cube, and their positions are determined by the value of the variable position parameter u. In the elapsolite structure cubic symmetry is maintained, but the anions are not constrained to lie on the unit cell edges, thus lowering the symmetry (space group $Pa3$) below that of the ideal cubic cryolite (space group $Fm3m$) and the ideal perovskite structure (space group $Pm3m$). The chemical composition of elpasolite is K_2NaAlF_6 with sodium once again occupying an octahedral site and the larger potassium atom a 12-coordinated site.

Several vanadium, titanium, niobium, molybdenum, and tungsten oxyfluoride salts of the general formula $A_3MO_xF_{6-x}$ have been prepared by solid-state reactions, and their structural and magnetic properties as well as visible and infrared (ir) spectra have been reported (Pausewang and Dehnicke, 1969; Pausewang and Rüdorff, 1969). The structural properties indicate the cubic cryolite structure, but the authors reported these compounds to crystallize with the elpasolite structure. With large A cations, such as K^+ and Rb^+, monoclinic and orthorhombic symmetry was observed.

Several zirconium compounds with formula A_2BZrOF_5 (A = Cs, Rb, Tl, or K; B = Cs, Rb, Tl, K, Na, or Li) were reported (Vedrine *et al.*, 1972). Many of these compounds were found to have cubic symmetry, and refinement of the crystallographic data on five compounds suggested the space group $Pa3$.

2. Corundum, Ilmenite, and $LiNbO_3$ Structures

The corundum structure is very common among the sesquioxides of transition metals. This structure requires that both metal ions be of similar size and charge; otherwise, ordering of cations occurs and a different, ordered structure is generated.

The corundum structure is a hexagonal close-packed structure of anions in which two-thirds of the octahedral sites are filled between each of the layers. The structure can best be illustrated in terms of these close-packed layers and the placement of metal atoms in certain positions of the structure, as illustrated in Fig. 10 by the (110) plane of the hexagonal unit cell. Each horizontal line represents a close-packed layer of anions.

When the metals order, due to charge size differences, structures related to corundum are formed, namely, ilmenite, $FeTiO_3$, or the lithium niobate structure $LiNbO_3$ (see Fig. 10).

Very few oxyfluoride examples have been reported for any of these structures, but those that have been prepared have interesting properties. The system $V_2O_{3-x}F_x$ was investigated (Ueda *et al.*, 1978) by the reaction of V_2O_3, V, and VF_3 at 1100°C and 15–20 kbars pressure. Single-phase products were obtained in the range $0 \leq x \leq 0.3$; thereafter, a two-phase region was observed. Pure V_2O_3 exhibits a metal–insulator transition at 170 K; fluorine substitution for oxygen tends to suppress this transition as well as reduce the magnitude of change at the lower transition temperature. The magnetic data indicate Pauli paramagnetism. The change in magnetic behavior was followed as a function of temperature for members of the series. The unit cell volume was observed to increase as the fluorine concentration increased.

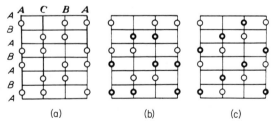

(a)   (b)   (c)

Fig. 10. Elevations for close-packed ABO_3 structures containing 3D systems of linked octahedra. (a) Corundum Al_2O_3 (○, Al), (b) ilmenite $FeTiO_3$ (○, Fe; ○, Ti), and (c) the $LiNbO_3$ structure, (○, Li; ○, Nb). Taken from Wells (1975) with permission of Oxford University Press.

Substitution of fluorine in ordered structures, based on corundum, has been investigated to a limited degree. Studies to date have concerned the compound $LiNbO_3$, which possesses ferroelectric properties. $LiNbO_2F$ was obtained (Rüdorff and Krug, 1964) in the general reactions of alkali metal fluorides with NbO_2. The product has a rhombohedral structure with hexagonal parameters $a = 5.14_4$ and $c = 13.83_5$ Å. The properties of this compound have already been presented in a previous section.

A study of the compound $Li(Fe_{0.5}Ta_{0.5})O_2F$, a pseudo-ilmenite derivative, was reported (Ismailzade et al., 1971) in which a ferroelectric–antiferromagnetic transition was observed at 610 K. This is the highest known transition temperature reported for a ferroelectric material. The product was found to be an insulator with resistivity of the order of 10^{14} at room temperature.

3. $KSbO_3$ Structure

The $KSbO_3$ structure has been known for a long time. This structure was first determined in 1940 by Spielberg and is composed of groups of two edge-shared octahedra interconnected by the equatorial vertices to other pairs of octahedra in a 3D framework of composition $Sb_{12}O_{36}^{12-}$, which possesses cubic symmetry. The large central cavity can accommodate large cations, polyhedral groups, or clusters. The space groups in $Pn3$, and the large unit cell contains 12 formula weights of $KSbO_3$. Controversy has arisen over the exact composition and the space group for this structure. Many believe that the true chemical composition of $KSbO_3$ is stabilized by SiO_2 or water as in $K_4Sb_6O_{16}(OH)_2$ and that this composition can better fit in the unit cell possessing cubic symmetry. Several crystallographic studies on AMO_3 compounds have generated the "SbO_3" framework of edge- and vertex-shared octahedra, but the space group is more consistent with $Im3$. The difference in space group occurs through random distribution of the large cations within the cavity of the framework.

Brower et al. (1974) and Waring et al. (1976) studied the synthesis of oxyfluoride compositions of $KSbO_3$ by flux-exchange methods. The $K_{1-x}SbO_{3-x}F_x$ system ($0 < x < 0.5$) was investigated and found to be primitive cubic with a unit cell parameter of $\sim 9.5\overline{7}8$ Å. Similarly, the $Na_{1-x}SbO_{3-x}F_x$ system was formed through an ion-exchange method utilizing $K_{1-x}SbO_{3-x}F_x$ and a molten $NaNO_3$ flux. This system generated a product with a unit cell parameter of ~ 9.328 Å.

Using a similar technique, Goodenough et al. (1976) prepared fast ion transport materials for use as solid electrolytes in fuel cells and other electric generators. Oxyfluorides of $KSbO_3$ and $NaSbO_3$ were found to possess the $Im3$ cubic space group, and the single-crystal data were suitably refined, $R = 0.077$ for $KSbO_{3-x}F_x$ and $R = 0.064$ for $NaSbO_{3-x}F_x$.

TABLE XI

Oxyfluoride Compositions with the Cubic $KSbO_3$ Structure

Composition	Unit Cell Parameter a (Å)	Reference[a]
$KSbO_{3-x}F_x$	9.605	1
$NaSbO_{3-x}F_x$	9.328	1
$LiSbO_{3-x}F_x$	9.397	1
$AgSbO_{3-x}F_x$	9.388	1
$RbSbO_{3-x}F_x$	9.662	1
$NaSbO_3 \cdot \frac{1}{4}NaF$	9.353 (at 220°C)	2
$NaSbO_{3-x}F_x$	9.334	2
$NaSbO_{3-x}F_x$ ($x = 0.17$)	9.306(1)	3
$KSbO_{3-x}F_x$ ($x = 0.17$)	9.606(5)	3
$Ca_7Sb_{12}O_{36}F_2$	9.245(3)	4
$Cd_7Sb_{12}O_{36}F_2$	9.195(3)	4

[a] 1, Brower et al. (1974); 2, Waring et al. (1976); 3, Goodenough et al. (1976); 4, Watelet et al. (1982).

The calcium and cadmium derivatives of $A_7Sb_{12}O_{36}F_2$ were prepared by flux cation exchange, and the structural details were determined (Watelet et al., 1982). Refinement of the crystallographic data generated low R factors of 4.4 and 6.7%, respectively, in the space group $Im3$ or $I23$. Table XI lists the reported unit cell parameters of the $AMO_{3-x}F_x$ derivatives.

D. AMO_3F COMPOUNDS: SCHEELITE STRUCTURE

There are several possible combinations of oxidation states for A and M cations that would generate an overall positive charge of 7 for the composition AMO_3F. Of the known oxyfluoride compounds, however, only $KCrO_3F$ and $CsCrO_3F$ have been characterized as possessing the scheelite structure. The mineral scheelite, chemically $CaWO_4$, contains discrete WO_4^{2-} tetrahedra. The coordination sites for the various atoms in the structure are represented by $^{XII}Ca^{IV}W^{IV}O_4$. The structure is related to the fluorite structure in which two fluorite unit cells are stacked on one another with calcium and tungsten atoms occupying the original calcium positions in fluorite and are so arranged that they alternate in the structure. The fluorine atoms are displaced toward the tungsten atoms to form a tetrahedron. The resulting structure is tetragonal with space group $I4_1/a$.

In $KCrO_3F$ there is a random arrangement of the anions in the crystal, which is supported by the fact that the CrO_3F^- tetrahedra are apparently quite regular in shape and that the anions are indistinguishable. The

TABLE XII

Oxyfluoride Compounds with the Tetragonal Scheelite Structure

Composition	Unit Cell Parameters		Reference[a]
	a (Å)	c (Å)	
$KCrO_3F$	5.46	12.89	1
$CsCrO_3F$	5.712	14.5	2

[a] 1, Ketelaar and Wegerif (1938); 2, Ketelaar and Wegerif (1939).

compound KUO_3F has been reported (Mitra, 1963), but its structure has not been determined. It is claimed that KUO_3F forms addition compounds with oxalic and acetic acids. Table XII lists the few known oxyfluorides with the scheelite structure.

E. M_3O_3F, MM'_2O_3F, AND A_2MO_3F COMPOUNDS

Compounds having the general formula M_3X_4, MM'_2X_4, or $M_2M'X_4$ can possess several different structures depending on the size of the M or M' ions, which occupy certain specific coordination sites in oxide and fluoride structures. Of the possible structures, only the spinel, $MgAl_2O_4$, and the K_2NiF_4 structures are discussed in detail here. These two structures are not directly related except by the general formula or by the total cation/anion ratio. K_2NiF_4 can best be regarded as an A_2MX_4 formula because the metal ion in the A position is normally a highly electropositive metal and not a transition metal ion as in the spinel case.

1. Spinel Structure

Magnetite, or Fe_3O_4, is a common mineral dispersed in many areas of the earth's crust. This mineral, known for centuries, possesses the spinel structure, which chemically is $MgAl_2O_4$, a rare mineral. The structure is important in that substitution of most of the transition elements in the cationic sites has been carried out in nature or in the laboratory, generating a large family of different compounds with varied chemical and physical properties. Compounds in which transition elements occupy the metal sites possess interesting magnetic properties, whereas those containing non-transition elements, such as spinel itself, can yield important laser host materials for optical applications.

The mineral spinel contain Mg^{2+} ions in tetrahedral sites, with Al^{3+} ions located in octahedral sites. The coordinations for the different ions are expressed by $^{IV}Mg^{VI}Al_2^{IV}O_4$. The structure is slightly complex but can be described in terms of either polyhedral arrangement or sphere packing of anions. The first description is based on the arrangement of tetrahedra and octahedra. The structure can be visualized as layers of edge-shared octahedra alternating with layers of mixed octahedra and tetrahedra. Neither layer is completely filled, and the octahedral layer can be further described as alternating between rows of oppositely edge-shared octahedra and rows that are only half-occupied by octahedra, as found in MX_3 layer structures. The mixed polyhedral layer once again contains rows of half-occupied octahedra connected to tetrahedra that alternate between vertex down and vertex up in a zigzag manner. The second description is based on the close packing of anions in a cubic arrangement: half of the octahedral sites and one-eighth of the tetrahedral vacancies are occupied by metal ions. The cubic space group is $Fd3m$, and there are eight MM'_2X_4 groups per unit cell.

The spinel structure is adopted by a large number of compounds, which are of three types in terms of combinations of cation charges: (1) M^{2+}, M^{3+}; (2) M^{2+}, M^{4+}; and (3) M^+, M^{6+}. The first two types are of greater importance because we are concerned with fluorine substitution for oxygen, which requires lower oxidation states for the metal ions, as in examples (1) and (2). As noted previously, when the M atom is the same chemical specie as the M' atom, one can obtain the formula M_3O_4. In such a case, however, the M atom must be present in two different oxidation states. A few first-row transition elements exist with this composition and structure, namely, Mn_3O_4, Fe_3O_4, and Co_3O_4. All of these binary oxides with the spinel structure have important ferrimagnetic properties.

Spinel oxyfluorides have been investigated as a new class of compounds because of the diverse properties expected to be exhibited by these compounds. In the synthesis of oxyfluoride spinels, metal oxides and metal fluorides are heated in sealed containers to form the desired $MM'_2O_{4-x}F_x$ compounds. Sealed systems are necessary to prevent the hydrolysis and/or volatilization of the metal fluoride in the reaction mixture.

One of the first studies on the preparation of spinel-type oxyfluorides was reported by Robbins et al. in 1963. The authors investigated the $Cu_{1+x}Fe_{2-x}O_{4-x}F_x$ system and obtained crystallographic and magnetic data on three different oxyfluoride phases. This study was followed by a series of investigations on other substituted spinel systems. Most of the research focused on determining changes in the magnetic properties of a ferrimagnetic oxide as a function of fluorine substitution. The maximum fluorine concentration was normally observed to be unity in the general formula $MM'_2O_{4-x}F_x$, even though one perfluoro spinel is known, namely, Li_2NiF_4. High pressure was found to assist in increasing the fluorine content to the limit of $x = 1.0$ in spinel phases (Austin and Sclar, 1968; B. L. Chamberland, unpublished results, 1968; Claverie and Georges, 1973), but for x values greater than unity disproportionation of the product to metal fluorides and metal oxides predominated.

The formation of $LiFe_2O_3F$ has been claimed by several investigators (Brixner, 1960; Frei et al., 1963; Hegyi, 1964; Okazaki et al., 1966), but several attempts to isolate a pure product (B. L. Chamberland, unpublished results, 1968; Harrison and Lang, 1968) using a variety of conditions have not been successful. Yanagida et al. (1966) also reported the formation of a lithium aluminum oxyfluoride spinel from the reaction of LiF with various aluminum oxides and hydrated aluminum oxides at 800 to 900°C. The composition of the spinel is not known, but some fluorine was detected from nmr results. The unit cell parameter was 7.922 Å, versus 7.906 Å for the well-known oxide spinel $LiAl_5O_8$.

Several $M_{1-x}Li_xFe_2O_{4-x}F_x$ compounds (M = Cd, Co, Fe, Mg, Mn, Ni, or Zn) were prepared, and preliminary magnetic data were presented (Okazaki et al., 1966). In most cases the substitutions caused an increase in Curie temperature, but the saturation magnetization was not drastically altered. These workers also claimed the formation of $LiFe_2O_3F$ as the end member of the series, where $x = 1.0$. The crystal chemistry of compounds formed in the $M_{2-x}^{2+}Li_x^+M^{4+}O_{4-x}F_x$ series were also reported by Okazaki in 1966. The divalent metal was either magnesium or zinc and the M^{4+} species was either tin or titanium. No magnetic data were presented, but the various cation distributions in the spinel were determined. The substitution of divalent metal ions and fluorine in the spinel structure was also studied (Schieber, 1964). Some magnetic data on these phases were reported. Substitution of fluorine in the $MgGa_2O_4$ spinel was investigated (Baffier and Huber, 1969). The investigators noted vacancy formation at cationic sites as the fluorine concentration increased. A thorough study of the $Fe_3O_{4-x}F_x$ system was presented by Portier et al. (1970b) and by Rigo (1976). The crystallographic, electrical, and magnetic properties are summarized in these two major reports. A separate study of the $Zn_xFe_{3-x}O_{4-x}F_x$ system ($0 \le x \le 0.50$) was reported by Claverie et al. (1971). Other investigations of the $ZnFe_2O_{4-x}F_x$ system were conducted (Claverie and Georges, 1973; Claverie et al., 1972a,b, 1977) wherein zinc was replaced by other divalent transition metals (iron, cobalt, and nickel), and charge compensation for the fluorine substitution occurred at the Fe^{3+} site. A major discussion of the cationic coupling giving rise to the different magnetic and electrical conductivity data is presented by Casalot et al. (1973) and Claverie et al. (1977). A more limited study of the $Zn_xNiFe_{2-x}O_{4-x}F_x$ system was reported in 1973 by the group at the University of Bordeaux (Claverie and Georges, 1973). Sobel (1967) studied the paramagnetism in fluorine-substituted spinels of the type $MgM_xAl_{2-x}O_{4-x}F_x$ (M = Ni, Co, or Mn) by means of ^{19}F-nmr experiments. He concluded that fluorine was randomly distributed in the crystal structure.

The crystallographic and some representative magnetic data on oxyfluoride spinels are presented in Table XIII.

MAGNETOPLUMBITE STRUCTURE. This structure, having the chemical composition $PbFe_{12}O_{19}$, is built of spinel $(MgAl_2O_4)$ blocks that alternate with PbO_3 layers. Similar structures can be derived by varying the number of spinel blocks and alternating metal oxide layers in the repeat unit. These compounds have been extensively studied because, through proper transition metal ion substitution in the spinel block or the metal oxide layer, they form a large family of anisotropic ferrimagnetic materials suitable for electronic devices. These compounds are referred to as layered hexagonal ferrites, and many polytypes in which the c axis can be as large as 990 Å are known.

Early work on oxyfluoride compositions with the magnetoplumbite structure was reported in a paper by Frei et al. (1960). Compounds of general formula $AFe_{10}M_2O_{17}F_2$ (A = Pb or an alkaline-earth element; M = Mg, Zn, Fe, or Co) have been prepared. The iron and cobalt compounds having barium as the countercation were both found to have larger saturation magnetization values than that of the parent oxide compound. Robbins et al. (1963) also studied the fluorine substitution in magnetoplumbite; charge compensation on the Fe^{3+} metal sites was effected by incorporating lower-valent Ni^{2+} and Co^{2+} ions. These authors found that the divalent ions preferentially occupied the tetrahedral sites in the structure and that the fluorine substitution for oxygen facilitated this exchange. The compositions $BaNi_{0.5}Ga_{11.5}O_{18.5}F_{0.5}$ and $BaCo_{0.5}Ga_{11.5}O_{18.5}F_{0.5}$, both deep blue in color, were found to crystallize with the magnetoplumbite structure. The diffuse reflectance spectra of the two compounds were investigated and analyzed in great detail. The magnetic results on cobalt and fluorine substitution in the magnetoplumbite structure are presented in conference proceedings (Banks et al., 1962; Robbins and Banks, 1963).

2. K_2NiF_4 Structure

The parent compound for this structural type is a fluoride into which oxygen can be partly or totally substituted. The original structure determination on K_2NiF_4 was made in 1955 (Balz and Plieth, 1955), and since then several isostructural oxides of formula A_2MO_4 have been prepared and characterized.

The K_2NiF_4 structure is derived principally from the perovskite structure in which layers of vertex-sharing octahedra are separated by individual potassium atoms. The alternate perovskite layers are displaced so that vertically the cationic positions are Ni–K–K–Ni (Fig. 11). The crystal class for this structure is tetragonal, and the space group is $I4/mmm$. The coordinations of

TABLE XIII

Oxyfluoride Compositions with the Spinel Structure and Representative Magnetic Properties

Composition	Unit Cell Parameter a (Å)	Saturation Magnetization σ (emu/g) (at K)	Curie Temp T_c (K)	Reference[a]
$Mg_{0.8}Fe_2O_{3.3}F_{0.7}$	8.367	64 (90)	693	1
$Mn_{0.5}Fe_{2.15}O_{3.3}F_{0.7}$	8.462	110 (90)	723	1
$Co_{0.8}Fe_2O_{3.3}F_{0.7}$	8.383	91 (90)	773	1
$Ni_{0.8}Fe_{1.95}O_{3.3}O_{0.7}$	8.347	56 (90)	703	1
Cu_2FeO_3F	8.418	103 (0)	658	2
$Cu_{1.4}Fe_{1.6}O_{3.6}F_{0.4}$	8.419	61 (0)	643	2
$Cu_{1.65}Fe_{1.35}O_{3.35}F_{0.65}$	8.412	86 (0)	603	2
"$LiFe_2O_3F$"	8.33	~70 (0)	903	3
Fe_3O_3F	8.408(10)	78	772	4
$Fe_3O_{3.9}F_{0.1}$	8.404(5)	—	—	5
$Fe_3O_{4-x}F_x (0 \leq x \leq 0.5)$	8.39–8.42	—	843–853	6
$MnFe_2O_3F$	8.45	96	692	4
$CoFe_2O_3F$	8.36	57	746	4
$NiFe_2O_3F$	8.36	49	802	4
VFe_2O_3F	8.46	3.4	593	4
VMn_2O_3F	—	32	336	4
$Zn_xFe_{3-x}O_{4-x}F_x (0 \leq x \leq 0.5)$	8.395–8.430	—	350–580	7

[a] 1, Robbins et al. (1963); 2, Schieber (1964); 3, Brixner (1960); 4, Chamberland (1968); 5, Austin and Sclar (1968); 6, Portier et al. (1970b); 7, Claverie et al. (1972b).

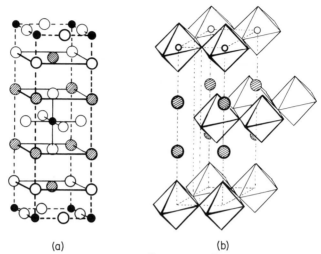

(a)   (b)

Fig. 11. K_2NiF_4 structure. (a) Unit cell (⊘, K; ●, Ni; ○, F); (b) polyhedral model (note the staggered layers of vertex-shared octahedra). Adapted from Wells (1975) with permission of Oxford University Press.

the atoms are given in the formula $^{IX}K_2{}^{VI}Ni^{VI}F_4$. The A site in the structure is normally occupied by a large, low-valent metal ion such as K^+, Rb^+, Ca^{2+}, Sr^{2+}, or La^{3+}, whereas the octahedral M site is occupied by diverse divalent to tetravalent transition metal ions of approximate size 0.60(5) Å. The ideal c/a value for spheres (anions) in direct contact within the tetragonal structure should be $2 + 2^{1/2}$, or 3.41; however, in most oxides and fluorides the c/a ratio is less than the ideal value (Table XIV). In K_2NbO_3F and stoichiometric Sr_2FeO_3F the c/a ratios are equal to or larger than the theoretical value, and these two compounds are believed to contain ordered anions. The model proposed for these two compounds has fluorine atoms located in opposite apical positions along the c direction of the octahedra surrounding the transition metal, causing an elongation along the c crystallographic axis.

Many studies have been undertaken on perfluoro and peroxo compounds, but it was not until 1962 that an oxyfluoride analog was prepared with the K_2NiF_4 structure. The preparation of these oxyfluoride compounds is accomplished by a direct reaction of alkali or alkaline-earth fluorides with transition metal oxides. The first compound with this structure, K_2NbO_3F, was isolated by Galasso and Darby (1962). Shortly thereafter, Sr_2FeO_3F was reported (Galasso and Darby, 1963). Thermal expansion and thermoelectric power measurements indicated ordering of anions in both compounds. The Sr_2FeO_3F product was brown to black depending on the presence of variable

TABLE XIV

Oxyfluoride Compounds with the K_2NiF_4 Structure

Composition	Unit Cell Parameters		c/a Ratio	Reference[a]
	a (Å)	c (Å)		
K_2NbO_3F	3.956	13.670	3.46	1
"$Sr_2FeO_{3.2}F_{0.8}$"	3.84	12.98	3.38	2
$Sr_2FeO_{3.20}F_{0.80}$	3.844(1)	13.064(5)	3.40	3
$Sr_2FeO_{3.11}F_{0.89}$	3.857(2)	13.10(1)	3.40	3
Sr_2FeO_3F	3.864(2)	13.17(1)	3.41	3
$Ca_2MnO_{3.90}F_{0.10}$	3.688(5)	12.04(1)	3.26	4
$Ca_2MnO_{3.85}F_{0.15}$	3.694(5)	12.03(1)	3.26	4
$Ca_2MnO_{3.80}F_{0.20}$	3.702(5)	12.02(1)	3.25	4
$Ca_2MnO_{3.75}F_{0.25}$	3.712(5)	12.01(1)	3.24	4
$Ca_2MnO_{3.70}F_{0.30}$	3.719(5)	12.00(1)	3.22	4

[a] 1, Galasso and Darby (1962); 2, Galasso and Darby (1963); 3, Menil *et al.* (1981); 4, LeFlem *et al.* (1976).

oxidation states of iron as the oxygen/fluorine ratio was varied. The 2D feature of this structure is of interest for magnetic and optical effects. Anisotropic properties are intrinsic to these materials because of the unique structure and bonding.

The ir transmission on nonoptical-grade K_2NbO_3F has been reported (Galasso and Darby, 1965). Thin platelets exhibited ir transmission up to 12 μm wavelength and as such showed some promise for long-wavelength transmission of ir radiation. This compound is an insulator possessing a positive temperature coefficient of resistance in the temperature range 0–160°C.

Thermoelectric power measurements on nonstoichiometric Sr_2FeO_3F indicated a p-type semiconductor (Galasso and Darby, 1963) with antiferromagnetic ordering near 50 K (Schelleng, 1973). A pure stoichiometric sample of Sr_2FeO_3F was isolated by Fournes *et al.* (1980), and it was found (Menil *et al.*, 1981) to be an insulator with an antiferromagnetic ordering temperature of 358 K. Mössbauer studies supported the magnetic data and suggested that the fluorine atoms were ordered on opposite apical positions of the octahedra, as suggested earlier. These authors carried out a complete study on stoichiometric and nonstoichiometric phases in the system $Sr_2FeO_{3-x}F_x$ (0.80 $\leq x \leq$ 1.0). The nonstoichiometric phases contain mixtures of Fe^{2+} and Fe^{3+}, which introduced conductivity and lowered the magnetic ordering temperature and Curie constant. The relationship between structure and physical properties for compounds with the K_2NiF_4 structure has been developed (LeFlem *et al.*, 1982).

Several compositions in the $Ca_2MnO_{4-x}F_x$ system were prepared by LeFlem et al. (1976). The unit cell parameters showed an increase in the a parameter and a decrease in the c parameter as a function of increasing fluorine concentration. On heating to $500°C$, the $Ca_2MnO_{3.8}F_{0.2}$ sample showed a contraction along the c axial direction and then exhibited an elongation above this temperature. This feature is similar to that observed in Ca_2MnO_4 and $Ca_{1.8}Y_{0.2}MnO_4$, which also possess the K_2NiF_4 structure. It was concluded that fluorine preferentially occupied anion sites in the equatorial positions of the octahedra within the perovskite layers. The presence of Mn^{3+} gave rise to a Jahn–Teller distortion. Magnetic studies (LeFlem et al., 1980) indicated low-temperature, weak ferromagnetic interactions ($\sigma_s = 0.19\ \mu_B$ when $x = 0.25$), which decreased in intensity as the fluorine concentration increased (the canting decreased). The magnetic interactions at higher temperature were antiferromagnetic ($\theta_p = -150$ K and $T_N = 65$ K for $x = 0.25$). The electrical resistivity data suggest a hopping mechanism for the conduction of electrons ($E_a = 0.40$ eV for $x = 0.2$). These results are consistent with fluorine atoms in equatorial positions of the octahedra within the perovskite layers.

F. $A_2M_2O_{7-x}F_x$ COMPOUNDS: PYROCHLORE STRUCTURE

The mineral pyrochlore has a complex chemical composition: $NaCa(Nb,Ta)_2O_6(OH,F)$. A substantial amount of titanium, zirconium, thorium, uranium, and the lanthanide elements are also found in some mineral samples. There is also considerable variation in the sodium and calcium content.

The pyrochlore structure can be regarded as an anion-deficient fluorite structure with a doubled unit cell in which the anions assume an ordered arrangement. The cations are located along face-centered cubic positions with eightfold coordination for the larger A cations and octahedral coordination for the M ions. The hexagonal bipyramidal site for A accommodates ions of size $0.9 \leq r_A \leq 1.2$ Å, whereas the octahedral site is best suited for smaller transition metal ions ($0.6 \leq r_M \leq 0.7$ Å). The cubic space group is $Fd3m$ with $Z = 8$. The structure can also be regarded as two interpenetrating networks. The A_2O network consists of tetrahedral groups (anti-cristobalite structure) in which the oxygen atom is 4 coordinated and the M atom is 2 coordinated. The main framework is an M_2O_6 network of vertex-shared octahedra, four such polyhedra being arranged in a tetrahedral cluster. The six oxygen atoms forming hexagonal tunnels in the M_2O_6 framework are also locations of the A cations in the interpenetrating A_2O network, giving rise to the eightfold coordination of the A atoms in the final structure. The formula and coordinations of the two descriptions are $^{II}A_2{}^{IV}O + {}^{VI}M_2{}^{II}O_6$ and $^{VIII}A_2{}^{VI}M_2{}^{IV}O_6{}^{IV}O$. The oxygen atoms are not equivalent in the structure but are of two types. The unique oxygen position can be replaced by several different atoms, such as fluorine, sulfur, or chlorine, or can be a vacancy in the structure, giving rise to the formula $A_2M_2O_6$ or AMO_3—a defect pyrochlore.

In certain cases the oxygen atoms in the M_2O_6 framework can also be replaced by fluorine atoms, generating a random arrangement in anion sites. Ordering of M atoms in $Cd_2M_{2-2x}M'_{2x}O_{7-2x}F_{2x}$ ($0.25 \leq x \leq 0.55$) has been observed (Calage et al., 1971) in systems containing zirconium and niobium, or tin and niobium, for M and M', respectively.

Cationic substitution in oxyfluoride pyrochlores was first carried out by Mazelsky and Ward in 1961. These authors studied the synthesis of a series of pyrochlores of general formula $NaA(Nb)_2O_6F$, wherein they substituted cations in the A site, and other transition metals for some of the niobium atoms. They reported representative magnetic and electrical resistivity results on many of the compounds prepared. Shortly after the publication of this article, Aleshin and Roy (1962) presented a general paper on the crystal chemistry of the pyrochlore structure. These authors attempted to prepare and isolate pyrochlores of the type $A_2M_2O_6F$, $A_2M_2O_5F_2$, and $AM_2O_5F_2$. In many of their preparations they obtained two phases. Their products were not further characterized. The attempts to prepare $Pb_2Ru_2O_6F$ by Beyerlein et al. (1984) under a variety of experimental conditions were also unsuccessful.

$AgSbO_3$ was reported (Sleight, 1969a) as a defect pyrochlore, and attempts to prepare the two analogous compounds $Ag_2Sb_2O_5(OH)_2$ and $Ag_2Sb_2O_5F_2$ were unsuccessful. However, the 5:2 oxygen:fluorine ratio was achieved (Bradley, 1969) in the synthesis of $Pb_2Ti_2O_5F_2$. The bright orange polycrystalline product was found to have low resistivity, which prevented measurement of a dielectric constant or hysteresis behavior. $Cd_2Ti_2O_5F_2$ was also prepared (Laguitton and Lucas, 1969b), and the solid solution between this compound and the oxide end member $Cd_2Nb_2O_7$ was studied. A large family of pyrochlores containing lanthanide ions has been studied (Grannec et al., 1974). According to the dielectric constant measurements, these materials were found to possess ferroelectric–paraelectric transitions. Fluorine substitution in the oxide pyrochlores caused a reduction in the Curie temperature: low fluorine substitutions were followed by increases in the dielectric constant and the spontaneous polarization. Fluorine substitution for oxygen and Na^+ replacement of Pb^{2+} in $Pb_2Nb_2O_7$ has been reported (Campet et al., 1974). Cubic pyrochlores were obtained in $Pb_{2-x}Na_xNb_2O_{7-x}F_x$ for the region $0.75 \leq x \leq 1.85$. At other fluorine concentrations a monoclinic distortion of the structure was observed. All of these materials were found to have high dielectric constants, possibly arising from ionic polarizability and F^- mobility at high temperature.

Phase equilibria and crystal growth in the $Sb_2O_4/NaSbO_3/NaF$ system have been studied (Waring et al., 1976). Oxyfluoride pyrochlores were obtained but in several cases were contaminated with a body-centered cubic phase of composition $4NaSbO_3 \cdot NaF$. Goodenough et al. (1976) studied the $KSbO_3/KF$ and $NaSbO_3/NaF$ systems in a similar manner and also observed oxyfluoride compounds of the type $6MSbO_3 \cdot MF$, but these had the cubic $KSbO_3$ structure (space group $Im3$) and not the pyrochlore structure. Defect pyrochlores of the type $RbTa_2O_5F$ and KTa_2O_5F, however, were synthesized in the latter study. These compounds were investigated as fast ion transport agents for use in fuel cells.

The pyrochlore $Pb_{1.5}U_2O_6F$ was prepared (Kemmler-Sack, 1968a), and its magnetic properties were measured to determine the oxidation state of the uranium ion. Substitution of alkali metals for lead in $Pb_2U_2O_7$ with charge compensation provided by fluorine led (Kemmler-Sack, 1969) to a series of compounds with composition $A_xPb_{2-x}U_2O_{6-y}F_y$ (A = K, Rb, or Tl; $x \sim \frac{1}{3}$, $y \sim 1$). Magnetic data on several phases were reported, and these data suggest the presence of UO_2^+ groups in the cubic pyrochlore compounds. A polymorph of $Na_2Ta_2O_5F_2$ having a large monoclinic cell (Vlasse et al., 1974) was found to contain both layers of pyrochlore and weberite structures in the same structure. It should be noted that both structures have the same general formula $A_2M_2X_7$ but that the linking of octahedral groups is quite different in these two structures. Distorted pyrochlore structures with monoclinic symmetry are obtained (Chaminade and Pouchard, 1975; Chaminade et al., 1971, 1973) for the compositions $A_2Ta_2O_5F_2$ (A = Na or Ag).

A rhombohedral distortion of the pyrochlore structure was noted (Laguitton and Lucas, 1969a) in $Cd_2(NbZr)O_6F$ and $Cd_2(NbHf)O_6F$. This distortion is believed to arise from the regular hexacoordination of zirconium and hafnium, which normally prefer higher coordination.

The crystallographic properties of several pyrochlore oxyfluorides are given in Table XV.

Thin films of oxyfluoride pyrochlores have been prepared for electrical conductivity and dielectric constant measurements (Campet et al., 1979; Pompei et al., 1977). A fluorine- substituted $Cd_2Sb_2O_7$ compound was studied (Mooney and Aia, 1964) as a dopant in a calcium halophosphate phosphor and was found to increase the overall luminous efficiency in the process.

G. $A_3M_2M_3'O_{12-x}F_x$ COMPOUNDS: GARNET STRUCTURE

The garnet structure can be formulated as $\{Ca\}_3[Al]_2(Si)_3O_{12}$, with the notation defined as follows: $\{\ \}$ = octacoordinated or dodecahedral sites;

TABLE XV

Oxyfluoride Compounds with the Pyrochlore Structure

Composition	Unit Cell Parameter a (Å)	Reference
$NaCaNb_2O_6F$	10.431(1)	1
$NaLa_{2/3}Nb_2O_6F$	10.489(1)	1
$NaSrNb_2O_6F$	10.525(1)	1
$Ca_2Nb_2O_6F$	10.364(2)	1
$Pb_2Ti_2O_5F_2$	10.372	2
$Cd_2Ti_2O_5F_2$	10.179(1)	3
$SrLiTa_2O_6F$	10.470(2)	4
$CaNaSb_2O_6F$	10.30	5
$NaYTiNbO_6F$	10.262(8)	6
$NaCeTiNbO_6F$	10.374(8)	6
$CdYTi_2O_6F$	10.140(8)	6
$CdNdTi_2O_6F$	10.240(8)	6
$RbTa_2O_5F$	10.496(3)	7
$Pb_{1.5}U_2O_6F$	11.10(1)	8
$Cd_2Nb_{2-2x}Ti_{2x}O_{7-2x}F_{2x}(0 \le x \le 1)$	10.37–10.18(1)	3
$Cd_2Nb_{2-2x}M^{4+}O_6F(M = Ti, Ge, Sn)$	10.275–10.397(2)	9
$Cd_2Nb_{2-2x}M_{2x}{}^{4+}O_{7-2x}F_{2x}$		
$\quad M = Sn, Ge \ (0 \le x \le 0.5)$	—	10
$\quad M = Zr, Hf \ (0 \le x \le 0.875)$	Varies	10
$A_{0.33}^+PbU_2O_{5.67}F_{0.33}$ (A = K, Rb, Tl)	11.10–11.20(1)	11
$Pb_{2-x}Na_xNb_2O_{7-x}F_x$ $(0.75 \le x \le 1.85)$	10.481–10.521(2)	12

[a] 1, Mazelsky and Ward (1961); 2, Bradley (1969); 3, Laguitton and Lucas (1969b); 4, Tutov et al. (1972); 5, Aleshin and Roy (1962); 6, Grannec et al. (1974); 7, Goodenough et al. (1976); 8, Kemmler-Sack (1968a); 9, Laguitton and Lucas (1969a); 10, Calage et al. (1971); 11, Kemmler-Sack (1969); 12, Campet et al. (1974).

[ ] = octahedral sites; and ( ) = tetrahedral sites. The fluorine equivalent of this compound is found in the mineral cryolithionite $Na_3Al_2Li_3F_{12}$. The general formula and the coordination numbers are given by $^{VIII}A_3{}^{VI}M_2(^{IV}M'X_4)_3$ (M' = Si or Li; X = O or F). The space group is $Ia3d$, and there are eight formula weights per cubic unit cell with an approximate edge of ~ 12.5 Å. Garnets have long been known as an abundant class of orthosilicate minerals. The mineral compositions have calcium, magnesium, or iron in the $A(VIII)^{2+}$ sites and aluminum, chromium, or iron in the $M(VI)^{3+}$ sites. Various substitutions in the garnet structure have generated a large number of complex derivatives having technological importance. The utility of compounds with this structure is exemplified by the occurrence of interesting optical, magnetic, and magneto-optical properties. In some compounds the octahedral and tetrahedral sites are occupied by the same element in the same or different oxidation states, as in $Y_3Fe_5O_{12}$, yttrium iron garnet, commonly

referred to as YIG—an important ferrimagnet used in "bubble" magnetic technology. The corresponding aluminum garnet, $Y_3Al_5O_{12}$, called YAG, is a well-known and important laser host material.

The garnet framework can also exist without the M^{2+} ions, as in $Al_2(WO_4)_3$ or $[Al]_2(W)_3O_{12}$, where there is considerable distortion about the metal–oxygen–metal angles (143–175° versus the ideal value of 109° in garnet). Some oxidic garnets, such as $CaGeO_3$ (II) and $Mg_3Al_2(SiO_4)_3$ have been prepared at high pressure, but attempts (Austin and Sclar, 1968) to substitute fluorine in $Y_3Fe_5O_{12}$ at 90 kbars and 1000°C were unsuccessful in isolating $Y_3Fe_5O_7F_5$ or $Y_3Fe_5O_9F_3$. These workers concluded that the oxyfluoride garnets were unstable relative to the normal oxides, metal difluorides, and YOF.

Fluorine substitution in the garnet structure began with the work of Frei et al. (1960) and Banks et al. (1962) in the early 1960s. Robbins and Banks (1963) reported evidence that the divalent metal ion preferred a site where fluorine substitution had occurred. Diffuse reflectivity measurements indicated that Ni^{2+} or Co^{2+} substitution in the garnet $Y_3Al_5O_{12}$ did not stabilize the tetrahedral sites, but rather some of the divalent ions entered the yttrium sites. Divalent substitution of transition metals for Al^{3+} was charge compensated by fluorine substitution for oxygen.

This work was followed by fluorine substitution in YIG and the investigation of magnetic properties of the new compositions. The oxyfluoride garnet system $Y_{3-x}Ca_xFe_5O_{12-x}F_x$ was studied (Ichinose and Kurihara, 1965) in the range $0 \leq x \leq 1.0$; however, a second phase was observed in the products when x was greater than 0.7. The Néel temperature decreased as the fluorine concentration increased, from 550 to 540 K, whereas the magnetic moment remained essentially constant ($\sim 5 \mu_B$) throughout the series. Additional magnetic results on other garnet systems were reported by Ichinose (1966).

The $Gd_3Fe_5O_{12-x}F_x$ ($0 \leq x \leq 0.60$) system studied by Portier et al. (1970a) showed a decrease in the Néel temperature from 564 to 528 K as the fluorine concentration increased. Charge compensation was obtained in this system by the reduction of trivalent iron to the divalent state. Magnetic studies further showed an increase in saturation magnetization as fluorine was introduced and as the divalent iron was generated. The molar magnetization value (extrapolated to 0 K) increased from 15.0 μ_B for $x = 0$ to 17.0 μ_B for $x = 0.60$.

The magnetic properties in the systems $Y_3Fe_5O_{12-x}F_x$ and $Y_3Fe_{5-x}Zn_xO_{12-x}F_x$ were also investigated (Tanguy et al., 1971), and the results indicated decreasing Néel temperature as the fluorine concentration increased, as well as decreasing saturation magnetization as the divalent ions Fe^{2+} or Zn^{2+} were introduced in the structure. The YIG garnet was observed to possess a saturation magnetization value of $4.7 \pm 0.2 \mu_B$, which decreased to 3.7 ± 0.2 μ_B for $Y_3Fe_5O_{11}F$ and 3.37 ± 0.2 μ_B for $Y_3Fe_{4.5}Zn_{0.5}O_{11.5}F_{.5}$. Several rare-earth transition metal oxyfluoride compounds possessing the garnet-type structure were prepared by Morell $et\ al.$ (1973, 1973-1974). The systems successfully prepared include those shown in the following tabultion:

M^{2+}	$Y_3Fe_{5-x}M_xO_{12-x}F_x$	$Gd_3Fe_{5-x}M_xO_{12-x}F_x$
Mn	$0 \leq x \leq 0.40$	$0 \leq x \leq 0.50$
Co	—	$0 \leq x \leq 0.40$
Ni	$0 \leq x \leq 0.50$	$x = 0$
Cu	$0 \leq x \leq 0.75$	$0 \leq x \leq 0.75$
Zn	$0 \leq x \leq 0.70$	$0 \leq x \leq 0.35$

In all the systems the unit cell parameters decreased as x increased. For the $Y_3Fe_{5-x}Mn_xO_{12-x}F_x$ series, the Néel temperature decreased from 550 to 505 K at $x = 0.40$. The saturation magnetization value remained essentially constant at 5 μ_B. The magnetic moment values (± 0.2 μ_B) at 0 K for several members of the $Y_3Fe_{5-x}M_xO_{12-x}F_x$ series are given in the following tabulation:

				x				
M^{2+}	0	0.10	0.20	0.25	0.30	0.40	0.50	0.70
Mn	5.0	5.1	5.1	—	5.0	5.0	—	—
Ni	5.0	5.2	—	5.5	—	5.8	6.0	—
Zn	5.0	4.5	4.0	—	3.5	—	2.6	1.7

Similar magnetization data for the $Gd_3Fe_{5-x}M_xO_{12-x}F_x$ series are as follows:

			x			
M^{2+}	0	0.10	0.20	0.25	0.35	0.40	0.50
Mn	15.2	—	15.2	—	—	15.1	15.1
Zn	15.2	16.3	—	16.9	17.3	—	—

The Néel temperature decreased as the Mn^{2+} or Zn^{2+} concentration increased in both series. Mössbauer data for $Ln_3Fe_{4.6}Mn_{0.4}O_{11.6}F_{0.4}$ ($Ln = Y$ or Gd) have been obtained. The results indicate that Mn^{2+} preferentially occupies octahedral sites in the structure.

Crystallographic data for garnet oxyfluorides are presented in Table XVI.

TABLE XVI
Oxyfluorides with the Garnet Structure

Composition	Fluorine Content	Range of Unit Cell Parameter a (Å)	Reference[a]
$Y_3Al_{5-x}Ni_xO_{12-x}F_x$	$0 \leq x \leq 0.9$	12.020–12.032	1
$Y_3Ga_{5-x}Ni_xO_{12-x}F_x$	$0 \leq x \leq 0.4$	12.275–13.384	1
$Y_3Al_{5-x}Co_xO_{12-x}F_x$	$0 \leq x \leq 0.6$	12.020–12.033	1
$Y_{3-x}Ca_xFe_5O_{12-x}F_x$	$0 \leq x \leq 0.9$	12.376–12.369	2
$Y_{3-3x}Ca_{3x}Fe_{5-x}Sb_xO_{12-x}F_x$	$0 \leq x \leq 0.55$	—	3
$Y_{3-3x}Ca_{3x}Fe_{5-x}V_xO_{12-x}F_x$	$0 \leq x \leq 0.50$	—	3
$Gd_3Fe_5O_{12-x}F_x$	$0 \leq x \leq 0.60$	12.473–12.441	4
$Y_3Fe_5O_{12-x}F_x$	$0 \leq x \leq 1.0$	12.375–12.360	5
$Y_3Fe_{5-x}Zn_xO_{12-x}F_x$	$0 \leq x \leq 0.70$	12.375–12.358	5

[a] 1, Robbins *et al.* (1963); 2, Ichinose and Kurihara (1965); 3, Ichinose (1966); 4, Portier *et al.* (1970a); 5, Tanguy *et al.* (1971).

H. OTHER COMPOUNDS AND STRUCTURES

1. Chiolite Structure

Chiolite is a native fluoride mineral of chemical composition $Na_5Al_3F_{14}$ wherein all the aluminum atoms occupy octahedra of two sorts. One octahedron has four vertices (equatorial) bonded to octahedra having only two vertices (opposite) attached in a layerlike fashion (Fig. 12). The layers are separated by the sodium ions in 8 + 2 coordination. Chiolite has tetragonal symmetry, but distortion of the framework and multiple twinning of crystals can yield structures with orthorhombic or monoclinic symmetry.

Various M^{3+} substitutions have been carried out in the chiolite structure, and these transition metal ions include vanadium, chromium, iron, and cobalt. The structural features indicate that such compounds may possess optical and ferroelectric properties useful in nonlinear optical applications.

Ravez *et al.* (1980) have studied a variety of chiolite derivatives in order to form new ferroelastic–ferroelectric materials. Their work has focused on the substitution of oxygen for fluorine in the chiolite structure as a possible route to the synthesis of useful compounds.

In a study (Doumerc and Pouchard, 1970) of the NaF/WO_3 system, the oxyfluoride $Na_5W_3O_9F_5$ was isolated and found to be derived from the chiolite structure having orthorhombic symmetry. This compound was investigated (Doumerc *et al.*, 1979) as a ferroelectric material. $Na_5W_3O_9F_5$ is piezoelectric at room temperature and undergoes three phase transitions, at 190, 530, and 800 K. At the high temperature the structure changes to tetragonal symmetry, and the material is believed to undergo a ferroelectric–

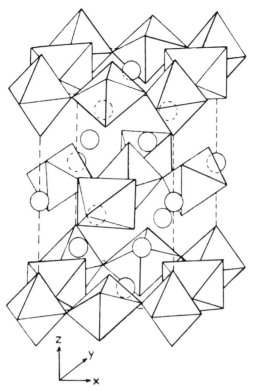

Fig. 12. Polyhedral representation of the unit cell in the chiolite $Na_5Al_3F_{14}$ structure ($\bigcirc$, Na^+). Adapted from Doumerc *et al.* (1979) with permission of Pergamon Press.

paraelectric transition as observed from a study of the dielectric constant behavior. A crystal growth and domain structure study on the compound $Na_5W_3O_9F_5$ was reported (Ravez *et al.*, 1979a). The optical properties, as a function of temperature, of single crystals are given. Ionic conduction, due to Na^+ mobility between the octahedral layers, has been determined at high temperature.

Similar studies were performed (Ravez *et al.*, 1979b) on a family derived from the $(Na_{5-x}Nd_x)(W_{3-2x}Nb_{2x})O_9F_5$ system ($0 \leq x \leq 1.50$), which has the chiolite-type structure. The authors noted that as the Nd^{3+} concentration increased the Curie temperature for the two observed transitions decreased.

A different series having the general formula $Na_4Ln(WNb_2)O_9F_5$ was investigated (Elaatmani *et al.*, 1980). The lanthanide ions substituted in the formula were yttrium, neodymium, europium, gadolinium, dysprosium, and lutetium. At room temperature these compounds possessed tetragonal symmetry, and a ferroelectric–paraelectric transition was observed in each case. The Curie temperature increased as the size of the lanthanide ion increased (between 100 and 300 K).

The oxygen/fluorine ratio in the chiolite-type compounds $Na_5(W_{3-x}Nb_x)O_{9-x}F_{5+x}$ was varied in the range $0 \leq x \leq 2$ to determine (Elaatmani et al., 1981) the effect of such anion substitution. The Curie temperature for the orthorhombic phases decreased very rapidly as the fluorine (and niobium) concentration increased. These oxyfluoride compounds with chiolite-type structure were found to be true ferroelectric materials with relatively high Curie temperatures.

A dielectric measurement of $Na_5W_3O_9F_5$ at varying temperatures under a fluorinating atmosphere was carried out (Tressaud et al., 1973) in a new type of cell. Natural chiolite crystals were similarly investigated, and the results were compared.

Oxyfluoride compounds derived by Li^+ and K^+ substitution for Na^+ in the chiolite structure have been investigated (Ravez et al., 1981a). The orthorhombic compounds of the type $Na_{5-x}Li_xW_3O_9F_5$ and $Na_{5-x}K_xW_3O_9F_5$ were all found to exhibit ferroelectric and ferroelastic behavior. The Curie temperature decreased slightly as the substituted alkali metal concentration increased. Single crystals were grown by slow cooling of the melt using two different techniques. Four different transitions were observed in most of the new compounds.

A complete listing of the observed ferroelectric transitions for many of these chiolite oxyfluoride derivatives has been presented (Ravez et al., 1981b).

A study on the "detwinning" of chiolite-type crystals by the application of compressive stress at high temperature yielded a $Na_5W_3O_9F_5$ crystal with monoclinic symmetry. The best fit to the diffraction data suggested space Group Pn with unit cell parameters $a = 7.3620(6)$, $b = 10.6345(22)$, $c = 7.3620(8)$ Å, and $\beta = 90.82(2)°$. A plot summarizing the Curie temperatures for derivatives in the $Na_5W_3O_9F_5$ family is given (Ravez et al., 1981c) as a function of chemical composition with Na^+ or W^{6+} substitution.

2. Copper Niobium Oxyfluorides

The system $Cu/Nb/O/F$ has been studied by a number of investigators, and new compounds with interesting physical properties have been obtained. The preparation and crystal structure of $CuNbO_3F$ were reported (Lundberg and Sävborg, 1978–1979) in 1979. The new compound was found to have triclinic symmetry, space group $P\bar{1}$, and its structure was determined from single-crystal studies. It is built of infinite zigzag chains of edge-shared octahedral $Nb(O,F)_6$ groups along the c axis of the structure. The copper atoms are bonded to three different octahedral chains. The $Nb(O,F)_6$ octahedra are distorted because of the adjacent edge sharing in the chain. The compound is structurally related to the oxide $Cu_3Nb_2O_8$. $CuNbO_3F$ was found to be a temperature-dependent paramagnet from magnetic experiments., having an effective magnetic moment of 2.04 μ_B, which arises from the Cu^{2+} and is consistent with square-planar coordination for this ion.

A similar, but variable-composition compound was later obtained (Lundberg and Ndalamba Wa Ilunga, 1981), having the formula $Cu_{0.6}NbO_{2.6}F_{0.4}$. The single-crystal structure was determined to an R factor of 5.4%. The orthorhombic unit cell had the space group $Pnam$ and was composed of $Nb(O,F)_6$ octahedral layers similar to that found in MoO_3. The compound crystallized as rod-shaped red crystals with an observed density of 4.50 g/cm^3. The unit cell parameters are $a = 17.694(5)$, $b = 3.944(1)$, and $c = 3.801(3)$ Å. The copper atoms in the structure bridge the niobium oxyfluoride layers to give nearly linear X—Cu—X bonds.

Another variable-composition oxyfluoride in the Cu/Nb system was obtained (Khazai et al., 1980) from the fluorination of $CuNb_2O_6$ at 800°C. The compound $CuNb_2O_{5.3}F_{0.7}$ was isolated as a dark brown orthorhombic phase, isostructural with the starting oxide material. The unit cell parameters are $a = 9.665(5), b = 10.395(3)$, and $c = 7.847(3)$ Å. Curie–Weiss behavior was observed, and the calculated moment was 0.68 μ_B, indicating the presence of a large amount of diamagnetic Cu^+. Resistivity data showed semiconductivity behavior with a room-temperature resistance of 0.5 Ω cm and an activation energy of 0.06 eV.

References

Aleshin, E., and Roy, R. (1962). *J. Am. Ceram. Soc.* **45**, 18–25.

Andersson, S. (1964). *Acta Chem. Scand.* **18**, 2339–2344.

Arend, H., Coufova, P., and Novak, J. (1967). *J. Am. Ceram. Soc.* **50**, 22–25.

Aurivillius, B. (1952), *Ark. Kemi* **5**, 39–47.

Austin, A. E. (1969). *J. Phys. Chem. Solids* **30**, 1282–1285.

Austin, A. E., and Sclar, C. B. (1968). Final Rep. **AD 673295**, 31 July 1968. U.S. Army Res. Office, Durham, North Carolina.

Aynsley, E. E., Peacock, R. D., and Robinson, P. L. (1950). *J. Chem. Soc.* pp. 1622–1624.

Baenziger, N. C., Holden, J. R., Knudson, G. E., and Popov, A. I. (1954), *J. Am. Chem. Soc.* **76**, 4734–4735.

Baffier, N., and Huber, M. (1969). *C. R. Hebd. Seances Acad. Sci., Ser. C* **269**, 231–234.

Balz, D., and Plieth, K. (1955). *Z. Elektrochem.* **59**, 545–551.

Banks, E., Robbins, M., and Tauber, A. (1962). *J. Phys. Soc. Jpn.* **17**, Suppl. B-1, 196–200.

Barker, W. W. (1968). *Acta Crystallogr., Sect. A* **A24**, 700.

Bartlett, N., and Lohmann, D. H. (1964). *J. Chem. Soc.* pp. 619–626.

Bayard, M. L. F., Reynolds, T. G., Vlasse, M., McKinzie, H. L., Arnott, R. J., and Wold, A. (1971). *J. Solid State Chem.* **3**, 484–489.

Bayard, M. L. F., Pouchard, M., and Hagenmuller, P. (1975a). *J. Solid State Chem.* **12**, 31–40.

Bayard, M. L. F., Pouchard, M., Hagenmuller, P., and Wold, A. (1975b) *J. Solid State Chem.* **12**, 41–50.

Bedford, R. G., and Catalano, E. (1970). *J. Solid State Chem.* **2**, 585–592.
Bevan, D. J. M., and Mann, A. W. (1969). *Proc. Rare Earth Res Conf., 7th, 1968*, Vol. 1, pp. 149–162.
Beyerlein, R. A., Horowitz, H. S., Longo, J. M., Leonowicz, M. E., Jorgensen, J. D., and Rotella, F. J. (1984). *J. Solid State Chem.* **51**, 253–265 (ref. 13).
Bhat, T. R., and Khan, I. A. (1965). *J. Less-Common Met.* **9**, 388–389.
Bhat, T. R., and Khan, I. A. (1966). *J. Less-Common Met.* **11**, 290–292.
Blasse, G. and Bril, A. (1967). *Philips Res. Rep.* **22**, 481–504.
Bradley, F. N. (1969). *J. Am. Ceram. Soc.* **52**, 114.
Brixner, L. H. (1960). U. S. Patent 2,962,345.
Brower, W. S., Minor, D. B., Parker, H. S., Roth, R. S., and Waring, J. L. (1974). *Mater. Res. Bull.* **9**, 1045–1052.
Brown, D. (1968). "Halides of the Transition Elements: Halides of the Lanthanides and Actinides." J. Wiley, New York.
Buessem, W. R., Forland, K. S., and Marshall, P. A. Jr. (1963). U. S. Patent 3,111,414.
Calage, Y., Pannetier, J., and Lucas, J. (1971). *J. Solid State Chem.* **3**, 425–428.
Campet, G., Claverie, J., Perigord M., Ravez, J., Portier, J., and Hagenmuller, P. (1974). *Mater. Res. Bull.* **9**, 1589–1596.
Campet, G., Claverie, J., and Hagenmuller, P. (1979). *Rev. Phys. Appl.* **14**, 415–420.
Carpy, A., and Galy, J. (1971). *Acta Chem. Scand.* **25**, 1918.
Casalot, A., Claverie, J., and Hagenmuller, P. (1973). *J. Phys. Chem. Solids* **34**, 347–354.
Chamberland, B. L. (1968). U. S. Patent 3,365,269.
Chamberland, B. L. (1970). U. S. Patent 3,532,641.
Chamberland, B. L. (1971a). *Mater. Res. Bull.* **6**, 311–316.
Chamberland, B. L. (1971b). *Mater. Res. Bull.* **6**, 425–432.
Chamberland, B. L., and Sleight, A. W. (1967), *Solid State Commun.* **5**, 765–767.
Camberland, B. L., Sleight, A. W., and Cloud, W. H. (1970), *J. Solid State Chem.* **2**, 49–54.
Chamberland, B. L., Frederick, C. G., and Gillson, J. L. (1973). *J. Solid State Chem.* **6**, 561–564.
Chaminade J.-P., and Pouchard, M (1975). *Ann. Chim. (Paris)* **10**, 75–99.
Chaminade, J.-P., Pouchard, M., and Hagenmuller, P. (1971), *C. R. Hebd. Seances Acad. Sci., Ser. C* **273**, 984–987.
Chaminade, J.-P., Pouchard, M., and Hagenmuller, P. (1972), *Rev. Chim. Miner.* **9**, 381–402.
Chaminade, J.-P., Vlasse, M., Pouchard, M., and Hagenmuller, P. (1973). *C. R. Hebd. Seances Acad. Sci., Ser. C* **277**, 1141–1143.
Chappert, J., and Portier, J. (1966a), *Solid State Commun.* **4**, 185–188.
Chappert, J., and Portier, J. (1966b), *Solid State Commun.* **4**, 395–398.
Claverie, J., and Georges, R. (1973), *Mater. Res. Bull.* **8**, 283–292.
Claverie, J., Dexpert, H., Portier, J., Pauthenet, R., and Hagenmuller, P. (1971). *Mater. Res. Bull.* **6**, 1125–1130.
Claverie, J., Portier, J., and Hagenmuller, P. (1972a). *Z. Anorg. Allg. Chem.* **393**, 314–320.
Claverie, J., Portier, J., and Pauthenet, R. (1972b). *J. Solid State Chem.* **5**, 207–211.
Claverie, J., Portier, J., and Hagenmuller, P. (1977). *J. Phys. (Orsay, Fr.)*, **38**, Cl, 169–173.
Corbin, O. (1982). U.S. NTIS no. DE82 703254. Les composes fluores, oxyfluores et les oxides de niobium, antimoine et tellure. CEA Centre d'Etudes Nucleaires de Fontenay-aux-Roses (France), Feb 1982. 81 pp. (in France, CEA-BIB-236).
Dance, J. M., Tressaud, A., Portier, J., and Hagenmuller, P. (1977). "Reactivity of Solids," p. 289–295. Plenum, New York.
Dehnicke, K. (1965a). *Naturwissenschaften* **52**, 660.
Dehnicke, K. (1965b). *Angew. Chem., Int. Ed. Engl.* **4**, 22–29.
Demazeau, G., Grannec, J., Marbeuf, A., Portier, J., and Hagenmuller, P. (1969). *C. R. Hebd. Seances Acad. Sci., Ser. C* **269**, 987–988.

de Pape, R., Gauthier, G., and Hagenmuller, P. (1968). *C. R. Hebd. Seances Acad. Sci., Ser. C* **266**, 803–805.

Derrington, C. E., Godek, W. S., Castro, C. A., and Wold, A. (1978), *Inorg. Chem.* **17**, 977–980.

D'Eye, R. W. M. (1958). *J. Chem. Soc.* pp. 196–199.

Doumerc, J.-P., and Pouchard, M. (1970). *C. R. Hebd. Seances Acad. Sci., Ser. C* **270**, 547–550.

Doumerc, J. P., Elaatmani, M., Ravez, J., Pouchard, M. and Hagenmuller, P. (1979). *Solid State Commun.* **32**, 111–113.

Ehrlich, P., Ploger, F., and Pietzka, G. (1955). *Z. Anorg. Allg. Chem.* **282**, 19–23.

Elaatmani, M., Ravez, J., and Hagenmuller, P. (1980). *Mater. Res. Bull.* **15**, 981–983.

Elaatmani, M., Ravez, J., Doumerc, J.-P., and Hagenmuller, P. (1981). *Mater. Res. Bull.* **16**, 105–108.

Finkelnberg, W., and Stein, A. (1950). *J. Chem. Phys.* **18**, 1296.

Fournes, L., Kinomura, N., and Menil, F. (1980). *C. R. Hebd. Seances Acad. Sci., Ser. C* **291**, 235–238.

Frei, E. H., Schieber, M., and Shtrikman, S. (1960). *Phys. Rev.* **118**, 657.

Frei, E. H., Schieber, M., and Shtrikman, S. (1963). U. S. Patent 3,093,453.

Frevel, L. K., and Rinn, H. W. (1956). *Acta Crystallogr.* **9**, 626–627.

Galasso, F., and Darby, W. (1962). *J. Phys. Chem.* **66**, 1318–1320.

Galasso, F., and Darby, W. (1963). *J. Phys. Chem.* **67**, 1451–1453.

Galasso, F., and Darby, W. (1965) *J. Opt. Soc. Am.* **55**, 332–333.

Gondrand, M., Joubert, J.-C., Chenevas, J., Capponi, J.-J., and Perroud, M. (1970). *Mater. Res. Bull.* 5, 769–773.

Goodenough, J. B., Hong, H. Y.-P., and Kafalas, J. A. (1976). *Mater. Res. Bull.* **11**, 203–220.

Gortsema, F. P., and Didchenko, R. (1965). *Inorg. Chem.* **4**, 182–186.

Grannec, J., Baudry, H., Ravez, J., and Portier, J. (1974). *J. Solid State Chem.* **10**, 66–71.

Gulick, J. C., and Sienko, M. J. (1970). *J. Solid State Chem.* **1**, 195–204.

Gutmann, V., and Jack, K. H. (1951). *Acta Crystallogr.* **4**, 244–246.

Hagenmuller, P., Portier, J., Cadiou, J., and de Pape, R., (1965). *C. R. Hebd. Seances Acad. Sci.* **260**, 4768–4770.

Harrison, F. W. and Lang, G. K. (1968). *J. Phys. Soc. Jpn.* **25**, 1609–1610.

Haussonne, J. M., Desgardin, G., Bajolet, Ph., and Raveau, B. (1983). *J. Am. Ceram. Soc.* **66**, 801–807.

Hegyi, I. J. (1964). U. S. Patent 3,146,205.

Holmberg, B. (1966). *Acta Chem. Scand.* **20**, 1082–1088.

Hubble, F. F., Gulick, J. M., and Moulton, W. G. (1971). *J. Phys. Chem. Solids* **32**, 2345–2350.

Hund, F. (1962a). *Angew. Chem., Int. Ed. Engl.* **1**, 41–45.

Hund, F. (1962b). U. S. Patent 3,022,186.

Ichinose, N. (1966). *J. Appl. Phys. Jpn.* **5**, 461–468.

Ichinose, N., and Kurihara, K. (1965). *J. Phys. Soc. Jpn.* **20**, 1530.

Ingraham, J. N., and Swoboda, T. J. (1962). U. S. Patent 3,068,176.

Ismailzade, I. H., and Ravez, J. (1978), *Ferroelectrics* **21**, 423–424.

Ismailzade, I. H., Yakupov, R. G., and Melik-Shanazarova, T. A. (1971). *Phys. Status Solidi A* **8**, K63–K64.

Kemmler-Sack, S. (1968a). *Z. Naturforsch., B: Anorg. Chem., Org. Chem., Biochem. Biophys., Biol.* **23B**, 1010.

Kemmler-Sack, S. (1969). *Z. Anorg. Allg. Chem.* **364**, 135–147.

Ketelaar, J. A. A., and Wegerif, E. (1938). *Recl. Trav. Chim. Pays-Bas* **57**, 1269–1275.

Ketelaar, J. A. A., and Wegerif, E. (1939), *Recl. Trav. Chim. Pays-Bas* **58**, 948.

Khazai, B., Dwight, K., Kostiner, E., and Wold, A. (1980). *Inorg. Chem.* **19**, 1670–1672.

Kutek, F. (1964). *Zh. Neorg. Khim. (Engl. Transl.)* **9**, 1499–1501.

Laguitton, D., and Lucas, J. (1969a). *C. R. Hebd. Seances Acad. Sci., Ser. C* **269**, 105–108.

Laguitton, D., and Lucas, J. (1969b). *C. R. Hebd. Seances Acad. Sci., Ser. C* **269**, 228–230.

La Valle, D. E., Steele, R. M., Wilkinson, M. K., and Yakel, H. L., Jr.(1960), *J. Am. Chem. Soc.* **82**, 2433–2434.

Le Flem, G., Colmet, R., Chaumont, C., Claverie, J., and Hagenmuller, P. (1976). *Mater. Res. Bull.* **11**, 389–396.

Le Flem, G., Colmet, R., Claverie, J., and Hagenmuller, P., (1980). *J. Phys. Chem. Solids* **41**, 55–59.

Le Flem, G., Demazeau, G., and Hagenmuller, P. (1982). *J. Solid State Chem.* **44**, 82–88.

Lucas, J., and Rannou, J.-P. (1968). *C. R. Hebd. Seances Acad. Sci., Ser. C* **266**, 1056–1058.

Lundberg, M., and Ndalamba Wa Ilunga, P. (1981). *Rev. Chim. Miner.* **18**, 118–124.

Lundberg, M., and Sävborg, Ö. (1978–1979). *Chem. Scr.* **13**, 197–200.

Magneli, A., and Nord, S. (1965). *Acta Chem. Scand.* **19**, 1510.

Malabry, G., Ravez, J., Fourquet, J.-L., and de Pape, R. (1973). *C. R. Hebd. Seances Acad. Sci., Ser. C* **277**, 105–108.

Mann, A. W., and Bevan, D. J. M. (1970). *Acta Crystallogr.*, **B26**, 2129–2131.

Mazelsky, R., and Ward, R. (1961). *J. Inorg. Nucl. Chem.* **20**, 39–44.

Mazza, L., and Jandelli, A. (1951). *Atti Acad. Ligure Sci. Lett. Pavia* **7**, 44–52.

Menil, F., Kinomura, N., Fournes, L., Portier, J., and Hagenmuller, P. (1981). *Phys. Status Solidi A* **64**, 261–274.

Mirishli, F. A., and Ismailzade, I. H. (1971). *Izv. Akad. Nauk SSSR, Ser. Fiz.* **35**, 1833.

Mitra, G. (1963). *Z. Anorg. Allg. Chem.* **326**, 98–100.

Mooney, R. W., and Aia, M. A. (1964). U. S. Patent 3,157,603.

Morell, A., Tanguy, B., Menil, F., and Portier J. (1973). *J. Solid State Chem.* **8**, 253–259.

Morell, A., Tanguy, B., Portier, J., Hagenmuller, P., and Nicolas, J. (1973–1974). *J. Fluorine Chem.* **3**, 351–359.

Moutou, J. M., Vlasse, M., Cerrera-Marzal, M., Chaminade, J. P., and Pouchard, M. (1984). *J. Solid State Chem.* **51**, 190–195.

Muetterties, E. L., and Castle J. E. (1961), *J. Inorg. Nucl. Chem.* **18**, 148–153.

Munoz, J. L., and Eugster, H. P. (1969). *Am. Miner,* **54**, 943–959.

Niihara, K., and Yajima, S. (1971). *Bull. Chem. Soc. Jpn.* **44**, 643–648.

Niihara, K., and Yajima, S. (1972). *Bull. Chem. Soc. Jpn.* **45**, 20–23.

Odenthal, R.-H., Grannec, J., Dance, J.-M., Portier, J., and Hagenmuller, P. (1974). *J. Solid State Chem.* **9**, 120–123.

Okazaki, C., Hirota, E., Neichi, Y., Okazaki, H., and Nakajima, S. (1966). *J. Phys. Soc. Jpn.* **21**, 199–200.

Okazaki, H. (1966). *J. Appl. Phys. Jpn.* **5**, 559–560.

Pannetier, J., and Lucas, J. (1969). *C. R. Hebd. Seances Acad. Sci., Ser. C* **268**, 604–607.

Pausewang, G., and Dehnicke, K. (1969). *Z. Anorg. Allg. Chem.* **369**, 265–277.

Pausewang, G., and Rüdorff, W. (1969). *Z. Anorg. Allg. Chem.* **364**, 69–87.

Peterson, J. R., and Burns, J. H. (1968). *J. Inorg. Nucl. Chem.* **30**, 2955–2958.

Pierce, J. W., and Vlasse, M. (1971). *Acta Crystallogr., Sect. B* **B27**, 158–163.

Pierce, J. W., McKinzie, H. L., Vlasse, M., and Wold, A. (1970). *J. Solid State Chem.* **1**, 332–338.

Pistorius, C. W. F. T. (1973). *J. Less-Common Met.* **31**, 119–124.

Pompei, J., Campet, G., Claverie, and Hagenmuller, P. (1977). *Bull. Soc. Fr. Ceram.* **117**, 15–18.

Popov, A. I., and Knudson, G. E. (1954). *J. Am. Chem. Soc.* **76**, 3921–3922.

Portier, J., and Roux, J. (1968). *Chim. Anal. (Paris)* **50**, 390–392.

Portier, J., Claverie, J., Dexpert, H., Olazcuaga, R., and Hagenmuller, P. (1970a). *C. R. Hebd Seances Acad. Sci., Ser. C* **270**, 2142–2145.

Portier, J., Tanguy, B., Morell, A., Pauthenet, R., Olazcuaga, R., and Hagenmuller, P. (1970b). *C. R. Hebd. Seances Acad. Sci., Ser. C* **270**, 821–824.

Pouchard, M., Torki, M. R., Demazeau, G., and Hagenmuller, P. (1971). *C. R. Hebd. Seances Acad. Sci., Ser. C* **273**, 1093–1096.

Rannou, J. P., and Lucas, J. (1969). *Mater. Res. Bull.* **4**, 443–450.

Ravez J., Tourneur, D., and Hagenmuller, P. (1972). *Mater. Res. Bull.* **7**, 473–478.

Ravez, J., Tourneur, D., Grannec, J., and Hagenmuller, P. (1973). *Z. Anorg. Allg. Chem.* **399**, 34–42.

Ravez, J., Elaatmani, M., and Chaminade, J.-P. (1979a), *Solid State Commun.* **32**, 749–754.

Ravez, J., Elaatmani, M., von der Mühll, R., and Hagenmuller, P. (1979b). *Mater. Res. Bull.* **14**, 1083–1087.

Ravez, J., Perandeau, G. Arend, H., Abrahams, S. C., and Hagenmuller, P. (1980). *Ferroelectrics* **26**, 767–769.

Ravez, J., Elaatmani, M., Cervera-Marzal, M., Chaminade, J.-P., and Pouchard, M. (1981a). *Mater, Res. Bull.* **16**, 1167–1175.

Ravez, J., Elaatmani, M. and Hagenmuller, P. (1981b). *Mater. Res. Bull.* **16**, 1253–1259.

Ravez, J., Elaatmani, M., Hagenmuller, P., and Abrahams, S. C. (1981c). *Ferroelectrics* **38**, 777–780.

Reynolds, T. G., and Wold, A. (1973). *J. Solid State Chem.* **6**, 565–568.

Richter, F. E., and Cook, R. L. (1966). *Am. Ceram. Soc. Bull.* **45**, 409 (Abstr. 30–E–66).

Rigo, M.-O. (1976). Ph.D. Thesis, University of Nancy I, France.

Robbins, M., and Banks, E. (1963). *J. Appl. Phys.* **34**, 1260–1261.

Robbins, M., Lerner, S., and Banks, E. (1963). *J. Phys. Chem. Solids* **24**, 759–769.

Robbins, M., Pierce, R. D., and Wolfe, R. (1971). *J. Phys. Chem. Solids* **32**, 1789–96.

Roy, R. (1954). *J. Am. Ceram. Soc.* **37**, 581–588.

Rüdorff, W., and Krug, D. (1964). *Z. Anorg. Allg. Chem.* **329**, 211–217.

Ruff, O., and Kwasnik, W. (1934). *Z. Anorg. Allg. Chem.* **219**, 65–81.

Ruff, O., and Lickfelt, H. (1911). *Chem. Ber.* **44**, 2539–2549.

Schäfer, H., Bauer, D., Beckmann, W., Gerken, R., Nieder-Vahrenholz, H.-G., Niehues, K.-J., and Scholz, H. (1964). *Naturwissenschaften* **51**, 241.

Schäfer, H., Schnering, H. G., Niehues, K.-J., and Nieder-Vahrenholz, H.-G. (1965), *J. Less-Common Met.* **9**, 95–104.

Schelleng, J. H. (1973). *AIP Conf. Proc.* **10**, 1054.

Schieber, M. (1964). *J. Appl. Phys.* **35**, 1072–1073.

Senegas, J., and Galy, J. (1972). *J. Solid State Chem.* **5**, 481–486.

Senegas, J., and Galy, J. (1973). *C. R. Hebd. Seances Acad. Sci., Ser. C* **277**, 1243–1246.

Shannon, R. D., and Prewitt, C. T. (1968). *Acta Crystallogr. Sect. B* **B25**, 925–946.

Shaplygin, I. S., and Lazarev, V. B. (1975). *Mater. Res. Bull.* **10**, 903–908.

Shinn, D. B., and Eick, H. A. (1969). *Inorg. Chem.* **8**, 232–235.

Siegel, B. (1968). *Inorg. Chim. Acta, Rev.* **2**, 137–146.

Sleight, A. W. (1969a) *Mater. Res. Bull.* **4**, 377–380.

Sleight, A. W. (1969b). *Inorg. Chem.* **8**, 1764–1767.

Smith, S. K., and Krause, D. W. (1968). *Anal. Chem.* **40**, 2034–2035.

Sobel, A. (1967). *J. Phys. Chem Solids* **28**, 185–196.

Spielberg, P. (1940). *Ark. Kemi, Miner. Geol.* **14A**, (5), 1–12.

Storm, A. R. (1970). *Mater. Res. Bull.* **5**, 973–976.

Tanguy, B., Portier, J., Morell, A., Ol Azcuaga, R., Francillon, M., Pauthenet, R., and Hagenmuller, P. (1971). *Mater. Res. Bull.* **6**, 63–68.

Tanguy, B., Vlasse, M., and Portier, J. (1973). *Rev. Chim. Miner.* **10**, 63–75.

Templeton, D. H. (1957). *Acta Crystallogr.* **10**, 788.

Templeton, D. H., and Dauben, C. H. (1954). *J. Am. Chem. Soc.* **76**, 5237–5239.

Tressaud, A., Dance, J. A., Menil, F., Portier, J., and Hagenmuller, P. (1973). *Z. Anorg. Allg. Chem.* **399**, 231–238.

Tressaud, A., Lozano, L., and Ravez, J. (1981–1982). *J. Fluorine Chem.* **19,** 61–66.

Tutov, A. G., Bader, V. I., and Myl'nikova, I. E. (1972). *Sov. Phys.–Crystallogr.* (*Engl. Transl.*) **17,** 345–349.

Ueda, Y., Ohtani, T., Kosuge, K., Kachi, S., Shimada, M., and Koizumi, M. (1978). *Mater. Res. Bull.* **13,** 305–310.

Vedrine, A., Berlin, D., and Besse, J.-P. (1972). *Bull. Soc. Chim. Fr.* pp. 76–78.

Vlasse, M., Massies, J.-C., and Demazeau, G. (1973). *J. Solid State Chem.* **8,** 109–113.

Vlasse, M., Massies, J.-C., Chaminade, J.-P. and Pouchard, M. (1974). *C. R. Hebd. Seances Acad. Sci., Ser. C* **278,** 1505–1507.

Vlasse, M., Saux, M., Echegut, P., and Villeneuve, G. (1979). *Mater. Res. Bull.* **14,** 807–812.

Vorres, K. S., and Donahue, J. (1955). *Acta Crystallogr.* **8,** 25–26.

Vorres, K. S., and Dutton, F. B. (1955). *J. Am. Chem. Soc.* **77,** 2019.

Waring, J. L., Roth, R. S., Parker, H. S., and Brower, W. S., Jr. (1976). *J. Res. Nat. Bur. Stand., Sect. A* **80,** 761–774.

Watelet, H., Baud, G., Besse, J.-P., and Chevalier, R. (1982). *Mater. Res. Bull.* **17,** 1155–1159.

Weaver, E. A., and Li, C. T. (1969). *J. Am. Ceram. Soc.* **52,** 335–338.

Wells, A. F. (1975). "Structural Inorganic Chemistry," 4th ed. Oxford Univ. Press, London and New York.

Yanagida, H., Yamaguchi, G., and Ono, Y. (1966). *Bull. Chem. Soc. Jpn.* **39,** 1346.

Zachariasen, W. H. (1949). *Acta Crystallogr.* **2,** 388–390.

Zachariasen, W. H. (1951). *Acta Crystallogr.* **4,** 231–236.

Zalkin, A., and Templeton, D. H. (1953). *J. Am. Chem. Soc.* **75,** 2453–2458.

Zalkin, A., Templeton, D. H., and Hopkins, T. E. (1966). *Inorg. Chem.* **5,** 1466–1470.

5

Defects in Solid Fluorides

C. R. A. CATLOW

Department of Chemistry
University College London
London, England

I. Introduction

Many of the most important properties of crystalline solids are controlled by defects, that is, deviations from the perfect translational periodicity of the solid. Transport properties, both electrical conductivity and diffusion, are almost invariably effected by the migration of point defects (Lidiard, 1957; Bénière and Catlow, 1983). In addition, much of the thermodynamic and mechanical behavior of solids is determined by their defect structure. The solid fluorides provide intriguing and varied case studies in defect structure. Despite the relative simplicity of their crystal structures, they range from highly perfect to highly disordered solids, and they have relevance to many topical problems in solid-state chemistry (e.g., superionic conduction and photochromism).

The materials with which we are concerned are summarized and classified in Table I. By far the greatest attention has been devoted to the alkaline-earth

INORGANIC
SOLID FLUORIDES

TABLE I

Defect Properties of Inorganic Fluorides[a]

Class	Compound	Structure Type	Intrinsic Disorder Type	Intrinsic Defect Formation Energy (eV)	Defect Migration Energies (eV) where Available
Alkali metal fluorides	LiF	Rock salt	Schottky pair	2.3–2.7	Cation vacancy, 0.7; Anion vacancy, 1.1
	NaF	Rock salt	Schottky pair	2.4	Cation vacancy, 0.9; Anion vacancy, 1.4
	KF	Rock salt	Schottky pair	2.64	Cation vacancy, 1.0; Anion vacancy, —
Alkaline-earth fluorides	MgF_2	Rutile	Uncertain	2–4	—
	CaF_2	Fluorite	Frenkel pair	2.7	Anion vacancy, 0.4; Anion interstitial, 0.8
	SrF_2	Fluorite	Frenkel pair	2.4	Anion vacancy, 0.55; Anion interstitial, 0.75
	BaF_2	Fluorite	Frenkel pair	2.2	Anion vacancy, 0.5; Anion interstitial, 0.75
	PbF_2	Fluorite	Frenkel pair	1.1	Anion vacancy, 0.25; Anion interstitial, 0.50
	MnF_2	Rutile	Uncertain	2–4	—
	LaF_3	Tysonite	Uncertain	2–4	—
	YF_3	—	Uncertain	—	—

[a] References for experimental data can be found in the text and Corish and Jacobs (1973, 1978).

fluorides (CaF_2, SrF_2, and BaF_2). These show exceptional variety in their defect chemistry. At low temperatures the pure materials are conventional ionic solids with low levels of intrinsic disorder due to the high electronegativity of fluorine. The high-temperature solids, however, show superionic properties (Catlow, 1980; Chadwick, 1983), that is, ionic conductivities of the same order of magnitude as those of molten salts. The materials may also be doped with high levels of trivalent impurity cations (La^{3+} and the rare-earth cations), which induce corresponding concentrations of charge-compensating defects. The structural properties of these heavily defective phases are among the most intensively studied in solid-state chemistry (see Catlow, 1981). The properties of the irradiated materials are also of considerable interest. The electronic defects that are created have been studied by a variety of spectroscopic and theoretical methods (Hayes and Stoneham, 1974), and the photochromism of the doped materials is of special interest (Hayes and Staebler, 1974).

There is growing interest in other disordered fluorides—for instance, the high-temperature rare-earth fluorides that have superionic properties. But, as will be apparent in the discussion that follows, much less information is available for these systems.

II. Basic Defect Structures

Table I summarizes the nature and energetics of intrinsic disorder in solid fluorides. The alkali metal fluorides all show Schottky disorder, which involves the generation of cation and anion vacancies by the migration of lattice ions to the surface. In contrast, for the alkaline-earth fluorides, all of which have the fluorite structure illustrated in Fig. 1, anion–Frenkel disorder predominates. This mode of disorder involves the displacement of lattice ions to interstitial sites, and the large interstitial sites in the fluorite structure as shown in Fig. 1 result in relatively low energies for the creation of anion interstitials. In contrast, the levels of cation disorder in the alkaline-earth fluorides are low; an important consequence is that the rates of cation diffusion, which are affected by cationic defects, are low (e.g., see Catlow et al., 1977).

Transport measurements have been made on the rutile-structured fluorides MgF_2 and MnF_2 (Park and Nowick, 1976). There is no clear evidence, however, as to the nature of the intrinsic disorder in these materials. A similar situation applies to the rare-earth fluorides for which both Frenkel and Schottky models are plausible.

Defect structure is generally deduced from analysis of transport data— electrical conductivity or diffusion studies. Ion transport is, as noted, effected

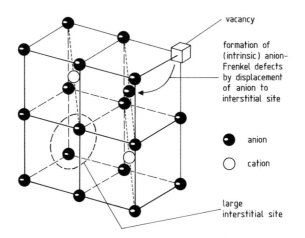

Fig. 1.   Fluorite structure. The cubic interstitial site is indicated.

by the migration of defects, which in ionic solids are electrically charged entities. Conductivity and diffusion are thus controlled by the concentrations and mobilities of defects. In addition to the intrinsic Frenkel and Schottky disorder, defects can be created as "charge compensators" for aliovalent impurity ions, that is, ions with a charge that is different from that of the host-lattice ions for which they substitute. For example, the replacement of Ca^{2+} by a rare-earth cation leads to the creation of charge-compensating fluoride interstitials (see Chapter 12). Indeed, at low temperatures, defects induced by impurity aliovalent ions generally dominate the defect population and hence the transport properties of the material, whereas at higher temperatures the intrinsic (Schottky or Frenkel disorder) dominates. This change in defect population is apparent in Arrhenius plots of conductivity or diffusion, an example of which is shown for BaF_2 in Fig. 2. From the analysis of such plots it is possible to deduce energies of formation and activation of defects (see Corish and Jacobs, 1973, 1978; Bénière and Catlow, 1983). Energies obtained from such analyses are given in Table I. There is generally good agreement between calculated and experimental parameters. The calculations that use modern computational defect theory have been discussed by Catlow and Mackrodt (1982).

In alkali metal fluorides, the migration mechanisms involve simple jumps of neighboring ions into vacancies. Calculations (Catlow *et al.*, 1979) have shown that, in the saddle point for these mechanisms, the migrating ion is midway between the two lattice sites between which the ion jump occurs. A similar type of mechanism operates for the anion vacancy in alkaline-earth fluorides. The low value of the activation energy for these defects (generally about 0.3–0.5 eV) should be noted; it is a general feature of the fluorite structure and is

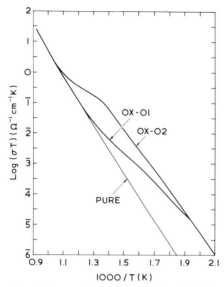

Fig. 2. Arrhenius plot of the conductivity of O^2-doped BaF_2. The low-temperature "extrinsic" region is separated from the higher-temperature "intrinsic" region (see Jacobs and Ong, 1980).

responsible for superionic properties of fluorite-structured oxides, for example, CeO_2 (see Kilner and Steele, 1981). A more complex mechanism seems to effect interstitial migration. This involves a concerted process in which the migrating interstitial displaces a lattice ion into a neighboring interstitial position; the saddle point for this mechanism is shown in Fig. 3.

The defect properties of alkali and alkaline-earth fluorides appear, therefore, to be well understood, although for rare-earth fluorides, there is, as noted in Table I, no definite knowledge of the predominant intrinsic defects or of the

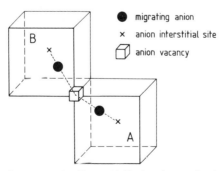

Fig. 3. Saddle point for concerted or "interstitial" migration mechanism for anion interstitials in alkaline-earth fluorides.

migration mechanism. Even for fluorite-structured fluorides, however, there are major areas of uncertainty when we consider the high-temperature behavior of these systems, as discussed in the next section.

III. Superionic Fluorides

All fluorite-structured materials appear to show a diffuse phase transition, manifested by a λ-type specific heat anomaly (Dworkin and Bredig, 1968; Schroter and Nolting, 1980), within a few hundred degrees of their melting point. Typical data are illustrated in Fig. 4 for PbF_2. Above the phase transition temperature T_C, the materials show exceptionally high ionic conductivity, with the Arrhenius plot, however, leveling off, as shown in Fig. 4. It is generally accepted that the diffuse phase transition and the superionicity are due to the generation of disorder on the anion sublattice, but the extent and nature of this disorder have been a matter of considerable controversy. Earlier work (e.g., Derrington et al., 1975) proposed a comprehensive disordering of the anion sublattice at T_C; the term *sublattice melting* was coined to describe this process. However, it has become clear that the extent of disorder is far more limited than implied by this term. Diffraction studies (Dickens et al., 1979) have shown that in the superionic phase most anions ($>90\%$) are still located at regular lattice sites. This conclusion was reinforced by dynamic computer simulation studies of Dixon and Gillan (1980a,b). In addition, Catlow (1980) showed that the energetics associated with the generation of high levels of disorder were incompatible with enthalpy of the diffuse transitions obtained from the specific heat data.

It seems, therefore, that the diffuse phase transition involves the generation of about 1 to 5 mol% of anionic defects, the high mobility of which is responsible for the superionic properties[†] of the high-temperature phase. Such models, however, raise the question of why the generation of disorder does not continue beyond the diffuse phase transition, as is suggested by both the specific heat data and the flatness of the Arrhenius plot of the conductivity above T_C illustrated in Fig. 4.

A possible solution to this problem was suggested by Dickens et al. (1982), who proposed that in the superionic phase there were significant concentrations of interstitial–vacancy clusters of the type illustrated in Fig. 5. These are similar to the dopant–interstitial clusters discussed in the next section, and their stability was demonstrated by energy calculations (Catlow and Hayes, 1982). Their formation, moreover, provides a good explanation of the

[†] Note that both simulation studies (Dixon and Gillan, 1980a,b) and conductivity measurements (Carr et al., 1978) show that interstitials make an appreciable contribution to anion transport in the superionic high-temperature fluorites.

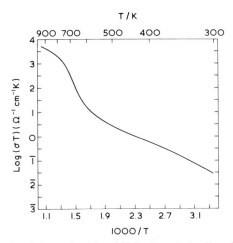

Fig. 4. Arrhenius plot of the conductivity of PbF_2. Note the leveling off of the plot above the phase transition temperature T_c.

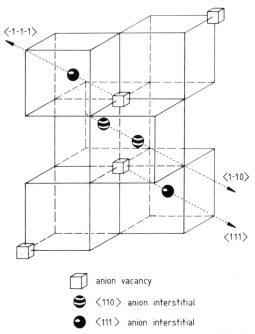

Fig. 5. Interstitial–vacancy complex proposed by Dickens *et al.* (1982) for high-temperature fluorites. Note the similarity to the dopant–interstitial complexes shown in Fig. 8.

observed features of quasi-elastic and Bragg scattering data. It may be argued, therefore, that in the vicinity of T_C the vacancy and interstitial concentrations have reached sufficient levels to lead to the generation of such clusters. Repulsive interactions between these large bulky defects, however, prevent further, large-scale generation of disorder above T_C. However, further work, both theoretical and experimental will be needed before such models may be taken as established.

There has been speculation that superionic behavior is present in high-temperature LaF_3 and YF_3, and in the former case there is also clear evidence for a specific heat excess (Lyon et al., 1978), which starts at $\sim 1100°C$ and continues up to the melting point (unlike the λ-type anomaly observed for the fluorites). There is, however, no definite knowledge of the defect structure in these superionic phases.

In addition to high-temperature superionic behavior, doped fluorite-structured fluorides may also show high ionic conductivity, a property arising from the very high levels of disorder that can be induced by doping of alkaline-earth fluorides by trivalent ions; this topic is examined in the next section.

IV. Doped Alkaline-Earth Fluorides

A remarkable feature of alkaline-earth fluorides is their capacity to form solid solutions, retaining the fluorite structure, with high concentrations (up to 40 mol %) of rare-earth fluorides. The rare-earth cations enter the lattice at cation sites; electroneutrality is, as noted, restored by the creation of anion interstitials, which in heavily doped materials are consequently present in high concentrations.

Much of the interest in these materials has focused on the intriguing structural features of the high interstitial concentrations. For low dopant concentrations (< 1 mol %), the defect structure seems to be simple. Spin resonance studies (see Baker, 1974) show that simple pair clusters of the type shown in Fig. 6 are formed; the interstitials are bound to sites neighboring the dopant, owing to the coulombic interaction between the substitutional dopant and its charge-compensating defect. More concentrated interstitial solutions (> 5 mol % dopant) have been investigated by diffraction techniques, which reveal a radically different interstitial structure. Refinements of neutron Bragg diffraction data (Cheetham et al., 1971; Catlow et al., 1983a,b) reveal the presence of interstitials at sites strongly displaced from the body-center positions of the cubic interstitial site, which are occupied in the low-concentration region. Two types of displaced interstitial are noted: the first is displaced along the [110] and the second along the [111] direction, as

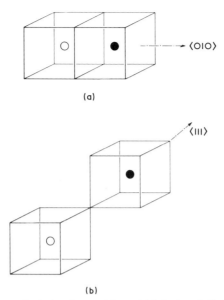

Fig. 6. Simple cluster of trivalent ions and interstitials in doped alkaline-earth fluorides: (a) nearest-neighbor pair; (b) next-nearest-neighbor pair.

illustrated in Fig. 7. In addition, the Bragg diffraction data showed the presence of vacancies.

Cheetham *et al.* (1971) argued that their results on Y^{3+}-doped CaF_2 could be explained in terms of the formation of interstitial clusters. The models they proposed are illustrated in Fig. 8; the smaller "2:2:2" cluster had been suggested by Willis (1964) in studies on the isostructural nonstoichiometric UO'_{2+x} phase. The cluster includes both [110] and [111] interstitials as well as vacancies and as such can rationalize the results of the Bragg scattering data. In addition, the ratios of the various types of species in the 2:2:2 cluster

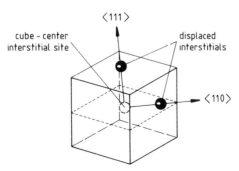

Fig. 7. Interstitial positions [110] and [111] in anion-excess fluorides.

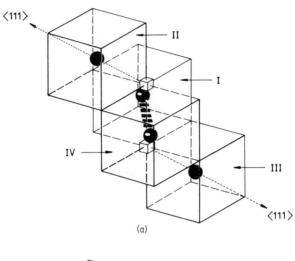

(a)

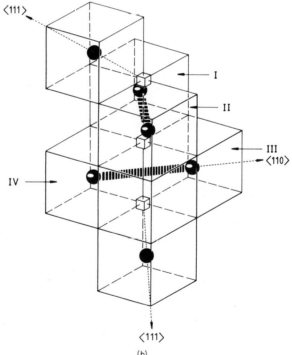

(b)

Fig. 8. Interstitial clusters "2:2:2" (a) and "4:3:2" (b).

are compatible with Bragg scattering data on 5 mol % Y^{3+}-doped CaF_2, whereas the $4:3:2$ model seems to explain the data for 10 mol % solution. It is doubtful whether any simple cluster model would be appropriate for higher dopant concentrations.

Support for the models proposed by Cheetham et al. (1971) was provided by calculations of the energies of these clusters by Catlow (1973, 1976). These studies showed that the clusters were strongly bound when dopant ions were in nearest-neighbor cation position with respect to the interstitial cluster. This work provided, moreover, a plausible rationalization of the structure of these unusual clusters. Thus, the $2:2:2$ cluster is essentially a dimer of the simple pair clusters shown in Fig. 6. The most obvious structure for such a dimer is shown in Fig. 9. Calculations showed that such clusters were unstable, however, and that stability required a coupled relaxation in which interstitial ions move inward (to become [110] interstitials) and lattice ions are displaced toward interstial sites (to become [111] "interstitials" and leaving vacancies). This process generates the $2:2:2$ cluster and was shown by the calculations to be highly favored energetically. Similar arguments apply to the $4:3:2$ cluster.

Other workers (e.g., see Gettman and Greiss, 1978), using x-ray techniques, usually in very heavily doped CaF_2, have suggested larger, more complex cluster structures in which six dopant ions are grouped around a central interstitial site. Lattice anion sites surrounding the interstitial position are vacant, but anions are situated above each edge of the cube to give the structure shown in Fig. 10. Support for these models is provided by the observation by Bevan et al. (1982) of an ordered system ($Ca_{14}Y_5F_{43}$, which is a naturally occurring mineral, tveitite) the structure of which is based on the cubo-octahedral arrangement of trivalent ions within a fluorite-structured host. Moreover, we note that mixed-fluoride systems (e.g., $RbBi_3F_{10}/KY_3F_{10}$) discussed in Chapter 12 contain this type of cubo-octahedral anion cluster.

Insight into the factors controlling the stability of the cubo-octahedral structures has been provided by calculations of their energies (Catlow et al., 1984). This work showed that the stability of such clusters relative to the $2:2:2$ and $4:3:2$ models was critically dependent on the size of the dopant ion

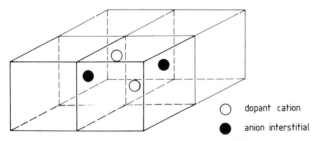

○ dopant cation

● anion interstitial

Fig. 9.   Planar dimer of two dopant–interstitial pairs.

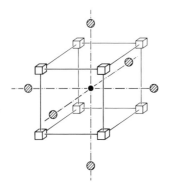

☐ anion vacancy

⊘ dopant ion

● interstitial site

Fig. 10.   Cubo-octahedral dopant–interstitial cluster.
Note that interstitials are present above each cube edge.

relative to that of the host-lattice cation. For large cations (e.g., La^{3+}) the 2:2:2 type of model is preferred, whereas smaller cations show greatest stability for the cubo-octahedral clusters. Neutron diffraction studies (Catlow *et al.*, 1983a,b) on La^{3+}- and Er^{3+}-doped CaF_2 support these conclusions, although the type of model proposed for La^{3+}-doped CaF_2 involves the capture by the 2:2:2 cluster of an additional interstitial to give the cluster shown in Fig. 11. The stability of these clusters was strongly supported by calculations (Catlow *et al.*, 1983a,b).

It seems, therefore, that there is a highly complex range of possible cluster structures in doped fluorite-structured fluorides. The type of structure adopted depends on the size of the dopant ion and probably on other factors, for example, the duration of the high-temperature anneal involved in the sample preparation, a factor that is clearly relevant owing to the low cation mobility in the fluorite structure.

In view of the high interstitial concentrations in heavily doped materials, it might be expected that these materials would show high ionic conductivities. Thus, Schoonman and co-workers (Schoonman, 1980) have performed extensive conductivity studies of heavily doped alkaline-earth fluorides. An interesting feature of their results is the decrease in the activation energy for conductivity with the dopant concentration. This they attribute to the occurrence of relatively easy conduction paths around the clusters. The conductivity of the materials, however, is not sufficient in general to justify use of the term *superionic* for these materials.

Finally, we note that it is possible to dope alkaline-earth fluorides with low-valence ions (e.g., Na^+), which induce a charge-compensating vacancy

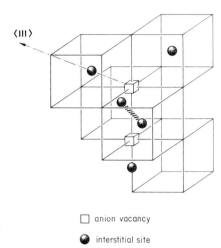

⟨III⟩

□ anion vacancy

● interstitial site

Fig. 11. Cluster formed by capture of interstitial by 2:2:2 cluster.

population, although the solubility of monovalent dopants is generally less than 1 mol %, very much less than that of the trivalent ions. Low solubilities of aliovalent dopant ions are also observed with the alkali metal fluorides; there appear to be high dopant solubilities in LaF_3, although the doped rare-earth fluorides have not been extensively investigated.

V. Electronic Defects

Irradiation of fluorides by x rays or electrons leads to the creation of electronic defects with properties that can be studied by spin resonance and a variety of other spectroscopic defects. General reviews are given by Stoneham (1975) and Hayes and Stoneham (1974). Again, the greatest amount of work has been reported for alkaline-earth fluorides, in which two principal types of defect, the H and F centers, have been identified. The former is an interstitial defect, essentially an interstitial fluorine atom, which forms a covalent bond with a neighboring lattice F^-, resulting in displacement of the atom from the body-center position. Information is available from epr studies on the distribution of unpaired spin density between interstitial and neighboring lattice ions. Details are given by Hayes and Stoneham (1984).

The F center consists of an electron trapped at an anion vacancy.[†] A large body of detailed spectroscopic data is available for these defects (e.g., see Hayes and Stoneham, 1974). Again, spin resonance data indicate appreciable

---

[†] This defect can also be created by additive coloration, that is, by exposure of the material to the appropriate metal vapor, as well as by radiation damage.

delocalization of the electron spin over neighboring anion sites. Extensive theoretical studies of these defects have also been reported (e.g., see Stoneham, 1975). Indeed, these defects provide good test cases for theories of localized electronic states in solids.

The mechanism whereby radiation leads to the formation of electron defects is uncertain. For alkali halide crystals, it appears that radiationless decay of self-trapped excitons provides the defect formation mechanism in irradiated crystals. Excitonic decay may be involved in alkaline-earth fluorides, but it seems likely that the mechanism differs significantly in the two classes of crystal. A detailed discussion is given by Catlow (1979).

Finally, we should note that alkaline-earth fluorides doped with rare-earth ions show the phenomenon of photochromism after x irradiation. Radiation results in the creation of F centers, a high proportion of which are in sites neighboring the rare-earth substitutional. Subsequent irradiation by visible or ultraviolet light leads to ionization of the electron in these vacancy traps; a proportion of the ionized electrons are trapped at the rare-earth cation sites. The reduced cations have a characteristic color; hence, the materials are photochromic. Other discussions of these interesting systems are given by Hayes and Staebler (1974) and Catlow (1979).

VI. Summary

This brief chapter has, we hope, justified our claim that solid fluorides offer a remarkably interesting and diverse range of defect properties. Our discussion has necessarily concentrated on alkaline-earth fluorides, for which the greatest amount of information is available. We believe that there may be a similarly rich and varied field of study in defective rare-earth fluorides. Studies of these systems should be encouraged.

References

Baker, J. M. (1974). In "Crystals with the Fluorite Structure" (W. Hayes, ed.). Oxford Univ. Press, London and New York.
Bénière, F., and Catlow, C. R. A. (1983). "Mass Transport in Solids." Plenum, New York.
Bevan, D. J. M., Strähle, J., and Greiss, O. (1982). J. Solid State Chem. 44. 75.
Carr, V. M., Chadwick, A. V., and Saghafian, R. (1978). J. Phys. Chem. 11, L637.
Catlow, C. R. A. (1973). J. Phys. C 6, L64.
Catlow, C. R. A. (1976). J. Phys. C 9, 1845.
Catlow, C. R. A. (1979). J. Phys. C 12, 969.
Catlow, C. R. A. (1980). Comments Solid State Phys. 9, 157.
Catlow, C. R. A., and Hayes, W. (1982). J. Phys. C 15, L9.

Catlow, C. R. A. (1981). *In* "Non-stoichiometric Oxides" (O. T. Sorensen, ed.). Academic Press, New York.

Catlow, C. R. A., and Mackrodt, W. C. (1982). *Lect. Notes Phys.* **166**.

Catlow, C. R. A., Norgett, M. J., and Ross, T. A. (1977). *J. Phys. C* **10**, 1627.

Catlow, C. R. A., Corish, J., Diller, K. M., Jacobs, P. W. M., and Norgett, M. J. (1979). *J. Phys. C* **12**, 451.

Catlow, C. R. A., Chadwick, A. V., and Corish, J. (1983a). *J. Solid State Chem.* **48**, 65.

Catlow, C. R. A., Chadwick, A. V., and Corish, J. (1983b). *Radiat. Eff.* **75**, 61.

Catlow, C. R. A., Corish, J., and Jacobs P. W. M. (1984). *J. Solid State Chem.* **51**, 159.

Chadwick, A. V. (1983). *Radiat. Eff.* **74**, 17.

Cheetham, A. K., Cooper, M. J., and Fender, B. E. F. (1971). *J. Phys. C* **4**, 3107.

Corish, J., and Jacobs, P. W. M. (1973). *Spec. Publ. Chem. Soc.* **2**, Chapter 7.

Corish, J., and Jacobs, P. W. M. (1978). *Spec. Publ.—Chem. Soc.* **8**, 128.

Derrington, C. E., Lindner, A., and O'Keeffe, M. (1975). *J. Solid State Chem.* **15**, 171.

Dickens, M. H., Hayes, W., Hutchings, M. T., and Smith, C. (1979). *J. Phys C* **12**, L97.

Dickens, M. H., Hayes, W., Hutchings, M. T., and Smith, C. (1982). *J. Phys. C* **15**, 4043.

Dixon, M. and Gillan, M. J. (1980a). *J. Phys. C* **13**, 1901.

Dixon, M., and Gillan, M. J. (1980b). *J. Phys. C* **13**, 1919.

Dworkin, A. S., and Bredig, M. A. (1968). *J. Phys. Chem.* **72**, 1277.

Gettman, P., and Greiss, O. (1978). *J. Solid State Chem.* **26**, 255.

Hayes, W., and Staebler, I. (1974). *In* "Crystals with the Fluorite Structure" (W. Hayes, ed.). Oxford Univ. Press, London and New York.

Hayes, W., and Stoneham, A. M. (1974). *In* "Crystals with the Fluorite Structure" (W. Hayes, ed.), p.472. Oxford Univ. Press, London and New York.

Jacobs, P. W. M., and Ong, S. (1980). *Cryst. Lattice Defects* **8**, 177.

Kilner, J. H., and Steele, B. C. H. (1981). *In* "Non-stoichiometric Oxides" (O. T. Sorensen, ed.). Academic Press, New York.

Lidiard, A. B. (1957). *In* "Handbuch der Physik" (S. Flügge, ed.), Vol. 20, p. 507. Springer-Verlag, Berlin and New York.

Lyon, W. G., Osborne, D. W., Flotow, H. E., and Grandjean, F. (1978). *J. Chem. Phys.* **69**, 167.

Park, D. S., and Nowick, A. S. (1976). *J. Phys. Chem. Solids* **37**, 607.

Schoonman J. (1980). *Solid State Ionics* **1**, 123.

Schroter, W., and Nolting, J. (1980). *J. Phys. (Paris)* **41**, C6-20.

Stoneham, A. M. (1975). "The Theory of Defects in Solids." Oxford Univ. Press, London and New York.

Willis, B. T. M. (1964). *Proc. Bri. Ceram. Soc.* **1**, 9.

6

High Oxidation States in Fluorine Chemistry

RUDOLF HOPPE

Institut für Anorganische und Analytische Chemie
Justus Liebig–Universität
Giessen, Federal Republic of Germany

I. What Is a High Oxidation State?

In inorganic chemistry, *oxidation state* is quite often a more precise expression than *valence*, especially for fluorides. Difficulties may arise,[†] however, for example, for SiH_2F_2. Concerning the rule that the oxidation state of fluorine should be always -1, it should be mentioned that fluorides such as $F_2^+[SbF_6]^-$ are fortunately still unknown. Such problems are only of minor or no concern for the compounds under consideration in this chapter. At any rate, one should know the constitution: $Mn_2O_2F_9$, for example, is not a mixed valence compound with formulation $Mn^{VI}Mn^{VII}O_2F_9$ but a derivative of MnF_4 (Hoppe *et al.*, 1962a); it is $O_2^+[Mn_2F_9]^-$.

Moreover, there is no generally accepted definition of *high oxidation state*. Should one attempt to put the term *high* on an absolute scale or rather use it in a relative connection? In order to avoid such more or less philosophical debates, a pragmatic definiton was chosen for this chapter:

> The metal ion in a fluoride will be considered of *high oxidation state* when there is no, or no stable, binary chloride with the metal ion in the same oxidation state.

If there are derivatives such as oxyfluorides that possibly exist [e.g., MnO_3F (Engelbrecht and Grosse, 1954)], these will be included even if the corresponding chloride is known and stable [see also CrO_2F_2/CrO_2Cl_2 (Engelbrecht and Grosse, 1952)]. The purpose of adding "and stable" was to make the definition a little less well defined but smoother so that it leaves, like many other chemical conceptions (such as atomic radii, coordination numbers, coordination polyhedra, and, last but not least, oxidation state itself), some space for a "personal" decision (e.g., by chemical intuition) on how to interpret the definition in doubtful cases.

A precise discussion of other definitions for the expression *high oxidation state* must be avoided mainly because it would be too space-consuming. However, one other definition will be considered briefly. With respect to the Born–Haber cycle, in Fig. 1 an attempt is made to put the "definition" on an absolute scale. One should bear in mind that this attempt is of little chemical importance because the stability and therefore (in a certain sense) the "existence" of a phase is subject to competition with the neighboring phases.

All oxidation states that are now known in chemistry correspond to $I_n/nI_1 < 2.3$. The consequences are the following:

1. All oxidation states that correspond to $I_n/nI_1 > 2.3$ are not only high but

---

[†] The oxidation state of fluorine is always -1 and that of oxygen mostly -2 (but -1 in the case of O_2^{2-}, $-\frac{1}{2}$ with O_2^-, $\frac{1}{3}$ regarding ozonides with O_3^-, and, e.g., $+\frac{1}{2}$ with the cation O_2^+).

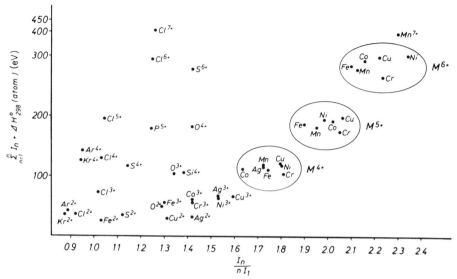

Fig. 1. Plot of heat of formation of E^{nd} versus ratio of nth (I_n) and first (I_1) energies of ionization divided by n.

could be called *exotically high oxidation states*. The hypothetical "CsF$_3$" (Bode, 1950) discussed by the author as early as 1948–1949 is certainly unstable against decomposition to CsF and F$_2$ because ΔH^0_{298}(CsF, solid) = -133 kcal/mol is exceedingly "larger" than the estimated value for CsF$_3$, with only ΔH^0_{298}(CsF$_3$, solid) = -30 kcal/mol.[†] Pathways of syntheses of such compounds outside the thermodynamic equilibrium [such as XeO$_4$ (Bartlett and Sladley, 1972), ΔH^0_{298} (XeO$_4$, solid) = 153 kcal/mol] are still unknown, but maybe the day will come.

2. The definition of a high oxidation state even in this sense is not possible without a certain arbitrariness. Should the borderline of *oxidation state* versus *high oxidation state* be drawn between Cl^{5+}/Cl^{6+} and the groups of fluorides MF$_4$/MF$_5$, or rather MF$_3$/MF$_4$? At any rate, Ar^{4+} would be as "normal" as is S^{4+}. But alas, the position of Kr^{2+} in Fig. 1, and more impressively of Ar^{2+}, excludes this attempt from serious discussion. Or would ArF$_2$ or even ArF$_4$, despite of all published unsuccessful attempts of synthesis be stable (or at least obtainable)?

[†] Discussed as early as 1949 with respect to calculations of the heat of formation of xenon fluorides by the present author, unpublished work. See for competition, Klemm (1982).

II. Consequences

Following the definition given above, there are chemical elements without a high oxidation state in fluorides, as well as some that have to be discussed here. Figure 2 shows the distribution of these within the periodic table. Clearly, we have five distinct groups:

Group A consists of IB elements:

> Copper, including Cu^{3+} and Cu^{4+}; silver, including Ag^{2+}, Ag^{3+}, and Ag^{4+} (or Ag^{3+} and Ag^{5+}); gold, including Au^{5+} (perhaps, in addition, Au^{4+})

Taxonomists could call Cu^{4+}, Ag^{3+}, Ag^{4+} and Au^{5+} "very high oxidation states" because these fluorides belong to the very few in which no corresponding oxide (whatever kind) of this or a higher oxidation state is known. This distinction is not used in the following sections.

Group B consists of *d* elements:

> V^{5+}, $Cr^{4+}-Cr^{6+}$, $Mn^{3+}-Mn^{7+}$, Co^{3+}, Co^{4+}, Ni^{3+}, and Ni^{4+}

from the series of 3*d* metals;

> Tc^{6+}, Tc^{7+}, $Ru^{4+}-Ru^{8+}$, $Rh^{4+}-Rh^{6+}$,
> Pd^{3+} (only obtained under high pressure), Pd^{4+}, (Pd^{5+}?)

from the series of 4*d* metals; and

> Re^{7+}, $Os^{5+}-Os^{8+}$, Ir^{4+}, Ir^{5+}, Ir^{6+}, Pt^{5+}, and Pt^{6+}

from the 5*d* metals.

Group C includes lanthanoids:

> Ce^{4+}, Pr^{4+} (perhaps Pr^{5+}), Nd^{4+}, Tb^{4+}, and Dy^{4+}

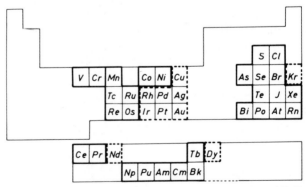

Fig. 2.   Part of the periodic table of chemical elements showing those that are dealt with in this chapter.

Group D contains actinoids:

$$Np^{5+}-Np^{7+}, Pu^{4+}-Pu^{7+}, Am^{4+}, Am^{6+}, Bk^{4+}, Cf^{4+}, \text{ and } Es^{4+}$$

(*without* any example of very high oxidation states known today).

Group E consists of nonmetallic elements:

As^{5+} and Bi^{5+} from VA
S^{6+}, Se^{6+}, and Te^{6+} from VIA
Cl^{3+}, Cl^{5+}, Cl^{7+}, Br^{3+}, Br^{5+}, Br^{7+}, I^{5+}, and I^{7+} from VIIA
Kr^{2+}, Xe^{2+}, Xe^{4+}, Xe^{6+} and Xe^{8+} from VIIIA

This classification is the basis for the following discussion.

III. High Oxidation States of Copper, Silver, and Gold

Table I summarizes the characteristics of the known high-oxidation-state compounds in this group.

A. OXIDATION STATE +2

Here, only AgF_2 must be cited. Surprisingly, it is structurally quite different from CuF_2 (Fig. 3). A series of ternary derivatives are known, some having been defined by single-crystal work (e.g., $CsAgFeF_6$).

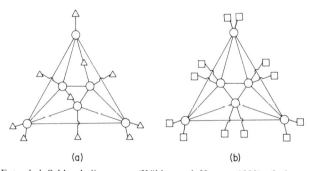

(a)   (b)

Fig. 3. Extended Schlegel diagramm (Köhler and Hoppe, 1983) of the polyhedron of coordination for (a) Cu^{2+} in CuF_2 ($\triangle = Cu$, $\bigcirc = F$) (typical for all rutiles) and (b) Ag^{2+} in AgF_2 ($\square = Ag$, $\bigcirc = F$) (showing that each F^- is a common corner of three polyhedras of coordination. This is possible for such remarkably elongated octahedra). The diagram shows whether additional cations share the polyhedron of coordination by common corners (terminal) or by common edges (angular; here the cation is placed symbolically on the very edge).

TABLE I

Characteristic Fluorides of High Oxidation State: Group IB Elements[a,b]

Oxidation State	Copper	Silver	Gold
+2	—	$\underline{AgF_2}$ (1) Dark blue ($\mu_{eff}/\mu_B = 1.0$, $T_C = 163$ K) $CsAgF_3$ (2) Brown ($\mu_{eff}/\mu_B = 0.65$, $T_C = 50$ K) ($GdFeO_3$) $\underline{CsAgAlF_6}$ (3) Dark blue ($\mu_{eff}/\mu_B = 1.3$)	—
+3	Cs_2NaCuF_6 (4) Green (Cs_2NaCrF_6) K_3CuF_6 (5) Green ($\mu/\mu_B = 3.03$) (K_3CrF_6) Rb_2KCuF_6 (8) Green ($\mu/\mu_B = 2.98$) (K_2NaAlF_6) $CsCuF_4$ (9) Orange (Diamagnetic) ($KBrF_4$)	AgF_3 (6) (?) Red-brown $KAgF_4$ (7) Yellow (diamagnetic) ($KBrF_4$) Cs_2KAgF_6 (10) Purple ($\mu/\mu_B = 2.6$) (K_2NaAlF_6)	—

+4		
Cs$_2$CuF$_6$ (11, 12)	Cs$_2$AgF$_6$ (13)	—
Orange	Orange	
($\mu/\mu_B = 1.55$) (K$_2$PtCl$_6$)	($\mu/\mu_B = 1.39$) (K$_2$PtCl$_6$)	

+5		
—	Cs$_2$Ag$_{1/2}$Ga$_{1/2}$F$_6$ (13)	AuF$_5$ (14)
	Orange	Red
	(K$_2$PtCl$_6$)	(diamagnetic)
		CsAuF$_6$ (15)
		Yellow-green
		(diamagnetic) (LiSbF$_6$)
		(O$_2^+$)AuF$_6$ (15)
		Yellow
		($\mu/\mu_B = 1.66$) [(O$_2^+$)PtF$_6$]

[a] Underscored are those formulas in which the crystal structure is known by single-crystal work; otherwise, the type of structure, if known, is given in parentheses.

[b] Numbers in parentheses are references, as follows: 1, Charpin et al. (1970); 2, Odenthal and Hoppe (1971); 3, Müller (1981a); 4, Grannec et al. (1975); 5, Klemm and Huss (1949); 6, Bougon and Lance (1983); 7, Klemm (1954); 8, R. Hoppe and H. Mattauch, unpublished, see Mattauch (1962); 9, Fleischer and Hoppe (1982d); 10, Hoppe and Homann (1966); 11, Harnischmacher and Hoppe (1973); 12, Hagenmuller et al. (1976); 13, Hagenmuller et al. (1978); 14, Waslie and Falconer (1976); 15, Bartlett and Leary (1976).

B. OXIDATION STATE +3

Neither CuF_3 nor AgF_3 (but see Bougon and Lance, 1983) are known; AuF_3 is definitely known. All ternary fluorides known so far isolated have been obtained only as powders. The following are noteworthy:

1. Orange $Cs[CuF_4]$. This new diamagnetic fluoride with square coordination [coordination number (C.N.) = 4] is a "missing link" between the paramagnetic complexes $A_3[CuF_6]$ and the diamagnetic fluorides $A[MF_4]$ (M = Ag or Au).
2. Deep red $Cs_2K[AgF_6]$. This is still the only other "link" (C.N. = 6).
3. Derivatives such as $Cs_2K[AuF_6]$. These are still unknown. In any case, they are excluded by declaration.

C. OXIDATION STATE +4

It is very difficult to characterize such extremely sensitive materials as $Cs_2[CuF_6]$ or $Cs_2[AgF_6]$ (Table II). This is why making a choice between formulations such as $Cs_2[Ag^{IV}]$ and $Cs_2(Ag^{III}_{0.5}Ag^{V}_{0.5})F_6$ is so delicate at the moment (despite the existence of a similar $Cs_2(Ga^{III}_{0.5}Ag^{V}_{0.5})F_6$ also diamagnetic).

D. OXIDATION STATE +5

$Cs[AuF_6]$ and $O_2^+ [AuF_6]^-$ are known examples. The latter fluoride was prepared in the form of single crystals (Müller, 1981b), which were unfortunately too sensitive to be handled for structural investigations. An exciting development has been the preparation of AuF_5.

TABLE II

Unit Cell Parameters of Various Preparations of Cs_2CuF_6[a]

Reference	Unit Cell Parameters (Å)
Harnischmacher and Hoppe (1973)	$a = 8.871; b = 6.272; c = 6.310$
	$a = 8.858$
	$a = 8.922$
	$a = 8.922; a = 8.852$ (two phases)
	$a = 8.862; b = 6.266; c = 6.327$
Hagenmuller et al. (1976)	$a = 8.876$

[a] Different preparations obtained under similar experimental conditions

E. STATE OF THE ART AND CONCLUSIONS

Without doubt, the existence of AuF_5 is the most surprising experimental result in the entire scope of this chapter. A more detailed discussion is not possible because there is a lack of information about energies of ionization (Au^{II}–Au^{V}).

The drastic drop from AuF_5 to HgF_2 has caused many (unpublished) attempts to prepare, for example, HgF_4, $CsHgF_4$, $Cs_2K(HgF_6)$, or $Cs_2(HgF_6)$, obviously without any known success. Once again, ignorance of the energies of ionization prohibits further discussion.

Finally, it should be remembered that the high-oxidation-state Cu^{4+}, Ag^{3+}, Ag^{4+} (Ag^{5+}), and Au^{5+} are still unknown in the case of oxides. This is puzzling, at least for Ag^{3+}. The reasons are unknown, but it is certainly related to the high electronegativity of fluorine.

IV. High Oxidation States of 3d Elements

Typical examples are compiled in Table III (pp. 284–286).

A. OXIDATION STATE +3: Mn(III), Co(III), AND Ni(III)

Here, Mn(III) is included (although complex chlorides are known). The very plausible crystal structure of dark red MnF_3 ($T_N = 43$ K) should be confirmed by single-crystal work. $BaMnF_5$ is one of the very few examples with proved structure. All known fluorides are of high spin state.

Although $CoCl_3$ is known in the gaseous state (Schäfer and Krehl, 1952), olive CoF_3 ($T_N = 460$ K) and derivatives are included. Single crystals of CoF_3 are needed for further determination. All compounds are of high spin state.

The few derivatives of the as yet unknown NiF_3 have interesting magnetic properties. For instance, $K_2Na[NiF_6]$ shows low spin state behavior at low temperature and high spin state behavior at room temperature.

B. OXIDATION STATE +4: Cr(IV), Mn(IV), Co(IV), AND Ni(IV)

Although unstable $CrCl_4$ was described (von Wartenberg, 1942), red CrF_4 is included. No single-crystal-proved structure of CrF_4 or of sky blue MnF_4 is known, although crystals of MnF_4 have been prepared (Müller, 1981b). Note the following:

1. The long unknown existence of fluorides such as yellow $Cs_3F[MnF_6]$, which are similar in behavior to SiF_4 in forming, for example, $K_3F[SiF_6]$

TABLE III

Characteristic Fluorides of High Oxidation State: $3d$ Elements[a,b]

Oxidation State	Vanadium	Chromium	Manganese	Cobalt	Nickel
+3	—	—	MnF_3 (1) Purple	CoF_3 (2) Blue (VF_3)	Na_3NiF_6 (5) Violet (Na_3AlF_6)
			$CsMnF_4$ (3) Brown	$CsCoF_4$ (4) Violet $(\beta\text{-RbAlF}_4)$	Cs_2KNiF_6 (8) Violet (K_2NaAlF_6)
			$(NH_4)_2MnF_5$ (6) Purple	$CdCoF_5$ (7) Blue	$CsZnNiF_6$ (11) Dark brown $(RbNiCrF_6)$
			$K_2MnF_5 \cdot H_2O$ (9) Purple	$(\mu/\mu_B = 3.8)\,(CaCrF_5)$	
			$PbMnF_5$ (3) Red-violet	$LiMgCoF_6$ (10) Light blue (trirutile)	
			Rb_2NaMnF_6 (12) Violet $(\mu/\mu_B = 4.98)$	Cs_3CoF_6 (13) Light blue $(\mu/\mu_B = 5.38)\,(K_3FeF_6)$	
				Cs_2KCoF_6 (14) Light blue $(\mu/\mu_B = 5.41)\,(K_2NaAlF_6)$	
				Ba_2ZnCoF_9 (15) Blue (Ba_2MnFeF_9)	
				$Sr_3(CoF_6)_2$ (16) Greenish blue $(Ba_3(FeF_6)_2)$	
				$Na_3Li_3Co_2F_{12}$ (17) Blue $(Ca_3Al_2(SiO_4)_3)$	

	V	Cr	Mn	Co	Ni
+4	—	CrF$_4$ (18) Dark greenish black ($\mu/\mu_B = 3.02$) CsCrF$_5$ (18) Brick red ($\mu/\mu_B = 3.09$) Li$_2$CrF$_6$ (25) Yellow BaCrF$_5$ (25) Yellow (BaGeF$_6$) Cs$_3$F[CrF$_6$] (26) Pink	MnF$_4$ (19) Bluish gray K$_2$MnF$_6$ (22) Yellow Cs$_3$F[MnF$_5$] (26) Yellow (K$_3$F[SiF$_6$]) BaMnF$_6$ (28) Yellow (BaGeF$_6$)	Sr$_5$Co$_3$F$_{19}$ (16) Blue (Sr$_5$Ga$_3$F$_{19}$) K$_2$CoF$_6$ (20) Yellow-orange (K$_2$PtCl$_6$) Cs$_2$CoF$_6$ (13) Yellow ($\mu/\mu_B = 2.97$) (K$_2$PtCl$_6$)	BaNiF$_6$ (21) Carmine Na$_2$NiF$_6$ (23, 24) Purple-red K$_2$NiF$_6$ (27) Carmine (K$_2$PtCl$_6$)
+5	VF$_5$ (29) Colorless (Edwards) VOF$_3$ (31) Yellowish NaVF$_6$ (32) (LiSbF$_6$) KVOF$_4$ (34) Light yellow K$_2$VO$_2$F$_3$ (35)	CrF$_5$ (30) Dark red CrOF$_3$ (12) Purple CsCrF$_6$ (33)	—	—	—

(cont.)

TABLE III (cont.)

Oxidation State	Vanadium	Chromium	Manganese	Cobalt	Nickel
+5 (cont.)	$Cs_2KVO_2F_4$ (35) (K_2NaAlF_6) $Na_3VO_2F_4$ (35) (Na_3AlF_6) $(C_5H_5)_4P[VO_2F_2]$ (36) Colorless				—
+6	—	CrF_6 (37) Lemon yellow CrO_2F_2 (38) Red-violet $CrOF_4$ (39) Red	—	—	
+7	—	—	MnO_3F (40) Dark green	—	

^{a} Underscored are those formulas in which the crystal structure is known by single-crystal work, otherwise, the type of structure, if known, is given in parentheses.

^{b} Numbers in parentheses are references, as follows: 1, Hepworth and Jack (1957); 2, Hepworth et al. (1957); 3, Köhler et al. (1978); 4, Fleischer and Hoppe (1982b); 5, Henkel and Hoppe (1969); 6, Sears and Hoard (1969); 7, Fleischer and Hoppe (1982c); 8, Alter and Hoppe (1974a); 9, Edwards (1971); 10, Fleischer and Hoppe (1982a); 11, Jesse and Hoppe (1974); 12, Siebert and Hoppe (1972); 13, Hoppe (1956); 14, Alter and Hoppe (1974b); 15, Fleischer and Hoppe (1982e); 16, Ravez et al. (1971); 17, de Pape et al. (1967); 18, Clark and Sadana (1964); 19, Hoppe et al. (1961); 20, Grannec et al. (1976); 21, Müller and Hoppe (1983); 22, Bukovec and Hoppe (1983); 23, Bougon (1968); 24, Henkel et al. (1969); 25, Siebert and Hoppe (1971); 26, Hofmann and Hoppe (1979); 27, Klemm and Huss (1949); 28, Hoppe and Blinne (1957); 29, Edwards and Jones (1969); 30, von Wartenberg (1941); 31, Edwards and Taylor (1970); 32, Kemmit et al. (1963); 33, Brown et al. (1976); 34, Rieskamp and Mattes (1973); 35, Pausewang and Dehnicke (1969); 36, Ahlborn et al. (1972), 37, Glemser et al. (1963); 38, Engelbrecht and Grosse (1952); 39, Edwards (1963); 40, Engelbrecht and Grosse (1954).

2. The failure of all attempts to prepare fluorides with Fe(IV), for example $Cs_2[FeF_6]$ or $O_2^+[FeF_9]$, well known with Mn(IV)
3. The relative instability (or sensitivity) of compounds $A_2[CoF_6]$ (A = K, Rb, or Cs) and their magnetic similarity to $K_2Na[NiF_6]$
4. The impressive stability of, for example, carmine red $Cs_2[NiF_6]$ as well as the existence of equally colored $Ba[NiF_6]$ and $Sr(NiF_6)$. [Surprisingly, $Ba[NiF_6]$ was obtained even as single crystals (Müller and Hoppe, 1983) and the proposed structure has been confirmed]

C. OXIDATION STATE +5: Cr(V) AND Mn(V)

Surprisingly, carmine red CrF_5 is isostructural with VF_5 (Edwards and Jones, 1969). Its red derivatives, such as $Cs[CrF_6]$, require further characterization. Note the following: (1) The absence of corresponding fluorides $A[MnF_6]$ (e.g., A = Cs) [the (literally "hidden") search for such compounds has so far been unsuccessful, even for $A = O_2^+$ or XeF^+]; (2) the absence of $CrOF_3$ and CrO_2F as well as $MnOF_3$ corresponding to VOF_3 (Edwards and Taylor, 1970) and VO_2F, which should be investigated more intensely.

D. OXIDATION STATE +6: Cr(VI)

The report on CrF_6 (Glemser et al., 1963) requires confirmation. Single crystals of CrO_2F_2 and $CrOF_4$ are needed. Surprisingly, nothing is known about the corresponding fluorides of Mn(VI). No report on fluorides such as $A[CrOF_5]$ has been published, although corresponding chlorides such as $(C_5NH_6)^+[CrOCl_5]^-$ (Weinland and Fiederer, 1907) are easily obtained!

E. OXIDATION STATE +7: Mn(VII)

MnO_3F (a green solid and liquid) is the only known example. The molecule is said to present nearly tetrahedral arrangement of the ligands. Further work is needed. It is difficult to predict anything about derivatives of $[MnF_6]^+$ at the moment.

F. STATE OF THE ART AND CONCLUSIONS

The chemistry of all these fluorides contains a great deal of mystery. The absence of $A[Mn^VF_6]$, for instance, seems unexplainable at the moment. A detailed discussion of the reasons for the observed instability of fluorides such as $A_2[FeF_6]$ is not possible; first, one should have more background information (i.e., the energy of ionization for $Fe^{3+} \rightarrow Fe^{4+}$). The search for $Cs_3F[CoF_6]$ is still under way. In an early stage of fluorine chemistry its existence as Cs_3CoF_7 (as well as K_3CoF_7) (Klemm and Huss, 1949) was erroneously reported.

Another problem is the absolute absence of complex oxide–fluorides such as (ruby red?) $Cs_2K[Mn^{IV}OF_5]$ or $Cs_2K[Fe^{IV}OF_5]$, as well as of mixed chlorides such as $CrClF_3$ and $MnClF_2$ or complexes such as (garnet red?) $Cs_2K[MnF_5Cl]$. It is apparently still an open field.

With the puzzling exception of colorless $Cs_3[Fe_2F_9]$ (Babel *et al.*, 1971) in which two coordination octahedra share a *common face*, nothing is known concerning "oligofluorides" such as $K_3[Mn_2^{IV}F_{11}]$ or $Cs_2[Mn_3^{IV}F_{14}]$. No one knows why. The composition of $NiF_{2.5}$ (Henkel and Klemm, 1935) has not been confirmed. It was shown to correspond to $NiF_{2.2}$ (Henkel and Hoppe, 1968). Since Cr_2F_5 and Fe_2F_5 are well known, it is hard to understand why Co_2F_5 and $Ni_2F_5 \hateq NiF_{2.5}$ are still unknown.

V. High Oxidation States of 4*d* and 5*d* Elements

Tables IV (pp. 290–291) and V (pp. 292–293) contain data for these compounds, with the exception of PdF_3 (known as a high-pressure form) and ternary fluorides such as $NaPdF_4$ (Hagenmuller *et al.*, 1982); only the aforementioned oxidation states are included.

In Chapters 3 and 4, how known compounds are combined with unknown compounds, and the consequences of such combinations, are discussed in greater detail. Thus, it may be sufficient here to mention only briefly those cases in which unusual differences occur:

1. Little is known about all tetrafluorides; this results from the still unsatisfactory state of the synthetic methods used in the preparation of fluorides of "medium high" oxidation states.
2. All pentafluorides are tetrameric, in striking contrast to those of the 3*d* metals (e.g., VF_5 or Nb_2F_{10}). The explanation of This phenomenon is difficult because we know that in low-temperature $Ba_3Al_2F_{12}$ (Domesle and Hoppe, 1980) that is, $Ba_6F_4[(AlF_{4/1}F_{2/2})_4]$ the same tetrameric unit is present, but no low-energy *d* orbitals are available.
3. All complex fluorides with M^{4+} or M^{5+} so far known are of common composition. Dinuclear entities such as $[M_2F_{11}]^+$ (unknown in the case of 3*d* metals) are well known.
4. Mixed-valency compounds are lacking (with the exception of "PdF_3", which is actually $Pd^{II}Pd^{IV}F_6$, and derivatives like $Pd^{II}Pt^{IV}F_6$).
5. All known hexafluorides adopt the UF_6 structural type as the low-

temperature variety and the insufficiently known OsF_6 type as the high-temperature form but with a few exceptions).

6. The reasons for the absence of PdF_6 are unknown and difficult to discuss because of the lack of experimental data (e.g., corresponding energies of ionization).

7. There is a striking difference between the behavior of WF_6, for example (forming compounds such as KWF_7 and K_2WF_8), and the hexafluorides such as OsF_6, IrF_6, and PtF_6.

8. Oxide–fluorides such as $Ba_2WO_3F_4$ (i.e., $Ba_2F_2\,[WO_{2/2}O_{2/1}F_{2/1}]$) and the isotypic $Ba_2MoO_3F_4$ (Wingefeld and Hoppe, 1984) as well as K_2ReO_4F, (Chakravorti and Chandhuri, 1973) are easily formed. There is a lack of information concerning corresponding compounds of the VIIIB elements.

The entire field must be examined further for more detailed information. Single-crystal work on the tetrafluorides is needed urgently. The penta-fluorides should be investigated for high-pressure forms to improve the basis for discussing the puzzling tetramerization. As in the case of the $3d$ metals, oxide–fluorides have to be synthesized. The elucidation of the important crystal structure of high-temperature OsF_6 is a pressing problem. It seems to be a key to deeper understanding of the crystal chemistry of fluorides called *umhüllte Verbindungen* ("enveloped" compounds) by Biltz (1934). More information is needed on OsF_7. The uncertainty connected with the possible existence of fluorides such as OsF_8 [reported erraneously in the earlier stages of fluorine chemistry (Ruff and Tschirch, 1913)] IrF_7, and TcF_7 is disheartening. We know nothing about their stability and we are not really certain how to prepare them. Curious experimental observations excluded from the tables may be illustrated by the observation that $NaRhF_6$ is much more easily obtained than $CsRhF_6$ (Wilhelm and Hoppe, 1976). In general, such complexes are obtained more easily with the higher alkaline metals.

VI. High Oxidation States of $4f$ and $5f$ Elements

Data characterizing the present state of knowledge are collected in Table VI (p. 294) and VII (pp. 295–296), respectively.

In the case of $4f$ elements, the tetrafluoride PrF_4 should be reexamined. Single-crystal work is also needed on TbF_4. The still unsuccessful search for Pr^{5+} started by Prandtl and Huttner (1925) in the field of oxides continues. Mössbauer measurements on fluorides such as Cs_3PrF_7 not only confirmed the presence of Pr^{4+} but indicated Pr^{5+} impurities (K. Feldner and R. Hoppe, unpublished; see Feldner, 1979). The author so far has failed to

TABLE IV

Characteristic Fluorides of High Oxidation State: 4d Elements[a,b]

Oxidation State	Technetium	Ruthenium	Rhodium	Palladium
+3	—	—	—	$Pd^{II}Pd^{IV}F_6$ (1) Black ($\mu/\mu_B = 2.88$) $NaPdF_4$ (2) Gray ($KBrF_4$) K_2NaPdF_6 (3, 4) Beige (dist. K_2NaPdF_6)
+4	—	RuF_4 (5) Yellow $BaRuF_6$ (9) Pale yellow ($\mu/\mu_B = 3.08$) ($BaGeF_6$) Cs_2RuF_6 (12) Pale yellow ($\mu/\mu_B = 2.98$) (K_2GeF_6)	RhF_4 (6) Sandy brown (paramagnetic) (PtF_4) $BaRhF_6$ (10) Lemon yellow ($\mu/\mu_B = 2.06$) ($BaGeF_6$) K_2RhF_6 (13) Yellow ($\mu/\mu_B = 1.97$) (K_2GeF_6)	PdF_4 (7, 8) Brick red (diamagnetic) (PtF_4) $MgPdF_6$ (11) Pale yellow ($LiSbF_6$) Cs_2PdF_6 (14) Pale yellow (K_2PtCl_6)
+5	—	$\underline{RuF_5}$ (15, 16) Emerald green	$\underline{RhF_5}$ (17, 18) Dark red (RuF_5)	

+6	KRuF$_6$ (9) Light blue ($\mu/\mu_B = 3.48$) (BaGeF$_6$) XeF$^+$[Ru$_2$F$_{11}$]$^-$ (21) Bright green (XeF$^+$[Sb$_2$F$_{11}$]$^-$) RuF$_6$ (I) (22) Dark brown (OsF$_6$) RuF$_6$ (II) (22, 24) Dark brown (UF$_6$)	NaRhF$_6$ (19) Bright orange ($\mu/\mu_B = 2.82$) (LiSbF$_6$) RhF$_6$ (I) (22, 23) Black (OsF$_6$) RhF$_6$ (II) (22, 23) Black (UF$_6$)	"NaPdF$_6$" (20) — —
+7	TcO$_3$F (25) Yellow	—	—

ᵃ Underscored are those formulas in which the crystal structure is known by single-crystal work; otherwise, the type of structure, if known, is given in parentheses.

ᵇ Numbers in parentheses are references, as follows: 1, Tressaud et al. (1976); 1a, Langlais et al. (1979); 2, Hagenmuller et al. (1982); 3, Khaïroun et al. (1983); 4, Tressaud et al. (1984); 5, Holloway and Peacock (1963); 6, Rao et al. (1976); 7, Hagenmuller et al. (1976); 8, Wright et al. (1978); 9, Weise and Klemm (1955); 10, Wilhelm and Hoppe (1974); 11, Henkel and Hoppe (1968); 12, Peacock (1956); 13, Weise and Klemm (1953); 14, Hoppe and Klemm (1952); 15, Holloway et al. (1964); 16, Ruff and Vidic (1925); 17, Morrell et al. (1973); 18, Holloway et al. (1965); 19, Wilhelm and Hoppe (1976); 20, Sokolov et al. (1976); 21, Sladky et al. (1969); 22, Siegel and Northrop (1966); 23, Chernik et al. (1961); 24, Claassen, et al. (1961); 25, Selig and Malm (1963).

TABLE V

Characteristic Fluorides of High Oxidation State: $5d$ Elements[a,b]

Oxidation State	Rhenium	Osmium	Iridium	Platinum
+4	—	OsF_4 (1) Yellow	IrF_4 (2, 3) Chestnut brown	PtF_4 (4) Yellow-brown (diamagnetic)
		K_2OsF_6 (5) (K_2GeF_6 (II))	Na_2IrF_6 (5) White (Na_2SiF_6)	K_2PtF_6 (6) Yellow (K_2GeF_6 (II))
			$BaIrF_6$ (7) ($KOsF_6$)	$BaPtF_6$ (8) Yellow ($BaSiF_6$)
+5	—	OsF_5 (1,3) Blue-gray ($\mu/\mu_B = 2.06$) (RuF_5)	IrF_5 (9) Yellow ($\mu/\mu_B = 1.32$) (RuF_5)	PtF_5 (1) Dark red (RuF_5)
		$KOsF_6$ (9,10) White ($\mu/\mu_B = 3.2$)	$CsIrF_6$ (11) ($KOsF_6$)	$KPtF_6$ (4,12) Mustard yellow ($KOsF_6$)
			$XeF^+[Ir_2F_{11}]^-$ (13) Orange-yellow ($XeF^+[Sb_2F_{11}]^-$)	$XeF^+[Pt_2F_{11}]^-$ (13) Dark red ($XeF^+[Sb_2F_{11}]^-$)

Oxidation state	Re	Os	Ir	Pt
+6	—	OsF$_6$ (I) (14–16) Yellow High-temperature form II $\longrightarrow$ I (1.4°C) OsF$_6$ (II) (14–16) Yellow Low-temperature form (T_C = 66 K) (UF$_6$)	IrF$_6$ (I) (15, 17) Yellow High-temperature form (OsF$_6$) IrF$_6$ (II) (15, 17) Yellow (μ/μ_B = 2.9) (UF$_6$) Low-temperature form II $\longrightarrow$ I (−1.2°C)	PtF$_6$ (I) (15) Purple (OsF$_6$) PtF$_6$ (II) (15) Dark red (UF$_6$)
+7	ReF$_7$ (18) Yellow ReOF$_5$ (20) Yellow ReO$_2$F$_3$ (20) Yellow ReO$_3$F (20) Yellow K$_2$ReO$_3$F$_3$ (22) K$_2$ReO$_4$F (23)	OsF$_7$ (19) Pale yellow OsOF$_5$ (21)	—	—

^{a} Underscored are those formulas in which the crystal structure is known by single-crystal work; otherwise, the type of structure, if known, is given in parentheses.

^{b} Numbers in parentheses are references, as follows: 1, Hargreaves and Peacock (1960); 2, Bartlett and Tressaud (1974); 3, Paine and Asprey (1975); 4, Bartlett and Lohmann (1964); 5, Hepworth et al. (1958); 6, Mellor and Stephenson (1951); 7, Hepworth et al. (1956); 8, Cox (1956); 9, Bartlett and Rao (1965); 10, Peacock (1956); 11, Klemmitt and Russel (1963); 12, Bartlett et al. (1962); 13, Sladky et al. (1969); 14, Siegel and Northrop (1966); 15, Weinstock (1964); 16, Hargreaves and Peacock (1959); 17, Figgis et al. (1961); 18, Malm and Selig (1961); 19, Glemser et al. (1966); 20, Sunder and Stevie (1975); 21, Bartlett et al. (1962); 22, Kuhlmann and Sawodny (1977a); 23, Chakravorti and Chandhuri (1973).

TABLE VI

Characteristic Fluorides of High Oxidation State: $4f$ Elements[a]

Oxidation State	Cerium	Praseodymium	Neodymium	Terbium	Dysprosium
+4	CeF_4 (1) Colorless (ZrF_4)	PrF_4 (2) Colorless ($\mu/\mu_B = 2.42$) (ZrF_4)	Cs_3NdF_7 (7) Orange	TbF_4 (3) Colorless (ZrF_4)	Cs_2RbDyF_7 (7) Orange ($\mu/\mu_B = 9.87$)
	K_2CeF_6 (4, 5) Colorless	Li_2PrF_6 (6) Colorless (Li_2ZrF_6)	Cs_2RbNdF_7 (7) Orange ($\mu/\mu_B = 3.14$)	Li_4TbF_8 (8) Colorless (Li_4UF_8)	Cs_3DyF_7 (7) Orange
	Rb_3CeF_7 (4) Colorless (diamagnetic)	Rb_2PrF_6 (8) Colorless (Rb_2UF_6)		$BaTbF_6$ (6) Colorless ($\mu/\mu_B = 8.22$)	
	$CaCeLi_2F_8$ (9) Colorless (scheelite type)	$BaPrF_6$ (6) Colorless ($RbPaF_6$)		Cs_2RbTbF_7 (6) Colorless ($(NH_4)_3ZrF_7$)	
	KCe_2F_9 (10) (KU_2F_9)	K_3PrF_7 (6) Colorless ($(NH_4)_3ZrF_7$)		Rb_2CsTbF_7 (6) Colorless ($(NH_4)_3ZrF_7$)	
		$CsRbKPrF_7$ (6) Colorless ($(NH_4)_3ZrF_7$)		$CsRbKTbF_7$ (6) Colorless ($(NH_4)_3ZrF_7$)	
		$CdPrLi_2F_8$ (6) Colorless (scheelite type)			

[a] Numbers in parentheses are references, as follows: 1, Klemm and Henkel (1934); 2, Asprey et al. (1967); 3, Cunningham et al. (1954); 4, Hoppe and Rödder (1961); 5, Hoppe (1959); 6, Feldner and Hoppe (1983); 7, Hoppe (1982); 8, Avignant and Cousseins

TABLE VII

Characteristic Fluorides of High Oxidation State: 5f Elements[a]

Oxidation State	Neptunium	Plutonium	Americium	Berkelium	Californium	Einsteinium
+4	—	PuF$_4$ (1) Pink-light brown (ZrF$_4$)	AmF$_4$ (2) Yellowish brown (ZrF$_4$)	BkF$_4$ (2) (ZrF$_4$)	CfF$_4$ (2) (ZrF$_4$)	—
		LiPuF$_5$ (3) Brown	LiAmF$_5$ (3) (LiUF$_5$)			
		Na$_2$PuF$_6$ (4) Pink	Rb$_2$AmF$_6$ (5) Pink (Rb$_2$UF$_6$)			
		Rb$_2$PuF$_6$ (5)				
		CaPuF$_6$ (6, 7) Red-brown (LaF$_3$)				
		Na$_3$PuF$_7$ (8)				
		Li$_4$PuF$_8$ (8–10)	(NH$_4$)$_4$AmF$_8$ (11) Red			
		(NH$_4$)Pu$_3$F$_{13}$ (12) Pink				
		Rb$_7$Pu$_6$F$_{31}$ (4)	Na$_7$Am$_6$F$_{31}$ (13, 14)			
+5	NpF$_5$ (15, 16) (α-UF$_5$)		—	—	—	—

(cont.)

$$\text{TABLE VII } (cont.)$$

Oxidation State	Neptunium	Plutonium	Americium	Berkelium	Californium	Einsteinium
+5 (cont.)	$CsNpF_6$ (13, 17) Pink-violet ($KOsF_6$) Rb_2NpF_7 (13) Pink-violet (K_2NbF_7) Na_3NpF_8 (19) Lilac (Na_3PaF_8)	$CsPuF_6$ (13) Green ($KOsF_6$) Rb_2PuF_7 (18) Green (K_2NbF_7) Na_3PuF_8 (20)				
+6	NpF_6 (21, 22) Bright orange (UF_6)	PuF_6 (21, 23) Reddish brown (UF_6)	—	—	—	—

[a] Numbers in parentheses are references, as follows: 1, Florin and Heath (1944); 2, Keenan and Asprey (1969); 3, Keenan (1966a); 4, Keller and Schmutz (1966); 5, Keenan (1967); 6, Salzer (1966); 7, Keller and Salzer (1967); 8, Riha and Trevorrow (1966); 9, Steindler (1968); 10, Steindler (1969); 11, Asprey and Penneman (1962); 12, Benz et al. (1963); 13, Penneman et al. (1967); 14, Keenan (1966b); 15, Baluka et al. (1980); 16, Drobyskevskii et al. (1975); 17, Asprey et al. (1966); 18, Penneman et al. (1965); 19, Brown et al. (1969); 20, Matcheret (1970); 21, Florin et al. (1956); 22, Zachariasen (1949); 23, Weinstock and Malm (1956).

prepare pure $Cs[PrF_6]$ or $Cs_2[PrF_7]$. In all cases, of known complex fluorides with Re^{4+} (Re = rare-earth element) single-crystal work is needed. No oxide–fluorides such as $PrOF_2$ have been reliably reported.

In the case of $5f$ elements, investigations are evolving so rapidly that any comment on the results given in Table VII may be out-of-date as soon as this chapter is printed.

VII. High Oxidation States of Main-Group Elements

By definition this chapter covers mainly nonmetals. Results are given in Table VIIIa–c (pp. 298–301). Because there is an excellent review article (Seppelt, 1979), we can focus our interest mainly on oddities:

1. In the case of *VA elements*:

The gap between Si_2F_6 and S_2F_{10} should be noted. Thus, P_2F_8 should be, and As_2F_8 could be, obtainable. In addition, we await $F_4P—O—PF_4$, $F_4As—O—AsF_4$, and related peroxo derivatives.

There is a lack of structural information for fluorides such as $K_2[AsF_7]$, which is proof of the somewhat mysterious relations between A and B elements of the same group, as well as for fluorides such as $Cs[AsF_6]$ and the still unknown analogues of the oligofluoroantimonates such as $Xe_2F_3[Sb_2F_{11}]$ with As(V) or Bi(V) [and P(V), Si(IV), etc.].

2. In the case of *VIA elements*:

The cation OF_3^+, stabilized by complexation (e.g., $OF_3^+[SbF_6]^-$), is as yet unknown, as is O_3^+ in fluorides such as $O_3^+[Sb_2F_{11}]^-$.

Te_2F_{10}, once reported and then identified to be $F_5TeOTeF_5$, is still to be expected.

3. For *halogen compounds*, it should be noted that

The reasons for the remarkable difference between fluorides (e.g., BrF, BrF_3, and BrF_5) known as binary compounds and corresponding unknown oxides (e.g., Br_2O_3 and Br_2O_5) are still open to discussion (especially with respect to the existence of unstable entities such as BrO_2 or ClO_3).

Fluorides such as $Cl_2^+[PtF_6]^-$ suggest the existence of metastable compounds such as $F_2^+[Sb_2F_{11}]^-$.

A precise prediction of vaporization (on sublimation) temperatures for binary fluorides such as "ClF_7" is difficult (Table IX, p. 302).

4. Concerning the fluorides of the *noble gases*, there is a striking contrast between the existence of $ClF_6^+[SbF_6]^-$ and the still unknown fluorides ArF_2

TABLE VIIIa

Characteristic Fluorides of High Oxidation State: Va and VIa Elements[a,b]

Oxidation State	Arsenic	Bismuth	Sulfur	Selenium	Tellurium
+5	AsF_5 (1) Colorless	BiF_5 (2) Colorless	S_2F_{10} (3) Colorless	Se_2F_{10} (4) Colorless	—
	$AsOF_3$ (5) Colorless	$BiOF_3$ (6) Colorless			
	S_8AsF_6 (7) Dark blue	$CsBiF_6$ (8) Colorless ($KOsF_6$)			
	Hg_3AsF_6 (9) Silvery golden				
	$KAsF_6$ (10) Colorless				
	$(C_5H_6N)_2As_2O_2F_8$ (11) Colorless				
	$(NEt_4)As_2F_{11}$ (12) Colorless				

298

+6			
—	SF_6 (3) Colorless	SeF_6 (4) Colorless	TeF_6 (4) Colorless
—	F_5SOSF_5 (13) Colorless	$F_5SeOSeF_5$ (14) Colorless	$F_5TeOTeF_5$ (15) Colorless
	$F_5SO_2SF_5$ (16) Colorless	$F_5SeO_2SeF_5$ (17) Colorless	$F_5TeO_2TeF_5$ (4) Colorless
	SOF_4 (17) Colorless	$F_4SeO_2SeF_4$ (17) Colorless	$F_4TeO_2TeF_4$ (17) Colorless
	$SF_4(OSF_5)_2$ (4) Colorless	$SeF_4(OSeF_5)_2$ (4) Colorless	$TeF_4(OTeF_5)_2$ (18) Colorless
	$CsOSF_5$ (13) Colorless		$Te(OTeF_5)_6$ (18) Colorless

[a] Underscored are those formulas in which the crystal structure is known by single-crystal work; otherwise, the type of structure, if known, is given in parentheses.

[b] Numbers in parentheses are references, as follows: 1, Rulf et al. (1932); 2, Hebecker (1971); 3, Seel (1974); 4, Seppelt (1979); 5, Dehnike and Weidlein (1966); 6, Gutmann and Emeléus (1950); 7, Davies et al. (1971); 8, Hebecker (1970); 9, Schultz et al. (1978); 10, Gutner and Krüger (1974); 11, Haase (1973); 12, Dean et al. (1971); 13, Seppelt (1973b); 14, Seppelt (1973a); 15, Engelbrecht et al. (1968); 16, Merril and Cady (1961); 17, Seppelt (1974); 18, Lentz et al. (1978).

299

TABLE VIIIb

Characteristic Fluorides of High Oxidation State: VIIa Elements[a,b]

Oxidation State	Chlorine	Bromine	Iodine
+1	—	—	—
+3	ClF_3 (1)	BrF_3 (2)	—
+5	ClF_5 (3)	BrF_5 (2)	IF_5 (4)
	$ClOF_3$ (5)	$BrOF_3$ (6)	IOF_3 (7)
	ClO_2F (8)	BrO_2F (8)	IO_2F (9)
		$KBrO_2F_2$ (10)	
+7	ClO_3F (9)		IF_7 (11)
	ClO_2F_3 (12)		IOF_5 (13)
	$(ClF_6)(PtF_6)$ (14) Yellow	BrO_3F (12)	IO_2F_3 (15) Yellow
		$(BrF_6)(SbF_6)$ (16)	IO_3F (9)
			$CsIF_8$ (17)

[a] Underscored are those formulas in which the crystal structure is known by single-crystal work. Unless otherwise indicated, crystals are colorless.

[b] Numbers in parentheses are references, as follows: 1, Burbank and Bensey (1953); 2, Burbank and Bensey (1957); 3, Smith (1963b); 4, Kuhlmann and Sawodny (1977b); 5, Christie et al. (1972); 6, Christie et al. (1978); 7, Donohue (1965); 8, Schmeisser and Pammer (1955); 9, Naumann (1980); 10, Gillespie and Schröbilgen (1974); 11, Gillespie and Quail (1963); 12, Christie and Wilson (1973); 13, Engelbrecht and Peterty (1969); 14, Christie (1973); 15, Seel and Pimpl (1977); 16, Schmeisser et al. (1970); 17, Gillespie and Spekkens (1976).

and ArF_4. Remember, there are only a small number of noble gas compounds at all. Moreover, there is a dearth of information about fluorides of radon. Finally, one of the highlights of inorganic chemistry is the preparation of $(NF_4)_2{}^+[XeF_8]^{2-}$.

VIII. Closing Remarks

The colored spectrum of known compounds and the enormous number of compounds still to be prepared could be only listed here. The author is depressed by his inability to understand what he knows and to realize what he is dreaming about. Young students may thus have a good chance to add new landmarks of chemical intuition to this field.

TABLE VIIIc

Characteristic Fluorides of High Oxidation State:
VIIIa Elements[a,b]

Oxidation State	Krypton	Xenon
+2	KrF_2 (1)	XeF_2 (2)
		$[XeF]^+[RuF_6]^-$ (3)
		Pale yellow-green
		$[XeF]^+[Sb_2F_{11}]^-$ (4)
		Yellow
		$FXeOSO_2F$ (5)
		$Xe(OSO_2F)_2$ (5)
		Pale yellow
		$Xe(OTeF_5)_2$ (6)
		$[Xe_2F_3]^+[AsF_6]^-$ (7)
		$XeF_2 \times XeF_4$ (8)
		$XeF_2 \times XeOF_4$ (9)
+4	—	XeF_4 (10)
		$XeOF_2$ (11)
+6	—	XeF_6 (12)
		$XeOF_4$ (13)
		XeO_2F_2 (14)
		$KXeO_3F$ (15)
		$[XeF_5]^+[RuF_6]^-$ (3)
		Pale yellow-green

[a] Underscored are those formulas in which the crystal structure is known by single-crystal work. Unless otherwise indicated, crystals are colorless.

[b] Numbers in parentheses are references, as follows: 1, Turner and Pimentel (1963); 2, Hoppe et al. (1962b); 3, Bartlett et al. (1973); 4, McRae et al. (1969); 5, Bartlett et al. (1969); 6, Sladky (1969); 7, Sladky et al. (1968); 8, Burns et al. (1965); 9, Bartlett and Wechsberg (1971); 10, Claassen et al. (1962); 11, Ogden and Turner (1966); 12, Malm et al. (1963); 13, Smith (1963a); 14, Huston (1967); 15, Hodgson and Ibers (1969).

TABLE IX

Boiling Points or Points of Sublimation of Some Binary Halogenides
EX_n, E_2X_n, and E_3X_n[a]

BF_3 58	CF_4 36	NF_3 48			
B_2F_4 60	C_2F_4 49	N_2F_4 50			
	C_2F_6 32				
	SiF_4 47	PF_5 38	SF_6 35[b]	ClF_5 52	
	Si_2F_6 42	PF_3 34	SF_4 58	ClF_3 95	
	Si_3F_8 39	P_2F_4 67	S_2F_{10} 30	ClF 172	
	GeF_4 59[b]	AsF_5 44	SeF_6 38[b]	BrF_5 63	
		AsF_3 110	SeF_4 95	BrF_3 133	
				BrF 293	
		SbF_5 83	TeF_6 39[b]	IF_7 40[b]	XeF_6 58
		SbF_3 197	TeF_4 117	IF_5 76	XeF_2 197[b]

[a] The values (K) correspond to bp/n or sp/n.
[b] Point of sublimation.

Acknowledgments

The author is indebted to his co-workers Dipl.-Chem. L. Grosse, Dipl.-Chem. D. Kissel, Dipl.-Chem. J. Scheffler, Dipl.-Chem. K. H. Wandner, and Dr. G. Wingefeld for helpful work in collecting data and constructing the tables. He thanks Priv.-Doz. Dr. B. Müller and Priv.-Doz. Dr. G. Meyer for critical remarks.

References

Ahlborn, E., Diemann, E., and Müller, A. (1972). *Z. Anorg. Allg. Chem.* **394,** 1.
Alter, E., and Hoppe, R. (1974a). *Z. Anorg. Allg. Chem.* **405,** 167.
Alter, E., and Hoppe, R. (1974b). *Z. Anorg. Allg. Chem.* **407,** 313.
Asprey, L. B., and Penneman, R. A. (1962). *Inorg. Chem.* **1,** 134.
Asprey, L. B., Keenan, T. K., Penneman, R. A., and Sturgeon, G. D. (1966). *Inorg. Nucl. Chem. Lett.* **2,** 19.
Asprey, L. B., Coleman, J. S., and Reisfeld, M. J. (1967) *Adv. Chem. Ser.* **71,** 122.
Avignant, D., and Cousseins, J. C. (1978). *Rev. Chim. Miner.* **15,** 360.
Babel, D. *et al.* (1971). *J. Less-Common Met.* **25,** 257.
Baluka, M., Yeh, S., Banks, R., and Edelstein, N. (1980) *Inorg. Nucl. Chem. Lett.* **16,** 75.
Bartlett, H. N., and Lohmann, D. H. (1964) *J. Chem. Soc.* p. 619.
Bartlett, N., Tha, N. K., and Testter, J. (1962). *Proc. Chem. Soc., London* p. 277.
Bartlett, N., and Leary, K. (1976). *Rev. Chim. Miner.* **13,** 82.
Bartlett, N., and Rao, P. R. (1965). *Chem. Commun.* p. 252.
Bartlett, N., and Sladky, F. O. (1972). "The Chemistry of Krypton, Xenon and Radon; Comprehensive Inorganic Chemistry." Pergamon, Oxford.
Bartlett, N., and Tressand, A. (1974). *C. R. Hebd. Seances Acad. Sci.* **278,** 1501.
Bartlett, N., and Wechsberg, M. (1971). *Z. Anorg. Allg. Chem.* **385,** 5.
Bartlett, N. *et al.* (1962). *J. Chem. Soc.* p. 5253.
Bartlett, N. *et al.* (1969). *Chem. Commun.* p. 703.
Bartlett, N., Gennis, M., Gibler, D. D., Morrell, B. K., and Zalkin, A. *Inorg. Chem.* (1973). **12,** 1717.
Benz, R., Douglass, R. M., Kruse, F. H., and Penneman, R. A. (1963). *Inorg. Chem.* **2,** 799.
Biltz, W. (1934). "Raumchemie der festen Stoffe." Leipzig.
Bode, H. (1950). *Naturwissenschaften* **37,** 477.
Bougon, R. (11968). *C. R. Hebd. Seances Acad. Sci.* **267,** 681.
Bougon, R., and Lance, M. (1983). *C. R. Hebd. Seances Acad, Sci., Ser. B* **297,** 117.
Brown, D., Easey, J. F., and Pickard, C. E. F. (1969). *J. Chem. Soc. A* p. 1161.
Brown, S. D., Loehr, T. M., and Gard, G. L. (1976). *J. Fluorine Chem.* **7,** 19.
Bukoveč, P., and Hoppe, R. (1983). *J. Fluorine Chem.* **23,** 579.
Burbank, R. D., and Bensey, F. N. (1953). *J. Chem. Phys.* **21,** 602.
Burbank, R. D., and Bensey, F. N. (1957). *J. Chem. Phys.* **27,** 982.
Burns, J. H., Ellison, R. D., and Levy, H. A. (1965). *Acta Crystallogr.* **18,** 11.
Chakravorti, M. C., and Chaudhuri, M. K. (1973). *Z. Anorg. Allg. Chem.* **398,** 221.
Charpin, P. *et al.* (1970) *Bull. Soc. Fr. Mineral. Cristallogr.* **93,** 7.
Chernik, C. L., Claassen, H. H., and Weinstock, B. (1961). *J. Am. Chem. Soc.* **83,** 3165.
Christie, K. O., (1973). *Inorg. Chem.* **12,** 1580.
Christie, K. O., and Wilson, R. D. (1973). *Inorg. Chem.* **12,** 1356.
Christie, K. O., Curtis, E. C., and Schack, C. J. (1972). *Inorg. Chem.* **11,** 2212.
Christie, K. O., Curtis, E. C., and Bougon, R. (1978). *Inorg. Chem.* **17,** 1533.

Claassen, H. H., Selig, H., Malm, J. G., Chernik, C. L., and Weinstock, B. (1961). *J. Am. Chem. Soc.* **83,** 2390.

Claassen, H. H., Selig, H., and Malm, J. G. (1962). *J. Am. Chem. Soc.* **84,** 3593.

Clark, H. C., and Sadana, Y. N. (1964). *Can. J. Chem.* **42,** 50.

Cox, B. (1956). *J. Chem. Soc.* p. 876.

Cunningham, B. B., Feay,D. C., and Rollier, M. A. (1954). *J. Am. Chem. Soc.* **76,** 3361.

Davies, C. G., Gillespie, R. J., Park, J. J., and Passmore, J. (1971). *Inorg. Chem.* **10,** 2781.

Dean, P. A. W., Gillespie, R. J., Hulme, R., and Humphreys, D. A. (1971). *J. Chem. Soc. A* p. 341.

Dehnike, K., and Weidlein, J. (1966). *Z. Anorg. Allg. Chem.* **342,** 225.

Delaigne, A., and Cousseins, J.-C. (1972). *Rev. Chim. Miner.* **9,** 789.

de Pape, R., Portier, J., and Gauthier, G. (1967). *C. R. Hebd. Seances Acad. Sci., Ser. C* **265,** 1244.

Domesle, R., and Hoppe, R. (1980). *Angew. Chem.* **92,** 499.

Donohue, J. (1965). *Acta Crystallogr.* **18,** 1071.

Drobyskevskii, Y. V., Serik, V. F., and Sokolov, V. B. (1975). *Dokl. Chem. (Engl. Transl.)* **225,** 675.

Edwards, A. J. (1963). *Proc. Chem. Soc.* p. 205.

Edwards, A. J. (1971). *J. Chem. Soc. A* p. 2653.

Edwards, A. J., and Jones, G. R. (1969). *J. Chem. Soc. A* p. 1651.

Edwards, A. J., and Taylor, P. (1970). *Chem. Commun.* p. 1474.

Engelbrecht, A., and Grosse, A. V. (1952). *J. Am. Chem. Soc.* **74,** 5262.

Engelbrecht, A., and Grosse, A. V. (1954). *J. Am. Chem. Soc.* **76,** 2042.

Engelbrecht, A., and Peterty, P. (1969). *Angew. Chem.* **81,** 753.

Engelbrecht, A., Loreck, W., and Nehoda, W. (1968). *Z. Anorg. Allg. Chem.* **360,** 88.

Feldner, K. (1979). Dissertation, D. 23. Giessen. (K. Feldner and R. Hoppe, unpublished)

Feldner, K., and Hoppe, R. (1983). *Rev. Chim. Miner.* **20,** 351.

Figgis, B. N., Lewis, L., and Mabbs, F. E. (1961). *J. Chem. Soc.* p. 3138.

Fleischer, T., and Hoppe, R. (1982a). *Z. Naturforsch., B: Anorg. Chem., Org. Chem.* **37B,** 988.

Fleischer, T., and Hoppe, R. (1982b). *Z. Naturforsch., B: Anorg. Chem., Org. Chem.* **37B,** 1132.

Fleischer, T., and Hoppe, R. (1982c). *Z. Anorg. Allg. Chem.* **490,** 111.

Fleischer, T., and Hoppe, R. (1982d). *Z. Anorg. Allg. Chem.* **492,** 76.

Fleischer, T., and Hoppe, R. (1982e). *Z. Anorg. Allg. Chem.* **493,** 59.

Florin, A. E., and Heath, R. G. (1944). Ck-1372.

Florin, A. E., Tannenbaum, J. A., and Lemons, J. F. (1956) *J. Inorg. Nucl. Chem.* **2,** 368.

Gillespie, R. J., and Quail, J. W. (1963). *Proc. Chem. Soc., London* p. 278.

Gillespie, R. J., and Schrobligen, G. J. (1974). *Inorg. Chem.* **13,** 1230.

Gillespie, R. J., and Spekkens, P. H. (1976). *J. Chem. Soc., Dalton Trans.* p. 2391.

Glemser, O., Roesky, H., and Hellberg, K.-H. (1963) *Angew. Chem.* **75,** 346.

Glemser, O., Roesky, H. W., Hellberg, K. H., and Werther, H. V. (1966). *Chem. Ber.* **99,** 2652.

Grannec, J., Sorbe, P., Portier, J., and Hagenmuller, P. (1975). *C. R. Hebd. Seances Acad. Sci.* **280,** 45.

Grannec, J., Sorbe, P., and Portier, J. (1976), *C. R. Hebd. Seances Acad. Sci.* **283,** 441.

Gutmann, V., and Eméleus, H. J. (1950). *J. Chem. Soc.* p. 1046.

Gutner, G., and Krüger, G. J. (1974). *Acta Crystallogr., Sect. B* **B30,** 250.

Haase, W. (1973). *Chem. Ber.* **106,** 734.

Hagenmuller, P. *et al.* (1976). *C. R. Hebd. Seances Acad. Sci., Ser. C* **282,** 663.

Hagenmuller, P. *et al.* (1978). *J. Fluorine Chem.* **11,** 243.

Hagenmuller, P. *et al.* (1982). *C. R. Hebd. Seances Acad. Sci.* **295,** 183.

Hargreaves, G. B., and Peacock, R. D. (1959). *Proc. Chem. Soc. London* p. 85.

Hargreaves, G. B., and Peacock, R. D. (1960). *J. Chem. Soc.* p. 2618.

Harnischmacher, W., and Hoppe, R. (1973), *Angew. Chem.* **85,** 590. [B. Wingefeld and R. Hoppe, unpublished; see Wingefeld, G. (1984). Dissertation. Giessen]

Hebecker, C. (1970). *Z. Anorg. Allg. Chem.* **376,** 236.

Hebecker, C. (1971). *Z. Anorg. Allg. Chem.* **384**, 111.

Henkel, H. (1968). Dissertation, Giessen.

Henkel, H., and Hoppe, R. (1968). *Z. Anorg. Allg. Chem.* **359**, 160.

Henkel H., and Hoppe, R. (1969). *Z. Anorg. Allg. Chem.* **364**, 253.

Henkel, P., and Klemm, W. (1935). *Z. Anorg. Allg. Chem.* **222**, 73.

Henkel, H., Hoppe, R., and Allen, G. C. (1969). *J. Inorg. Nucl. Chem.* **31**, 3855.

Hepworth, H. E., Jack, K. H., and Westland, G. J. (1956). *J. Inorg. Nucl. Chem.* **2**, 79–87.

Hepworth, M. A., and Jack, K. H. (1957). *Acta Crystallogr.* **10**, 345.

Hepworth, M. A., Jack, K. H., Peacock, R. D., and Westland, G. J. (1957). *Acta Crystallogr.* **10**, 63.

Hepworth, M. A., Robinson, P. L., and Westland, G. J. (1958). *J. Chem. Soc.* p. 611.

Hodgson, D. J., and Ibers, I. A. (1969). *Inorg. Chem.* **8**, 326.

Hofmann, B., and Hoppe, R. (1979). *Z. Anorg. Allg. Chem.* **458**, 151.

Holloway, J. H., and Peacock, R. D. (1963). *J. Chem. Soc.* p. 3892.

Holloway, J. H., Peacock, R. D., and Small, R. W. H. (1964). *J. Chem. Soc.* p. 644.

Holloway, J. H., Rao, P. R., and Bartlett, N. (1965). *Chem. Commun.* p. 393.

Hoppe, R. (1956). *Recl. Trav. Chim. Pays-Bas* **75**, 569.

Hoppe, R. (1959). *Angew. Chem.* **71**, 457.

Hoppe, R. (1982). *In* "The Rare Earths in Modern Science and Technology," (G. J. MacCarthy, H. B. Silber, and J. J. Rhynne, eds.), Vol. 3, p. 315. New York.

Hoppe, R., and Blinne, K. (1957). *Z. Anorg. Allg. Chem.* **291**, 269.

Hoppe, R., and Homann, R. (1966). *Naturwissenschaften* **53**, 501.

Hoppe, R., and Klemm, W. (1952). *Z. Anorg. Allg. Chem.* **268**, 364.

Hoppe, R., and Rödder, K.-M. (1961). *Z Anorg. Allg. Chem.* **313**, 154.

Hoppe, R., Dähne, W., and Klemm, W. (1961). *Naturwissenschaften* **48**, 429.

Hoppe, R., Dähne, W., and Klemm, W. (1962a). *Justus Liebigs Ann. Chem.* **658**, 1.

Hoppe, R., Dähne, W., Mattauch, H., and Rödder, K. M. (1962b). *Angew. Chem.* **74**, 903.

Huston, J. L. (1967). *J. Phys. Chem.* **71**, 3339.

Jesse, R., and Hoppe, R. (1974). *Z. Anorg. Allg. Chem.* **403**, 143.

Keenan, T. K. (1966a). *Inorg. Nucl. Chem. Lett.* **2**, 153.

Keenan, T. K. (1966b). *Inorg. Nucl. Chem. Lett.* **2**, 211.

Keenan, T. K. (1967). *Inorg. Nucl. Chem. Lett.* **3**, 463.

Keenan, T. K., and Asprey, L. B. (1969). *Inorg. Chem.* **8**, 235.

Keller, C., and Salzer, M. (1967). *J. Inorg. Nucl. Chem.* **29**, 2925.

Keller, C., and Schmutz, H. (1966). *Inorg. Nucl. Chem. Lett.* **2**, 355.

Kemmit, R. D. W., Russell, D. R., and Sharp, D. W. A. (1963). *J. Chem. Soc.* p. 4408.

Khaïroun, S., Dance, J. M., Grannec, J., Demazeau, G., and Tressaud, A. (1983). *Rev. Chim. Miner.* **20**, 871.

Klemm, W. (1954). *Angew. Chem.* **66**, 468.

Klemm, W. (1982). *Nachr. Chem., Tech. Lab.* **30**, 963.

Klemm, W., and Henkel, P. (1934). *Z. Anorg. Allg. Chem.* **220**, 180.

Klemm, W., and Huss, E. (1949). *Z. Anorg. Allg. Chem.* **258**, 221.

Klemmitt, R. D. W., Russel, D. R., and Sharp, D. W. A. (1963). *J. Chem. Soc.* p. 4408.

Köhler, J., and Hoppe, R. (1983). *Z. Naturforsch., B: Anorg. Chem., Org. Chem.* **38B**, 130.

Köhler, P., Massa, W., Reinen, D., Hofmann, B., and Hoppe, R. (1978). *Z. Anorg. Allg. Chem.* **446**, 131.

Kuhlmann, W., and Sawodny, W. (1977a). *J. Fluorine Chem.* **9**, 337.

Kuhlmann, W., and Sawodny, W. (1977b). *Eur. Fluorsymp. 6th, 1977* Abstract 6.

Langlais, F., Demazeau, G., Portier, J., Tressaud, A., and Hagenmuller, P. (1979). *Solid State Commun.* **29**, 473.

Lentz, D., Pritzkow, H., and Seppelt, K. (1978). *Inorg. Chem.* **17**, 1926.

McRae, V. M., Peacock, R. D., and Russel, D. R. (1969). *Chem. Commun.* p. 62.

Malm, J. G., and Selig, H. (1961). *J. Inorg. Nucl. Chem.* **20,** 189.

Malm, J. G., Sheft, I., and Chernick, C. L. (1963). *J. Am. Chem. Soc.* **85, 110.**

Matcheret, G. (1970). *Rep.* **CEA-R-4051.**

Mattauch, H. (1962). Dissertation, Münster.

Mellor, D. P., and Stephenson, N. C. (1951). *Aust. J. Sci. Res.* **4,** 406.

Merril, C. I., and Cady, G. H. (1961). *J. Am. Chem. Soc.* **83,** 298.

Morrell, B. K., Zatkin, A., Tressaud, A., and Bartlett, N. (1973). *Inorg. Chem.* **12,** 2640.

Müller, B. G. (1981a). *J. Fluorine Chem.* **17,** 317.

Müller, B. G. (1981b). *J. Fluorine Chem.* **17,** 409.

Müller, B. G., and Hoppe, R. (1983). *Z. Anorg. Allg. Chem.* **498,** 128.

Naumann, D. (1980). "Fluor und Fluorverbindungen." Steinkopf Verlag, Darmstadt.

Odenthal, R.-H., and Hoppe, R. (1971). *Monatsh. Chem.* **102,** 1340.

Ogden, I. S., and Turner, J. J. (1966). *Chem. Commun.* p. 693.

Paine, R. T., and Asprey, L. B. (1975). *Inorg. Chem.* **14,** 1111.

Pausewang, G., and Dehnicke, K. (1969). *Z. Anorg. Allg. Chem.* **369,** 265.

Peacock, R. D. (1956). *Recl. Trav. Chim. Pays-Bas* **75,** 576.

Penneman, R. A., Sturgeon, G. D., Asprey, L. B., and Kruse, F. H. (1965). *J. Am. Chem. Soc.* **87,** 5803.

Penneman, R. A., Keenan, T. K., and Asprey, L. B. (1967). *In* "Lanthanide–Actinide Chemistry" (P. R. Fields and T. Moeller, eds.), Washington, D. C. p. 248.

Prandtl. W., and Huttner, K. (1925). *Z. Anorg. Allg. Chem.* **149,** 235.

Rao, P. R., Tressaud, A., and Bartlett, N. (1976). *J. Inorg. Nucl. Chem.* **Suppl. Hyman,** 23.

Ravez, J., Grannec, R., and von der Mühll, R. (1971). *C. R. Hebd. Seances Acad. Sci., Ser. C* **272,** 1042.

Rieskamp, H., and Mattes, R. (1973). *Z. Anorg. Allg. Chem.* **401,** 158.

Riha, J., and Trevorrow, L. E. (1966). *Argonne Natl. Lab.* [Rep.] **ANL-7425.**

Rulf, O., Braida, A., Bretschneider, G. C., Menzel W., and Plant, H. (1932). *Z. Anorg. Allg. Chem.* **206,** 59.

Ruff, O., and Tschirch, F. N. (1913). *Ber. Dtsch. Chem. Ges.* **46,** 926.

Ruff, O., and Vidic, E. (1925). *Z. Anorg. Allg. Chem.* **143,** 163.

Salzer, M. (1966). **KFK-385.**

Schäfer, H., and Krehl, K. (1952). *Z. Anorg. Allg. Chem.* **268,** 25.

Schmeisser, M., and Pammer, E. (1955). *Angew. Chem.* **67,** 156.

Schmeisser, M., Sartori, P., and Naumann, D. (1970). *Chem. Ber.* **103,** 590.

Schultz, A. J., Williams, J. M., Miro, N. D., Diarmid, A. G., and Heeger, A. J. (1978). *Inorg. Chem.* **17,** 646.

Sears, D. R., and Hoard, J. L. (1969). *J. Chem. Phys.* **50,** 1066.

Seel, F. (1974). *Adv. Inorg. Chem. Radiochem.* **16,** 297.

Seel, F., and Pimpl, M. (1977). *J. Fluorine Chem.* **10,** 413.

Selig, H., and Malm, J. G. (1963). *J. Inorg. Nucl. Chem.* **25,** 349.

Seppelt, K. (1973a). *Chem. Ber.* **106,** 157.

Seppelt, K. (1973b). *Z. Anorg. Allg. Chem.* **399,** 87.

Seppelt, K. (1974). *Z. Anorg. Allg. Chem.* **406,** 287

Seppelt, K. (1979). *Angew. Chew.* **91,** 199.

Siebert, G., and Hoppe, R. (1971). *Naturwissenschaften* **58,** 95.

Siebert, G., and Hoppe, R. (1972). *Z. Anorg. Allg. Chem.* **391,** 117.

Siegel, S., and Northrop, D. A. (1966). *Inorg. Chem.* **5,** 2187.

Sladky, F. O. (1969). *Angew. Chem., Int. Ed. Engl.* **8,** 523.

Sladky, F. O., Bulliner, P. A., Bartlett, N., De Boer, B. G., and Zalkin, A. (1968). *Chem. Commun.* p. 1048.

Sladky, F. O., Bulliner, P. A., and Bartlett, N. (1969). *J. Chem. Soc. A* p. 2179.

Smith, D. F. (1963a). *Science* **140**, 899.

Smith, D. F. (1963b). *Science* **141**, 1039.

Sokolov, V. B., Drobyschevskij, Y. U., Prusakov, V. M., Ryzhkov, A. V., and Koroshev, S. S. (1976). *Dokl. Akad, Nauk SSSR* **229**, 641.

Steindler, M. J. (1968). *Argonne Natl. Lab.* [*Rep.*] **ANL-7438**.

Steindler, M. J. (1969). *Argonne Natl. Lab.* [*Rep.*] **ANL-7595**.

Sunder, W. A., and Stevie, F. A. (1975), *J. Fluorine Chem.* **6**, 449.

Tressaud, A., Winterberger, M., Bartlett, N., and Hagenmuller, P. (1976). *C. R. Hebd. Seances Acad. Sci., Ser. C.* **282**, 1069.

Tressaud, A., Kaïroun, S., Dance, J. M., and Hagenmuller, P. (1984) *Z. Anorg. Allg.* Chem. **508**, 93.

Turner, J. J., and Pimentel, G. C. (1963). *Science* **140**, 974.

Vedrine, A., Baradue, L., and Cousseins, J.-C. (1973). *Mater. Res. Bull.* **8**, 581.

von Wartenberg, H. (1941). *Z. Anorg. Allg. Chem.* **247**, 135.

von Wartenberg, H. (1942). *Z. Anorg. Allg. Chem.* **250**, 122.

Waslie, M. J., and Falconer, W. E. (1976). *J. Chem. Soc., Dalton Trans.* p. 351.

Weinland, R. F., and Fiederer, B. (1907). *Chem. Ber.* **40**, 2090.

Weinstock, B. (1964). *Chem. Eng. News* **42**, 86–100.

Weinstock, B., and Malm, J. G. (1956). *J. Inorg. Nucl. Chem.* **2**, 380.

Weise, E., and Klemm, W. (1953). *Z. Anorg. Allg. Chem.* **272**, 211.

Weise, E., and Klemm, W. (1955). *Z. Anorg. Allg. Chem.* **279**, 74.

Wilhelm, V., and Hoppe, R. (1974). *Z. Anorg. Allg. Chem.* **407**, 13.

Wilhelm, V., and Hoppe, R. (1976). *J. Inorg. Nucl. Chem.* **Suppl. Hyman**, 113.

Wingefeld, G., and Hoppe, R. (1984). *Z. Anorg. Allg. Chem.* **518**, 149.

Wright, A. F. Fender, B. E. F., Bartlett, N., and Leary, K. (1978). *Inorg. Chem.* **17**, 748.

Zachariasen, W. H. (1949). *In* "The Transuranium Elements" (G. T. Seaborg. J. J. Katz, and W. H. Manning, eds.), TL 2. p. 1462. New York.

7

Fluoride Glasses

JEAN JACQUES VIDEAU and JOSIK PORTIER
Laboratoire de Chimie du Solide du CNRS
Université de Bordeaux
Talence, France

I. Introduction

The use of fluorides in glass is several centuries old. But these compounds actually do not behave as glass "progenitor agents," but as vitroceramic "progenitors" for the fabrication of opaline glasses (see Chapter 20). With the extensive development of fluorine chemistry it has been discovered that, as is true of many oxides, the fluorides can give rise to glasses. This analogy may be related to a certain extent to the similar size of the O^{2-} and F^- anions: SiO_2 and BeF_2 have the same crystal structure, and both lead to glasses.

Studies have been devoted to fluoride glasses for more than half a century. Yet these investigations did not lead to significant applications until lately. Only new needs have brought about a renewed interest in nonoxide glasses, particularly fluoride glasses.

In this chapter we review the various glass formation domains with a structural approach and discuss the specific optical, thermomechanical, and electrical properties of the various fluoride glass families so far discovered. In addition, some prospective aspects are given for optical applications.

II. Glass Formation and Structural Features

Numerous studies devoted to fluoride glasses have induced a proliferation of new vitreous compositions. The best known are fluoroberyllate glasses, for which the Zachariasen (1932) model of glass formation agrees as well as it does for silicate glasses. Fluorozirconate glasses are another new glass family; ZrF_4 is the primary component, although it cannot actually be characterized as a "glass network former." The other new fluoride glasses generally do not have major components. They are grouped in one section, which includes glasses formed with AlF_3, ThF_4, $3d$ transition metal fluorides, etc.

Among the oxide–fluoride glasses, we have selected the fluorophosphates because of their practical interest. Chlorofluoroglasses have been discovered by several authors (Almeida and Mackenzie, 1982; Matecki et al., 1983) interested in their absorption in the infrared (ir) region. Although these glasses are very attractive, the subject is not addressed here.

A. FLUOROBERYLLATE GLASSES

The vitreous materials based on beryllium fluoride were discovered by Goldschmidt (1926). Sun and Callear (1949), Sun (1949a–c), and Sun and Higgins (1950) were the first authors interested in the optical properties. Since then, research concerning the glass formation regions in various composition diagrams has been reported (Counts et al., 1953; Roy et al., 1950; Imaoka, 1954a,b; Imaoka and Mizusawa, 1953; Vogel and Gerth, 1958a,b,c).

Beryllium difluoride, like SiO_2, easily forms a glass on cooling from the melted state. Due to the similarity of both crystal structures, models for the formation of silicate glasses can be adapted to the related fluoride systems. As already pointed out, Zachariasen's rules (1932) can be applied to BeF_2 glasses (Cooper, 1982). Baldwin and Mackenzie (1979a) used the bonding-energy approach for a rough classification of the glass-forming, modifying, and intermediate components. The most common fluorides can be grouped as follows:

Modifying fluorides: Li, Na, K, Rb, Cs
Ca, Sr, Ba
Zn, Cd, Ce, Nd, Pr, Th
Intermediate fluorides: Al, Mg, Y, Pb

Modifying fluorides do not enter the network structure and tend to depolymerize the $(Be_nF_{3n+1})^{(n+1)-}$ chains. Intermediate fluorides participate in the vitreous network, leading generally to higher chemical stability and, as a consequence, to a higher melting point.

The tendency of glass phase separation was shown to increase with potassium → sodium → lithium substitution by Vogel and Gerth (1958a–c). The authors did not point out that the binary LiF/BeF_2 glass system appears to be completely demixed through the entire glass formation region. The proportions of BeF_2 can vary from 100 to ~ 30 mol % according to the nature of the added compound. The presence of BeF_2 can even drop to a few mole percent when AlF_3 is present in the glass (Videau et al., 1979b).

Neutron diffraction and x-ray studies have been interpreted by assuming a model based on the structure of β-quartz. Interatomic distances and coordination numbers have been measured, calculated, and compared (Batsanova et al., 1968; Leadbetter and Wright, 1972; Narten, 1972; Zarzycki, 1971). By means of ir absorption and Raman spectroscopy it has been possible to determine the basic groups that form the network as a function of the nature and content of modifying fluorides (Galeener et al., 1978; Kondrat'eva et al., 1969; Piriou et al., 1981; Quist et al., 1972; Walrafen and Stolen, 1978). Both esr and optical spectroscopy have also yielded much structural information (Abrashitova and Petrovskii, 1967; Griscom et al., 1980; Kolobkov and Petrovskii, 1971; Wong and Angell, 1976; Yudin et al., 1976).

Those studies illustrate significant differences between the networks of BeF_2- and SiO_2-rich glasses, although formerly both glasses were assumed to be similar. This difference is enhanced when the BeF_2 content decreases. The "weakened" model of SiO_2 (Osborn, 1952) is not more suitable, and a chain structure is formed. Typical examples are the aluminum fluoroberyllate glasses, in which BeF_4 tetrahedra are present more or less as polymerized species $(Be_nF_{3n+1})^{(n+1)-}$. These chains are cross-linked by $(AlF_5)_n$ chains (Videau et al., 1982).

B. FLUOROZIRCONATE GLASSES

The discovery of the first ZrF_4-based fluoride glasses in 1975 was the starting point of numerous studies on vitreous fluorozirconates (Poulain et al., 1975). Standard ternary glasses have been found in the ZrF_4/BaF_2 and $/MF_x$ systems (M = Th, U, Al, La, Nd, Ca, or Na) (Poulain and Lucas, 1978), in which zirconium fluoride can be more accurately characterized as a "glass network former" than as a "glass progenitor" according to Angell's concept (1981). It does not seem possible to obtain vitreous materials if ZrF_4 is not associated with other fluorides, particularly BaF_2. Barium-free glasses may exist, nevertheless, in $ZrF_4/ThF_4/MF_x$ systems that contain thorium fluoride (M = La, Nd, Y, Lu, Sc, or Al) (Matecki et al., 1978, 1982a). The absence of barium lowers the glass transition temperature.

Fluorozirconate glasses have a relatively strong tendency to crystallize, and the area of glass formation in the different ternary systems is usually small. The introduction of a fourth element, such as aluminum, however, improves the stability (Lecoq and Poulain, 1980a).

Although these glasses are an innovation, so far they have not been studied extensively. We shall mention two investigations using Raman and Mössbauer spectroscopy by Almeida and Mackenzie (1981) and Coey et al. (1981). A structural model has been proposed by several authors based on a comparison with the corresponding crystallized materials (Poulain et al., 1977a) and the physical properties (Hu and Mackenzie, 1983; Matecki et al., 1978). The determination of fluorescence and absorption spectra after doping with transition and rare-earth elements has also been useful (Aliaga et al., 1978; Lucas et al., 1978; Poulain et al., 1977b).

The vitreous network is built from the association of various ZrF_6, ZrF_7, and ZrF_8 polyhedra sharing either corners or edges. The coordination number of zirconium varies between 6 and 8 with different geometries. The periodicity of the lattice is broken from place to place by the presence of barium when this element is present. In the $ZrF_4/ThF_4/LnF_3$ systems (Ln = lanthanide), lanthanum and thorium can be considered "intermediate"; they are often called "network stabilizers." In these glasses, which have a high fluorine content, the Ln^{3+} and Th^{4+} cations can be surrounded apparently by 9, 10, or even 11 fluorine atoms, inducing a progressive decrease in the coordination of zirconium from 8 to 7 or even 6 (Matecki et al., 1978). In BaF_2-containing systems, the addition of Al^{3+} and Y^{3+} seems to stabilize the glass because of the occupation of sites that otherwise would be empty, so that the vitreous network is built up by a larger number of polyhedra (aluminum octahedra close to ZrF_n polyhedra), giving more geometric possibilities to the local structural arrangements (Lecoq and Poulain, 1980b).

C. OTHER FLUORIDE GLASSES

The glass-forming capacity of AlF_3 was discovered by Sun (1949c) in the $SrF_2/MgF_2/PbF_2/AlF_3$ quaternary system. More recently, several authors have prepared glasses from the CaF_2/AlF_3 and BaF_2/AlF_3 systems by squeezing the corresponding liquid phase between stainless steel plates (Baldwin et al., 1981; Ehrt et al., 1982) and from the PbF_2/AlF_3 system by a splat quenching technique (Shibata et al., 1980).

The devitrification stability of these glasses can be improved by the addition of ThF_4, ZnF_2, or LnF_3 (Ln = rare-earth metal or yttrium) (Matecki et al., 1981; Poulain et al., 1981a, 1982a). The difference between the glass transition temperature T_g and the glass crystallization temperature T_c, generally lower for the binary systems, can exceed $100°C$. A glass of 20 BaF_2/40 CaF_2/40 AlF_3 (mole per cent) composition has been studied by Raman spectroscopy and by comparison with the crystallized material. A model based on chains of AlF_6 octahedra linked by some AlF_4 tetrahedra has been proposed (Videau et al., 1979c) and confirmed by Ehrt et al. (1983a). In the multicomponent glasses such as those of $BaF_2/ThF_4/YF_3/AlF_3$ composition the proportions of each component are quite close. Therefore, the role of each cation cannot be clearly defined. Poulain et al. (1981a) suggested an analysis of the glass formation in terms of an ionic model. The fluoride glasses would result from a random insertion of cations in a disordered anionic sublattice. This supposes that a every cation has several insertion possibilities and that the intercationic repulsion is minimized either by a screen effect due to the anions or by a lowering of cationic charge and an increase in interatomic distances.

Glass formation was detected in the $HfF_4/BaF_2/MF_x$ (M = Th, U, or La) systems (Lucas et al., 1977). Drexhage et al. (1980) pointed out that the tendency of HfF_4 systems to form glass is greater than that of ZrF_4 systems. If one supposes a structural analogy between the vitreous and the crystallized forms, just as crystalline HfF_4 and ZrF_4 are isostructural, one may also expect similar glass-forming regions and structures. Comparative Raman spectroscopic investigations of HfF_4- and ZrF_4-based glasses have confirmed that prediction (Banerjee et al., 1983).

A new series of fluoride glasses containing $3d$ transition elements has been detected in $M^IF/M^{II}F_2/M^{III}F_3$, systems ($M^I$ = Li, Na, K, or Ag; M^{II} = Ca, Sr, Pb, Mn, Fe, Co, Ni, Cu, Zn, or Ag; M^{III} = V, Cr, Fe, or Ga) by Miranday et al. (1979, 1981). The $PbF_2/MnF_2/FeF_3$ ternary glass domain was found by the same authors to be particularly wide, but the difference between T_g and T_c is weak and cannot exceed $90°C$ in the best case. From spectroscopy investigations a structural model with a random corner sharing of $M^{III}F_6$ octahedra, PbF_2 being a network modifier, has been proposed.

Another noteworthy group of fluoride glasses based on transition metals such as manganese or zinc but associated with thorium or rare-earth fluorides can be obtained, but they also have a strong tendency to crystallize (Fonteneau *et al.*, 1980a,b; Le Page *et al.*, 1982). Their stability has been improved by the use of a fourth fluoride, generally BaF_2, which plays the role of a modifier, the other cation present being randomly distributed in a three-dimensional lattice (Fonteneau *et al.*, 1982).

Large amounts of lanthanide or actinide fluorides such as ThF_4 or UF_4 can be included in various fluoride glasses. Most likely, they have some glass-forming capacity. Ternary glasses containing BaF_2 have been synthesized (Lucas *et al.*, 1981; Poulain and Poulain, 1983; Poulain *et al.*, 1981b).

Other vitreous materials have been prepared from scandium, cadmium, and indium fluorides. Vitreous domains have been obtained in the following ternary systems: $ScF_3/YF_3/BaF_2$, $CdF_2/BaF_2/ZnF_2$, and $InF_3/BaF_2/YF_3$ with glasses of formulation $Cd_{0.50}Ba_{0.50}F_{2.00}$ or $In_{0.55}Ba_{0.45}F_{2.55}$ (Matecki *et al.*, 1982b; Poulain *et al.*, 1982b; Videau *et al.*, 1983b). The difference between T_g and T_c is weak ($\Delta T \le 80°C$).

D. FLUOROPHOSPHATE GLASSES

A chapter on fluoride glasses should include fluorophosphate glasses, although they cannot be considered pure fluoride glasses. Fluorine plays a specific role, however. Numerous investigations on fluorophosphate glasses have actually shown a very clear evolution of structural, physical, and chemical properties with fluorine content.

Most research has been related to the following pseudosystems:

1. $NaPO_3/MF_x$ (Murphy, 1963; Murphy and Mueller, 1963; Williams, 1959)
2. $Ba(PO_3)_2/MF_x$ and $BaPO_3F/MF_x$ (Chalilev *et al.*, 1977; Evstrop'ev *et al.*, 1971a; Pronkin *et al.*, 1978)
3. $Al(PO_3)_3/MF_x$ (Kolobkov and Petrovskii, 1971)
4. $Sr(PO_3)_2/MF_x$, $Zn(PO_3)_2/MF_x$, and $Pb(PO_3)_2/MF_x$ (Ehrt, 1983b)

Here M is an alkali or alkaline-earth element: zinc, cadmium, gallium, or lead. The vitreous areas can vary greatly from one system to another. In the case of $NaPO_3$, $Ba(PO_3)_2$, or $BaPO_3F$ the content of the fluoride introduced seldom exceeds 50 mol %, but for $Al(PO_3)_3$ the phosphate concentration can decrease to a few mole percent. Extensive work has been done on those materials by Gan Fuxi (1981) and Yasi *et al.* (1982).

Whatever the nature of the starting compound, during the melt most fluorides volatilize partially. This phenomenon is enhanced in phosphate glasses because of the tendency to form gaseous PF_5 or POF_3 (Urusovskaya, 1968a; Volkov et al., 1974). The result is a considerable composition change. This problem can be minimized by the use of a fluorinating agent such as NH_4HF_2 which is added to the starting compounds (see Chapter 2, Section V,A,1). It allows the fluorine content to be maintained or even increased (Stević et al., 1978).

Several structural models have been proposed. The presence of fluorine anions brings about cutting of the infinite $(P_nO_{3n+1})^{(n+2)-}$ chains in the phosphate structure and the formation of shorter chains, the extremity of which can be monofluorinated or difluorinated (Stević et al., 1982; Veksler et al., 1974; Videau et al., 1983a). Raman spectroscopy, epr, and Mössbauer resonance studies have shown that, in $NaPO_3/AlF_3$, $NaPO_3/CaF_2/AlF_3$, and $NaPO_3/FeF_3$ pseudosystems, increasing fluorine content induces the appearance of $(AlF_6)^{3-}$, $(Fe^{III}F_6)^{3-}$, $(AlF_4)^-$, $(Fe^{III}O_2F_2)^{3-}$, or $(F_3Al/O/AlF_3)^{2-}$ groups and the progressive disappearance of the phosphate chains with the formation of $P(O,F)_4$ tetrahedra, which can continue eventually to share oxygen atoms. The strongly charged Al^{3+} cation may link two or several chains, playing the role of a bridge (Menil et al., 1979; Videau et al., 1982). Similar fluoro- and oxyfluoroaluminate groups are present in glasses formed from $Al(PO_3)_3$ (Petrovskii et al., 1973). When BaF_2 is associated with $Al(PO_3)_3$, the metaphosphate network collapses; AlF_3 and $Ba(PO_3)_2$ form in the melt. $Ba(PO_3)_2$ goes progressively over to the $BaPO_3F$ fluorophosphate. $BaPO_3F$ can actually be used to form glasses when mixed with fluorides. The addition of MgF_2 and/or AlF_3 fills the interchain space and stabilizes the vitreous structure in forming copolymers, particularly with PO_3F and MgO_2F_2 tetrahedra (Evstrop'ev et al., 1971a).

III. Optical Properties and Applications

A. REFRACTIVE INDEX AND DISPERSION

Fluoride glasses very often have a low refractive index n_D, a low nonlinear index n_2, and a weak dispersion (high Abbe number v) (Table I). Index n_2 is obtained from the expression giving the variation of the refractive index for an intense optical beam (Weber, 1976).

Values differ most for vitreous BeF_2-based materials. The addition of modifying fluorides to BeF_2 increases n_D and decreases v. For binary alkali and alkaline-earth fluoroberyllate glasses, n_D increases by sodium → lithium → potassium → rubidium or magnesium → calcium → strontium substitutions (Vogel and Gerth, 1958a,b). For fluoroberyllate glasses

TABLE I

Refractive Index, Optical Dispersion, and Infrared Transmission of Fluoride Glasses

Composition	Refractive Index n_D	Nonlinear Index n_2 (10^{-13} esu)	Abbe Number ν	Transmission, >50%	Theoretical Loss (cm^{-1})	Minimum Attenuation (μm)	Reference
BeF$_2$ based	1.28–1.48	0.23–0.40	85–107	0.15–4.00	10^{-7}	2.0	Baldwin and Mackenzie (1979b); Cline and Weber (1978); Stokowski et al. (1978)
ZrF$_4$ based	1.48–1.54	0.90	68–80	1.25–7	10^{-8}–10^{-9}	3.0–4.0	Brown et al. (1982); Drexhage et al. (1982b); Poulain et al. (1977a); Shibata et al. (1981)
HfF$_4$ based	1.52	—	84	0.3–7.5	10^{-8}	4.6	Brown et al. (1982); Drexhage et al. (1982a); Gannon (1980)
AlF$_3$ + additions	1.42–1.44	0.47	95	0.23–5.20	10^{-7}–10^{-8}	2.5–3.0	Kanamori et al. (1981); Shibata et al. (1981); Stokowski et al. (1978)
AlF$_3$, ZnF$_2$, and heavy metals ($4f$–$5f$)	1.49–1.54	—	86	0.25–10	10^{-8}	3.5–4.0	Drexhage et al. (1982a,b,c); Maze et al. (1982)
3d Fluorides + additions	1.59–1.67	3.40	40	0.25–11	—	—	Miranday et al. (1981); Stokowski et al. (1978)
Other fluorides with Sc, Cs, or In	1.50–1.60	—	—	–11	—	—	Matecki et al. (1982b); Poulain et al. (1982b); Videau et al. (1983)
Fluorophosphates	1.33–1.60	1.50	56–128	0.28–4.5	—	—	Cook and Mader (1982); Ehrt et al. (1983b); Stokowski et al. (1978); Yasi et al. (1982)

containing a large amount of lead, n_D reaches 1.48. Zirconate and heavy-metal fluoride glasses have a high n_D, whereas aluminum fluoride has a relatively low one.

For fluorophosphate glasses n_D narrowly depends on the fluorine content and the cations present. In $AlF_3/Al(PO_3)_2$ pseudobinary glasses with low phosphate content, n_D can be as low as 1.33. In the more complex fluorophosphate glasses, n_D varies with the cations present. When they contain $Al(PO_3)_3$, n_D generally decreases as the starting phosphate content increases. Heavy cations also tend to enhance n_D, but LiF, despite the small size of Li^+, has the same influence, apparently due to the small polarizability of lithium.

The observed zero-dispersion wavelengths are about ~ 1.62 μm for the lanthanide fluorozirconate glasses and 1.48 μm for aluminofluoride glasses (Jinguji et al., 1982).

B. OPTICAL TRANSMISSION RANGE

Due to the strong electronegativity of fluorine, the charge-transfer band is generally in the ultraviolet (uv) region. The ir absorption edge is related to the atomic vibration spectrum, so it is shifted toward long wavelengths when the glass contains heavy atoms and as the metal fluorine becomes more ionic.

Many investigations have shown that the uv and ir absorption edges are strongly dependent on the preparation conditions (contamination by crucible, inert or reactive atmosphere, etc.). As an example, Dumbaugh and Morgan (1980) have shown that by appropriate selection of the modifying cation and, minimizing the pollution, it is possible to prepare an excellent uv-transmitting fluoroberyllate glass. For fluorozirconate, fluorohafnate, and heavy-metal fluoride glasses uv and ir absorption edges are obtained in a reactive atmosphere of CCl_4 (see Table I). The main feature of such glasses, in addition to a very wide optical transmission range, is a high transmittance in the whole transmission range.

In fluorozirconate glasses, $3d$ transition elements such as iron, cobalt, nickel, and copper show weak absorption coefficients due to ionic bonding with fluorine, which prohibits same-parity transitions (Poulain and Lucas, 1978). Their influence is less than one-tenth of that due to rare-earth elements whose transitions between $4f$ levels are not influenced by the anionic surrounding (Ohishi et al., 1981).

In contrast, in the fluorophosphate glasses the influence of fluorine on $3d$ cation absorption is very significant. Nevertheless, the attenuation resulting from the $3d$ transition metal contamination in $NaPO_3$/fluoride glasses is smaller than that observed in silicate glasses (Videau et al., 1979a). The strength of the $3d$ element–anion bonding is intermediate between those occuring in pure fluorides and in silicates.

These optical properties make the fluoride glasses good candidates for most optical applications except those involving wavelengths longer than 8 μm, which seems to be an upper limit in the ir region (Dubois *et al.*, 1983; Miyashita and Manabe, 1982).

The most exciting application of fluoride glasses at present is their use in optical fibers for optical transmissions. Fluorozirconate glasses seem to form the best low-loss ir glass fibers. Fibers with a diameter of 250 μm and a length of 100 m with an attenuation of 21 dB km^{-1} at 2.55 μm have been prepared (Mitachi and Miyashita, 1982), despite a narrow glass stability region and high activation energy for viscous flowing inappropriate for fiber elaboration (Tran *et al.*, 1982). Other fluoride glasses such as aluminum fluoride–based materials have a large wavelength transmission range, but they show, unfortunately, a strong tendency to devitrify (Kananmori, 1983). Many advanced applications can be considered for optical fibers, including monomode transmission (Bendow and Mitra, 1981) and short fibers for surgery using an HF/DF laser. Infrared heat detection and various other military applications such as power guides and long-range radiometry imaging detection systems can also make use of either fluoride fibers or fluoride transmission windows (Patel, 1981).

C. FLUORESCENCE PROPERTIES

Various studies using the lanthanide fluorescence in fluoride glasses have shown the unusual optical properties of these materials (Blanzat *et al.*, 1980; Wong and Angell, 1976; Kolobkov and Petrovskii, 1971; Kolobkov *et al.*, 1977; Videau *et al.*, 1979a, 1982).

Laser fluorescence studies of Nd^{3+}-doped fluoride glasses have been reported, for example, by Stokowski *et al.* (1978). Table II summarizes typical values of spectroscopic parameters for some fluoride glasses compared with classical oxide glasses.

It is worth emphasizing the important difference between oxide glasses and fluoride glasses. Neodynium ion absorption peaks in fluoride glasses are generally narrower than those observed in silicates, so that the relative intensities are higher. The presence of fluorine instead of oxygen shifts the peak of the $^4F_{3/2} \rightarrow {}^4I_{11/2}$ emission toward shorter wavelengths due to lower covalency. The long lifetime on the $^4F_{3/2}$ Nd^{3+} emission level results from the ionic character of the bonding (small cation–anion mixing) but is also due partially to the small value of the refractive indexes, which enters as n_D (n_D + 2)^2 in the spontaneous emission probability. Fluoroberyllate glasses are the best due to competition between Nd—F and strong Be—F bonds.

TABLE II

Comparison of Refractive Index and Spectroscopic Properties of the $^4F_{3/2} \longrightarrow {}^4I_{11/2}$ Nd^{3+} Transition in Some Representative Oxide and Fluoride Glasses[a]

Property	Silicate (1)	Phosphate (2)	Fluoro-phosphate (3)	Fluoro-zirconate (4)	Lead Fluoride (5)	Alumino-fluoride (6)	Fluoro-beryllate (7)	Alumino-fluoro-beryllate (8)
Refractive index n_D	1.567	1.507	1.460	1.560	1.636	1.428	1.346	1.390
Nonlinear index n_2 (10^{-13} esu)	1.40	0.91	0.57	0.90	3.4	0.47	0.34	0.4
Emission cross section σ (10^{-20} cm^2)	2.9	4.7	2.5	3.0	2.4	2.2	3.2	3.5
Linewidth $\Delta\lambda_{\mathrm{eff}}$ (nm)	29[b]	19[b]	31	26	25	32	23	17
Fluorescence peak (nm)	1062	1064	1052	1049	1052	1049	1047	1048
Radiative lifetime τ (μsec)	340	330	510	440	516	568	611	630
Branching ratio ($^4F_{3/2} \longrightarrow {}^4I_{11/2}$)	0.47	0.47	0.51	0.50	0.50	0.50	0.50	0.50

[a] Glass systems are as follows: 1, 60 SiO$_2$/27.5 Li$_2$O/10 CaO/25 Al$_2$O$_3$ (Weber et al., 1976); 2, 50 P$_2$O$_5$/33 K$_2$O/17 BaO (Weber et al., 1976); 3, 5 Al(PO$_3$)$_3$/25 AlF$_3$/12 MgF$_2$/35 CaF$_2$/12 SrF$_2$/12 BaF (Stokowski et al., 1978); 4, 61 ZrF$_4$/34 BaF$_2$/5 LaF$_3$ (Lucas et al., 1978); 5, 50 PbF$_2$/25 MnF$_2$/25 GaF$_3$ (Stokowski et al., 1978); 6, 45 AlF$_3$/35 CaF$_2$/20 BaF$_2$ (Stokowski et al., 1978); 7, 47 BeF$_2$/27 KF/16 CaF$_2$/10 AlF$_3$ (Stokowski et al., 1978); 8, 31 AlF$_3$/15 BaF$_2$/27 CaF$_2$/27 BeF$_2$ (Videau et al., 1979b).

[b] Linewidth (FWHM).

The emission linewidths in fluoride glasses are narrow. For the fluoroberyllate glasses, they decrease with increasing size of the alkali or alkaline-earth ions present, as also observed in phosphate glasses (Cline and Weber, 1979).

The low linear and nonlinear refractive indexes, the low dispersion (high Abbe number), the high-frequency uv transmission edge, and the strong emission of fluoride glasses, particularly in BeF_2-based glasses, would make some of them attractive materials for host structures of high-power fusion lasers because the hardness and moisture resistance would be improved. The purpose of such materials is to simulate instantaneous fusion reactions (Weber, 1976; Weber et al., 1976).

Weber (1982) collated Faraday rotation studies on fluoride glasses and related crystals doped with paramagnetic ions. The comparison shows the good performances of Tb^{3+} fluorophosphate glasses with a Verdet constant of -10.8 radians $T^{-1} m^{-1}$ and a figure of merit of 4.8×10^{20} radians $W T^{-1} m^{-3}$ at 1064 nm; Tb^{3+} fluoroberyllate and Nd^{3+} zirconate glasses are less attractive, because their Verdet constants are only -2.9 and -3.3 radians $T^{-1} m^{-1}$, respectively (Pye et al., 1983).

IV. Other Properties

A. GLASS TRANSITION TEMPERATURE

Glass transition (T_g), recrystallization (T_c), and melting (T_m) temperatures obtained by differential thermal analysis (DTA) for various types of glasses are summarized in Table III. Often, T_g values are in the vicinity of 300°C; they are always lower than those of the silicate glasses ($T_g > 500$°C). The classical two-thirds rule expressed as $T_g/T_m \approx 2/3$ (K) can be applied in most cases.

For BeF_2-based and fluorophosphate glasses T_g and T_c are strongly dependent on composition. The large T_g range of these glasses results from the wide existence domains in the composition diagrams. Fluoroaluminate and heavy-metal fluoride glasses have a higher T_g. The T_g is lower for alkali fluorozirconate glasses and becomes higher when the glass contains tri- and tetravalent cations (Baldwin and Mackenzie, 1980b; Lecoq and Poulain, 1980a; Matecki et al., 1978, 1982a).

Most fluoride glasses have a relatively strong tendency to recrystallize at rising temperature. This feature results from the small difference generally observed between T_c and T_g (ΔT). Only a few fluoride glasses, such as BeF_2-based ones, have a ΔT exceeding 120 K. As a rule the transition temperature of

TABLE III

Physical Characteristics of Fluoride Glasses[a]

Glass	Composition	T_g (°C)	T_c (°C)	T_m (°C)	$\alpha_{20-200°C}$ ($K^{-1} \times 10^{-7}$)	d (g cm^{-3})	H (kg mm^{-2})	Reference
Fluoroberyllate	Vitreous BeF$_2$	250	—	545	68	2.0	180	Cline and Weber (1979)
	Multicompounds	160–350	500	—	70–350	2.0–3.3	180–315	Videau (1979)
Fluorozirconate	Multicompounds	240–455	300–540	500–600	100–187	4.5–5.3	250	Baldwin et al. (1981); Poulain (1983)
Fluorohafnate	Ba, Hf, La, Al	310	400	560	173	6.27	250	Drexhage et al. (1982a, 1983)
Fluoroaluminate	Ba, Ca, Y, Al	430	535	710	165	4.00	360	Kanamori et al. (1981); Videau et al. (1979)
Heavy-metal fluoride	Ba, Th, Lu, Zn	355	445	670	151	6.45	300	Drexhage et al. (1982a,1983); Poulain et al. (1982a)
	Ba, Th, Ln, Al	450	550	710	150	5.10	—	Le Page et al. (1982)
3d Transition metal fluoride	Ba, Ln, Mn	320	340	720	—	4.5–5.5	—	Matecki et al. (1981)
	Ba, Th, Al, Zn	347	460	650	154	5.3	—	Miranday et al. (1981)
	Pb, M_t^{II}, M_t^{III}	250–350	280–400	500–700	—	5.5–6.0	—	Matecki and Poulain (1983); Videau 1979, also unpublished results, 1982)
Fluorophosphate	NaPO$_3$ + Fluorides	125–420	300–500	—	180–300	2.0–4.0	—	
	M^{II}(PO$_3$) + Fluorides	410–540	—	—	140–170	2.7–4.06	—	Ehrt et al. (1983b); Urusovskaya (1968b)
	Al(PO$_3$)$_3$ + Fluorides	300–460	—	—	130–190	2.5–4.0	300	Stokowski et al. (1978); Yasi et al. (1982)

[a] Abbreviations and symbols: T_g, T_c, and T_m are the glass transition, recrystallization, and melting temperatures, respectively, α is the thermal expansion coefficient, d the density, and H the Knoop hardness; Ln = lanthanides; M_t = 3d transition metals; M^{II} = Ba, Sr, Pb, or Zn.

the fluorophosphate glasses decreases with fluorine content and increases with cationic oxidation state (Stokowski et al., 1978; Videau, 1979).

B. THERMAL EXPANSION

Fluoride glasses can be considered "soft glasses" because they expand more than silicate glasses due to weaker bonds. Their thermal expansion α is higher than $70 \times 10^{-7} \, \text{K}^{-1}$ (Table III). When they contain 40 mol % CsF, BeF_2/CsF glasses show a maximum α value of $350 \times 10^{-7} \, \text{K}^{-1}$ (Baldwin and Mackenzie, 1980b). For barium fluorozirconate glasses, α depends on the barium concentration, varying from $100 \times 10^{-7} \, \text{K}^{-1}$ for barium-free glasses up to $180 \times 10^{-7} \, \text{K}^{-1}$ for 35 mol % BaF_2 (Poulain and Grosdemonge 1982). The presence of barium obviously softens the phases. In fluorophosphate glasses with low $Al(PO_3)_3$ content, α strongly decreases with the electrostatic force Z/a^2 (Yasi et al., 1982). In sodium metaphosphate, the thermal expansion coefficient also decreases with fluoride incorporation (Matecki and Poulain, 1983).

C. DENSITY

The density of vitreous fluorides strongly changes with the nature of the metals contained in the various glasses, as seen in Table III. The addition of alkali fluoride modifiers to BeF_2 increases the density. Although the lithium and sodium compositional molar volumes follow an ideal mixture law, this is not the case for potassium. This behavior emphasizes the different packing mode of K^+ in fluoroberyllate glasses. Whereas Li^+ and Na^+ may be incorporated into the glass network with a minimum of distortion, K^+, Rb^+, and Cs^+ seem to shrink it (Vogel and Gerth, 1958a). For binary alkaline-earth fluoroberyllate glasses, a linear composition–density dependence has been reported. Density increases with magnesium $\rightarrow$ calcium $\rightarrow$ strontium substitutions (Vogel and Gerth, 1958b).

Fluorozirconate glasses containing only modifying fluorides have a density lower than $4.5 \, \text{g cm}^{-3}$. In contrast, modifier-free fluorozirconate glasses are more dense. In aluminum-substituted fluorozirconate glasses the density decreases linearly, inducing greater compactness and therefore higher thermodynamic stability than in corresponding aluminum-free ZrF_4 glasses (Lecoq and Poulain, 1980b).

The density of fluorophosphate glasses rises with fluoride content, whatever the nature of the starting phosphate due to chain breaking, but the presence of intermediate fluorides such as MgF_2 or AlF_3 hardly modifies the density (Urusovskaya, 1968b).

D. VISCOSITY

The viscosity of fluoroberyllate glasses has been the subject of many investigations (Cantor et al., 1969; Mackenzie, 1960; Moynihan and Canter, 1968; Petrovskii, 1967). The viscosity–temperature behavior above the glass transition temperature follows an Arrhenius law. Melted BeF_2 is almost as viscous as melted SiO_2 near the melting point: 10^6 and 10^7 P, respectively, for BeF_2 and SiO_2 (Hu and Mackenzie, 1983).

In contrast, melted fluorozirconates are characterized by a low viscosity (10^3 P near a melting point of $540°C$) and a quick increase in the flowing activation energy with decreasing temperature. As a consequence the strong tendency to recrystallize (10^4 P $< \eta < 10^9$ P) as well as the narrow glass existence range makes fiber preparation difficult.

In general, the addition of modifying fluorides such as LiF or KF to BeF_2 reduces the viscosity (Cantor et al., 1969; Nemilov et al., 1968). This tendency is less sensitive in the presence of intermediate fluorides such as AlF_3 (Nemilov et al., 1968). The influence of fluoride dopants, especially of PbF_2, on ZrF_4-based glasses results in a sharp drop in the flowing activation energy, which becomes very close to that of the corresponding BeF_2-based glasses (Tran et al., 1982).

E. CHEMICAL DURABILITY

Generally speaking, the chemical inertness of fluoride glasses is poorer than that of classical vitreous materials based on silica. Table IV summarizes the approximate solubility of fluoride glasses compared with that of silicate and phosphate glasses.

TABLE IV

Approximate Solubility of Some Glasses in Water at $23°C$[a]

Material	Solubility (g cm^{-2} min^{-1})
Vitreous BeF_2	18×10^{-6}
BeF_2-based glasses	2.7×10^{-6}
ZrF_4-based glasses	4.7×10^{-6}[b]
Fluorophosphates	0.09×10^{-6}
Phosphates	0.04×10^{-6}
Silicates	0.02×10^{-6}
Borosilicates	0

[a] After Cline et al. (1979).
[b] Solubility at $100°C$. After Baldwin et al. (1981).

The stability of many beryllium fluoride glasses in the presence of water is weak. Vitreous BeF_2 itself, for instance, is extremely hygroscopic (Heyne, 1933). Moreover, beryllium fluoride is soluble in acidic solutions such as HF or HCl and also in basic solutions, but it appears to be insoluble in inorganic liquids (Grebenshchikova and Petrovskii, 1963; Margaryan and Evstrop'ev, 1968). For multicomponent fluoroberyllate glasses, it has been shown that the durability can be improved by reducing the amount of BeF_2 and alkali fluoride and by substituting alkaline-earth fluorides and other modifiers or intermediate fluoride such as AlF_3 or YF_3 (Grebenshchikova and Petrovskii, 1963; Margaryan and Evstrop'ev, 1968; Nikolina et al., 1970). For example, the alkali-free fluoroberyllate glasses listed in Table IV have a solubility in water nearly 250 times lower than that of pure vitreous BeF_2.

Fluorozirconate glasses are stable in the ambient atmosphere, but the surface of the glasses in the $ZrF_4/BaF_2/ThF_4$ system or of the ZrF_4-based glasses containing large amounts of alkali fluorides is attacked by water after a few days at 20°C. The dissolution rate in boiling water is 10^3 faster than that of typical silicate glasses (Baldwin et al., 1981). The mechanism of aqueous corrosion of fluorozirconate glasses has been investigated. An F^-/OH^- surface exchange occurs due to hydrolysis (Simmons et al., 1982).

The dissolution rate of glasses containing ZnF_2 is about 10 times weaker than that of fluorozirconate or fluorohafnate glasses (Robinson and Drexhage, 1983). Fluoroaluminate glasses are not very sensitive to water; the solubility is similar to that of CaF_2 (Kanamori et al., 1981).

The stability of fluorophosphate glasses varies according to the nature of the phosphate and the fluorine/oxygen ratio. $NaPO_3$-based glasses with high fluorine content are sensitive to atmospheric moisture (Stević et al., 1978). Vitreous materials containing little $Al(PO_3)_3$ and alkaline-earth fluorides are as stable as heavy-metal fluoride glasses in the presence of aqueous solutions (Stokowski et al., 1978).

F. ELECTRICAL PROPERTIES

Several investigations in the field of fluoroberyllate or fluorozirconate glasses have shown that such materials are fast F^- conductors (Aleksandrova and Batsanova, 1972; Baldwin, and Mackenzie, 1980a; Kondrat'eva and Petrovskii, 1967; Leroy and Ravaine, 1978; Petrovskii et al., 1965). The F^- anion motion, however, is lower than that of most fluorite-type fluorides (see Chapter 11).

In the case of BeF_2 glasses the conductivity ($10^{-10}\ \Omega^{-1}\ cm^{-1}$ at 200°C) seems to be very sensitive to hydroxyl impurities; it can increase by four orders of magnitude (Baldwin and Mackenzie, 1979b, 1980b). On the addition of CsF to BeF_2 at low alkali concentration, the conductivity of the glasses decreases, but at higher concentrations it increases again. Such an anomaly does not appear for LiF fluoroberyllate glasses (Baldwin and Mackenzie, 1980b). If lithium is present, the conductivity could be, according to the authors, either cationic (Hitch and Baes, 1969; Ohmichi et al., 1976; Ohno et al., 1978; Vallet and Braunstein, 1975) or of the mixed type (Evstrop'ev et al., 1972).

The conductivity of fluorozirconate glasses is higher due to weaker bonding of fluorine. It seems to be related to a network built up by polyhedra of various shapes and sizes giving rise to nonbinding F^-. When moving, these fluorine atoms could give rise to C.N. exchange between neighboring polyhedra (Ravaine et al., 1982).

Fluorophosphate-based glasses are poor fluorine conductors due to higher compactness ($10^{-10} \Omega^{-1}$ at 200°C), but they can eventually show a high cationic conductivity ($2.10^{-4} \Omega^{-1}$ cm^{-1} at 150°C for a lithium fluorophosphate glass). Such materials have been proposed as solid electrolytes (Evstrop'ev et al., 1971b). Reviews of systematic investigation of alkali fluorophosphate glasses with appropriate references have been published by Evstrop'ev et al. (1978) and Gan Fuxi (1981).

V. Conclusions

Fluoride glasses constitute a new class of materials; they can be divided into several families according to the nature of the predominant cation: fluoroberyllates, fluorozirconates, fluorohafnates, fluoroaluminates, etc.

Although fluoroberyllate glasses roughly obey Zachariassen's rules as well as to silicate glasses due to the formation of corner-sharing tetrahedra, this is not the case for other fluoride glasses, for which the notion of "network former" does not apply. As a consequence, such glasses have physical properties that differ from those of beryllium-based fluorides.

Most fluoride glasses show a stronger tendency to recrystallize than corresponding oxide glasses, diminished hardness related to a larger thermal expansion coefficient, and lower chemical stability but a lower refractive index, nonlinear index, and dispersion. They are transparent in a large spectral domain involving uv as well as ir regions with a good optical transmittance.

Because of their excellent optical properties and relatively good resistance to optical damage, fluoroberyllate glasses seem to be outstanding candidates for use in high-power neodymium lasers.

With a theoretical minimum of absorption losses of 10^{-3} dB km^{-1} at 3.6 μm and low attenuation due to impurities, especially transition cations, some fluorozirconate glasses and others related to them could constitute the second generation of glasses for optical fibers. These materials could be quite efficient once the fiber preparation difficulties resulting from recrystallization and small existence range of the glass formation domain were solved.

Unfortunately, despite high transparency in the ir region, there is always a threshold at which fluoride glasses become absorbing, which excludes ir applications, as in the CO_2 laser. New glass families, which are mixed halides, have been discovered, however. Despite higher hygroscopicity they seem to be promising for fiber and window applications at long wavelengths.

References

Abdrashitova, E. I., and Petrovskii, G. I. (1967). *Dokl. Akad. Nauk SSSR* **175,** 1305.

Aleksandrova, I. P., and Batsanova, L. R. (1972). *Zh. Strukt. Khim.* **13,** 232.

Aliaga, N., Fonteneau, G., and Lucas, J. (1978). *Ann. Chim. Sci. Nat.* **3,** 51.

Almeida, R. M., and Mackenzie, J. D. (1981). *J. Chem. Phys.* **74,** (11), 5954.

Almedia, R. M., and Mackenzie, J. D. (1982). *J. Non-Cryst. Solids* **51,** 187.

Angell, C. A., and Ziegler, D. C. (1981). *Mater. Res. Bull.* **16,** 279.

Baldwin, C. M., and Mackenzie, J. D. (1979a). *J. Am. Ceram. Soc.* **62,** 573.

Baldwin, C. M., and Mackenzie, J. D. (1979b). *J. Non-Cryst. Solids* **31,** 441.

Baldwin, C. M., and Mackenzie, J. D. (1980a). *J. Non-Cryst. Solids* **40,** 135.

Baldwin, C. M., and Mackenzie, J. D. (1980b). *J. Non-Cryst. Solids* **42,** 455.

Baldwin, C. M., Almeida, R. M., and Mackenzie, J. D. (1981). *J. Non-Cryst. Solids* **43,** 309.

Batsanova, L. R., Yur'ev, G. S., and Doronina, V. P. (1968). *Zh. Strukt. Khim.* **9,** 79.

Bendow, B., and Mitra, S. S. (1981). "Physics of Fiber Optics." Am. Ceram. Soc., Columbus, Ohio.

Bendow, B., Banerjee, P. K., Drexhage, M. G., Goltman, J., Mitra, S. S., and Moynihan, C. T. (1982). *J. Am. Ceram. Soc,* **65**(1), C8.

Blanzat, B., Bochon, L., Jorgensen, C. K., Reisfeld, R., and Spector, N. (1980). *J. Solid State Chem.* **32,** 185.

Brown, R. N., Bendow, B., Drexhage, M. G., and Moynihan, C. T. (1982). *Appl. Opt.* **21**(3), 361.

Cantor, S., Ward, W. T., and Moynihan, C. T. (1969). *J. Chem. Phys.* **50,** 2874.

Chalilev, V. D., Wasylak, J. P., Igitchanjan, J. G., Oganesjan, R. M., and Pogosjan, M. A. (1977). *Proc. Int. Congr. Glass, 11th, 1977* Vol. I, p. 133.

Cline, C. F., and Weber, M. J. (1979). *Wiss. Z. Friedrich-Schiller-Univ. Jena, Math. Naturmiss Reine* **28,** 351.

Coey, J. M. D., McEvoy, A., and Shafer, N. W. (1981). *J. Non-Cryst. Solids* **43,** 387.

Cook, L., and Mader, K. M. (1982). *J. Am. Ceram. Soc.* **655**(12), 597.

Cooper, A. R., Jr. (1982). *J. Non-Cryst. Solids* **49,** 1.

Counts, W. E., Roy, R., and Osborn, E. F. (1953). *J. Am. Ceram. Soc.* **36,** 12.

Drexhage, M. G., Moynihan, C. T., and Saleh, M. (1980). *Mater. Res. Bull.* **15,** 213.

Drexhage, M. G., El-Bayoumi, O. M., Moyniham, C. T., Bruce, A. T., Chung, K. M., Gavin, D. L., and Loretz, T. J. (1982a). *J. Am. Ceram. Soc.* **65**(10), C168.

Drexhage, M. G., El-Bayoumi, O. M., and Moynihan, C. T. (1982b). *SPIE Proc.* p. 320.

Drexhage, M. G., Bendow B., Brown, R. N., Banerjee, P. K., Lipson, H. G., Fonteneau, G., Lucas, J., and Moynihan, C. T. (1982c). *Appl. Opt.* **21**(6), 971.

Drexhage, M. G., El-Bayoumi, O. M., Lipson J., Moynihan, C. T., Bruce, A. J., Lucas, J., and Fonteneau, G. (1983). *J. Non-Cryst. Solids* **56,** 51.

Dubois, B., Portier, J., and Videau, J. J. (1984). *J. Optics* **15**(5), 351.

Dumbaugh, W. M., and Morgan, D. W. (1980). *J. Non-Cryst. Solids* **38/39,** 24.

Ehrt, D., Kraub, M., Erdmann, C., and Vogel, W. (1982). *Z. Chem.* **22**, 315.

Ehrt, D., Erdmann, C., and Vogel, W. (1983a). *Z. Chem.* **23**, 37.

Ehrt, D., Atzrodt, R., and Vogel, W. (1983b). *Wiss. Z.—Friedrich-Schiller-Univ. Jena, Math. Naturwiss. Reine* **32**, 509.

Evstrop'ev, K. K., Petrovskii, G. T., and Chalilev, W. D. (1971a). *C. R. Trav. Congr. Int. Verre, 9th, 1971* p. 485.

Evstrop'ev, K. K. Veksler, G. I., Kondrat'eva, B. S., Khalilev, V. D., and Kiryanova, T. N. (1971b). *Invent Certif.* No. 313, p. 794 (Byull. Izobret. No. 27).

Evstrop'ev, K. K., Ivanov, I. A., Kondrat'eva, G. I., Petrovskii, G. T., and Shvedov, V. P. (1972). *Zh. Prikl. Khim.* **45**, 1475.

Evstrop'ev, K. K., Ivanov, I. A., Petrovskii, G. T., and Pronkin, A. A. (1978). *Zh. Prikl. Khim.* **51**, 985.

Fonteneau, G., Lahaie, F., and Lucas, J. (1980a). *Mater. Res. Bull.* **15**, 1143.

Fonteneau, G., Slim, H., Lahaie, F., and Lucas, J. (1980b). *Mater. Res. Bull.* **15**, 1425.

Fonteneau, G., Slim, H., and Lucas, J. (1982). *J. Non-Cryst. Solids* **50**, 61.

Galeener, F. L., Lucovsky, G., and Geils, R. H. (1978). *Solid State Commun.* **25**, 405.

Gan Fuxi (1981). "Calculation of Properties and Design of Composition of Inorganic Glasses." Shangai Science Press.

Gannon, J. R. (1980). *J. Non-Cryst. Solids* **42**, 239.

Goldschmidt, V. M. (1926). *Skr. Nor. Vidensk.-Akad.* **8**, 127.

Grebenshchikova, N. I., and Petrovskii, G. T. (1963). *Zh. Prikl. Khim.* **36**, 1199.

Griscom, D. L., Stapebrook, N., and Weber, W. J. (1980) *J. Non-Cryst. Solids* **41**, 329.

Heyne, G. (1933). *Angew. Chem.* **46**, 473.

Hitch, B. F., and Baes, C. F. (1969). *Inorg. Chem.* **8**, 201.

Hu, H., and Mackenzie, J. D. (1983). *J. Non-Cryst. Solids* **54**, 241.

Imaoka, M. (1954a). *J. Res. Inst. Univ. Tokyo* **20**, 14.

Imaoka, M. (1954b). *Yogyo Kyokaishi* **62**, 24.

Imaoka, M., and Mizusawa, S. (1953). *Yogyo Kyokaishi* **61**, 21.

Jinguji, K., Horiguchi, M., Shibata, S., Kanamori, T., Mitachi, S., and Manabe, T. (1982). *Electron. Lett.* **16**(4). 164.

Kanamori, T. (1983). *J. Non-Cryst. Solids* **57**, 443.

Kanamori, T., Oikawa, K., Shibata, S., and Manabe, T. (1981). *Jpn. J. Appl. Phys.* **20**(5), 326.

Kolobkov, V. P., and Petrovskii, G. T. (1971). *Sov. J. Opt. Technol.* (*Engl. Transl.*) **38**(3), 175.

Kolobkov, V. P., Khalilev, V. D., Vasylyak, Y. P., Vakhrameev, V. I., Zhmyreva, I. A., and Kovaleva, I. V. (1977). *Fiz. Khim. Stekla* **3**(3), 249.

Kondrat'eva, B. S., and Petrovskii, G. T. (1967). *Russ. J. Inorg. Chem.* (*Engl. Transl.*) **12**, 1643.

Kondrat'eva, Y. N., Petrovskii, G. T., and Raaben, E. L. (1969). *Zh. Prikl. Spektrosk.* **10**, 69.

Leadbetter, A. J., and Wright, A. L. (1972). *J. Non-Cryst. Solids* **7**, 156.

Lecoq, A., and Poulain, M. (1980a). *Verres Refract.* **34**, 333.

Lecoq, A., and Poulain, M. (1980b). *J. Non-Cryst. Solids* **41**, 209.

Le Page, Y., Fonteneau, G., and Lucas, J. (1982). *Mater. Res. Bull.* **17**, 647.

Leroy, D., and Ravaine, D. (1978). *C. R. Acad. Sci. Paris. Ser. C* **286**, 413.

Leroy, D., Lucas, J., Poulain, M., and Ravaine, D. (1978). *Mater. Res. Bull.* **13**, 1125.

Lucas, J., Poulain, M., and Poulain, M. (1977). German Patent 2,726,170.

Lucas, J., Chanthanasinh, M., Poulain M., Brun, M., and Weber, M. J. (1978). *J. Non-Cryst. Solids* **27**, 273.

Lucas, J., Slim. J., and Fonteneau, G. (1981). *J. Non-Cryst. Solids* **44**, 31.

Mackenzie, J. D. (1960). *J. Chem. Phys.* **32**, 1150.

Margaryan, A. A., and Evstrop'ev, K. S. (1968). *Izv. Akad. Nauk. SSSR, Neorg. Mater.* **6**, 403.

Matecki, M., and Poulain, M. (1983). *J. Non-Cryst. Solids* **56**, 111.

Matecki, M., Poulain, M., Poulain, M., and Lucas, J. (1978). *Mater. Res. Bull.* **13,** 1039.
Matecki, M., Poulain, M., and Poulain, M. (1981). *Mater. Res. Bull.* **16,** 749.
Matecki, M., Poulain, M., and Poulain, M. (1982a). *Mater. Res. Bull.* **17,** 1035.
Matecki, M., Poulain, M., and Poulain, M. (1982b). *Mater. Res. Bull.* **17,** 1275.
Matecki, M., Poulain, M., and Poulain, M. (1983). *J. Non-Cryst. Solids* **56,** 81.
Maze, G., Cardin, V., and Poulain, M. (1982). OPTO **1,** 18.
Menil, F., Fournés, L., Dance, J. M., and Videau, J. J. (1979) *J. Non-Cryst. Solids* **34,** 209.
Miranday, J. P., Jacobini, C., and de Pape, R. (1979). *Rev. Chim. Miner.* **16,** 277.
Miranday, J. P., Jacobini, C., and de Pape, R. (1981). *J. Non-Cryst. Solids* **43,** 393.
Mitachi, S., and Miyashita, T. (1982). *Electron. Lett.* **18**(4), 170.
Miyashita, T., and Manabe, T. (1982). *IEEE J. Quantum Electron.* **QE-18,** 1432.
Moynihan, C. T., and Cantor, S. (1968). *J. Chem. Phys.* **48,** 114.
Murphy, M. K. (1963). *J. Am. Ceram. Soc.* **46,** 558.
Murphy, M. K., and Mueller, A. (1963). *J. Am. Ceram. Soc.* **46,** 530.
Narten, A. H. (1972). *J. Chem. Phys.* **56,** 1905.
Nemilov, S. V., Petrovskii, G. T., and Krylova, L. A. (1968). *Izv. Akad, Nauk SSSR, Neorg. Mater.* **4,** 1664.
Nikolina, G. P., Khalilev, V. D., and Evstrop'ev, K. S. (1970). *Ivz. Akad. Nauk. SSSR, Neorg. Mater.* **6,** 582.
Ohishi, Y., Mitachi, S., and Kanamori, T. (1981). *Jpn. J. Appl. Phys.* **20**(11), 787.
Ohmichi, T., Ohno, H., and Furukawa, K. (1976). *J. Phys. Chem.* **80,** 1628.
Ohno, H. Furukawa, K., Tsunawaki, T., Umesaki, N., and Iwamoto, N. (1978). *J. Chem. Res.* **5,** 158.
Osborn, E. F. (1952). *Ceram. Age* **60,** 36.
Patel, C. K. N. (1981). *SPIE Proc.* **266,** 22.
Petrovskii, G. T. (1967). *Zh. Neorg. Khim.* **12,** 485.
Petrovskii, G. T., Leko, E. K., and Mazurin, O. V. (1965). *In* "The Structure of Glass" (Mazurin, ed.), Vol. 4, p. 88. Consultants Bureau, New York.
Petrovskii, G. T., Urusovskaya, L. N., and Yudin, D. M. (1973). *Izv. Akad. Nauk SSSR, Neorg. Mater.* **9,** 1615.
Piriou, B., Videau, J. J., and Portier, J. (1981). *J. Non-Cryst. Solids* **46,** 105.
Poulain, M. (1981). *Nature* (*London*) **293,** 279.
Poulain, M. (1983). *J. Non-Cryst. Solids* **56,** 1.
Poulain, M., and Grosdemonge, M. A. (1982). *Verres Refract.* **36,** 853.
Poulain, M., and Lucas, J. (1978). *Verres Refract.* **32**(4), 505.
Poulain, M. A., and Poulain, M. I. (1983). *J. Non-Cryst. Solids* **56,** 57.
Poulain, M. A., Poulain, M. I., Lucas, J., and Brun, P. (1975). *Mater. Res. Bull.* **10,** 243.
Poulain, M., Chanthanasinh, M., and Lucas, J. (1977a). *Mater. Res. Bull.* **12,** 151.
Poulain, M., Lucas, J., Brun, P., and Driffort, M. (1977b). *Colloq. Int. C. N. R. S.* **255,** 257.
Poulain, M. A., Poulain, M. I., and Matecki, M. (1981a). *Mater. Res. Bull.* **16,** 555.
Poulain, M. A., Poulain, M. I., and Maze, G. (1981b). European Patent 4,004,090.
Poulain, M. A., Poulain, M. I., and Matecki, M. (1982). *J. Non-Cryst. Solids* **51,** 201.
Poulain, M. A., Poulain, M. I., and Matecki, M. (1982b). *Mater. Res. Bull.* **17,** 661.
Pronkin, A. A., Igitchanjan, J. G., Chalilev, V. D., Zavukijan, S. G., and Tarlakov, J. P. (1978). *Fiz. Khim. Stekla* **4,** 118.
Pye, L. D., Cherukeri, S. C., Mansfield, J., and Loretz, T. (1983). *J. Non-Cryst. Solids* **56,** 99.
Quist, A. S., Bates, J. B., and Boyd, G. E. (1972). *Spectrochim. Acta Part A* **28A,** 1103.
Ravaine, D., Perera, W. G., and Minier, M. (1982). *J. Phys. Colloq.* (*Orsay, Fr.*) **C9,** Supp. 12(43), 407.
Robinson, M., and Drexhage, M. G. (1983). *Mater. Res. Bull.* **18,** 1101.

Roy, D. M., Roy, R., and Osborn, E. F. (1950). *J. Am. Ceram. Soc.* **33,** 85.

Shibata, S., Kanamori, T., Mitachi, S., and Manabe, T. (1980). *Mater. Res. Bull.* **15,** 129.

Shibata, S., Horiguchi, M., Jinguji, K., Mitachi, S., Kanamori, T., and Manabe, T. (1981). *Electron, Lett.* **17,** 775.

Simmons, C. J., Sutter, H., Simmons, J. H., and Tran, D. C. (1982). *Mater. Res. Bull.* **17,** 1203.

Stević, S., Videau, J. J., and Portier, J. (1978). *Rev. Chim. Miner,* **15,** 529.

Stević, S., Radosaljević, S., and Poleti, D. (1982). *Rev. Chim. Miner.* **19,** 192.

Stokowski, S. E., Saroyan, R. A., and Weber, M. J. (1978). "Nd-doped Laser Glass Spectroscopy and Physical Properties." Lawrence Livermore Laboratory, Livermore, California.

Sun, K. H. (1949a). U. S. Patent 2,466,507.

Sun, K. H. (1949b). U. S. Patent 2,466,508.

Sun, K. H. (1949c). U. S. Patent 2,466,509.

Sun, K. H., and Callear, T. E. (1949). U. S. Patent 2,466,506.

Sun, K. H., and Huggins, M. L. (1950). U. S. Patent 2,511,224.

Tran, D. C., Guither, R. J., and Sigel, G. M., Jr. (1982). *Mater. Res. Bull.* **17,** 1177.

Urusovskaya, L. N. (1968a). *Opt.-Mekh.Prom-st.* **35**(7), 41.

Urusovskaya, L. N. (1968b). *Zh. Prikl. Khim.* **41,** 2361.

Vallet, C. E., and Braunstein, J. (1975). *J. Am. Ceram. Soc.* **58,** 209.

Veksler, G. I., Evstrop'ev, K. K., and Kondrat'eva, B. S. (1974). *Izv. Akad. Nauk SSSR, Neorg. Mater.* **10,** 171.

Videau, J. J. (1979). Thesis, Univ. of Bordeaux, France.

Videau, J. J., Portier, J., and Fouassier, C. (1979a). *Mater. Res. Bull.* **14,** 177.

Videau, J. J., Fava, J., Fouassier, C., and Hagenmuller, P. (1979b). *Mater. Res. Bull.* **14,** 499.

Videau, J. J., Portier, J., and Piriou, B. (1979c). *Rev. Chim. Miner.* **16,** 393.

Videau, J. J., Portier, J., and Piriou, B. (1982). *J. Non-Cryst. Solids* **48,** 385.

Videau, J. J., Portier, J., Tanguy, B., and Stević, S. (1983a). *Glass Technol.* **24**(3), 171.

Videau, J. J., Dubois, B., and Portier, J. (1983b). *C. R. Hebd. Seances Acad. Sci.* **297**(II), 483.

Vogel, W., and Gerth, K. (1958a). *Glastech. Ber.* **31,** 15.

Vogel, W., and Gerth, K. (1958b). *Silikattechnik* **9,** 353.

Vogel, W., and Gerth, K. (1958c). *Silikattechnik* **9,** 495.

Volkov, O. S., Novozhilova, L. D., Sizova, E. D., and Urusovskaya, L. N. (1974). *Izv. Akad. Nauk SSSR, Neorg. Mater.* **10**(12), 2249.

Walrafen, G. E., and Stolen, R. H. (1978). *Solid State Commun.* **21,** 417.

Weber, M. J. (1976). *In* "Critical Materials Problems in Energy Production" (C. Stein, ed.), Chapter 8, p. 261. Academic Press, New York.

Weber, M. J. (1982). "Faraday Rotator Materials." Lawrence Livermore Laboratory, Livermore, California.

Weber, M. J., Layne, C. B., Saroyan, R. A., and Milam, D. (1976). *Opt. Commun.* **18,** 171.

Willians, D. J. (1959). *J. Soc. Glass Technol.* **43,** 308.

Wong, J., and Angell, C. A. (1976). "Glass Structure by Spectroscopy." Dekker, New York.

Yasi, J., Fusong, J., and Gan Fuxi (1982). *J. Phys. Colloq.* (*Orsay. Fr.*) *C9*, 5, 12(43), 315.

Yudin, D. M., Tsurikova, G. A., Petrovskii, G. T., and Starostin, N. V. (1968). *Opt. Spectrosc.* **24**(6), 518.

Zachariasen, W. H. (1932). *J. Am. Chem. Soc.* **54,** 3841.

Zarzycki, J. (1971). *Phys. Chem. Glasses* **12,** 97.

8

Fluorine Intercalation Compounds of Graphite[†]

NOBUATSU WATANABE, HIDEKAZU TOUHARA,
and TSUYOSHI NAKAJIMA
Department of Industrial Chemistry
and Division of Molecular Engineering
Kyoto University
Kyoto, Japan

NEIL BARTLETT
Department of Chemistry
University of California
Berkeley, California

THOMAS MALLOUK
Materials and Molecular Research Division
Lawrence Berkeley Laboratory
Berkeley, California

H. SELIG
Department of Inorganic and Analytical Chemistry
The Hebrew University
Jerusalem, Israel

[†] Section I by Watanabe, Touhara, and Nakajima, Section II by Bartlett and Mallouk, and Section III by Selig.

I. Graphite Fluorides and Graphite Intercalation Compounds of Fluorine and Metal Fluorides

A. INTRODUCTION

Graphite fluoride is a nonstoichiometric solid fluorocarbon of empirical formula CF_x ($0.5 < x \leq \sim 1.25$), the value x strongly depending on the fluorination temperature of the carbon material. Two kinds of graphite fluoride have been prepared. They are called poly(carbon monofluoride) and poly(dicarbon monofluoride), $(CF)_n$ and $(C_2F)_n$, respectively; they differ in structure. The materials are covalent compounds of graphite; in a somewhat broader sense, however, they can be regarded as intercalation compounds because of the layer structure, which is derived from graphite by insertion of covalently bound fluorine atoms.

Poly(carbon monofluoride) has been known since 1934, when Ruff *et al.*, prepared a gray compound of composition $CF_{0.92}$. Later, Rüdorff *et al.* published several papers (Rüdorff and Rüdorff, 1947a; Rüdorff and Brodersen, 1957) on fluorides the composition range of which varied from $CF_{0.676}$ to $CF_{0.98}$. These were obtained by reaction of elemental fluorine with graphite at temperatures of between 410 and 550°C. Later, Palin and Wadsworth (1948) prepared $CF_{1.04}$ by a similar method.

Poly(tetracarbon monofluoride) was first prepared by Rüdorff and Rüdorff (1947a) by the reaction of a mixture of fluorine–hydrogen fluoride with graphite and later by Lagow *et al.* (1974). The material is a black solid, is unstable above 100°C in air, and contains a certain amount of residual unsaturated carbon atoms, which is the basis for the difference between this material and the highly stable white solids $(CF)_n$ or $(C_2F)_n$.

Despite these pioneering works, systematic studies on graphite fluorides were not carried out before the 1960s, presumably because their important characteristics had not been recognized and because no application was apparent at that time. Our own interest in graphite fluorides originated from works on the notorious "anode effects" (e.g., see Ring and Royston, 1973) associated with the production of elemental fluorine or aluminum by electrolysis of molten salts. The mechanism of the anode effect had been debated since 1893, but a satisfactory theory was not developed until Watanabe *et al.*, found in 1963 that the compound formed on the carbon anode surface was graphite fluoride and that its extremely low surface energy was responsible for this troublesome phenomenon (Watanabe *et al.*, 1963a,b,c). Since that time interest in the compound has led to systematic research on the fluorination reaction of carbon materials, crystal structure, physicochemical properties, and applications of graphite fluoride (Watanabe and Nakajima, 1982) and has resulted in a new graphite fluoride, $(C_2F)_n$ (Kita *et al.*, 1979). Margrave's group at Rice University has also been active in fundamental investigations of graphite fluorides (Lagow *et al.*, 1974; Kammarchik and Margrave, 1978).

Because comprehensive reviews on these materials have been published (Watanabe and Nakajima, 1982; Selig and Ebert, 1980; Kammarchik and Margrave, 1978; Watanabe *et al.*, 1974), this chapter concentrates on investigations performed since the discovery of $(C_2F)_n$ up to the present, particular emphasis being given to advances in characterization, electrochemical applications, and surface properties. In addition to graphite fluoride, a group of graphite intercalation compounds of fluorine with metal fluorides, $C_xF(MF_n)_y$ ($MF_n = LiF$, MgF_2, AlF_3, or CuF_2), is also described. These compounds are extremely stable even in moist air and have important potential applications as cathodes or conductive materials.

B.   GRAPHITE FLUORIDES

1.   Preparation and Characterization

Preparative methods are given in detail elsewhere; these involve direct fluorination of graphite (Watanabe and Nakajima, 1982; Kammarchik and Margrave, 1978) using a fluidized bed, a static bomb, and a rotating reactor. The chemical composition, crystal structure, and crystallinity depend primarily on the reaction temperature; crystallinity also depends on that of the starting carbon. The fluorination reaction is strongly exothermic (Wood *et al.*, 1969), so that careful temperature control and also relevant position of the sample are required to achieve reproducible results without combustion or thermal decomposition.

The results described here concern mainly natural graphite from Madagascar (Kita *et al.*, 1979) because its high crystallinity gives a relatively good crystallized graphite fluoride, which is convenient for characterization. The composition and color of graphite fluoride are a function of the reaction temperature. The value of x in $(CF_x)_n$ increases with rising temperature, the color changing from black, through gray, to white (the material may also be transparent) above 600°C under 1 atm fluorine pressure. It should be noted that the higher x value of graphite fluoride prepared at 375°C no longer increases even after further treatment at 600°C in fluorine atmosphere, while its color changes from black to white. The same is true for graphite fluorides with x values of less than 1. On the basis of these results, together with x-ray and esca analyses, the formation reaction and the characterization of two graphite fluorides, $(CF)_n$ and $(C_2F)_n$, can be summarized as follows:

1. Only $(CF)_n$ is formed in the temperature range 600–640°C by the following reaction: $2n\ C(s) + n\ F_2(g) \rightarrow 2\ (CF)_n(s)$.
2. Graphite fluoride prepared at 350 to 400°C is essentially $(C_2F)_n$, according to the following reaction: $4n\ C(s) + n\ F_2(g) \rightarrow 2\ (C_2F)_n(s)$.
3. Nonstoichiometric graphite fluoride $(CF_x)_n$ $(0.5 < x < 1)$ forms in the temperature range 400–600°C; the material is actually a mixture of $(CF)_n$ and $(C_2F)_n$. The difference in the value of x between the observed and theoretical stoichiometries can be ascribed to the existence of some CF_2 and CF_3 groups at grain boundaries or crystal lattice defects.
4. In addition to this phenomenon, $(C_2F)_n$ has its own features that increase the value of x. In considering the formation mechanism of $(C_2F)_n$, it has been reasonably concluded that some $(CF)_n$ layers are forced to form on the surface of crystal grains, which is responsible for x values higher than that of the theoretical stoichiometry of $(C_2F)_n$. This factor becomes more significant with increasing size of the crystal grains (Touhara *et al.*, 1983a).

These results have been roughly confirmed by investigations of fluorination reactions of other forms of carbon, but it should be noted that the formation temperature is lowered when the crystallinity of the starting carbons is low; for example, $(CF)_n$ is formed at 400°C by fluorination of petroleum coke (Kita *et al.*, 1979). No residual graphite or unsaturated carbon must be detected, at least by x-ray diffraction, for a good characterization of graphite fluoride.

2. Nature of the C—F Bond and Structure

a. NATURE OF THE C—F BOND.   Graphite fluoride is a covalent compound. The valence bond structure of an isolated two-dimensional infinite layer of $(CF)_n$ was calculated by a modified extended Hückel model (Perry *et al.*, 1974; Cadman *et al.*, 1975), and a good agreement between theoretically computed and experimentally measured density of states was shown for $(CF_{1.13})_n$. The polarization of the C—F bond was found to be

$q_C = q_F = 0.4|e|$. The esca and solid-state ^{13}C-nmr spectra confirm the covalent C—F bonding in graphite fluorides (Kita *et al.*, 1979; Clark and Peeling, 1976; Watanabe *et al.*, 1979; Wilke *et al.*, 1979). The aromatic planarity of the previous graphite layers disappears by formation of the covalent C—F bonds, and buckled, sp^3-hybridized sheets are created. In the case of $(C_2F)_n$, in addition to C—F covalent bonds, C—C covalent bonds are also formed between adjacent carbon layers. Concerning the nature of the bonding, the only difference is observed by broad-band ^{19}F-nmr spectroscopy between $(CF)_n$ and $(C_2F)_n$. The fluorine resonance absorption line of $(C_2F)_n$ consists of broad and narrow lines with linewidths of 8.5 and 2.0 G, respectively. This shows that a small number of fluorine atoms are rather weakly combined with carbon atoms in $(C_2F)_n$ at grain boundaries or lattice defects. The C—F bond energy of $(CF)_n$ is reported to be ~ 480 kJ mol^{-1} (Wood *et al.*, 1969).

b. STRUCTURE. Although it is buckled, the layered structure remains after fluorination, as clearly shown in Fig. 1, where the concentric shape of the original graphite fiber is also preserved around the fiber axis (Touhara *et al.*, 1983a). Figure 2 shows the $(00\,l)$ lattice image for $(CF)_n$ prepared from natural graphite. Noteworthy is the observation of an interlayer *d* spacing of ~ 6 Å, comparable with $d(00\,l)$ spacing determined by x-ray diffraction. Note that the

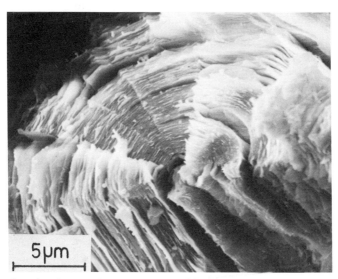

Fig. 1. Scanning electron micrograph showing the layer structure of graphite fluoride $(CF)_n$ prepared by fluorination of highly ordered graphite fibers at 614°C. Concentric layers around the fiber axis are also shown.

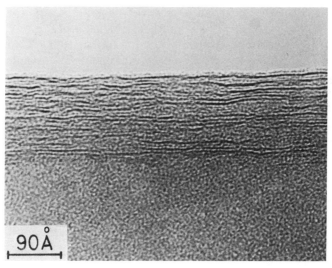

90 Å

Fig. 2. Bright field image of a (00*l*) plane of graphite fluoride (CF)$_n$ prepared by fluorination of natural graphite from Madagascar ores at 600°C.

lattice rays become much weaker than those of the original graphite, because a large strain accompanies the fluorine diffusion process between the graphite layers and also at the reaction interface, bringing about mechanical destruction of the crystal grains of the starting graphite. The same phenomenon is observed for $(C_2F)_n$ by high-resolution transmission electron microscopy (Touhara *et al.*, 1983a).

Possible structures have already been proposed for both (CF)$_n$ (Kita *et al.*, 1970a; Mahajan *et al.*, 1974; Ebert *et al.*, 1974; Takashima and Watanabe, 1975) and $(C_2F)_n$ (Kita *et al.*, 1979; Watanabe, 1980), but there remains some ambiguity because determining the structure of graphite fluorides is a difficult task, mainly due to the formidable problems of obtaining single crystals as mentioned earlier. On the basis of structural data from x-ray powder diffraction, electron diffraction, and nmr and infrared (ir) spectra, however, the structure of (CF)$_n$ has been proposed to be a layer-type lattice of weakly coupled nongraphitic sheets, each formed by an infinite array of trans-linked cyclohexane-type chairs in which fluorine atoms occupy the fourth bonding site on each of the sp^3-hybridized carbons. The hexagonal lattice parameters for (CF)$_n$ prepared from natural graphite at 600°C are $a = 2.57$ Å and $c = 5.85$ Å, in which the c parameter depends slightly on the crystallinity of the original carbon material and the fluorination temperature.

The structure of $(C_2F)_n$ has been supposed to be of hexagonal symmetry, with $a = 2.5$ Å and $c = 8.2$ Å obtained from the interpretation of structural data of x-ray diffraction, as well as ^{19}F-nmr and ir spectra; the unit cell contains four carbon and two fluorine atoms. The successive layers along the c axis consist of trans-linked cyclohexane rings in which three carbons combine with those of adjacent cyclohexane rings by diamondlike sp^3 bonds perpendicular to the a axis, and three others combine with fluorine atoms. This structure is not quite conclusive, because such features as stacking sequences of layers and space-group symmetry have not been examined. A study in progress to refine the structure is based on detailed structural analyses by x-ray diffraction photographs, calculation of structure factor, and determination of nmr second moments (Touhara et al., 1984a).

Fluoride $(CF)_n$ can be considered a first-stage compound from the structural viewpoint of graphite intercalation compounds, whereas $(C_2F)_n$ is a second-stage compound.

3. Properties and Uses

a. LITHIUM BATTERIES BASED ON $(CF)_n/Li$ and $(C_2F)_n/Li$ SYSTEMS. A primary battery based on a fluorine and lithium combination is theoretically the best one from the standpoint of the high-energy-density power source. This possibility has been realized by the use of $(CF)_n$ as a cathode material for lithium batteries. Because an extensive original and review literature is available (Braeuer, 1968; Watanabe and Fukuda, 1970, 1972; Watanabe and Nakajima, 1982; Watanabe, 1980), only the main and more recently developed aspects are mentioned here.

$(CF)_n/Li$-based dry batteries have been developed successfully by the Matsushita Company, Osaka, Japan (Watanabe and Nakajima, 1982) and are now used on a large scale. The working voltage of a $(CF)_n/Li$ cell is about twice that of a carbon/zinc cell, and the energy density is ~ 300 W hr kg^{-1} (~ 550 W hr liter^{-1}), which is only $\sim 15\%$ of the theoretical energy density but still five times that of a carbon/zinc cell. The cell has other outstanding characteristics, such as a flat discharge potential and long shelf life over a wide temperature range (-20 to $60°C$).

The new graphite fluoride $(C_2F)_n$ has received attention (Watanabe, 1980). Figure 3 gives the discharge characteristics of graphite fluorides in 1 M LiClO$_4$–propylene carbonate at $20°C$ (Watanabe et al., 1983a). Samples were prepared from natural graphite from Madagascar ores at different temperatures in the range 380–600°C. They have different compositions, and the $(C_2F)_n$ content in each sample increases with decreasing fluorine/carbon ratio. It should be noted that the sample, the fluorine/carbon atomic ratio of which is 0.6, is essentially $(C_2F)_n$. Obviously, the discharge voltage increases with increasing $(C_2F)_n$ content, and the $(C_2F)_n$ electrode shows a discharge voltage of 2.4 V versus Li/Li$^+$, which is higher by 0.4 V than that obtained for $(CF)_n$

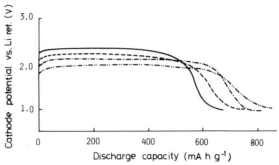

Fig. 3. Discharge characteristics of graphite fluorides with different composition in 1 M LiClO$_4$–propylene carbonate, 0.5 mA cm^{-2}. Key: ——, CF$_{0.62}$; ---, CF$_{0.70}$; ····, CF$_{0.90}$; ─··─··, CF$_{1.00}$.

at the current density of 0.5 mA cm^{-2}. The higher discharge potential of $(C_2F)_n$ can be attributed to the lower overpotential of its electrode, because the open circuit voltage (ocv) values for each graphite fluoride are almost similar (i.e., in the range 3.2–3.3 V).

The superiority of $(C_2F)_n$ is more clearly shown in Fig. 4, which gives galvanostatic polarization curves for similar samples. The discharge potential of the $(C_2F)_n$ electrode is considerably higher than that of $(CF)_n$ for current densities between 10^{-5} and 10^{-6} A cm^{-2} but becomes almost the same as that of $(CF)_n$ at very low current densities ($<5 \times 10^{-7}$ A cm^{-2}). The presence of appropriate graphite fibers due to the use of a new starting graphite, which results from a benzene-derived precursor material, can yield $(C_2F)_n$ at relatively lower fluorination temperature, $\sim 350°$C (Touhara *et al.*, 1983a).

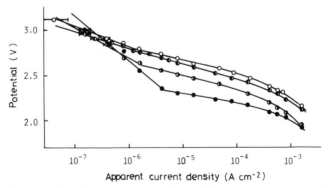

Fig. 4. Galvanostatic polarization curves of graphite fluoride electrodes with different compositions in 1 M LiClO$_4$–propylene carbonate. Key: ○, CF$_{0.62}$; ◐, CF$_{0.70}$; ◑, CF$_{0.90}$; ●, CF$_{1.00}$.

The electrode based on this $(C_2F)_n$ gives the higher ocv, 3.3–3.5 V, and a discharge potential of 2.4 to 2.5 V (Touhara et al., 1983b). It is thus obvious that $(C_2F)_n$ is potentially a new cathode material for primary cells with lithium anodes.

The discharge reaction mechanism for $(CF)_n$/Li and $(C_2F)_n$/Li cells has been investigated (Watanabe et al., 1983b; Whittingham, 1975; Touhara et al., 1984b. The ocv for the $(CF)_n$/Li cell is ~ 3.2 V, which is appreciably less than the value of the electromotive force (emf), 4.57 V, calculated from the free energies of formation of $CF_{1.0}$ and LiF (Wood et al., 1969). The difference in the values of and emf has been interpreted as being the result of the formation of a nonstoichiometric phase during discharge of the graphite fluoride cathode. Intermediate compounds formed during discharge have been investigated in detail by x-ray diffraction, esca, nmr, and transmission electron microscopy, and it has been revealed that the discharge reaction proceeds through the formation of an intermediate solvated ternary compound of composition $(C \cdot F \cdot Li) \cdot PC$ or $(C_2 \cdot F \cdot Li) \cdot PC$ in the $LiClO_4$–propylene carbonate system. They decompose later, leading to graphite with low crystallinity and to LiF (Watanabe et al., 1982a; Touhara et al., 1984b. The discharge reaction can be described as follows:

Anode reaction: $Li + z\,PC \longrightarrow Li^+ \cdot zPC + e^-$

Cathode reaction: $CF + Li^+ \cdot zPC + e^- \longrightarrow (C \cdot F \cdot Li) \cdot PC \longrightarrow C + LiF + z\,PC$

The decomposition process is not involved in the electrochemical reaction, which is why the ocv value is lower than that of the emf. Studies of the influence of solvent on the electrochemical behavior and the change in thermodynamic functions of the discharge reactions also support a model involving the formation of an intermediate phase (Ueno et al., 1982; Watanabe et al., 1982).

b. SURFACE PROPERTIES.   The surface properties of graphite fluorides $(CF)_n$ and $(C_2F)_n$ have been investigated by calorimetric, contact angle, and water adsorption isotherm methods (Watanabe and Touhara, 1981). The heats of immersion ΔH_i for fine-powder graphite fluorides are given in Table I together with those of Teflon and graphite (Watanabe and Touhara, 1981). The $-\Delta H_i$ of $(CF)_n$ is practically independent of the polarity of wetting liquids and lower than that of Teflon. This means that the interaction of the surface with a liquid is due entirely to dispersion forces. Surface energy values ϵ_s at 298 K are evaluated by the following equation:

$$\epsilon_s = (\epsilon_1 - \Delta H_i)^2 / 4\epsilon_1 \Phi^2,$$

where ϵ_1 and Φ are, respectively, the surface free energy of a liquid and a constant, both characteristic of a given system. The value used for Φ is

TABLE I

Heats of Immersion and Surface Energies of Graphite Fluorides, Teflon, and Graphite at 298 K

| Solvent | $-\Delta H_i$ (ergs cm^{-2}) | | | | ϵ_1 (ergs cm^{-2}) | ϵ_s (ergs cm^{-2}) | | | |
| | | | | | | $\Phi = 0.95$ | | | $\Phi = 1.0$ |
	$(CF)_n$	$(C_2F)_n$	Teflon	Graphite		$(CF)_n$	$(C_2F)_n$	Teflon	Graphite
1-Butanol	36	54	56	114	55.3	42	59	62	129
Methanol	42	—	—	102	51.0	47	—	—	131
Cyclohexane	32	—	58[a]	—	60.7[a]	39	—	65	107
CCl$_4$	35	54	—	115	63.0	42	60	—	126
1-Nitropropane	—	—	54	—	60.5	—	—	60	—
1-Chlorobutane	—	63	56	106	57.4	—	69	62	116
Water	<0	>0	6	32	119.8	—	—	—	—

[a] Value in n-heptane.

0.95, which is the average value calculated from data on fluorocarbon–hydrocarbon systems. The surface free energy of $(C_2F)_n$ is of the same order of magnitude as that of Teflon, whereas the value of $(CF)_n$ is lower. Graphite fluoride may have the lowest surface energy among hydrophobic solids, owing to the high density of covalently bound fluorine atoms per unit surface area, which is confirmed by measurements of contact angles for various liquids (Watanabe and Nakajima, 1982).

c. LUBRICANT PROPERTIES. The lubricant properties of graphite fluorides have been examined in detail (Kammarchik and Margrave, 1978). For a solid lubricant, the characteristics desired are thermal stability, low shear strength, lamellar crystal structure, surface protection, and surface adherence. Graphite fluoride, either $(CF)_n$ or $(C_2F)_n$, has all these properties except the last and lubricates as well in dry as in moist air. Its use as an additive in conventional grease, mechanical carbons, or Teflon and related composites suppresses the increase in surface temperature due to friction, which allows use under high pressure and high-temperature conditions. A typical example of the comparative performances of solid lubricants is shown in Fig. 5 (Hiratsuka et al., 1966). The influence of a $(CF)_n$ addition (10 wt %) is remarkable in every respect (weight loss, temperature increase, and friction coefficient).

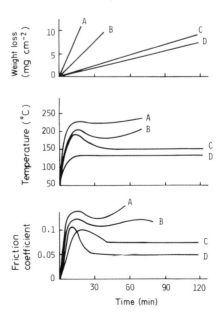

Fig. 5. Influence of the addition of solid lubricant (10 wt %) to a phenol resin at a PV value of 882 kg cm^{-2} min^{-1}. A, Graphite; B, Teflon; C, MoS$_2$; D, $(CF)_n$.

C. GRAPHITE INTERCALATION COMPOUNDS OF FLUORINE AND METAL FLUORIDES[†]

1. Preparation and Analyses

Intercalation compounds were prepared in ternary systems of graphite, fluorine gas, and metal fluoride (LiF, MgF_2, AlF_3, or CuF_2) below $300°C$, where the reaction of fluorine with graphite is so slow that graphite fluoride is not formed. Carbon materials used as hosts were flaky natural graphite, highly oriented pyrolytic graphite (HOPG), graphitized petroleum coke, and PAN-based carbon fibers. The purity of fluorine gas was 99.4–99.7% (N_2: 0.3–0.6%; HF < 0.01%). It was confirmed before the experiments that no intercalation compound was formed in the two-component system consisting of graphite/fluorine and graphite/metal fluoride. The reaction system is easily contaminated by traces of hydrogen fluoride, which is probably produced by reaction of fluorine gas with some water adsorbed on the metal fluoride and/or on the internal wall of the reactor. Hydrogen fluoride is intercalated into graphite with fluorine if it is not removed. The presence of traces of hydrogen fluoride can be checked by ir spectra.

x-Ray diffractometry indicated that the periodicity along the c axis was 9.4 and 12.7 Å. Graphitized petroleum coke and HOPG intercalated fluorine with a small amount of metal fluoride more easily than other carbon materials, giving x-ray diffraction patterns in which the (003) and (004) diffraction lines had the highest intensities for the compounds with periods of 9.4 and 12.7 Å, respectively. This suggests that they are second- and third-stage compounds.

From the chemical analysis, the composition was described as $C_{8-20}F(MF_n)_{0.01-0.0001}$. The esca peak of the F_{1s} electron appeared at 688 eV for the surface, which is near the position of the F_{1s} peak of graphite fluoride and at 686 eV for the bulk of the compound. The chemical bond between carbon and intercalated fluorine is nearly covalent near the surface but slightly ionic in the bulk. The ^{19}F-nmr spectra showed narrow derivatives, indicating the presence of mobile fluorine atoms. Intercalation of fluorine with metal fluoride was observed between $\sim 100°C$ and room temperature. Because pure fluorine gas cannot be intercalated into graphite, the reaction of fluorine with metal fluorides would eventually produce intermediate fluorides of higher oxidation state, which would be first intercalated and would then act as catalysts for further intercalation of fluorine. It was very difficult to detect a new chemical species produced from fluorine and lithium fluoride at room temperature; when the temperature increased, however, lithium fluoride melted around $530°C$ in fluorine atmosphere.

[†] See Nakajima *et al.* (1981, 1982, 1983).

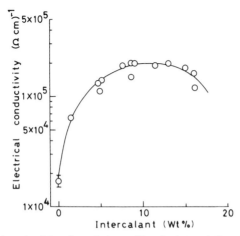

Fig. 6. Electrical conductivity of pyrolytic graphite after intercalation of fluorine and lithium fluoride at $22 \pm 2°C$.

2. Electrical Conductivity

The electrical conductivity along the c axis was measured by means of a contactless bridge for the HOPG host material (Fig. 6). The electrical conductivity grew with increasing intercalation and reached a maximum, $2.0 \times 10^5 \, \Omega^{-1} \, cm^{-1}$, for ~ 10 wt % intercalant. The value of this maximum was higher by one order of magnitude than that of HOPG. The compound obtained from PAN-based carbon fibers gave $8-9 \times 10^3 \, \Omega^{-1} \, cm^{-1}$ as the highest conductivity, which was about seven times that of the original fiber. The material made from carbon fibers showed high stability in air. The conductivity decrease in air was only 7% after several months.

3. Electrochemical Behavior

This type of intercalation compound discharges as a cathode in a lithium battery with 1 M LiClO$_4$–propylene carbonate solution as the electrolyte. Open-circuit voltage was between 3.4 and 4 V relative to the lithium reference electrode. Discharge potentials were in the range 2.8–2.5 V at current densities of 40 to 400 μA cm^{-2} (i.e., higher than those of graphite fluorides below 300 μA cm^{-2}) (Fig. 7). Utility decreased to $\sim 50\%$ with increasing current density. The x-ray diffraction pattern and esca spectra of discharged products showed that the discharge reaction was due to the intercalation of Li$^+$ into the cathode, with the formation of lithium fluoride:

$$C_x F(MF_n)_y + Li \longrightarrow C_x(LiF)(MF_n)_y$$

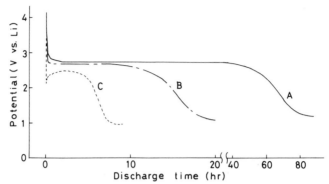

Fig. 7. Galvanostatic discharge curves of $C_{7.3}F(MgF_2)_{0.002}$. A, 40 μA cm^{-2}; B, 160 μA cm^{-2}; C, 400 μA cm^{-2}.

II. Graphite Fluorides Prepared in the Presence of Hydrogen Fluoride

A. INTRODUCTION

Although elemental fluorine does not interact with graphite unless the temperature is 375°C or higher (see Section I) Rüdorff and Rüdorff (1947b) discovered that in the presence of gaseous hydrogen fluoride, at room temperature, graphite is spontaneously intercalated by fluorine. They described the product as "tetracarbon monofluoride" but observed a composition range $C_xF_{1-\delta}(HF)_\delta$ (3.57 < x < 4.03). They assumed that δ was small. Lagow and Margrave and co-workers (Lagow *et al.*, 1972, 1974) determined that δ lay in the range 0.27–0.34, but other authors (Wilke *et al.*, 1979; Clarke and Peeling, 1976; Whittingham, 1975; Brauer and Moyes, 1970) referred to the material simply as "C_4F." The Rüdorffs noted the preservation of the graphite sheet structure and proposed a structure in which fluorine ligands are linked to the carbon in an ordered array. The diffraction data are sparse, however, and one cannot be certain of the structure.

Mallouk and Bartlett (1983) found that the composition range for the first-stage product of the interaction of graphite with fluorine (in the presence of hydrogen fluoride) is greater than previously recognized, with x in the formula $C_xF_{1-\delta}(HF)_\delta$ varying between 1.95 and ~6. These first-stage materials are preceded by second- and higher-stage hydrofluorides. The latter are salts of general formula $C_x{}^+F(HF)_y{}^-$. Such salts were not found in first-stage form.

It seems that graphite is not intercalated by fluorine alone, because the small fluorine ligand cannot catenate (to form species such as $F_3{}^-$) and does bond strongly to carbon atoms. There is, then, no mechanism for fluorine migration into the graphite galleries. The formation of hydrofluorides such as (F—H—F—H—F)$^-$ and (F—H—F)$^-$ provides a larger species that is less likely to interact strongly with a carbon atom, so reducing the kinetic barrier. In addition, vibrations within the hydrofluoride species cause variations in the fluorine ligand charging and so provide for the movement of the species relative to the sheets of carbon. It seems probable that the first-stage hydrofluorides do not exist, because the localization of positive charges at carbon by hydrofluoride ions on both sides of the atom causes a loss of fluoride ion to that positive carbon, thus establishing a C—F bond.

B. GRAPHITE DIFLUORIDE, $C_{11-14}^{+}HF_2^{-}$, AND RELATED SALTS

1. Preparation

When graphite (HOPG chips or powder or SP $-$ 1 powder) is shaken together with fluorine (2 atm) in the presence of excess anhydrous, liquid hydrogen fluoride, immediate fluorine uptake and a slight warming of the hydrogen fluoride occurs. The reaction can be followed tensimetrically: when 1 atom of fluorine has been taken up for every 18 to 20 carbon atoms, a pure third-stage compound, $C_{\sim 18}HF_2$, is obtained after the removal of volatiles. With more fluorine uptake, the pure second-stage compound of composition $C_{11-14}HF_2$ is formed. Hydrogen analysis consistently gives hydrogen/fluorine ratios in the range 0.46–0.60 for such preparations.

2. Crystal Structure

The structure of $C_{12}HF_2$ is depicted in Fig. 8. Fluorine atoms lie nearly midway between the enclosing carbon sheets, the elevation of the fluorine atoms above the nearest carbon atom plane being about 2.8 Å. Because the sum of the van der Waals' radii of carbon and fluorine is about 3.0 Å, the structures establish that the ionic formulation $C_{12}^{+}HF_2^{-}$ is appropriate. The observed d spacings require that the c axis of the cell be 19.47 Å, twice that of the primitive cell required by the $(00l)$ data. The ordering of the $(hk0)$ intensities is the same as in hexagonal graphite, which suggests that the superlattice is a consequence of the carbon layer stacking AB/BA/AB... (the stacking sequence is ABAB... in hexagonal graphite and ABCABC... in rhombohedral graphite). This sequence provides an equal number of A and B layers in the $(hk0)$ projection, and also accounts for the doubled unit cell with systematically absent $(00l)$, $l \neq 2n$ and (hhl), $l \neq 2n$. The fluorine atom array is likely to be highly disordered in the ab plane.

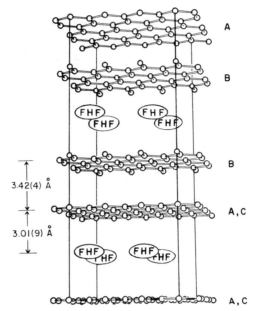

Fig. 8.   Crystal structure of graphite bifluoride $C_{12}^{+}HF_2^{-}$.

3. Electrical Conductivity of C_xHF_2

The *ab* plane electrical conductivities of several graphite bifluoride samples are presented in Table II. The per-plane conductivity for second-stage samples is only 1.4–2.0 times that of the pristine graphite and perhaps signifies that some fluorine is also bound to carbon. In the case of third-stage bifluoride the conductivity (per plane) is roughly five times that of graphite.

C. GRAPHITE FLUORIDES $C_xF_{1-\delta}(HF)_\delta$

1. Preparation

Treatment of graphite powder with F_2 and liquid or gaseous hydrogen fluoride produced, after 1 day, first-stage hydrofluorides of the general formula $C_xF_{1-\delta}(HF)_\delta$ ($x/\delta \approx 12$ and $2 < x < 5$–6). These materials were also prepared electrochemically (see Section II,D,2).

2. Powder Patterns

These compounds are hexagonal with nearly constant *a*-axis and smoothly varying *c*-axis dimensions (see Table II). As *x* approaches 2, only (*hk*0) and

TABLE II

Some Physical Properties of $C_xF_{1-\delta}(HF)_\delta$

C/F Ratio	a (Å)	c (Å)	Resistivity (Ω cm)
5.2[a]	2.456(3)	5.22(2)	$>8 \times 10^{-3}$
3.67	2.459	5.36	8×10^{-1}
3.20	2.456	5.45	3×10^{0}
2.70	2.460	5.71	2×10^{1}
2.49	2.457	5.90	1×10^{3}
2.42	2.459	6.02	1×10^{3}
2.15	2.468	6.22	1×10^{7}
1.94	2.466	6.45	2×10^{7}

[a] Contains some $C_{12}{}^+HF_2{}^-$, which is highly conductive.

(00l) lines are observed, indicating an absence of registry between one sheet and its neighbors. The c distances therefore represent only an average interlayer distance in each case. The (hk0) lines are sharp and persist to high Bragg angle [in contrast with (00l), which are broad and lost at higher angles]. The derived a values, listed in Table II, are remarkably constant over a large composition range and depart significantly from the graphite cell value only as x approaches 2.

3. Properties

Compounds in the series $C_xF_{1-\delta}(HF)_\delta$ are black in color and are insensitive to air and moisture, at least over a period of several days. If not thoroughly pumped, they lose a small amount of hydrogen fluoride on standing and so can attack glass. Analysis for hydrogen fluoride, by pyrolysis, shows carbon/hydrogen atomic ratios of 11 to 12. Examples are $C_{2.05}FH_{0.18}$ (C/H = 11.6), $C_{2.70}FH_{0.22}$ (C/H = 12.3), and $C_{3.75}FH_{0.33}$ (C/H = 11.4). Warming $C_{2.05}FH_{0.18}$ at 140°C *in vacuo* for 2 days brings about the loss of some hydrogen fluoride to give a composition $C_{2.25}FH_{0.09}$. The c-axis spacing of this heated material shrinks from 6.25 to 5.92 Å, but the a spacing remains constant at 2.464 Å. Heating to 350°C causes further c-axis contraction (to 5.65 Å) with no significant change in a.

The electrical resistivities of the members of this series are shown in Table III. The resistivity shows a dramatic increase (by 10 orders of magnitude) between carbon/fluorine = 5.2 and carbon/fluorine = 2, the marked upturn in ρ occurring near carbon/fluorine ≈ 2.7. Near the high ρ end of the series, the a spacing expands by ~ 0.01 Å, indicating a slight weakening of the graphite–sheet bonding at these compositions.

TABLE III

ab Plane Electrical Conductivity of Graphite
Bifluorides

Composition	Stage present	$\sigma/\sigma_{\text{graphite}}$
$C_{15.9}HF_2$	II, III	~2.0
$C_{13.5}HF_2$	II	2.0
$C_{12.5}HF_2$	II	2.8
$C_{14.5}HF_2$	II	2.5
$C_{18.1}HF_2$	II	6.1

4. x-Ray Photoelectron Spectrum of $C_{2.5}F_{1-\delta}(HF)_\delta$

x-Ray photoelectron spectra Mallouk and Bartlett, 1983 recorded using a freshly cleaved sample of $C_{2.5}F_{1-\delta}(HF)_\delta$, which had been prepared by exhaustive fluorination (2 weeks) of an HOPG chip in the presence of liquid hydrogen fluoride, are given in Table IV. The binding energies obtained (reported here relative to the carbon 1s level of a graphite standard, 284.2 eV) are compared with those of other compounds. The binding energy of fluorine in this sample is intermediate between that of fluorine in LiF and fluorine in Teflon of $(C_2F)_n$. This suggests that the fluorine ligand in $C_xF_{1-\delta}(HF)_\delta$ bears a partial negative charge and that the C—F bond has semiionic character. The carbon binding energy indicates two kinds of carbon, one resembling that of graphite, but neither the C in $(C_2F)_n$ nor Teflon, where the carbon coordination geometry is tetrahedral.

TABLE IV

x-Ray Photoelectron Spectra Binding Energies of $C_{2.5}F_{1-\delta}(HF)_\delta$ and
Comparison Materials

Compound	C_{1s} (eV)	F_{1s} (eV)	Reference
$C_{2.5}F_{1-\delta}(HF)_\delta$	284.6	686.6	Present work
	287.5		Takenaka and Bartlett (1984)
$(C_2F)_n$	288.5	689.3	Kita et al. (1979)
	290.4		
$(CF_2)_n$	291.8	689.1	Muilenberg (1978)
LiF	—	684.9	Muilenberg (1978)
Graphite	284.2	—	Muilenberg (1978)

5. ^{1}H-nmr Spectrum of $C_{2.15}F_{1-\delta}(HF)_\delta$

A proton spectrum[†] of a representative sample of $C_xF_{1-\delta}(HF)_\delta$ of low conductivity reveals a proton line, the width of which (37 ppm) can be ascribed to dipolar interactions that are, at 300 K, beginning to be averaged away by hindered rotation or exchange of the proton-containing species. The activation energy for the process is about 40 kJ mol^{-1}.

Evidently, the hydrogen fluoride in $C_xF_{1-\delta}(HF)_\delta$ is interacting strongly with the host lattice. This interaction is likely to take place via hydrogen bonding of the hydrogen fluoride with the rather negative fluorine ligands bound to the carbon.

6. ^{13}C-nmr Spectrum of $C_xF_{1-\delta}(HF)_\delta$

^{13}C-nmr spectra of powders for which carbon/fluorine = 3.70 and = 2.05 have been obtained.[*] The technique used was that of $^{19}F \rightarrow {}^{13}C$ cross-polarization (which greatly enhances the ^{13}C signal and at the same time removes the ^{19}F scalar and dipolar couplings) and magic angle spinning (which averages out chemical shift anisotropy). The pulse sequences used were standard (Yannoni, 1982): a 90° pulse followed by a ^{19}F spin-locking (and -decoupling) pulse causes the ^{19}F magnetization to decay with a characteristic time $T_{1\rho}$. The ^{13}C spins are brought into contact with the ^{19}F spins, and are so magnetized, by suppling a pulse that satisfies the Hartmann–Hann double-resonance condition (Hartmann and Hahn, 1962) for a time t (the cross-polarization interval). Because the characteristic cross-relaxation time of the ith carbon spin T^i_{CF} depends on the strength of the dipolar coupling with its ^{19}F nearest neighbors, the observed spectral intensity, which depends on $T_{1\rho}$, t, and T^i_{CF}, will be sensitive to the distance between C_i and the nearest fluorine atoms.

A representative spectrum is shown in Fig. 9. For both carbon/fluorine ratios (3.70 and 2.05), two peaks, separated by 47 ppm, are observed; the similarity of the spectra suggests that the same bonding environments are found in both compounds. The low-field peak ($\delta = +135$ ppm relative to TMS) shows spinning side bands displaced ±97 ppm from the central peak, which indicates a large chemical shift anisotropy for these carbon atoms. This chemical shift and shift anisotropy are characteristic of aromatic (i.e., graphite-like) carbon. The high-field peak is attributed to carbon atoms that are attached directly to fluorine. In addition to these, a small peak at 112 ppm

---

[†] The proton spectrum of $C_{3.15}F_{1-\delta}(HF)_\delta$ was recorded by Dr. K. Zilm at the University of California, Berkeley.

[*] The ^{13}C-nmr spectra were obtained by Dr. B. L. Hawkins and Dr. G. E. Maciel of the Regional NMR Facility, Colorado State University, Fort Collins, Colorado.

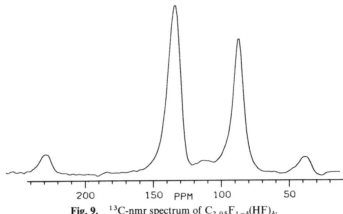

Fig. 9. ^{13}C-nmr spectrum of $C_{2.05}F_{1-\delta}(HF)_\delta$.

may be resolved in some of the spectra. This peak is attributed to a small concentration of —CF_2 carbon atoms, which are probably located at the edges of the powder grains. The ir spectrum (see next section) also shows evidence of these —CF_2 groups.

7. Vibrational Spectra of $C_xF_{1-\delta}(HF)_\delta$

Transmission ir spectra[†] of $C_{2.5}F_{1-\delta}(HF)_\delta$ were recorded (Mallouk and Bartlett, 1983) with a sample prepared by grinding a chip of graphite in such a way that the c axis lay in the plane of the resulting thin flake. This arrangement exploits the high conductivity anisotropy of graphite and its compounds; electromagnetic radiation may be transmitted through the graphite, provided that the electric vector of the light is aligned parallel to the c axis. The flake was intercalated by treatment with fluorine and hydrogen fluoride vapor. A very strong absorption, centered at about 1090 cm^{-1}, is attributed to C–F stretching. Comparison with the C–F stretching frequencies in $(CF)_n$ and $(C_2F)_n$, which are 1219 and 1221 cm^{-1}, respectively, indicates a C—F bond order of slightly less than unity. Bands situated at 1587 and 850 cm^{-1} are assigned to C–C stretching and carbon-out-of-plane motions; these vibrations, which also occur in the Raman spectra, are known for graphite itself and occur at almost the same frequency (Solin, 1980). The 1270 cm^{-1} band, which lies in the C–F stretching region, may arise from a small concentration of >CF_2 groups at the edges of the chip.

Raman spectra recorded from powdered samples for several compositions are all virtually identical to the spectrum of graphite powder (Solin, 1980) except that the Raman relative of the 850 cm^{-1} ir band is seen (at 839 cm^{-1}),

[†] The ir spectra of $C_{2.5}F_{1-\delta}(HF)_\delta$ were obtained by M. P. Conrad.

and a broad brand at 1355 cm^{-1} is also observed. Such a band is seen in graphite powder that is poorly ordered or has been disordered by grinding. It has been attributed (Tuinstra and Koenig, 1970) to a Brillouin zone boundary mode that is Raman inactive in crystallites of large extent; it is essentially an in-plane ring-breathing vibration. The intensity of this band relative to the graphite E_{2g}-like band at ~ 1600 cm^{-1} indicates the extent of the ordered domains in the ab plane. They must be approximately 50–100 Å in diameter. A slight shifting (~ 30 cm^{-1}) toward lower frequency of the ~ 1600 cm^{-1} band is found for the highest fluorine concentrations, indicating a very slight weakening of the C—C bond, in accord with the increased bond distance (of ~ 0.006 Å) inferred from the crystallographic a spacings. The Raman spectra establish, by the presence of only graphite-like modes, even at a carbon/fluorine ratio of 2, that the planarity of the carbon atom in its sheets, in $C_xF_{1-\delta}(HF)_\delta$, is maintained.

8. Bonding and Electronic Structure in $C_xF_{1-\delta}(HF)_\delta$

The Raman spectra confirm what the x-ray powder data already hinted at: the carbon atom sheets in $C_xF_{1-\delta}(HF)_\delta$ are as flat and as π bonded as in graphite itself. Moreover, the photoelectron spectroscopic data and the ^{13}C-nmr data show that the fluorine ligand, in its linkage to the carbon, is semi ionic. This suggests that the fluorine atom is sharing an electron from carbon $C\cdot \rightarrow F$, much as in the classical dative binding of oxygen atom (e.g., $R_3N:\rightarrow O$).

To account for the preservation of the π bonding of the graphite sheets, it is essential that the attachment of the fluorine ligands to the sheet not have an impact on the π bonding. That is, there cannot be net addition nor substraction of electrons from the π-bonding role.[†]

From inspection of the data in Table III, it will be seen that the intersheet separation (c_0) in $C_xF_{1-\delta}(HF)_\delta$ increases markedly for carbon/fluorine ratios

[†] The obvious bonding model that accommodates these requirements is that which attaches two fluorine ligands simultaneously to a carbon atom: one above and one below it. The three p_z orbitals of such a linear array would give rise to three molecular orbitals. Only the bonding (σ_u) and nonbonding (σ_g) orbitals of this set would be filled. The bonding σ_u orbital, however, having the appropriate symmetry, would still be part of the π system. Hence, the pair of electrons provided by that system would be restored to it. The nonbonding σ_g orbital, having no contribution from the carbon, is a fluorine-only orbital. Thus, the semiionic character of the C—F bond can be accounted for. It will be noted that the bonding molecular orbital (σ_u), which has the same symmetry as a bonding π orbital of the graphite, cannot be simply a three-center orbital. Rather, it must be akin to multicenter orbitals of the valence–band set. There is therefore no requirement for the two fluorine ligands to be attached (top and bottom) to the same carbon atom. We can regard these two fluorine ligands as being combined with an extended π-orbital system. As before, the nonbonding combination does not involve carbon participation in the orbital—it is of fluorine character only.

of less than 3. This can be understood if we allow that the fluorine ligands on one side of a carbon sheet are symmetrically distributed. The ready introduction and removal of fluorine ligands by electrochemical means (see next section) encourages that supposition. The symmetric arrangement of the fluorine ligands of one sheet meshes, in a close-packed fashion, with that of another sheet, as illustrated in Fig. 10a. At composition C_2F such close packing cannot occur (Fig. 10b). For $x < 3$, therefore, a marked increase in c_0 is anticipated.

Band structure calculations have not yet been made for the structures represented in Fig. 10, but one is available (Holzwarth *et al.*, 1981) for the idealized structure postulated by Rüdorff and Rüdorff (1947) for C_4F. A C—F bond distance of 1.4 Å was assumed, and with this perturbation an energy gap of ~ 2 eV develops between the valence and conduction bands. It is possible that the C—F bond in $C_xF_{1-\delta}(HF)_\delta$ is longer than 1.4 Å and has a less perturbing influence on the π system, but it is to be expected that this perturbing influence will grow with the concentration of fluorine. The rapidly increasing resistivity as a function of diminishing x may, to some extent, derive from an increasing band gap. There must, however, be a marked increase in resistivity with increase in the number of C—F bonds, simply as a consequence of the very low mobility of a current carrier through such carbon centers.

D. CHEMICAL AND ELETROCHEMICAL REACTIONS OF GRAPHITE HYDROFLUORIDES

1. Intercalation Reactions

Both first- and second-stage graphite hydrofluorides undergo intercalation reactions with strong Lewis acids such as AsF_5, GeF_4, and PF_5. The reaction of graphite bifluoride ($C_{12}{}^+HF_2{}^-$) with GeF_4 (1 atm) is sluggish, and a mixture of second- and third-stage compounds is formed. If the pressure is increased to about 10 atm, a pure first-stage compound, $a = 2.455(3)$,[†] $c = 7.83(2)$ Å is obtained, in which the carbon/germanium ratio is ~ 8. Reaction with PF_5 under similar conditions gives only compounds of high stage ($> III$). The compound AsF_5 is intercalated by $C_{12}HF_2$ (Okino and Bartlett, 1984) to give $C_{12}AsF_6 \cdot HF$, a first-stage salt, with $c = 8.0$ Å. Apparently, quite a strong Lewis acid is required to abstract a fluoride ion from $HF_2{}^-$ and overcome the favorable lattice energy of the bifluoride.

First-stage $C_xF_{1-\delta}(HF)_\delta$ reacts with GeF_4 and AsF_5 under 10 atm pressure to produce first-stage compounds, $c \approx 8.0$ Å. The amount of Lewis acid taken

---

[†] Numerals in parentheses indicate the standard deviation in the last figure given.

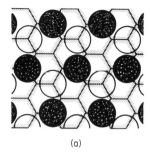 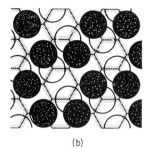

(a)   (b)

Fig. 10. Structural models for (a) C_3F and (b) C_2F. The unshaded circles represent the fluorine atoms on one side of a carbon atom sheet, and the shaded circles represent the fluorine atoms of an adjacent sheet in positions of closest packing.

up is limited by the value of x, and for $x \simeq 2$ no reaction is observed, even with AsF_5. For $2 < x < 3$, no reaction with AsF_3 was observed, and the reaction with water to produce hydrogen fluoride, graphite, and (presumably) O_2 is extremely slow. $C_{2.0}F_{1-\delta}(HF)_\delta$ appears to be inert to hydrolysis. Compounds in this series swell in anhydrous, liquid hydrogen fluoride, the c-axis spacing increasing by 0.5 to 0.6 Å, even at $C_{2.0}F_{1-\delta}(HF)_\delta$. This increase represents an additional volume of 1.3 to 1.4 Å^3 per carbon atom, or about 1 hydrogen fluoride molecule for every 11–14 carbons. The diffraction patterns show no discernible change with the increased hydrogen fluoride HF content, except for this increase in c.

2. Electrochemical Reversibility

The enthalpy of formation of tetracarbon monofluoride has been estimated to be about -44 kcal mol^{-1} of fluorine atoms (Wood et al., 1969); thermodynamically, it should be a potent oxidizer, readily reduced by even the most electronegative metals. For example, ΔH_f^0 for PbF_2 is -160 kcal mol^{-1}. Assuming that the entropy changes are small the C/C_4F couple is then expected to be 2.1 V below the Pb/PbF_2 couple in the electrochemical series. Previous work has established (Whittingham, 1975; (Beck et al., 1982) that the potential of this couple is near the thermodynamically expected value for compounds with carbon/fluorine ratios of ≥ 4.

The authors have found that for carbon/fluorine ratios of > 2.3, graphite hydrofluorides and bifluorides may be reduced electrochemically. The open-circuit potentials (using HF/NaF as the electrolyte) against Pb/PbF_2 for $C_xF_{1-\delta}(HF)_\delta$ were determined as follows (x in parentheses): (3.7), 2.03; (2.7), 2.2; (2.5), 2.38 V. The cell used for these measurements consisted of a Teflon T union connected via compression fittings to two glassy carbon electrodes approximately 1 mm in diameter and to the electrolyte reservoir. The two electrode compartments were separated by a wad of porous Teflon filter membrane.

With a resistive load in series, it was possible to assess the internal resistance of the cell using a variety of electrolytes and counterelectrodes. It is lowest when the electrolyte is hydrogen fluoride. This is undoubtedly a consequence of the fact that hydrogen fluoride enters the graphite galleries readily. It probably carries fluoride as HF_2^-, $H_2F_3^-$, etc. It is superior in this respect to H_2O, which is a weaker fluoride ion acceptor.

The intercalation of fluorine is completely reversible for carbon/fluorine ratios in excess of 2.3. Compounds in the series $C_xF_{1-\delta}(HF)_\delta$ can be prepared electrochemically from graphite, using PbF_2 as a source of fluoride ion, and a driving potential of 3.0 V. (For $x < 2.6$, the electrochemical synthesis is impracticable, possibly because an insulating layer is formed at the edges of the powder grains.) Compounds so synthesized have diffraction patterns identical to those prepared from graphite, hydrogen fluoride, and F_2. Chips of $C_{2.5}F_{1-\delta}(HF)_\delta$ and $C_{12}^+HF_2^-$, when placed in a large excess of powdered graphite and hydrogen fluoride, are reduced completely to graphite. (This arrangement is essentially a shorted electrochemical cell.)

III. Graphite Intercalation Compounds with Binary Fluorides

A. INTRODUCTION

The past few decades have witnessed intense activity in the field of graphite intercalation compounds (GICs). Particular impetus was provided by a report that GICs with SbF_5 and AsF_5 exhibit electrical conductivities exceeding that of copper (Foley *et al.*, 1977; Vogel, 1977). Although since then this report has been partially discounted, these compounds having a limiting conductivity that is only about half that of copper (Thompson, *et al.*, 1981; Interrante *et al.*, 1979), interest in the field has remained high. Many of these new developments have been summarized in reviews emphasizing different aspects of a field that is becoming increasingly interdisciplinary. Particularly to be noted are reviews on the chemistry (Bartlett and McQuillan, 1982; Forsman *et al.*, 1983; Selig and Ebert, 1980), catalytic properties (Ebert, 1982), and physics (Dresselhaus and Dresselhaus, 1981; Fischer, 1979) of GICs.

Although covalent fluorides are acceptors, they vary a great deal in the ease with which they can intercalate. Some fluorides intercalate spontaneously (e.g., AsF_5, SbF_5, and OsF_6), whereas others (e.g., IF_5, PF_5, BF_3, GeF_4, and WF_6) require the use of an oxidative coreagent such as fluorine. Still others intercalate with partial fluorination of the graphite network to yield low-conducting compounds (e.g. XeF_6, IF_7, and KrF_2).

Particular attention has been focused on GICs with AsF_5 in attempts to identify the nature of the intercalated species, the extent of charge transfer, and the correlation of these with enhanced conductivity.

B. PENTAFLUORIDE INTERCALANTS

1. Preparation

Most first-stage compounds are easy to prepare, this involving exposure of the graphite to sufficient vapor pressure of the intercalant for a long enough time. Pure stages higher than the first are more difficult to obtain except in the case of AsF_5. Compounds of the general formula $C_{8n}AsF_5$ (n = stage) are formed (Falardeau et al., 1978). Stages I and II are obtained by maintaining equilibrium pressures of ~ 700 and ~ 170 torr of AsF_5, respectively. Stage III can be obtained by heating lower stages to 200°C (Interrante et al., 1979). The c-axis thickness is a reliable indicator for stages I to III, plateaus occurring at expansions corresponding to specific stages. A given stage can therefore be obtained by terminating the intercalation at the appropriate value of $\Delta t/t_0$. Upon prolonged contact, there is evidence for partial fluorination of the graphite (Pentenrieder and Boehm, 1982). This is supported by the appearance of ir absorptions due to C–F vibrations. However, the fact that this is a function of particle size may indicate that this is not an intrinsic property of this GIC system.

In the case of SbF_5 intercalation this effect is more pronounced (Streifinger et al., 1979). Here, intercalation proceeds much more slowly, a reaction time of 20 days being required to reach stage I (Pentenrieder and Boehm, 1982). Compositions vary greatly, depending on graphite particle size. The intercalate richest in antimony had an approximate composition of $C_{6.5}SbF_6$.

A question arises as to the fate of the oxidation by-product SbF_3 and, even if this compound can be detected, whether it is found within the layers or on the surface (Forsman et al., 1983).

Compounds containing fluoride anions *only* can be prepared by reacting graphite with appropriate nitronium salts (Forsman and Mertwoy, 1980; Billaud et al., 1980; 1982; Moran et al., 1984) in inert solvents. The compounds $C_{23n}^+X^-\cdot1.7CH_3NO_2$ (X^- = PF_6^- or SbF_6^-) have been obtained.

2. Physical Properties

a. THE NATURE OF THE INTERCALANT.   Graphite acceptor compounds have been termed "synthetic metals" because they possess metallic properties such as high ir reflectivity and high electrical conductivity in the basal plane. The nature of the intercalated species in the system $C_{8n}AsF_5$ has been the subject of lively controversy. This controversy has focused on a number of outstanding but interrelated questions:

1. Does the intercalation reaction with AsF_5 go to completion, and is it reversible?
2. Is there an ionic salt limit?
3. Is the limiting conductivity due to reduced mobility arising from scattering sites (i.e., C—F bonds) or due to increased charge localization?

Answering these questions is complicated not only by different interpretations of experimental observations, but in some cases by conflicting experimental results.

i. THE SYSTEM $C_{8n}AsF_5$.   The GICs with SbF_5 (Lalancette and LaFontaine, 1973) and AsF_5 (Lin Chun-Hsu *et al.*, 1975) originally reported were assumed to contain largely neutral molecules. Mass spectra as a function of temperature (Selig *et al.*, 1977) showed only evolution of AsF_5. In addition, ^{19}F nmr gave a signal consistent with pentavalent arsenic (or antimony) but not AsF_3 (Ebert and Selig, 1977). Nevertheless, the occurrence of some oxidation was shown by the presence of AsF_3 ($\leq 5\%$ on the basis of arsenic) in the off-gases. Arsenic K-shell absorption spectra (Bartlett *et al.*, 1978a,b) were cited as evidence that the reaction

$$3\,AsF_5 + 2\,e^- \longrightarrow 2\,AsF_6^- + AsF_3 \tag{1}$$

goes to completion in the presence of excess AsF_5. The appearance of AsF_5 in the mass spectra was attributed to the reversibility of Eq. (1). Fluorination of $C_{10}AsF_5$ caused a component ascribed to AsF_3 to disappear, the spectrum being supplanted by one identical to that of $C_8^+AsF_6^-$ obtained by reaction of graphite with $O_2^+AsF_6^-$. These results were interpreted in terms of the reactions

$$12n\,C + \tfrac{3}{2}\,AsF_5 \longrightarrow C_{12n}^+[AsF_6^-]\cdot\tfrac{1}{2}AsF_3 = \tfrac{3}{2}\,C_{8n}AsF_5 \tag{2}$$

$$C_8AsF_5 + \tfrac{1}{2}\,F_2 \longrightarrow C_8^+AsF_6^- \longrightarrow 8C + O_2^+AsF_6^- \tag{3}$$

Upon pumping, the material loses initially mostly AsF_5 but later only AsF_3, in accordance with the following reaction:

$$\tfrac{3}{2}\,C_{8n}AsF_5 \rightleftharpoons C_{12n}^+AsF_6^- + \tfrac{1}{2}\,AsF_3(g) \tag{4}$$

Equation (4) is reversible. [A limiting composition of $C_{14}AsF_6$ was later suggested for the stage I salt (Bartlett *et al.*, 1983).]

Thus, two parallel, interconvertible series of compounds, $C_{8n}AsF_5$ and $C_{12n}^+AsF_6^-$, are obtained (McCarron and Bartlett, 1980), no changes in stage being observed during a given conversion.

This interpretation introduced difficulties, for a number of reasons. Certain acceptor compounds can be obtained by oxidation with nitronium salts in nitromethane solution (Billaud *et al.*, 1982; Forsman and Mertwoy, 1980; Moran *et al.*, 1984):

$$NO_2{}^+MF_6{}^- + 23n\,C + CH_3NO_2 \text{ (excess)} \longrightarrow C_{23n}^+MF_6^- \cdot 1.7CH_3NO_2 + NO_2(g)$$
$$(M = P, Sb) \tag{5}$$

The $NO_2{}^+$ acts as oxidizer, the $MF_6{}^-$ balancing the positive charge on the graphite. Neutral nitromethane is incorporated into the intercalate rather than there being complete oxidation to $C_8{}^+MF_6{}^-$. This seems to suggest that the maximum electronic charge that can be withdrawn from graphite is about 0.04 per carbon atom. The ^{19}F-nmr spectra of these ternary compounds are consistent with the presence of $AsF_6{}^-$ only. Subsequent intercalation of AsF_3 into this gave superimposed spectra of AsF_3 and $AsF_6{}^-$ with no evidence for fluorine exchange. This is in stark contrast to $C_{8n}AsF_5$ spectra, which do show exchange, implying the presence of AsF_5 (Moran *et al.*, 1984).

Further physical measurements conflict with the charge transfer implied by Eq. (1). Fischer (Moran *et al.*, 1980) has distinguished between the chemical charge transfer f_{chem}, as the fraction of electron charge transferred from the graphite per intercalant molecule (for acceptors), and the electronic charge transfer f_{elec}, defined as the fraction of free carrier added to the valence band per intercalated acceptor. The latter can be inferred from transport properties of the material. Normally, f_{chem} is equal to f_{elec}, unless part of the transferred charge is localized in the form of covalent C—F bonds. Equation (1), if it goes to completion, implies $f_{chem} = 0.67$. The blue color of C_8AsF_5 indicates that the charge transfer is much lower, namely, $f_{elec} \approx 0.33$, close to that of the metallic blue $C_{24}{}^+HSO_4{}^- \cdot 2H_2SO_4$ having a similar plasma frequency (Ebert, 1982).[†] In addition, magneto-oscillation experiments (Markiewicz *et al.*, 1980) lead to $f_{elec} = 0.37$ (stage I) and $f_{elec} = 0.41$ (stage II), whereas measurements of the Pauli spin susceptibility (Weinberger *et al.*, 1978) give $f_{elec} = 0.24$ for C_8AsF_5.

Further x-ray photoelectron spectroscopy measurements (Moran *et al.*, 1980), x-ray diffraction, and chemical analyses correlated with conductivity and reflectivity spectra indicate that AsF_5 is indeed reduced according to Eq. (1), thus creating holes in the graphite valence band, but the reaction does not go to completion.

[†] The charge transfer can be calculated from the Drude relation, $\omega_p^2 = 4\pi ne^2/m\epsilon$, where n is carrier density (i.e., charge transferred per unit volume), e the electronic charge, m the optical effective mass, ϵ the core dielectric constant, and ω_p the plasma frequency, which is the frequency of radiation corresponding to a minimum in the reflectivity. For both C_8AsF_5 and $C_{24}{}^+HSO_4{}^- \cdot 2H_2SO_4$; $\omega_p \approx 1.8$eV. In the latter, $f_{chem} \approx 0.33$, as determined from coulometry.

An upper limit of $\sim 30\%$ ionization was indicated for stage I compound. This compound should thus be formulated $C_8^{+0.27}(AsF_6^-)_{0.27}(AsF_3)_{0.13}$-$(AsF_5)_{0.60}$ with an approximate charge per carbon of 0.03.

The major species removed by initial pumping of stage I is AsF_5 (with little AsF_3). If this arises from reversibility of Eq. (1), as suggested by Bartlett, the resulting removal of AsF_6^- should cause a *substantial* reduction in the electrical conductivity, contrary to what is observed. The AsF_5 removed must thus have existed originally as the molecular species.

These apparently divergent interpretations have been reconciled partially with respect to the stage I compound, C_8AsF_5. The actual conductivity per graphite plane[†] for stage I is only 64% of that calculated from an extrapolation from higher stages. Thus, an upper limit of 36% has been set on the amount of un-ionized AsF_5 in stage I (Thompson *et al.*, 1981).

x-Ray photoelectron spectral data (Moran *et al.*, 1980) show a trend to more complete oxidation toward higher stages. At stage III, oxidation is thought to be complete. Whereas pumping on stage I seems to remove mostly AsF_5, stages II and III are reported to liberate mostly AsF_3 (Thompson *et al.*, 1981). In the latter case, the fluorine/arsenic ratio should change from 5 to 6, as has indeed been reported (Thompson *et al.*, 1981). This is not in disagreement with minor changes in conductivity upon pumping. More precise chemical analyses of the species evolved upon pumping and their correlation with fluorine/arsenic ratios in the residue would be most useful.

Thus, although there seems to be no conflict regarding stage III and a convergence of views regarding stage I, for stage II the discrepancy remains unresolved. Part of the problem may be due to different types of graphite used. The x-ray absorption work (Bartlett *et al.*, 1978a,b) was done on powdered graphite, whereas x-ray photoelectron spectral studies were carried out on HOPG. Although the presence of AsF_3 by the x-ray absorption work cannot be gainsaid, a quantitative interpretation on the basis of peak intensities may not be justified (Ebert, 1982).

It has been proposed that still another equilibrium may be involved (Ebert *et al.*, 1981), namely,

$$C_n^+ AsF_6^- + AsF_5 \rightleftharpoons C_n^+ As_2F_{11}^- \tag{6}$$

The $As_2F_{11}^-$ ion can be stabilized by bulky cations such as Et_4N^+, so its existence in the presence of the graphite macrocation is not unreasonable.

ii. THE FLUORINATED SYSTEM $C_n^+AsF_6^-$. The ionic salt limit has been defined as the charge density per carbon atom beyond which the conductivity decreases

---

[†] The specific conductivity per graphite plane is defined as $\sigma_s/\sigma_g \times t_s/t_g = \Delta V_s/\Delta V_g$, where σ is the conductivity, t the thickness, and ΔV a directly measurable parameter. The subscripts s and g refer to GIC and original HOPG, respectively.

(Moran $et\ al.$, 1980). The previous section alluded to the existence of such a limit, which may reach $\sim C_{30}^{+}$ for stage I (i.e., ~ 0.03 electron per carbon atom extracted by oxidation). According to Bartlett the charge per carbon atom may reach 0.125 for $C_8{}^+AsF_6{}^-$. The maximum conductivity attainable has been shown to be an intrinsic property of these compounds unrelated to the electron affinity of the intercalant (Moran $et\ al.$, 1981). In the case of $C_{8n}AsF_5$ the conductivity varies little between $n = 1-3$ and reaches a maximum σ_{max}/σ_g of 12.[†]

The fate of the incorporated fluorine in the highly fluorinated material has been a subject of controversy. There are some discrepancies regarding the extent of fluorine uptake for stage I. On the one hand, little or no reaction was said to occur (Thompson $et\ al.$, 1981), whereas others reported up to 20 mol % uptake at maximum conductivity based on total AsF_5 present (Milliken and Fischer, 1983).

The relative fluorine uptake of second or higher stages is substantial. $C_{12n}AsF_6$ ($n \geq 2$) salts are said to take up excess fluorine with a limiting composition $C_{12n}AsF_6 \cdot F$. This excess F_2 cannot be removed by pumping, but it can be titrated with AsF_3 to form an $AsF_6{}^-$ salt according to

$$3\,C_{4y}AsF_6 \cdot F + AsF_3 \longrightarrow 4\,C_{3y}AsF_6 \qquad (7)$$

The uptake of excess fluorine to yield fluorine/arsenic ratios of greater than 6 has also been noted by others (Münch and Selig, 1979; Pentenrieder and Boehm, 1982).

Although it is generally agreed that the highly fluorinated materials exhibit substantially lower electrical conductivities than the starting materials, it was also found that there is an $initial\ increase$ in conductivity upon exposing $C_{8n}AsF_5$ ($n = 1, 2$) to fluorine. In the case of $C_{12}AsF_6 \cdot F$, the extra fluorine is postulated to occupy, as fluoride ion, vacant sites normally filled by AsF_3 molecules in the $C_{12n}AsF_6 \cdot \frac{1}{2}AsF_3$ system (Bartlett and McQuillan, 1982) (see proposed structure for $C_{12n}AsF_6$). If the material is indeed $C_{12n}^{2+}AsF_6{}^- \cdot F^-$, it would parallel other highly charged systems such as $C_{12}{}^{2+}PtF_6{}^{2-}$ and $C_8{}^+MF_6{}^-$ (M = As, Os, or Ir), all of which are poor conductors. It has been pointed out (Bartlett $et\ al.$, 1979–1980) that poor conductivity is associated with high positive charge on the graphite layers, the cooperative effect of two anions on each side of the graphite sheet serving to $localize$ high positive charge at a carbon atom. Such highly charged systems would be expected to cause distortion of the regular trigonal carbon layers. In fact, loss of long-range order has been observed in $C_8{}^+MF_6{}^-$ systems, as shown by loss of crystallinity when samples are kept at room temperature for several hours.

[†] The basal plane conductivity of graphite, $\sigma_g = (26 \pm 3) \times 10^3\ \Omega^{-1}\ cm^{-1}$.

An alternative explanation has been proposed (Milliken and Fischer, 1983) based on analogies with the $C_m{}^+HSO_4{}^-\cdot nH_2SO_4$ system. Electrochemical oxidation of $C_{27}{}^+HSO_4{}^-\cdot 5H_2SO_4$ leads to C_{21}^+ compound with a concomitant increase in conductivity ("overcharging") indicating partial conversion of H_2SO_4 to $HSO_4{}^-$ (Metrot and Fischer, 1981; Fischer *et al.*, 1981). Further oxidation reduces the conductivity, broadens the reflectivity spectrum without shifting the plasma edge, and introduces covalent C—O bonds. Similarly, the initial increase in conductivity upon exposure of $C_{8n}AsF_5$ to fluorine is attributed to "overcharging" from C_{29}^+ to C_{20}^+, whereas further fluorination causes a substantial decrease in conductivity ("overoxidation"). The overcharging step, accompanied by a blue shift of the plasma edge, is attributed to the formation of additional $AsF_6{}^-$. "Overoxidation" with fluorine is presumed to result in the formation of C—F bonds, which act as scattering centers for the charge carriers. It does not alter the number of charge carriers (same plasma edge). Instead, the sharpness of the reflectivity edge, a parameter related to the density of scattering centers, decreases substantially. Results for stage II are similar, with overcharging causing an increase in charge density from C_{40}^+ to an average of C_{20}^+, the limit reached with stage I. In terms of the previous notation, f_{chem} is equal to f_{elec} up to the beginning of overfluorination, beyond which f_{chem} is greater than f_{elec}. At this point, f_{elec} equals ~ 0.4.

Concomitant with overfluorination, x-ray photoelectron spectra show the appearance of a new peak identified with covalent C—F bonds in analogy with Teflon and graphite fluoride (Streifinger *et al.*, 1979). Further support for defect formation is adduced from changes in the temperature dependence of the conductivity as a function of fluorination.

It is interesting that ir spectra of fluorinated $C_{8n}AsF_5$ show marked increases in absorptions attributed to C–F vibrations (Pentenrieder and Boehm, 1982). A simultaneous disappearance of AsF_3 absorptions and an increase in $AsF_6{}^-$ absorptions was noted as well.

Although a number of questions remain unresolved, it seems that the best room-temperature conductivities have been achieved. Further attempts to obtain better synthetic metals by the use of stronger acceptors seem doomed to failure.

b. STRUCTURE. Structural investigations have been limited mainly to the characterization of the stage by (00*l*) diffractometry. Single-crystal work on C_8AsF_6 (McCarron and Bartlett, 1980) established that it is hexagonal with $a = 4.92(2)$ Å, $c = 7.86(2)$ Å, and $V = 165$ Å^3. The *c*-axis contraction from $C_{12}AsF_6$ is attributed to the increased coulombic attraction.

Proposed model structures for $C_{8n}MF_5$ and $C_{12n}AsF_6$ are shown in Figs. 11 and 12, respectively (Bartlett and McQuillan, 1982). In the $C_{8n}AsF_5 =$

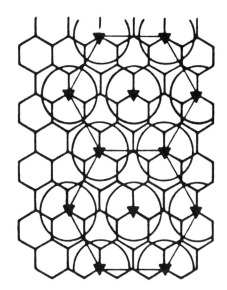

Fig. 11. Proposed structure for C_8MF_6. Note that each intercalated species has six nearest neighbors. From McCarron and Bartlett (1980).

$C_{12n}AsF_6 \cdot \frac{1}{2}MF_3$ structure each intercalant has six nearest neighbors. Upon pumping off AsF_3, $C_{12n}AsF_6$ is obtained in which each intercalant has three nearest neighbors. This void can be filled with AsF_3 or F^- (for $n \geq 2$) and possibly other molecules such as $trans$-N_2F_2, SF_4, and IOF_5 (Münch and Selig, 1979).

An x-ray diffraction study of $C_{16}AsF_5$ (stage II) yielded patterns corresponding to a hexagonal unit cell with $a = 2.46$ Å (as in graphite) and

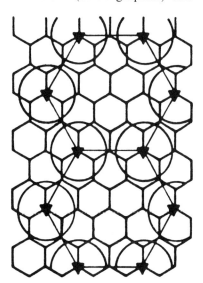

Fig. 12. Proposed structure for $C_{12}MF_6$. Note that each intercalated species has three nearest neighbors. From McCarron and Bartlett (1980).

$c = 11.50$ Å. Least-squares fitting of calculated and observed intensities can be accounted for only by the presence of both AsF_6^- and AsF_5, the amount of AsF_3 being indeterminate (Markiewicz *et al.*, 1980).

  c. CHEMICAL PROPERTIES.   Lowest-stage compounds retain some of the properties of the neutral intercalant molecules. They fume in air, indicating hydrolysis of the intercalated species by atmospheric moisture (Interrante *et al.*, 1979). Higher stages are considerably more stable. The reactivity of the intercalant molecules is substantially moderated, and as such these GICs have found some use as convenient fluorinating agents (Whittingham and Ebert, 1979). C/SbF_5 is a commercially available reagent.

Tetrafluorohydrazine reacts with C/AsF_5 according to

$$C_{10}AsF_5 + N_2F_4 \longrightarrow C_{10}AsF_7 + \text{trans-}N_2F_2 \qquad (8)$$

Because neither AsF_3 nor AsF_6^- reacts with N_2F_4, this has been taken as evidence for the presence of molecular AsF_5 in this GIC (Münch and Selig, 1979). Other fluorides that form complexes with AsF_5 ($\text{trans-}N_2F_2$, SF_4, and IOF_5) are taken up reversibly as well.

C.   COMPOUNDS WITH TRANSITION METAL HEXAFLUORIDES

1.   Introduction

  Attempts to identify the nature of the intercalant species and the extent of charge transfer have focused on the transition metal hexafluorides. These comprise a series of related compounds that have many physical properties that change monotonically in a given series. For spontaneous reaction of the hexafluorides with graphite to occur, the driving force must be sufficient to offset the energy necessary to separate the graphite layers as well as the highly unfavorable entropy term arising from ordering of guest species between the layers. Presumably, this driving force derives from the considerable coulombic forces between positively charged carbon layers and anionic guest species formed in the redox reaction accompanying intercalation (Bartlett and McQuillan, 1982). For a given degree of charge transfer these coulombic forces vary little for different MF_6, the latter being nearly isodimensional. Thus, the only term affecting the relative ease of intercalation is the electron affinity of the hexafluoride. It has been estimated that the free energy $(-\Delta G_{298}^0)$ of the reaction

$$MF_6(g) + e^- \longrightarrow MF_6^-(g) \qquad (9)$$

must exceed 120 kcal mol^{-1} for spontaneous intercalation to occur (Bartlett *et al.*, 1983). Electron affinities of the hexafluorides are given in Table V. Thus,

TABLE V

Electron Affinities of Transition Metal Hexafluorides[a]

MoF_6				
131				
WF_6	ReF_6	OsF_6	IrF_6	PtF_6
104	129	154	179	204
UF_6				
133				

[a] Based on table given in Bartlett and McQuillan (1982). The values (kilocalories per mole) have been revised because of a systematic error of ~ 20 kcal mol^{-1} (N. Bartlett, private communication, 1982).

all hexafluorides except WF_6 should intercalate spontaneously. The latter has been found to intercalate only in the presence of oxidizing agents (Hamwi et al., 1977). All other hexafluorides indeed intercalate spontaneously, undergoing either partial reduction (UF_6; Ebert et al., 1977; Binenboym et al., 1976) or complete reduction to a lower oxidation state (OsF_6, PtF_6; Bartlett et al., 1979–1980).

As can be seen from Table VI, compositions can vary considerably for a given stage, indicating incomplete filling of layers. This seems to be the rule

TABLE VI

Intercalation Compounds of Graphite with MF_6

Hexafluoride	Composition	Stage	I_c (Å)	Reference[a]
MoF_6	$C_{7.64}MoF_6$	I	—	1
	$C_{11\pm1}MoF_6$	I	—	2
	$C_{22\pm2}MoF_6$	II	11.70	2
	$C_{8-11}MoF_6$	I	8.35	3
UF_6	$C_{9.1}UF_6$	I	—	4
	$C_{13}UF_6$	—	—	5
OsF_6	C_8OsF_6	I	8.10	6
	$C_{12n}OsF_6$	II, etc.	—	7
IrF_6	C_8IrF_6	I	8.05	6
	$C_{12n}IrF_6$	II, etc.	—	7
PtF_6	$C_{12}PtF_6$	I	7.56	6
ReF_6	$C_{10}ReF_6$	I	8.35	3

[a] Key to references: 1, Opalovskii et al. (1974); 2, Hamwi et al. (1977); 3, Selig et al. (1984); 4, Binenboym et al. (1976); 5, Ebert et al. (1977); 6, Bartlett et al. (1979–1980); 7, Vaknin et al. (1983).

rather than the exception, and the stoichiometry C_8MF_6 (M = Os or Ir) reported previously is an ideal composition only rarely attained.

Kinetic considerations play an important role, these being influenced by the type of graphite used. Thus, UF_6 can intercalate easily with graphite powder or Grafoil but only with great difficulty with HOPG. The compound ReF_6 also intercalates into powder, albeit extremely slowly into HOPG (Selig *et al.*, 1984). Stage I $C_{12}PtF_6$ can only be attained with powdered graphite (Bartlett *et al.*, 1979–1980).

2. Structural Properties

Structural investigations have been limited almost exclusively to the determination of stage by (00*l*) diffractograms. Single-crystal studies on C_8OsF_6 indicate a primitive hexagonal cell with $a_0 = 4.92(2)$ Å, twice that of graphite, and $c_0 = 8.10(3)$ Å. The latter is consistent with orientation of $OsF_6{}^-$ ions with their threefold axes parallel to the graphite c axis. No complete crystal structure determination has been made, apparently because over the long term (> 1–2 days) *hk*0 spots become diffuse, spreading into rings typical of powder photographs. x-Ray diffraction spectra indicate that C_8IrF_6 and C_8AsF_6 are isostructural. The shorter c-axis repeat distance (Table VI) for $C_{12}PtF_6$ is thought to be due to higher coulombic attraction between the highly charged graphite sheets and the $PtF_6{}^{2-}$ guest species (Bartlett *et al.*, 1979–1980).

3. Magnetic Properties

Magnetic susceptibility data on powdered samples of C_8OsF_6 and C_8IrF_6 have been adduced as evidence for reduction of the hexafluorides to the pentavalent state (Bartlett, *et al.*, 1979–1980). Subsequent measurements on HOPG intercalated with OsF_6 show strong anisotropic effects with magnetic moments comparable to those reported by Bartlett. These are consistent with $OsF_6{}^-$ ions having uniform orientation within the graphite layers, leading to well-defined crystal field effects. Measurements of magnetization as a function of orientation with respect to the external magnetic field and temperature have made it possible to calculate the zero-field splitting (~ 9 cm^{-1}) separating the Kramer doublets, i.e. $\pm\frac{3}{2}$ and $\pm\frac{1}{2}$ spin states (Vaknin *et al.*, 1983).

The esr signals are also very anisotropic. They yield $g_{||} = 1.65$ and $g_{\perp} = 3.25$, in agreement with values calculated for a $5d^3$ system.

In the case of C_nMoF_6 for stages I and II, two signals are observed: one at $g = \sim 1.6$ arising from the localized moment and the other at ~ 1.9 from conduction electrons. The latter is observed because of the smaller spin–orbit coupling constant in the $4d$ series (Selig *et al.*, 1984).

The platinum compound $C_{12}PtF_6$ was found to be nonmagnetic. This was taken as evidence for the presence of the diamagnetic $PtF_6{}^{2-}$ species (Bartlett *et al.*, 1979–1980). Second-moment calculations based on a room-temperature ^{19}F spectrum, however, can be reconciled with a spin-paired $Pt_2F_{11}{}^-$ (Ebert, 1982). Because a highly charged system such as $C_{12}{}^{2+}PtF_6{}^{2-}$ is quite unusual, further investigations are warranted.

4. Electrical Properties

Basal plane conductivities of HOPG exposed to OsF_6 or IrF_6 increase smoothly, reaching a maximum at approximately second stage (Bartlett *et al.*, 1979–1980). The σ_{max}/σ_g ratio is ~ 8. First-stage materials show conductivities equal to or less than that of pristine HOPG. These results parallel those for the C/AsF_5 system and correspond to conductivities obtained for "overcharged" and "overfluorinated" compounds, respectively. These lower conductivities have been ascribed to distortion of the graphite layers in the highly charged system. The loss of crystallinity as a function of time is thought to be a consequence as well. By contrast, even second-stage $C_{24}PtF_6$ has low basal plane conductivity. The graphite/MoF_6 system, however, shows strikingly different behavior. The conductivity per graphite plane increases stepwise with time. The step heights do not correspond with those expected from changes of thickness in the c direction, indicating that charge transfer in this system does not increase smoothly. In addition, conductivity of stage I C/MoF_6 is larger than that of HOPG (Selig *et al.*, 1984).

D.   COMPOUNDS THAT INTERCALATE WITH OXIDATIVE COREAGENTS

Certain acceptors intercalate only in the presence of an oxidant. In order for this reaction to occur spontaneously in the presence of fluorine, the free energy $(-\Delta G^0_{298})$ of the reaction

$$MF_5(g) + \tfrac{1}{2} F_2(g) + e^- \longrightarrow MF_6{}^-(g) \tag{10}$$

must exceed 120 kcal mol^{-1} (Bartlett *et al.*, 1983). Other oxidants work as well. Thus, boron trifluoride and phosphorus pentafluoride can be intercalated in solution by the use of nitronium salts (Billaud *et al.*, 1982; Forsman and Mertwoy, 1980) [Eq. (5)].

Another oxidative coreagent used is chlorine monofluoride (Ebert *et al.*, 1981) (Eq. (11)]. In the dilute compound "$C_{28}PF_6$," both ^{31}P- and ^{19}F-nmr

$$m\,C + MF_x + ClF \longrightarrow C_m^+ MF_{x+1}^- + \tfrac{1}{2} Cl_2$$
$$(MF_x = BF_3, PF_5) \tag{11}$$

spectra show the presence of PF_6^-, but the more concentrated "$C_{14}PF_6$" (Ebert *et al.*, 1981) shows only a singlet, albeit with ^{31}P chemical shift closer to that of PF_6^- than PF_5. These results indicate an upper limit to the amount of charge transfer and either exchange between PF_5 and PF_6^- or formation of an oligomer:

$$PF_6^- + PF_5 \rightleftharpoons P_2F_{11}^- \qquad (12)$$

In the case of BF_3, the fluorine chemical shift (70 ± 10 ppm with respect to CF_3COOH) is more compatible with the presence of BF_4^- (71 ppm) than with BF_3 (54 ppm). Second-order quadrupole splitting of the ^{11}B resonance suggests the possible formation of $B_2F_7^-$, a known dimeric species (Ebert et al., 1981).

Germanium tetrafluoride can be intercalated in the presence of fluorine (McCarron *et al.*, 1980). At the intercalation limit a first-stage compound is obtained which loses F_2 reversibly according to

$$C_{12}GeF_5(s) + \tfrac{1}{2}F_2(g) \rightleftharpoons C_{12}GeF_6(s) \qquad (13)$$

Although IF_5 can be intercalated only in the presence of fluorine (Selig *et al.*, 1978), a ^{127}I Mössbauer study of $C/IF_5/F_2$ has shown that most of the intercalated species is neutral IF_5 (Wortman *et al.*, 1984). An upper limit of 5% was found for the presence of IF_6^-.

References

Bartlett, N., and McQuillan, B. W. (1982). *In* "Intercalation Chemistry" (M. S. Whittingham and A. J. Jacobson, eds.), p. 19. Academic Press, New York.

Bartlett, N., Biagoni, R. N., McQuillan, B. W., Robertson, A. S., and Thompson, A. C. (1978a). *J. Chem. Soc., Chem. Commun.* p. 200.

Bartlett, N., McQuillan, B. W., and Robertson, A. S. (1978b). *Mater. Res. Bull.* **13**, 1254.

Bartlett, N., McCarron, E. M., McQuillan, B. W., and Thompson, T. E. (1979–1980). *Synth. Met.* **1**, 221.

Bartlett, N., Mallouk, T., Okino, F., Rosenthal, G., and Verniolle, J. (1983). *J. Fluorine Chem.* **23**, 409.

Beck, F., Kaiser, W., and Krohn, H. (1982). *Angew. Chem., Suppl.* **57**.

Billaud, D., Pron, A., and Vogel, L. V. (1980). *Synth. Met.* **2**, 177.

Billaud, D., Flanders, P. J., Fischer, J. E., and Pron, A. (1982). *Mater. Sci. Eng.* **54**, 31.

Binenboym, J., Selig, H., and Sarig, S. (1976). *J. Inorg. Nucl. Chem.* **38**, 2313.

Braeuer, K. (1968). *Electrochem. Soc. Ext. Abstr., Spring Meet.* **210**, 495.

Brauer, K., and Moyes, K. R. (1970). U. S. Patent **3**, 514, 337.

Cadman, P., Scott, J. D., and Thomas, J. M. (1975). *J. Chem. Soc., Chem. Commun.* p. 654.

Clark, D. T., and Peeling J. (1976). *J. Polym. Sci., Polym. Chem. Ed.* **14**, 2941.

Dresselhaus, M. S., and Dresselhaus, G. (1981). *Adv. Phys.* **30**, 139.

Ebert, L. B. (1982). *J. Mol. Catal.* **15**, 275.

Ebert, L. B., and Selig, H. (1977). *Mater. Sci. Eng.* **31,** 177.

Ebert, L. B., Brauman, J. I., and Huggins, R. A. (1974). *J. Am. Chem. Soc.* **96,** 7841.

Ebert, L. B., DeLuca, J. P., Thompson, A. H., and Scanlon, J. C. (1977). *Mater. Res. Bull.* **12,** 1135.

Ebert, L. B., Mills, D. R., Scanlon, J. C., and Selig, H. (1981). *Mater. Res. Bull.* **16,** 831.

Falardeau, E. R., Hanlon, L. R., and Thompson, T. R. (1978). *Inorg. Chem.* **17,** 301.

Fischer, J. E. (1979). *In* "Intercalated Layered Materials" (F. Levy, ed.), p. 481. Reidel Publ., Dordrecht, Netherlands.

Fischer, J. E., Metrot, A., Flanders, P. J., Salaneck, W. R., and Brucker, C. F. (1981). *Phys. Rev. B: Condens. Matter* [3] **23,** 5576.

Foley, G. M. T., Zeller, C., Falardeau, E. R., and Vogel, F. L. (1977). *Solid State Commun.* **24,** 371.

Forsman, W. C., and Mertwoy, H. E. (1980). *Synth. Met.* **2,** 171.

Forsman, W. C., Dziemianowicz, T., Leong, K., and Carl, D. (1983). *Synth. Met.* **5,** 77.

Hamwi, A., Touzain, P., and Bonnetain, L. (1977). *Mater. Sci. Eng.* **31,** 95.

Hartmann, S. R., and Hahn, E. L. (1962). *Phys. Rev.* **128,** 2042.

Hiratsuka, T., Shimada, T., and Akatsuka, Y. (1966). *Symp. Carbon Mat., 19th, Jpn.,* p. 18.

Holzwarth, N. A. W., Louie, S. G., and Rabii, S. (1981). *Phys. Rev. Lett.* **47,** 1318.

Interrante, L. V., Markiewiz, R. S., and McKee, D. W. (1979). *Synth. Met.* **1,** 287.

Kammarchik, P., and Margrave, J. L. (1978). *Acc. Chem. Res.* **11,** 296.

Kita, Y., Watanabe, N., and Fujii,Y. (1979). *J. Am. Chem. Soc.* **101,** 3832.

Lagow, R. J., Badachhape, R. B., Ficalora, P., Wood, J. L., and Margrave, J. L. (1972). *Synth. Inorg. Met.-Org. Chem.* **2,** 145.

Lagow, R. J., Badachhape, R. B., Wood, J. L., and Margrave, J. L. (1974). *J. Chem. Soc., Dalton Trans.* p. 1268.

Lalancette, J. M., and LaFontaine, J. (1973). *J. Chem. Soc., Chem. Commun.* p. 815.

Lin Chun-Hsu, Selig, H., Rabinowitz, M., Agranat, I., and Sarig, S. (1975). *Inorg. Nucl. Chem. Lett.* **11,** 601.

McCarron, E. M., and Bartlett, N. (1980). *J. Chem. Soc., Chem. Commun.* p. 404.

McCarron, E. M., Grannec, Y. J., and Bartlett, N. (1980). *J. Chem. Soc., Chem. Commun.* p. 890.

Mahajan, V. K., Badachhape, R. B., and Margrave, J. L. (1974). *Inorg. Nucl. Lett.* **10,** 1103.

Mallouk, T., and Bartlett, N. (1983). *J. Chem. Soc., Chem. Commun.* p. 103.

Markiewicz, R. S., Hart, H. R., Jr., Interrante, L. V., and Kasper, J. S. (1980). *Synth. Met.* **2,** 331.

Metrot, A., and Fischer, J. E. (1981). *Synth. Met.* **3,** 201.

Milliken, J. W., and Fischer, J. E. (1983). *J. Chem. Phys.* **78,** 5800.

Moran, M. J., Fischer, J. E., and Salaneck, W. R. (1980). *J. Chem. Phys.* **73,** 629.

Moran, M. J., Milliken, J. W., Zeller, C., Grayeski, R. A., and Fischer, J. E. (1981). *Synth. Met.* **3,** 269.

Moran, M. J., Miller, G. R., DeMarco, R. A., and Resing, H. A. (1984). *J. Phys. Chem.* **88,** 158.

Muilenberg, G. E., ed. (1978). "Handbook of Photoelectron Spectroscopy." Perkin-Elmer Corp., Eden Prairie, Minnesota.

Münch, V., and Selig, H. (1979). *Synth. Met.* **1,** 407.

Nakajima, T., Kawaguchi, M., and Watanabe, N. (1981). *Z. Naturforsch., B: Anorg. Chem., Org. Chem.* **36B,** 1419.

Nakajima, T., Kawaguchi, M., and Watanabe, N. (1982). *Carbon* **20,** 287.

Nakajima, T., Kawaguchi, M., Kawasaki, T., and Watanabe, N. (1983). *Nippon Kagaku Kaishi* p. 283.

Okino, F., and Bartlett, N. (1984). Submitted for publication.

Opalovskii, A. A., Kuznetsova, Z. M., Chicagov, Yu. V., Nazarov, A. S., and Uminskii, A. A. (1974). *Zh. Neorg. Khim.* **19,** 2071.

Palin, D. E., and Wadsworth, K. D. (1948). *Nature (London)* **162,** 925.

Pentenrieder, R., and Boehm, H. P. (1982). *Rev. Chim. Miner.* **19,** 371.

Perry, D. E., Thomas, J. M., Bach, B., and Evans, E. L. (1974). *Chem. Phys. Lett.* **29**, 128.

Ring, R. J., and Royston, D. (1973). "A Review of Fluorine Cells and Fluorine Production Facilities." AAEC Rep. E-281.AAE Commission, Australia.

Rüdorff, W., and Brodersen, K. (1957). *Z. Naturforsch., B: Anorg. Chem., Org. Chem., Biochem., Biophys., Biol.* **12B**, 595.

Rüdorff, W., and Rüdorff, G. (1947a). *Z. Anorg. Allg. Chem.* **253**, 281.

Rüdorff, W., and Rüdorff, G. (1947b). *Chem. Ber.* **80**, 413.

Ruff, O., Bretschneider, O., and Ebert, F. (1934). *Z. Anorg. Allg. Chem.* **217**, 1.

Selig, H., and Ebert, L. B. (1980). *Adv. Inorg. Chem. Radiochem.* **23**, 281.

Selig, H., Vasile, M. J., Stevie, F. A., and Sunder, W. A. (1977). *J. Fluorine Chem.* **10**, 99.

Selig, H., Sunder, W. A., Vasile, M. J., Stevie, F. A., Gallagher, P. K., and Ebert, L. B. (1978). *J. Fluorine Chem.* **12**, 347.

Selig, H., Vaknin, D., Yeshurun, Y., and Davidov, D. (1984). *Proc. Materials Res. Soc. Symp., 1984, Boston.*

Solin, S. A. (1980). *Physica B (Amsterdam)* **99B**, 443.

Streifinger, L., Boehm, H. P., Schögl. R., and Pentenrieder, R. (1979). *Carbon* **17**, 195.

Takashima, M., and Watanabe, N. (1975). *Nippon Kagaku Kaishi* p. 432.

Takenaka, H., and Bartlett, N. (1984). To be published.

Thompson, T. E., McCarron, E. M., and Bartlett, N. (1981). *Synth. Met.* **3**, 255.

Touhara, H., Kadono, K., Endow, M., and Watanabe, N. (1983à). *Abstr. 50th Annu. Meet. Electrochem. Soc. Jpn.*, p. 150 (*J. Chim. Phys.* **85**).

Touhara, H., Kadono, K., Watanabe, N. and Endor, M. (1983b). *Abstr. 50th Annu. Meet. Electrochem. Soc.* 58 (*Electrochem. Acta*, submitted).

Touhara, H., Kadono, K., and Watanabe, N. (1984a). *Tanso (Carbon)* **117**, 98.

Touhara, H., Fujimoto, H., Watanabe, N., and Tressaud, A. (1984b). *Solid State Ionics* **14**, 163.

Tuinstra, F., and Koenig, J. L. (1970). *J. Chem. Phys.* **53**, 1126.

Ueno, K., Watanabe, N., and Nakajima, T. (1982). *J. Fluorine Chem.* **19**, 323.

Vaknin, D., Davidov, D., Yeshurun, Y., and Selig, H. (1983). *J. Fluorine Chem.* **23**, 425.

Vogel, F. L. (1977). *J. Mater. Sci.* **12**, 982.

Watanabe, N. (1980). *Solid State Ionics* **1**, 87.

Watanabe, N., and Fukuda, M. (1970). U. S. Patent 3,536,532. Watanabe, N., and Fukuda, M. (1972). U. S. Patent 3,700,502.

Watanabe, N., and Nakajima, T. (1982). *In* "Preparation Properties and Industrial Applications of Organofluorine Compounds" (R. E. Banks, ed.), p. 297, and publications quoted therein. Ellis Horwood, Chichester, England.

Watanabe, N., and Touhara, H. (1981). *Hyomen* **20**, 502.

Watanabe, N., Inoue, M., and Yoshizawa, S. (1963a). *J. Electrochem. Soc. Jpn.* **31**, 113.

Watanabe, N., Fujii, Y., and Yoshizawa, S. (1963b). *J. Electrochem. Soc. Jpn.* **31**, 131.

Watanabe, N., Inoue, M., and Yoshizawa, S. (1963c). *J. Electrochem. Soc. Jpn.* **31**, 168.

Watanabe, N., Takashima, M., and Kita, Y. (1974). *Nippon Kagaku Kaishi* p. 1033.

Watanabe, N., Takenaka, K., and Kimura, S. (1979). *Nippon Kagaku Kaishi* p. 1027.

Watanabe, N., Hagiwara, R., Nakajima, T., Touhara, H., and Ueno, K. (1982b). *Electrochim. Acta* **27**, 1615.

Watanabe, N., Hagiwara, R., Kumagai, N., Sakai, T., and Nakajima, T. (1983a). *Denki Kagaku* **51**, 183.

Watanabe, N., Touhara, H., and Nakajima, T. (1983b). *In* "Intercalated Graphites" (M. S. Dresselhause, G. Dresselhause, J. E. Fischer, and M. J. Moran, eds.), p. 247. North-Holland, New York.

Weinberger, B. R., Kaufer, J., Heeger, A. J., Fischer, J. E., Moran, M. J., and Holzwarth, N. A. W. (1978). *Phys. Rev. Lett.* **41**, 1417.

Whittingham, M. S. (1975). *J. Electrochem. Soc.* **122,** 526.
Whittingham, M. S., and Ebert, L. B. (1979). *In* "Intercalated Layered Materials" (F. Levy, ed.), p. 533. Reidel, Dortdrecht, Netherlands.
Wilke, C. A., Yu Lin, G., and Howarth, D. T. (1979). *J. Solid State Chem.* **30,** 197.
Wood, J. L., Badachhape, R. B., Lagow, R. J., and Margrave, J. L. (1969). *J. Phys. Chem.* **73,** 3139.
Wortman, G., Nowik, I., Kaindl, G., Selig, H., and Palchan, I. (1984–1985). *Synth, Met.* **10,** 141.
Yannoni, C. A. (1982). *Acc. Chem. Res.* **15,** 201.

9

Ferro- and Ferrimagnetism in Fluorides

JEAN-MICHEL DANCE and ALAIN TRESSAUD
Laboratoire de Chimie du Solide du CNRS
Université de Bordeaux
Talence, France

I. Introduction

In the early 1960s, speaking simultaneously of fluorine chemistry and strongly magnetic materials seemed to be a real paradox. Only one exception could be detected: the unusual ferrimagnet $Na_5Fe_3F_{14}$ of the chiolite type (Knox and Geller, 1958). Since that time, various structural types of fluorides showing spontaneous magnetization have been brought forward.

In addition to the practical interest of solid-state physicists in new materials showing both strong magnetization and optical transparency (Wolfe *et al.*,

INORGANIC
SOLID FLUORIDES

1970), the investigation of the magnetic couplings in ferro- and ferrimagnetic fluorides allows, on the basis of the semiempirical models proposed by Anderson, Goodenough, and Kanamori, a better approach to the nature of magnetic interactions in insulators.

II. Ferromagnetism in Fluorides

Most ionic ferromagnets are oxides (CrO_2, EuO, ferromagnetic manganites, etc.), chalcogenides [$CuCr_2X_4$ (X = S, Se, or Te), CoS_2, EuS, etc.], and halides containing chlorine, bromine, or iodine atoms (CrX_3, M_2CrX_4, M_2CuX_4, and $RbFeCl_3$).

Within the halide group three main classes of ferromagnetic fluorides have been investigated: rare-earth fluorides with scheelite or YF_3-type structure, layer-type A_xMF_4 series containing Jahn–Teller ions, and Pd_2F_6-type fluorides.

We deal here only with those series showing a bulk ferromagnetism. Compounds in which ferromagnetic units, that is, isolated clusters, infinite chains (of the $CsNiF_3$ type) or layers, are antiferromagnetically coupled below the three-dimensional (3D) ordering temperature are not considered, nor are those showing a weak ferromagnetic component associated with canted antiferromagnetism (FeF_3 and CoF_3).

It should be mentioned that ferromagnetism has been observed in amorphous FeF_2, which up to now has been the only example (Litterst, 1975). It orders ferromagnetically below 21 K, and a possible explanation of this property could be the change in bonding angles: the 135° Fe–F–Fe angle, which is responsible for strong antiferromagnetic interactions in crystalline rutile FeF_2, could approach 90° in the amorphous phase, thus favoring ferromagnetic interactions.

A. UNIAXIAL FERROMAGNETISM IN YTTRIUM AND RARE-EARTH FLUORIDES

Ferromagnetic properties have been observed in two different structural types: $LiRF_4$ compounds (R = rare earth) of the scheelite type and TbF_3. Numerous theoretical and applied studies of the first class of compounds have been particularly developed due to practical interest in these compounds as efficient laser materials (Chicklis et al., 1971). They can be used as converters of laser radiations in the infrared (ir)–visible region. Nd-doped $LiGdF_4$ and Ho- and Pr-doped $LiYF_4$ and $LiYbF_4$ have been thus considered (Esterowitz et al., 1977; Kuppenheimer and Baer, 1982; Fernelius et al., 1981; Antipenko et al., 1980). (These spectroscopic properties are further discussed in Chapter 14.)

1. Ferromagnetism in Scheelite-Type Fluorides

The LiRF$_4$ compounds crystallize in the tetragonal $I4_1/a$ space group with scheelite-type structure. Large, transparent, and optically uniaxial single crystals have been grown since the early 1960s generally using either the Stöckbarger or Czochralski method (Guggenheim, 1963; Harris *et al.*, 1983) (for the lattice constants, see Chapter 3, Table XLIV, and for further details on synthesis methods, see Chapter 2).

The relative simplicity of the structure (Fig. 1) allowed Misra and Felsteiner (1977) to make theoretical calculations of the possible magnetic ordering using a generalization of the Luttinger–Tisra method and considering only dipolar interactions between rare-earth elements. These authors predicted ferromagnetism for $g_\parallel > g_\perp$ and antiferromagnetism for $g_\parallel < g_\perp$, where $g_\parallel$ and $g_\perp$ are the Landé factors relative to the c axis.

LiTbF$_4$ and LiHoF$_4$ present a strong magnetization at low temperature, as shown in Table I. LiErF$_4$, which was initially predicted to be ferromagnetic (Hansen *et al.*, 1975), has been shown to be actually an antiferromagnet with $T_N = 0.381$ K and with the magnetic moments lying in the ab plane (Beauvillain *et al.*, 1977). These results are consistent with the theoretical proposals as $g_\perp \gg g_\parallel$ for LiErF$_4$ and $g_\perp \simeq 0$, $g_\parallel = 17.85$ and 13.95 for LiTbF$_4$ and LiHoF$_4$, respectively (Beauvillain *et al.*, 1978).

LiTbF$_4$ and LiHoF$_4$ are highly anisotropic, with the easy axis along the c direction; the transverse magnetization ($M_\perp[001]$) of LiTbF$_4$, for instance, is only 4% of the parallel magnetization ($M_\parallel[001]$) at $H = 60$ kOe and $T = 1.1$ K (Holmes *et al.*, 1973). These materials appear to be nearly pure uniaxial ferromagnets (Ising-type materials with $n = 1$) with a magnetic coupling

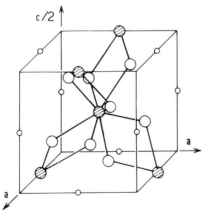

Fig 1. Half of the unit cell of a scheelite-type structure showing the four magnetic neighbors of the rare-earth (R) ions. Key: ○, Li; ◓, R; ○, F.

TABLE I

Magnetic Data Relative to Ferromagnetic Rare-Earth Fluorides

Compound	T_C (K)[a]	Experimental Technique	Reference
LiTbF$_4$	~2.86	Magnetization ($M_{0K} \simeq 8.9\ \mu_B$ along c axis)	Holmes et al. (1973)
	2.885(1)	Specific heat	Ahlers et al. (1975)
	2.8707(1)	Spontaneous magnetization	Griffin et al. (1977); Frowein et al. (1979); Frowein and Kötzler (1982)
	2.8850(2)	Magnetic susceptibility	Beauvillain et al. (1980b)
LiHoF$_4$	~1.30	Magnetization ($M_{0K} = 6.98\ \mu_B$ along c axis)	Hansen et al. (1975)
	1.527(1)	Magnetic susceptibility	Beauvillain et al. (1978)
TbF$_3$	~3.95	Magnetization ($M_{0K} = 8.09\ \mu_B$ along c axis)	Holmes et al. (1970)
	3.967(1)	Spontaneous magnetization, zero-field susceptibility	Brinkmann et al. (1978)

[a] Numerals in parentheses indicate standard deviation in the last figure given.

between R^{3+} ions mostly dipolar in nature. The exchange contribution to the magnetic interaction was shown by Beauvillain et al. (1978) to be three times smaller than the dipolar one in LiHoF$_4$ ($zJ_{exch}/4k = -0.820$ K). This uniaxial anisotropy is also responsible for the structure of the ferromagnetic domains, which are alternating cylinders confined to the c axis.

The variation of the physical data in the critical region (specific heat, magnetization, and magnetic susceptibility) has also been widely studied. The results provide a quantitative comparison between the predicted logarithmic corrections to mean-field theory and the experimental results. The spontaneous magnetization of LiTbF$_4$ has been measured in the critical region using the magneto-optic features. Over the range $0.002 \le t \le 0.09$ its variation is best described using the following logarithmic corrections:

$$M_c(T)/M_c(0) = B|t|^{1/2}|\ln(t_0/t)|^{1/3},$$

with $t = (T_C - T)/T_C$, $B = 1.77$, $T_C = 2.870$, and $t_0 = 0.53$ (Fig. 2) (Griffin et al., 1977; Griffin and Litster, 1979). Similar results were obtained with data on specific heat (Ahlers et al., 1975), correlation length (Als-Nielsen, 1976), and crossover of the susceptibility (Frowein et al., 1982).

The crossover between pure critical behavior and random critical behavior was studied in the randomly diluted ferromagnetic series LiTb$_x$Y$_{1-x}$F$_4$ for concentrations in the range $0.155 \le x < 1$. The T_C decreases linearly with x. The slope $[dT_C(x)/T_C(x = 1)]/dx = 1.137$ K extrapolates to a percolation concentration $x \simeq 0.12$, a value that could not be predicted using either of the calculations for dipolar coupled magnets (Folkins et al., 1982; Brierley and

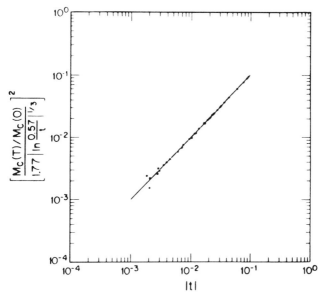

Fig. 2. Comparison of the measured spontaneous magnetization of $LiTbF_4$ with the predicted logarithmic corrections to mean-field theory (——) ($T_C = 2.8702$ K). From Griffin *et al.* (1977).

Griffin, 1982). For higher contents of nonmagnetic ions ($x = 0.30$ and 0.50), the critical behavior is strongly modified: in contrast with pure $LiTbF_4$, the parallel susceptibility, for instance, cannot be described using logarithmic corrections to a classical law but is better obtained by a power law $t^{-\gamma}$ with $\gamma = 1.80$ and 1.21 for $x = 0.30$ and 0.50, respectively (Beauvillain *et al.*, 1980a,c, 1983). This phenomenon is still not completely understood.

2. Ferromagnetism in TbF₃

In the orthorhombic TbF_3 (space group *Pnma*, Zalkin and Templeton, 1953), the magnetic behavior is also characterized by a high anisotropy approaching the Ising limit. The compound may be described as a uniaxial canted ferromagnet with the moment along the *a* axis or as a canted antiferromagnet with a staggered magnetic moment along the *c* axis. At 1.1 K the saturation magnetization is 8.09 μ_B along the *a* axis and only 3.84 μ_B along the *c* axis (Holmes *et al.*, 1970).

In the critical region, the physical data (spontaneous magnetization, zero-field susceptibility, and isothermal critical magnetization along the *a* axis) again confirm the renormalization group predictions and Larkin and Khmel'nitskii calculations (1969). The variations can best be described by logarithmic corrections to mean-field law (Brinkmann *et al.*, 1978).

B. FERROMAGNETISM IN LAYER-TYPE FLUORIDES

1. Jahn–Teller Orderings

For some d-element ions in E_g ground state (high-spin d^4, low-spin d^7, d^9) a cooperative Jahn–Teller effect generally yields important crystallographic distortions. In the case of a d^9 ion in an octahedral coordination, for example, $d_{x^2-y^2}$ orbitals are half-filled and d_{z^2} orbitals are totaly filled, leading to elongated MF_6 octahedra, the direction of elongation of which corresponds to the d_{z^2} orbital. Reinen and Friebel (1979) have shown that two arrangements preferentially occur. In the antiferrodistortive ordering the directions of the longer bonds are alternately perpendicular to each other, whereas in the ferrodistortive mode all these directions are parallel.

The magnetic couplings deriving from these two crystallographic arrangements differ considerably. For d^9 ions, for instance, showing an elongated octahedral coordination, a ferrodistortive crystallographic ordering will lead to antiferromagnetic couplings, whereas an antiferrodistortive mode will favor ferromagnetism. When the polyhedra containing the transition element are 3D linked by a common element (corner, edge, or face), the ferromagnetic units are generally coupled antiparallel, both positive and negative exchange constants having roughly the same order of magnitude. This is true, for instance, in rutile-related CuF_2, AgF_2, ReO_3-related MnF_3, and perovskite-related $KCrF_3$, in which a bulk 3D antiferromagnetism is observed. In contrast, when the structure is formed of layers containing the Jahn–Teller element, the intralayer ferromagnetic couplings are much stronger than the interlayer ones, and a 3D ferromagnetic behavior is maintained at low temperature.

This is the situation for compounds of formulations $A_2^I M^{II} F_4$ and $A^I M^{III} F_4$, which can be derived from $K_2 NiF_4$ and $TlAlF_4$, respectively. It may be pointed out that the $Ba_2 ZnF_6$-type structure, which can be alternatively written $(BaF)_2 ZnF_4$, is also favorable to 2D ferromagnetism (Reinen and Weitzel, 1977). A positive value of θ_p (15 K) had been detected by Schnering (1967) in $Ba_2 CuF_6$, which is actually ferromagnetic below 8 K (Renaudin and Ferey, 1984). $Pb_2 CuF_6$ is also ferromagnetic below $T_C = 5$ K (Dance, 1985).

2. $A_2 CuF_4$ Ferromagnetic Fluorides

Ferromagnetism in the $A_2 CuF_4$ series ($A = K, Rb, Cs, NH_4$, or Tl) has been widely investigated by means of a large number of physical techniques. Large, transparent single crystals were obtained by the melting-zone technique in a fluorinating atmosphere (Hirakawa and Ikeda, 1972). The structure determination of $K_2 CuF_4$ was initially carried out by analogy with that of $K_2 NiF_4$; further results required the presence of a larger unit cell (Haegele and

Babel, 1974). The CuF_6 octahedra present a tetragonal distortion, with the elongation axis being alternately directed along a and b axes (Friebel and Reinen, 1974). A multidomain structure has been more recently proposed; it is based on displacements of fluorine atoms (Hidaka and Walker, 1979). These displacements could be attributed to the two possible orientations of Cu^{2+} e_g orbitals relative to the crystallographic a and b axes. However, epr investigations of Reinen and Krause (1981) do not support these results and a crystallographic determination by Herdtweck and Babel (1981) has shown no superstructure reflections.

Ferromagnetism results from an electronic ordering of $d_{z^2-x^2}$ and $d_{z^2-y^2}$ orbitals initially proposed by Khomskii and Kugel (1973). The arrangement of orbitals shown in Fig. 3 allows for electronic transfers with unchanged spin direction from filled orbitals to half-filled orbitals. This model takes into account the crystallographic results and has been confirmed by an nmr study (Gupta et al., 1975).

Magnetization, neutron or Brillouin scattering, specific heat, and resonance measurements have been performed on the A_2CuF_4 series, mainly on K_2CuF_4 (Hirakawa and Ikeda, 1974; Le Dang Khoi and Veillet, 1976; Dupas and Renard, 1975; Yamada and Suzuki, 1976; Moussa et al., 1978). Isostructural Rb_2AgF_4 and Cs_2AgF_4 are ferromagnetic below $T_C \simeq 20$ K (Odenthal et al., 1974).

The magnetic interactions are of the ferromagnetic Heisenberg type with a 1% XY-square contribution arising from both anisotropic exchange and dipole–dipole interactions. The easy plane is $x0z$, and the spins are weakly bound along the a axis (Hirakawa and Ikeda, 1973b). The intralayer exchange

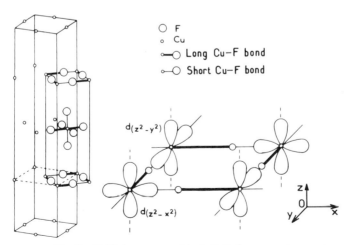

○ F
ₒ Cu
●—○ Long Cu–F bond
ₒ—○ Short Cu–F bond

$d_{(z^2-y^2)}$

$d_{(z^2-x^2)}$

Fig. 3. Orbital arrangement in A_2CuF_4 ferromagnets.

TABLE II

Magnetic Data of A_2CuF_4 Ferromagnets

Compound	T_C (K)	θ_p (K)	J/k (K)	J'/k (K)
K_2CuF_4	6.25(3)	16^a	11.2 (high-temperature expansion series)^a 8.8 (specific heat)^a 11.4 (spin–wave)^c	0.024 (Oguchi method)^b 0.034 (parallel pumping)^d
Rb_2CuF_4	17.5^e	13^f	—	—
Cs_2CuF_4	9.8^g	14^g	8.3 (high-temperature expansion series)^g	—

[a] Yamada (1970, 1971, 1972).
[b] Hirakawa and Ikeda (1973a, 1974).
[c] Funahashi et al. (1976).
[d] Yamazaki (1973, 1974).
[e] Gupta et al. (1975).
[f] Rüdorff et al. (1963).
[g] Dance et al. (1976b).

constant J/k has been obtained from high-temperature expansion series calculations and from the linear dependence of the spin–wave contribution to the heat capacity (Table II). The interlayer constant of K_2CuF_4 is $J'/k \simeq 0.03$ K. The corresponding value of J'/J ($\simeq 3 \times 10^{-3}$) is larger than those generally found in 2D antiferromagnetic systems (10^{-6}–10^{-4}) and a 3D critical behavior is thus expected. In addition, the temperature dependence of the magnetization follows in the low-temperature range ($0.12 < kT/J < 0.36$) a $T^{3/2}$ variation law predicted by the spin–wave theory for a 3D ferromagnet.

Another estimate of the dimensionality of the interactions is given by the value of the critical exponent β. For K_2CuF_4 it is 0.22 in the temperature range $0.85 < T/T_C < 0.99$ (Hirakawa and Ikeda, 1972); however, this value, which is close to those found for $AFeF_4$ 2D antiferromagnetic series (Dance et al., 1976a), does not seem to be sufficiently conclusive: $\beta(2D) = 0.125$; $\beta(3D$ Heisenberg$) = 0.36$.

The effect of dilution on the exchange constant has been studied in the $K_2Cu_{1-x}Mn_xF_4$ system (Dance et al., 1978; Ferré and Régis, 1978). The critical concentration for a ferromagnetic → paramagnetic transition is $x \simeq 0.50$ for the $K_2Cu_{1-x}Zn_xF_4$ system (Okuda et al., 1980).

3. A^IMnF_4 Ferromagnetic Fluorides

Mn(III) generally has a high-spin d^4 configuration in fluorides and is therefore associated with an important Jahn–Teller effect. Compounds A^IMnF_4 (A = K, Rb, Cs, or NH_4) have a layer structure deriving from the

TABLE III

Magnetic and Ligand Field Parameters (at 4.5 K) of $AMnF_4$ Compounds[a]

Compound	T_C (K)	θ_p (K)	$4\delta_1$ (cm^{-1})	$3\delta_2$ (cm^{-1})
$KMnF_4$	6	−45	15,200	4300
$RbMnF_4$	< 4.5	−14	~15,500	~4000
$CsMnF_4$	21	27	15,300	3900
	18.9[b]			
NH_4MnF_4	10	−12	~14,700	~3800

[a] $4\delta_1$ and $3\delta_2$ are the splittings of the ^5E_g ground state and the $^5T_{2g}$ excited state, respectively. From Köhler et al. (1978).

[b] Massa and Steiner (1980).

$TlAlF_4$ type. Here again, distorted MnF_6 octahedra are connected to each other by four vertices (Massa, 1977).

Although these phases have been investigated less extensively than the copper homologs, some of their magnetic and spectroscopic properties have been characterized (Table III). The thoroughly studied $CsMnF_4$ crystallizes in the tetragonal $P4/nmm$ space group ($a = 7.94$ Å, $c = 6.34$ Å), which would imply a ferrodistortive ordering of compressed MnF_6 octahedra. Because this hypothesis seems highly improbable, Massa and Steiner (1980) have proposed an antiferrodistortive order of elongated octahedra with an average behavior.

C.   FERROMAGNETISM IN Pd_2F_6-TYPE FLUORIDES

A large number of $M^{II}M'^{IV}F_6$ fluorides are isostructural with $Pd^{II}Pd^{IV}F_6$ ($LiSbF_6$ type). The structure derives from the ReO_3 type by a rhombohedral distortion and a cationic ordering. Additional crystallographic information is given in Chapter 3, Section II,D.

The magnetic data of Pd_2F_6-type ferromagnets are collected in Table IV. The values of saturation magnetization and T_C were determined from $M = f(H^{-1})$ and $M^2 = f(H/M)$ curves, respectively, using the methods developed by Belov and Goryaga (1956) and Kouvel (1957). Ferromagnetism may be associated with the cationic ordering. Neutron diffraction experiments are often required to detect this ordering, especially when both M and M' are d elements with close form factors. The ferromagnetic behavior can therefore be explained by the following mechanism. The spins of the electrons occupying the half-filled e_g orbitals of Ni(II), Pd(II), and Pt(II) in the high-spin $t_{2g}^6 e_g^2$ state are ferromagnetically coupled via a superexchange interaction involving the $2p$ fluorine orbitals and the empty e_g orbitals of tetravalent cations [Mn(IV) in the high-spin $t_{2g}^3 e_g^0$ configuration; Pd(IV) and Pt(IV) in the low-spin t_{2g}^6

TABLE IV

Magnetic Data of Pd_2F_6-Type Ferromagnets

Compound	T_C (K)	M_{sat} (μ_B)		θ_p (K)	C_{mol}		Reference
		Experimental (4.2 K)	Theoretical		Experimental	Theoretical	
$Ni^{II}Mn^{IV}F_6$	39	4.55	5	44	3.10	2.875 ⎫	
$Zn^{II}Mn^{IV}F_6$	9.5	2.87	3	6	2.43	1.875 ⎬ Lorin et al. (1981)	
$Cd^{II}Mn^{IV}F_6$	8	2.34	3	3[a]	1.79[c]	1.875 ⎭	
$Ni^{II}Pd^{IV}F_6$	6	1.42	2	4	1.32	1	A. Tressaud, R. C. Sherwood, and N. Bartlett (unpublished results, 1977); Lorin (1980)
$Pd^{II}Pd^{IV}F_6$	10	1.80	2	13	—	1	Tressaud et al. (1976a)
$Ni^{II}Pt^{IV}F_6$	8	1.09	2	4	1.12	1 ⎫ A. Tressaud, R. C. Sherwood, and N. Bartlett (unpublished results, 1977); Lorin (1980)
$Pd^{II}Pt^{IV}F_6$	25	1.70	2	22	1.038	1 ⎬
$Pt^{II}Pt^{IV}F_6$	16	1.60	2	18	1.16	1	Tressaud et al. (1976b)

[a] Hoppe and Siebert (1970).

configuration]. For Pd_2F_6 the best agreement between the observed and calculated magnetic intensities is obtained for magnetic moments oriented perpendicularly to the ternary axis; the calculated moment of ⁎Pd(II) is equal to 1.75 μ_B at 4.2 K (Tressaud et al., 1976a).

For ordered $ZnMnF_6$ and $CdMnF_6$, Mn(IV) ions are coupled according to the sequence $Mn_1-F_1-M^{II}-F_2-Mn_2$. The ferromagnetic behavior can be explained by a double-correlation superexchange mechanism in which two electrons, respectively located on F_1 and F_2 orbitals, are transferred without spin changing on Mn_1 and Mn_2 empty e_g orbitals or half-filled t_{2g} orbitals. The low value of T_C is consistent with the large Mn–Mn distance (Lorin et al., 1981).

One can point out that two conditions seem to be required to obtain a Pd_2F_6-type ferromagnet: (1) a cationic ordering [$PdMnF_6$, which has a disordered VF_3-type structure, is antiferromagnetic (Lorin et al., 1981)] and (2) the presence of empty e_g orbitals on the tetravalent species [$Pd^{II}Ge^{IV}F_6$ and $Pd^{II}Sn^{IV}F_6$ are paramagnetic down to 4.2 K (A. Tressaud, R. C. Sherwood, and N. Bartlett, unpublished results, 1977)].

III. Ferrimagnetism in Fluorides

In 1958 Knox and Geller discovered the first ferrimagnetic fluoride, $Na_5Fe_3F_{14}$, which for 10 years remained a unique example. Since the early 1970s, extensive work has been carried out on these materials, and ferrimagnetism has been found in several classes of fluorides. General information on preparative methods, crystal growth techniques, and magnetic properties have been given (Tressaud and Dance, 1977). Results concerning the main series are summarized in the following sections.

A. HEXAGONAL POLYTYPES OF PEROVSKITE

Most investigations on ferrimagnetic fluorides have been devoted to the study of the hexagonal polytypes of perovskite. The structures of the AMF_3 hexagonal phases can be deduced from the various arrangements of the cationic and anionic layers. If in the corresponding oxides the number of combinations is quite large, a limited number of structural types is found for fluorides, that is, 2H, 6H, 9R, 10H, and 12R types (Babel, 1967, 1969; Dance et al., 1984) (for terminology, see Chapter 3). Among these types the only structures liable to display ferrimagnetism are the 6H and 12R types in which at least two different and nonequivalent crystallographic sites are available for the transition element.

The ferrimagnetism of $6H$-type fluorides was first discovered for $RbNiF_3$ (Shafer et al., 1967; Smolensky et al., 1968). It has also been observed for $TlNiF_3$ (Kohn et al., 1967), $CsFeF_3$ (Portier et al., 1968; Eibschutz et al., 1969), and NH_4NiF_3 (Shafer and McGuire, 1968a). Ferrimagnetism results from the presence of two nonequivalent ($2a$) and ($4f$) sites for the divalent cation. Because the multiplicity of these sites is different, a net magnetization results due to ferromagnetic interactions within the ($4f$) sites and antiferromagnetic interactions between the ($4f$) and ($2a$) sites (Fig. 4). Numerous physical studies have been carried out on these phases by means of such methods as susceptibility measurements, neutron diffraction, neutron elastic scattering, Mössbauer resonance, Faraday rotation, far-ir, and Raman diffusion (Eibschutz et al., 1969; Pickart et al., 1964; Pickart and Alperin, 1971; McGuire and Shafer, 1971; Pisarev et al., 1967; Kohn and Nagawa, 1970; Chinn and Zeiger, 1968). Most of these materials are perfectly transparent over a large part of the visible spectrum. Between 20,000 and 150,000 cm^{-1}, $CsFeF_3$ and $RbNiF_3$, for instance, show an optical absorption factor of less than 1 cm^{-1}. In addition, magneto-optical resonance has been obtained in these compounds using Faraday rotation measurements (McGuire and Shafer, 1971; Pisarev et al., 1967). Shafer and McGuire (1969) found the existence of ferrimagnetic compositions of the $6H$ type in the $RbMg_{1-x}Co_xF_3$ and $RbNi_{1-x}Co_xF_3$ systems (McGuire and Shafer, 1968; Suits et al., 1968). The Curie temperatures vary regularly with the composition rate. Similarly, a large domain with the $6H$ type was found in the $Cs_{1-x}Rb_xCoF_3$ system (Dance et al., 1979). For the composition $Cs_{0.5}Rb_{0.5}CoF_3$, the value of T_C (62 K) corresponds to that predicted for a hypothetical $6H$ form of $RbCoF_3$ (Shafer and McGuire, 1968b). The determination of the magnetic structure of

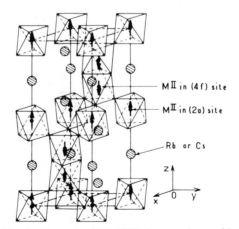

Fig. 4.   Structure of $6H$ AMF_3 hexagonal perovskite.

this phase by neutron diffraction indicates that the moments are parallel to the c axis (Dance *et al.*, 1981).

Ferrimagnetism can also be obtained by applying pressure on a structural type showing no net magnetization. Longo and Kafalas (1969) showed that pressure favors the formation of structures containing the largest number of compact cubic units (Kafalas and Longo, 1968, 1969; Longo *et al.*, 1970) and that the sequence is $2H \rightarrow 9R \rightarrow 6H \rightarrow 3C$. It is therefore possible to transform a fluoride showing antiferromagnetism ($9R$ $CsCoF_3$) or 1D ferromagnetism ($2H$ $CsNiF_3$) into a ferrimagnetic phase (Syono *et al.*, 1970). Magnetic parameters of $6H$ ferrimagnetic fluorides are collected in Table V.

In the $12R$ structure the ferrimagnetic properties arise from the cationic ordering between two transition elements. Dance *et al.* (1975) found large composition ranges with $12R$ structure in the $CsM_{1-x}Mn_xF_3$ systems (M = Ni or Co; $0.5 \leq x \leq 0.75$). The transition elements occupy three different sites, thus leading to ferrimagnetism (Dance and Tressaud, 1979; Yamaguchi and Sakuraba, 1977). In $CsNi_{0.5}Mn_{0.5}F_3$, the alternative formula of which (Cs_2MnNiF_6) may be compared to the $12R$ Cs_2NaCrF_6 type, a particular distribution of the cations has been observed by neutron diffraction (Fig. 5) (Dance *et al.*, 1977). Table VI gives the magnetic parameters of $12R$ ferrimagnetic fluorides.

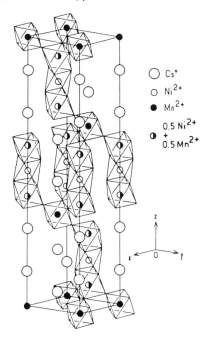

Fig. 5. Cationic distribution in $12R$ Cs_2MnNiF_6.

TABLE V

Magnetic Data of 6H Ferrimagnetic Fluorides

6H Compound	T_C (K)	M_{sat} (μ_B)	θ_p (K)	Reference
CsFeF$_3$	60	1.31	−100	Portier et al. (1968)
	62	1.40	−85	Eibschutz et al. (1969)
RbNiF$_3$	145	—	−130	Smolensky et al. (1967)
	139	0.76	−300	Shafer et al. (1967)
TlNiF$_3$	150	0.42 (at 77 K)	—	Kohn et al. (1967)
NH$_4$NiF$_3$	150	0.1	—	Shafer and McGuire (1968a)
RbNi$_{1-x}$Co$_x$F$_3$ ($0 \leq x \leq 0.25$)	139 (x = 0), 115 (x = 0.25)	Varies irregularly	—	Suits et al. (1968)
RbMg$_{1-x}$Co$_x$F$_3$ ($0.35 \leq x \leq 0.68$)	10 (x = 0.35), 35 (x = 0.68)	Increases with x $\sigma_{sat} = 0.48$ (x = 0.35) $\sigma_{sat} = 0.75$ (x = 0.68)	—	Shafer and McGuire (1968b)
(Rb$_{0.5}$Cs$_{0.5}$)CoF$_3$	62	0.70	−135	Dance et al. (1979)

TABLE VI

Magnetic Data for $12R$ Ferrimagnetic Fluorides

Compound	T_C (K)	M_{sat} (μ_B)	C_{exp}	θ_p (K)
Cs_2MnCoF_6	53	1.99	8.19	-61
Cs_2MnNiF_6	61	2.40	5.96	-78

B. CHIOLITE-TYPE COMPOUNDS $Na_5M_3F_{14}$ (M = 3d ELEMENT)

A precise determination of the structure of γ-$Na_5Fe_3F_{14}$ has been achieved by Vlasse *et al.* (1976). The distorted FeF_6 octahedra form layers of $(Fe_3F_{14})_n^{5n-}$ formula (Fig. 6). Within each layer, the difference of occupation of the two sublattices by Fe^{3+} [$(2a)$ and $(4d)$] sites leads to a spontaneous

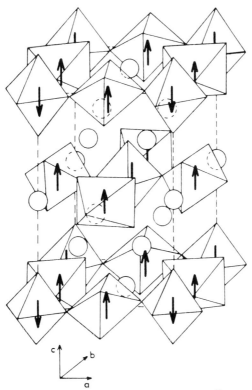

Fig. 6. Magnetic interactions in γ-$Na_5Fe_3F_{14}$ ($\bigcirc$, Na^+).

magnetization, the value of which is given by the relation $M = M(4d) - M(2a)$, yielding a 5-μ_B magnetization in the case of Fe^{3+}. The magnetic data (McKinzie et al., 1972) and the magnetic structure determination (Wintenberger et al., 1975b) showed that, in spite of a large distance between the layers ($\simeq 5.2$ Å), the net moments of each layer were aligned parallel to the c axis, leading to ferrimagnetism. Mössbauer and ferrimagnetic resonance data (Vlasse et al., 1976; Spencer et al., 1963) confirmed the previous results. The magnetic data of d-transition chiolites are reported in Table VII. The Curie temperatures of the vanadium and chromium compounds are low due to the absence of electrons in the e_g orbitals.

The only chiolite containing potassium instead of sodium ($K_5V_3F_{14}$) also exhibits spontaneous magnetization below 18 K; anisotropy measurements showed that the magnetic moments are parallel to the c axis (Cros et al., 1977).

C. WEBERITE-TYPE COMPOUNDS $NA_2M''M'^{III}F_7$ (M, M' = 3d ELEMENTS)

The weberite structure can be described as deriving from both hexagonal tungsten bronze (HTB) and pyrochlore types (see Chapter 3). The octahedra form layers that contain tunnels with hexagonal or triangular section. These layers are connected to each other by $M'^{III}F_6$ octahedra (Fig. 7). The two sites available for the transition elements are present in equivalent proportions. Ferrimagnetism results when both sites are occupied by paramagnetic ions in a different electronic configuration. A large number of weberite-type compounds have been prepared (Cosier et al., 1970; Tressaud et al., 1974). If only one of the sites is occupied by a paramagnetic species and the other by a diamagnetic one, the resulting magnetic property is either a chain antiferromagnetism (i.e., Na_2NiAlF_7) or an isolated-unit paramagnetism (i.e., Na_2MgFeF_7). Table VIII gives the magnetic data for ferrimagnetic weberites. The magnetic structure of Na_2NiFeF_7 has been determined and shows that the spins lie along the a axis (Heger and Viebahn-Hänsler, 1972; Tressaud et al., 1976c). Ferrimagnetism is due to strong antiferromagnetic interactions between Ni^{2+} and Fe^{3+}, which align the spins of Ni^{2+} ions ferromagnetically within chains as shown in Fig. 7a. These structural features may also favor phenomena of spin frustration (see Chapter 10).

D. FERRIMAGNETIC FLUORIDES WITH Na_2SiF_6 STRUCTURE

This structure presents three different crystallographic sites for the transition elements. So far, only two compounds with this structure have shown ferrimagnetic properties: $NaMnCrF_6$ and $LiMnCrF_6$ (Courbion et al., 1978).

TABLE VII

Magnetic Data for $Na_5M_3F_{14}$ Phases

$Na_5M^{III}_3F_{14}$ Compound	Ferrimagnetic Region			Paramagnetic Region			
	T_C (K)	M_{sat} (Experimental) (μ_B mol^{-1})	M_{sat} (Theoretical) (μ_B mol^{-1})	θ_p (K)	C_M (Experimental)	C_M (Theoretical) (Spin-Only Values)	Reference
$Na_5V_3F_{14}$	21	1.94	2	−48	2.44	3	McKinzie et al. (1972)
$Na_5Cr_3F_{14}$	18	2.97	3	−32	5.44	5.61	Miranday et al. (1975)
$Na_5Fe_3F_{14}$	90	4.98	5	−95	11.55	13.14 ⎫	McKinzie et al. (1972)
$Na_5Fe_2CoF_{14}$	94	4.33	4.5 (random) 6 (ordered)	−100	10.53	11.76 ⎭	
$Na_5Co_3F_{14}$	108	3.10	4	−110	8.50	9	McKinzie et al. (1972)
$K_5V_3F_{14}$	18	2.15	2	−77	3.22	3	Cros et al. (1977)

TABLE VIII

Magnetic Data for Ferrimagnetic Weberites

$Na_2M^{II}M^{III}F_7$ Compound	T_C (K)	Ferrimagnetic Region M_{sat} (Experimental) (μ_B mol^{-1})	M_{sat} (Theoretical) (μ_B mol^{-1})	Paramagnetic Region θ_p (K)	C_M (Experimental)	C_M (Theoretical) (Spin-Only Values)	Reference
Na_2FeFeF_7[a]	84	0.75	1	−104	7.42	7.38	Tressaud et al. (1974)
Na_2CoFeF_7	80	1.2	2	−100	7.35	6.26	Tressaud et al. (1974)
Na_2NiCrF_7	4	—[b]	1	−35	2.90	2.87	Tressaud et al. (1974)
Na_2NiFeF_7	90	1.5	3	−50	4.60	5.38	Cosier et al. (1970)
Na_2NiFeF_7	90	2.3 (single crystal)	3				Tressaud et al. (1976c)
Na_2NiCoF_7	126	0.9	2	−88	4.27	4	Tressaud et al. (1974)
Ag_2NiCrF_7	10	—[b]	1	−28	3.08	2.87	Tressaud and Dance (1977)
Ag_2NiFeF_7	103	2.1	3	−61	4.51	5.38	Tressaud and Dance (1977)

[a] Weberite-derived.

[b] Impossible to determine accurately.

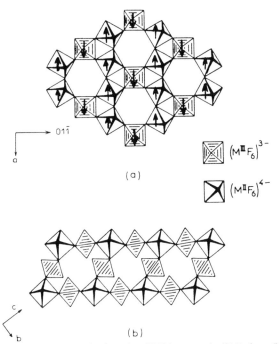

$$\boxed{} \quad \left(M^{III}F_6\right)^{3-}$$

(a)

$$\boxed{} \quad \left(M^{II}F_6\right)^{4-}$$

(b)

Fig. 7. Weberite structure: (a) projection of an HTB layer on the (011) plane, (b) connection of the layers.

In $NaMnCrF_6$ the paramagnetic ions occupy $(3f)$, $(1a)$, and $(2d)$ sites with Mn^{2+} in the first one and Cr^{3+} in the two others. MnF_6 and CrF_6 octahedra are connected by one or two fluorine atoms. They constitute a 3D network in which the nearest neighbors are of a different nature (see Chapter 3, Fig. 5). This disposition leads to ferrimagnetism, and if only a colinear model is taken into account, the magnetic structure consists of a ferromagnetic arrangement of Mn^{2+} ions in $(3f)$ and Cr^{3+} ions in $(2d)$ sites within a $(z = \frac{1}{2})$ layer (Fig. 8). The coupling between the layers is antiferromagnetic via Cr^{3+} ions in $(1a)$ sites. The theoretical saturation moment can therefore be calculated using the following formula: $[(3 \times 5) + (2 \times 3)] - (1 \times 3) = 18\mu_B$ per unit cell (or $6\,\mu_B$ mol^{-1}), in good agreement with the observed value (see Table IX).

$LiMnCrF_6$ has a different type of ordering: Cr^{3+} ions are still located in $(1a)$ and $(2d)$ sites, whereas Mn^{2+} ions are situated in $(3e)$ sites (those of Na^+ in $NaMnCrF_6$) and Li^+ ions occupy the $(3f)$ sites.

Table IX gives the magnetic data for both compounds. In fact, a "modulated" magnetic structure has been proposed more recently (Marmeggi and Courbion, 1979) and seems to indicate an irrational propagation vector for the structure.

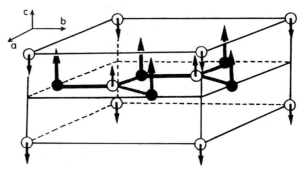

Fig. 8. Magnetic structure of $NaMnCrF_6$ ($\bullet$, Mn; $\circ$, Cr). From Courbion *et al.* (1978), copyright 1978, Pergamon Press Ltd.

TABLE IX

Magnetic Data for $NaMnCrF_6$ and $LiMnCrF_6$

		Ferrimagnetic Domain		Paramagnetic Domain		
Compound	T_C (K)	M_{sat} (Experimental) (μ_B)	M_{sat} (Calculated) (μ_B)	θ_p (K)	C_{exp}	C_{calc}
$NaMnCrF_6$	21	6	6	8	5.98	6.25
$LiMnCrF_6$	23	4.08	4	-26	6.55	6.25

E. $M^{II}M'^{III}F_5$ COMPOUNDS AND RELATED HYDRATES

$MM'F_5$ compounds crystallize in different but related structures (Dance and Tressaud, 1973). A common feature of these structures is the occurrence of parallel $(M'^{III}F_5)_n^{2n-}$ chains of octahedra sharing trans corners; these chains are held together by the M(II) cations. Two classes of ferrimagnetic fluorides have been found among a large number of $MM'F_5$ compounds.

Whereas $MnAlF_5$ is an antiferromagnet below 2.6 K (Wintenberger *et al.*, 1975a), the solid solution $MnAl_{1-x}Fe_xF_5$ exhibits ferrimagnetic properties ($0 < x \leq 0.58$). These properties have been interpreted by assuming the presence of ferrimagnetic clusters for low iron concentrations. With increasing x, the clusters are coupled to form a multimagnetic phase system consisting of antiferromagnetic and ferrimagnetic regions, yielding a spontaneous magnetization. Curie temperatures increase with x, and the magnetization first increases (up to $x \simeq 0.2$) and then decreases (Tressaud *et al.*, 1973a).

Among the $Cr^{II}M'^{III}F_5$ compounds (M' = Al, Ti, V, or Cr), the prototype of which is Cr_2F_5, only $CrTiF_5$ and $CrVF_5$ are ferrimagnetic (Table X). This

TABLE X

Magnetic Data for $MM'F_5$ Compounds

Compound	T_C (K)	M_{sat} (μ_B)	C_{exp}	C_{calc}	θ_p (K)
$MnAl_{0.90}Fe_{0.10}F_5$	18	2.3	4.55	4.82	-20
$MnAl_{0.45}Fe_{0.55}F_5$	34	1.5	6.25	6.78	-136
$CrTiF_5$	26	1.78	3.59	3.88	-78
$CrVF_5$	40	0.86	4.02	4.40	-95

property has been explained by competition between exchange interactions. In both cases a nonlinear alignment of the spins is assumed due to the anisotropy, which tends to align, for instance, Ti^{3+} moments in the c direction and not Cr^{2+} moments. In $CrVF_5$, both spin–orbit coupling and anisotropy align V^{3+} moments in the b direction of the unit cell. From this analysis the anisotropy has to be larger in the case of $CrVF_5$, as observed experimentally (Tressaud et al., 1973b).

Two hydrates with formula $Fe^{III}M^{II}F_5 \cdot 2H_2O$ (M^{II} = Fe or Co) also have ferrimagnetic properties. Their structure differs from that of the anhydrous $MM'F_5$ compounds insofar as the water molecules participate in the formation of $FeF_4(H_2O)_2$ octahedra containing divalent iron; Fe^{3+} ions are surrounded by fluorine atoms only (Hall et al., 1977).

Ferrimagnetism is due to strong antiferromagnetic interactions between neighboring Fe^{2+} and Fe^{3+} ions. For $Fe_2F_5 \cdot 2H_2O$, $T_C = 48$ K and $M_{obs} = 0.82$ μ_B ($M_{calc} = 1$ μ_B) (Walton et al., 1977; Jones and Dawson, 1978). For $FeCoF_5 \cdot 2H_2O$, $T_C = 27$ K and the net magnetization happens to be field dependent, suggesting a large anisotropy energy (Jones and Dawson, 1979).

References

Ahlers, G., Kornblit, A., and Guggenheim, H. J. (1975). *Phys. Rev. Lett.* **34**, 1227.

Als-Nielsen, J. (1976). *Phys. Rev. Lett.* **37**, 1161.

Antipenko, B. M., Podkolzina, I. G., and Tomashevich, Y. V. (1980). *Kvantovaya Elektron. (Moscow)* **7**, 647.

Babel, D. (1967). *Struct. Bonding (Berlin)* **3**, 1.

Babel, D. (1969). *Z. Anorg. Allg. Chem.* **369**, 117.

Beauvillain, P., Renard, J. P., and Hansen, P. E. (1977). *J. Phys. C* **10**, L709.

Beauvillain, P., Renard, J. P., Laursen, I., and Walker, P. J. (1978). *Phys. Rev. B* **18**, 3360.

Beauvillain, P., Chappert, C., Renard, J. P., and Griffin, J. A. (1980a). *J. Phys. C* **13**, 395.

Beauvillain, P., Chappert, C., and Laursen, I. (1980b). *J. Phys. C* **13**, 1481.

Beauvillain, P., Seiden, J., and Laursen, I. (1980c). *Phys. Rev. Lett.* **45**, 1362.

Beauvillain, P., Chappert, C., Renard, J. P., Seiden, J., and Laursen, I. (1983). *J. Magn. Magn. Mater.* **31–34**, 1103.

Belov, K. P., and Goryaga, A. N. (1956). *Fiz. Met. Metalloved.* **2**, 3.

Brierley, S. K., and Griffin, J. A. (1982). *Phys. Rev. B. Condens. Matter* [3] **26**, 315.

Brinkmann, J., Courths, R., and Guggenheim, H. J. (1978). *Phys. Rev. Lett.* **40**, 1286.

Chicklis, E. P., Naiman, C. S., Folweiler, R. C., Gabbe, D. R., Jenssen, H. P., and Linz, A. (1971). *Appl. Phys. Lett.* **19**, 119.

Chinn, B. R., and Zeiger, H. J. (1968). *Phys. Rev. Lett.* **21**, 1589.

Cosier, R., Wise, A., Tressaud, A., Grannec, J., Olazcuaga, R., and Portier, J. (1970). *C. R. Hebd. Seances Acad. Sci.* **271**, 142.

Courbion, G., Ferey, G., and de Pape, R. (1978). *Mater. Res. Bull.* **13**, 967.

Cros, C., Dance, J. M., Grenier, J. C., Wanklyn, B. M., and Garrard, B. J. (1977). *Mater Res. Bull.* **12**, 415.

Dance, J. M. (1985). To be published in *Mater. Res. Bull.*

Dance, J. M., and Tressaud, A. (1973). *C. R. Hebd. Seances Acad. Sci.* **279**, 379.

Dance, J. M., and Tressaud, A. (1979). *Mater. Res. Bull.* **14**, 37.

Dance, J. M., Grannec, J., and Tressaud, A. (1975). *C. R. Hebd. Seances Acad. Sci.* **281**, 91.

Dance, J. M., Sabatier, R., Ménil, F., Cousseins, J. C., Le Flem, G., and Tressaud, A. (1976a). *Solid State Commun.* **19**, 1059.

Dance, J. M., Grannec, J., and Tressaud, A. (1976b). *C. R. Hebd. Seances Acad. Sci.* **283**, 115.

Dance, J. M., Grannec, J., Tressaud, A , and Perrin, M. (1977). *Mater. Res. Bull.* **12**, 989.

Dance, J. M., Yoshizawa, H., and Hirakawa, K. (1978). *Mater. Res. Bull.* **13**, 1111.

Dance, J. M., Kerkouri, N., and Tressaud, A. (1979). *Mater. Res. Bull.* **14**, 869.

Dance, J. M., Soubeyroux, J. L., Kerkouri, N., and Tressaud, A. (1981). *C. R. Hebd. Seances Acad. Sci.* **293**, 279.

Dance, J. M., Darriet, J., Tressaud, A., and Hagenmuller, P. (1984). *Z. Anorg. Allg. Chem.* **508**, 93.

Dupas, A., and Renard, J. P. (1975). *Phys. Lett. A* **53A**, 141.

Eibschutz, M., Holmes, L., Guggenheim, H. J., and Levinstein, H. V. (1969). *J. Appl. Phys.* **40**, 1312.

Esterowitz, L., Allen, R., Kruer, M., Bartoli, F., Goldberg, L. S., Jenssen, H. P., Linz, A., and Nicolai, V. O. (1977). *J. Appl. Phys.* **48**, 650.

Fernelius, N. C., Dempsey, D. V., Walsh, D. A., O'Quinn, D. B., and Knecht, W. L. (1981). *NBS Spec. Publ. (V.S.)* **620**, 129.

Ferré, J., and Régis, M. (1978). *Solid State Commun.* **26**, 225.

Folkins, J. J., Griffin, J. A., and Gubser, D. U. (1982). *Phys. Rev. B* **25**, 405.

Friebel, C., and Reinen, D. (1974). *Z. Anorg. Allg. Chem.* **407**, 193.

Frowein, R., and Kötzler, J. (1982). *Phys. Rev. B* **25**, 3292.

Frowein, R., Kötzler, J., and Assmus, W. (1979). *Phys. Rev. Lett.* **42**, 739.

Frowein, R., Kötzler, J., Schaub, B., and Schuster, H. G. (1982). *Phys. Rev. B* **25**, 4905.

Funahashi, S., Moussa, F., and Steiner, M. (1976). *Solid State Commun.* **18**, 433.

Griffin, J. A., and Litster, J. D. (1979). *Phys. Rev. B* **19**, 3676.

Griffin, J. A., Litster, J. D., and Linz, A. (1977). *Phys. Rev. Lett.* **38**, 251.

Guggenheim, H. J. (1963). *J. Appl. Phys.* **34**, 2482.

Gupta, L. C., Vijayaraghavan, R., Damle, S. D., Rao, U. R. K., Le Dang Khoi, and Veillet, P. (1975). *J. Magn. Reson.* **17**, 41.

Haegele, R., and Babel, D. (1974). *Z. Anorg. Allg. Chem.* **409**, 11.

Hall, W., Kim, S., Zubieta, J., Walton, E., and Brown, D. B. (1977). *Inorg. Chem.* **16**, 1884.

Hansen, P. E., Johansson, T., and Nevald, R. (1975). *Phys. Rev. B* **12**, 5315.

Harris, I. R., Safi, H., Smith, N. A., Altunbas, M., Cockayne, B., and Plant, J. G. (1983). *J. Mater. Sci.* **18**, 1235.

Heger, G., and Viebahn-Hänsler, R. (1972). *Solid State Commun.* **11**, 1119.

Herdtweck, E., and Babel, D. (1981). *Z. Anorg. Allg. Chem.* **474**, 113.

Hidaka, M., and Walker, P. J. (1979). *Solid State Commun.* **31**, 383.

Hirakawa, K., and Ikeda, H. (1972). *J. Phys. Soc. Jpn.* **33**, 1483.
Hirakawa, K., and Ikeda, H. (1973a). *J. Phys. Soc. Jpn.* **35**, 1328.
Hirakawa, K., and Ikeda, H. (1973a). *J. Phys. Soc. Jpn.* **35**, 1608.
Hirakawa, K., and Ikeda, H. (1974). *Phys. Rev. Lett.* **33**, 374.
Holmes, L., Guggenheim, H. J., and Hull, G. W. (1970). *Solid State Commun.* **8**, 2005.
Holmes, L., Johansson, T., and Guggenheim, H. J. (1973). *Solid State Commun.* **12**, 993.
Hoppe, R., and Siebert, G. (1970). *Z. Anorg. Allg. Chem.* **376**, 261.
Jones, E. R., Jr., and Dawson, R. (1978). *J. Chem. Phys.* **69**, 3289.
Jones, E. R., Jr., and Dawson, R. (1979). *J. Chem. Phys.* **71**, 202.
Kafalas, J. A., and Longo, J. M. (1968). *Mater. Res. Bull.* **3**, 501.
Kafalas, J. A., and Longo, J. M. (1969). *J. Appl. Phys.* **40**, 1601.
Khomskii, D. I., and Kugel, K. I. (1973). *Solid State Commun.* **13**, 763.
Knox, K., and Geller, S. (1958). *Phys. Rev.* **110**, 771.
Köhler, P., Massa, W., Reinen, D., Hoffmann, B., and Hoppe, R. (1978). *Z. Anorg. Allg. Chem.* **446**, 131.
Kohn, K., and Nagawa, I. (1970). *Bull. Soc. Chim. Jpn.* **43**, 3780.
Kohn, K., Fukuda, R., and Iida, S. (1967). *J. Phys. Soc. Jpn* **22**, 333.
Kouvel, J. S. (1957). *Gen Electr. Res. Lab. Rep.* **RL1799**, No. 57, 4.
Kuppenheim, J. D., and Baer, J. W. (1982). U.S. Patent 4, 352, 186.
Larkin, A. I., and Khmel'nitskii, D. E. (1969). *Sov. Phys.–JETP (Engl. Transl.)* **29**, 1123.
Le Dang Khoi and Veillet, P. (1976). *Phys. Rev. B* **13**, 1919.
Litterst, F. J. (1975). *Jn. Phys. Lett.* **36**, L197.
Longo, J. M., and Kafalas, J. A. (1969). *J. Solid State Chem.* **1**, 103.
Longo, J. M., Kafalas, J. A., O'Connor, J. R., and Goodenough, J. B. (1970). *J. Appl. Phys.* **41**, 935.
Lorin, D. (1980). Thesis, Univ. of Bordeaux.
Lorin, D., Dance, J. M., Soubeyroux, J. L., Tressaud, A., and Hagenmuller, P. (1981). *J. Magn. Magn. Mater.* **23**, 92.
McGuire, T. R., and Shafer, M. W. (1968). *J. Appl. Phys.* **39**, 1130.
McGuire, T. R., and Shafer, M. W. (1971). *J. Phys. Colloq. C1,* **32**, 627.
McKinzie, H., Dance, J. M., Tressaud, A., Portier, J., and Hagenmuller, P. (1972). *Mater. Res. Bull.* **7**, 673.
Marmeggi, M., and Courbion, G. (1979). *Mater. Res. Bull.* **14**, 1057.
Massa, W. (1977). *Inorg. Nucl. Chem. Lett.* **13**, 253.
Massa, W., and Steiner, M. (1980). *J. Solid State Chem.* **32**, 137.
Miranday, J. P., Ferey, G., Jacoboni, C., Dance, J. M., Tressaud, A., and de Pape, R. (1975). *Rev. Chim. Miner.* **12**, 187.
Misra, S. K., and Felsteiner, J. (1977). *Phys. Rev. B* **15**, 4309.
Moussa, F., Hennion, B., Mons, J., and Pepy, G. (1978). *Solid State Commun.* **27**, 141.
Odenthal, R. H., Paus, D., and Hoppe, R. (1974). *Z. Anorg. Allg. Chem.* **407**, 144.
Okuda, Y., Tohi, Y., Yamada, I., and Haseda, T. (1980). *J. Phys. Soc. Jpn.* **49**, 936.
Pickart, S. J., and Alperin, H. A. (1971). *J. Appl. Phys.* **42**, 1617.
Pickart, S. J., Alperin, H. A., and Nathans, R. (1964). *J. Phys.* **25**, 565.
Pisarev, R. V., Siny, I. G., and Smolensky, G. A. (1967). *Fiz. Tverd. Tela* **9**, 3149.
Portier, J., Tressaud, A., Pauthenet, R., and Hagenmuller, P. (1968). *C. R. Hebd. Seances Acad. Sci.* **267**, 1329.
Reinen, D., and Friebel, C. (1979). *Struct. Bonding (Berlin)* **37**, 1.
Reinen, D., and Krause, S. (1981). *Inorg. Chem.* **20**, 2750.
Reinen, D., and Weitzel, M. (1977). *Z. Naturforsch., B: Anorg. Chem., Org. Chem.* **32B**, 476.
Renaudin, J., and Ferey, G. (1984). *Conf. Rep. Solid State Rep., Bordeaux.*
Rüdorff, W., Lincke, G., and Babel, D. (1963). *Z. Anorg. Allg. Chem.* **320**, 150.

Shafer, M. W., and McGuire, T. R. (1968a). *Proc. 133rd Meet. Electrochem. Soc., Boston.*

Shafer, M. W., and McGuire, T. R. (1968b). *Phys. Lett. A* **27A** (10), 676.

Shafer, M. W., and McGuire, T. R. (1969). *J. Phys. Chem. Solids* **30**, 1989.

Shafer, M. W., McGuire, T. R., Argyle, B. E., and Fan, G. J. (1967). *Appl. Phys. Lett.* **10**, 202.

Smolensky, G. A., Yudin, V. M., Syroikov, P. P., and Sherman, A. B. (1967). *JETP Lett. (Engl. Transl.)* **8**, 2368.

Spencer, E. G., Berger, S. B., Linares, R. C., and Lenzo, P. V. (1963). *Phys. Rev. Lett.* **10**, 236.

Suits, J. C., McGuire, T. R., and Shafer, M. W. (1968). *Appl. Phys. Lett.* **12**, 406.

Syono, Y., Akimoto, S., and Kohn, K. (1970). *Colloq. Int. C. N. R. S.* **188**, 415.

Tressaud, A., and Dance, J. M. (1977). *Adv. Inorg. Chem. Radiochem.* **20**, 133.

Tressaud, A., Parenteau, J. M., Dance, J. M., Portier, J., and Hagenmuller, P. (1973a). *Mater. Res. Bull.* **8**, 565.

Tressaud, A., Dance, J. M., Ravez, J., Portier, J., Hagenmuller, P., and Goodenough, J. B. (1973b). *Mater. Res. Bull.* **8**, 1467.

Tressaud, A., Dance, J. M., Portier, J., and Hagenmuller, P. (1974). *Mater. Res. Bull.* **9**, 1219.

Tressaud, A., Wintenberger, M., Bartlett, N., and Hagenmuller, P. (1976a). *C. R. Hebd. Seances Acad. Sci., Ser. C* **282**, 1069.

Tressaud, A., Pintchovski, F., Lozano, L., Wold, A., and Hagenmuller, P (1976b). *Mater. Res. Bull.* **11**, 689.

Tressaud, A., Dance, J. M., Vlasse, M., and Portier, J. (1976c). *C. R. Hebd. Seances Acad. Sci.,* **282**, 1105.

Vlasse, M., Ménil, F., Morilière, C., Dance, J. M., Tressaud, A., and Portier, J. (1976). *J. Solid State Chem.* **17**, 291.

von Schnering, H. G. (1967). *Z. Anorg. Allg. Chem.* **353**, 1, 13.

Walton, E. G., Brown, D. B., Wong, H., and Reiff, W. M. (1977). *Inorg. Chem.* **16**, 2425.

Wintenberger, M., Dance, J. M., and Tressaud, A. (1975a). *Solid State Commun.* **17**, 185.

Wintenberger, M., Dance, J. M., and Tressaud, A. (1975b). *Solid State Commun.* **17**, 1355.

Wolfe, R., Kurtzig, A. J., and Le Craw, R. C. (1970). *J. Appl. Phys.* **41**, 1218.

Yamada, I. (1970). *J. Phys. Soc. Jpn.* **28**, 1585.

Yamada, I. (1971). *J. Phys. Soc. Jpn* **30**, 896.

Yamada, I. (1972). *J. Phys. Soc. Jpn.* **33**, 979, 1334.

Yamada, I., and Suzuki, H. (1976). *Solid State Commun.* **18**, 237.

Yamaguchi, Y., and Sakuraba, T. (1977). *J. Phys. Chem. Solids* **38**, 957.

Yamazaki, H. (1973). *J. Phys. Soc. Jpn.* **34**, 270.

Yamazaki, H. (1974). *J. Phys. Soc. Jpn.* **37**, 667.

Zalkin, A., and Templeton, D. H. (1953). *J. Am. Chem. Soc.* **75**, 2453.

10

Competing Spin Interactions
and Frustration Effects
in Fluorides

G. FEREY, M. LEBLANC, and R. DE PAPE
Laboratoire des Fluorures et Oxyfluorures Ioniques
Faculté des Sciences
Le Mans, France

J. PANNETIER
ILL Grenoble
Grenoble, France

I. Competing Interactions: Concept of Frustration

The concept of frustration was introduced by G. Toulouse in 1977 (Toulouse, 1977; Vannimenus and Toulouse, 1977) as an attempt to justify the spin glass behavior of numerous compounds. In a general way, it describes the impossibility for a system to satisfy simultaneously all the interactions to which it is submitted.

Initially, Toulouse defined the frustration in the case of a bidimensional network of square platelets of Ising spins ($S_i = \pm 1$). If it is assumed that magnetic interactions exist only between nearest neighbors i and j, with $|J_{ij}| = 1$, the model is ferromagnetic for $J_{ij} = +1$ (Fig. 1a) and antiferromagnetic for $J_{ij} = -1$ (Fig. 1b). If $J_{ij} = \pm 1$ randomly, one can define for a contour line C the function

$$\theta_C = \prod_C J_{ij},$$

which measures the frustration effect; C, which has an area S, is formed of k square platelets. The frustration function is then written

$$\theta_C = \prod_{k \in S} \theta_k.$$

The platelet is frustrated if $\theta_k = -1$. This is the case summarized in Fig. 1c,d. If an antiferromagnetic interaction is introduced into a ferromagnetic platelet, it is impossible to satisfy simultaneously $4 - 1$ and $4 - 3$ antagonistic interactions; one of them is frustrated.

The mathematical treatment of the average of the function of frustration $\langle \theta \rangle$ predicts, for example, that above the critical threshold of 9% of antiferromagnetic bonds in a ferromagnetic set, long-range magnetic ordering cannot appear.

This frustration function can also be applied to a network of triangular platelets in which $J_{ij} = -1$ (Fig. 2). Here also $\theta_k = -1$, and the system is frustrated. If spins 1 and 2 interact antiferromagnetically, the third spin cannot adopt an antiferromagnetic coupling with both spins 1 and 2, and one bond must be frustrated.

Chemically, the two previous situations imply two types of system. For the

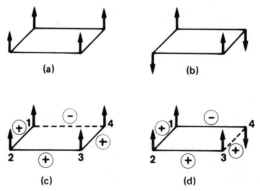

Fig. 1.   Square platelets. (a) Ferromagnetic; (b) antiferromagnetic; (c, d) frustrated.

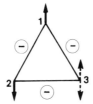

Fig. 2. Antiferromagnetic frustrated triangular platelet.

frustrated square platelets, it is necessary to introduce a cationic disorder that induces competing interactions. In this case, the scarceness of ferromagnetic interactions in $3d$ metallic fluorides explains the present lack of published results. A close example exists in chlorine chemistry, however; it concerns the system $Rb_2Cr^{II}_{1-x}Mn_xCl_4$ (Munninghoff, 1980; Munninghoff *et al.*, 1980, 1981, 1982).

When the platelets are triangular, many possibilities occur, which depend both on the topology of cationic sublattice and on the strength of antiferromagnetic exchange interactions J_{ij} between nearest metallic neighbors:

1. There is only one type of cation, and the J_{ij} are equal; an antiferromagnetic structure with three $120°$ sublattices is observed. An illustration is provided by hexagonal tungsten bronze (HTB) FeF_3 (e.g., $2:1$).

2. There are two cationic species M^{2+} and M^{3+}, with a crystallographic order between them. The magnetic structure is then governed by the strength of $M^{2+}-M^{3+}$ interactions:

    a. If they predominate, they oblige the weakly bound M^{2+} spins to adopt an antibonding parallel orientation (e.g., $2:2:1$ = weberite Na_2NiFeF_7).

    b. If they are weak, M^{2+} and M^{3+} form quasi-independent magnetic sublattices (e.g., $2:2:2$ $MnCrF_5$ and $2:2:3$ $NH_4Fe^{II}Fe^{III}F_6$).

    c. One of the two sublattices can remain paramagnetic below the ordering temperature of the other [e.g., $2:2:4$ $Fe^{II}Fe^{III}_2F_8(H_2O)_2$].

3. There is a structural disorder; it can concern the cationic sublattice in a crystalline compound or the whole network in an amorphous compound. This leads to spin glasses (e.g., $3:1$ $CsMnFeF_6$) or speromagnetic behavior (e.g., $3:2$ amorphous FeF_3).

The relative influence of the topology and the interactions on the frustration is not yet understood owing to the small number of examples. This chapter thus describes the experimental aspects of the magnetic frustration deduced from neutron diffraction experiments, Mössbauer spectroscopy, and susceptibility measurements and attempts to correlate them with the crystal structure.

II. Frustration and Cationic Ordering

As noted in the preceding section, frustration can exist in ordered structures if they exhibit triangular platelets in the metallic sublattice. The simplest corresponding network is probably given by the layers of the HTB structure (Magneli, 1953). Among the large number of compounds that contain these layers (Fig. 3), the low-temperature form of FeF_3 (Leblanc *et al.*, 1983) is of particular interest because only one type of cation (Fe^{3+}) occupies the metallic sites. All interactions are then equivalent in a layer and allow one to examine exclusively the influence of the structural topology on the frustrated character.

A. FRUSTRATION AND TOPOLOGY

The new HTB FeF_3 form is obtained by dehydration of $(H_2O)_{0.33}FeF_3$ at 150°C; it is synthesized by the hydrothermal method (Ferey *et al.*, 1975; Leblanc *et al.*, 1984b; Passaret *et al.*, 1974) and also exhibits an HTB structure with water molecules in the channels. The loss of water leads to HTB FeF_3 (space group $P6/mmm$).

Its susceptibility curve is very flat and indicates very strong antiferromagnetic interactions $[T_N = 97(2)\ K, \quad \theta_p \simeq -600\ K]$. The magnetic structure of HTB FeF_3 at 4.2 K can be described by three antiferromagnetic sublattices at 120° (Fig. 3); between two layers the coupling of Fe^{3+} spins is strictly antiferromagnetic (Leblanc *et al.*, 1985).

This situation was studied theoretically by Marland and Betts (1979) for an infinite planar triangular net of Heisenberg spins $S = \frac{1}{2}$. They compared its ground-state energy to the corresponding nonfrustrated ferromagnetic system. With an isotropic Heisenberg model, the ground-state entropy is

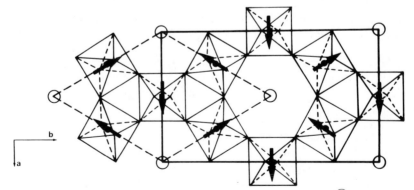

Fig. 3. Layer of the hexagonal tungsten bronze structure (■, Fe_1; ●, Fe_2; ○, H_2O). $z = 0.5$.

zero, and the high energy value in both cases led the authors to conclude curiously that such a system with three $120°$ sublattices is quite nonfrustrated. In spite of this result, it is obvious that HTB FeF_3 is magnetically frustrated if it is compared with the α-rhombohedral form of FeF_3 (Hepworth et al., 1957; Wollan et al., 1958), which derives from the ReO_3 type. In this structure, the topology is not frustrating; α-FeF_3 is an antiferromagnet of the G type (Bertaut, 1963) with moments lying in the (111) plane and a Néel temperature (365 K) four times that of HTB FeF_3.

This unique example illustrates that, if all magnetic interactions are equivalent, the triangular topology of the network is an important factor of frustration. It induces an important lowering of T_N and, correlatively, a large increase in the $|\theta_p|/T_N$ ratio. Moreover, molecular field calculations show that the magnetic binding energy E_T turns out to be only a fraction (say, $\leq \frac{1}{3}$) of the energy E_F in the ferromagnetic case. It is now important to take into account the strength of antiferromagnetic interactions when two different cationic species occupy the sites of the triangular net.

B. FRUSTRATION AND STRENGTH OF MAGNETIC INTERACTIONS IN ORDERED STRUCTURES

Fortunately, the literature provides more examples in this case. The species are divalent and trivalent $3d$ cations M(II) and M(III), arranged either in an ordered or in a disordered way. Four examples, leading to different frustrated magnetic behaviors, are described in the following subsections.

1. Na_2NiFeF_7

Fluoride Na_2NiFeF_7 (Cosier et al., 1970; Haegele et al., 1978; Heger and Viebahn-Hansler, 1972; Tressaud and Dance, 1977; Tressaud et al., 1974) is isostructural with the mineral weberite Na_2MgAlF_7 (Byström, 1944; Giuse-petti and Tadini, 1978; Knop et al., 1982) (Fig. 4a). This structure can be described either in terms of trans chains of $M^{II}F_6$ octahedra linked by isolated $M^{III}F_6$ octahedra or in terms of M(II)–M(III) HTB layers connected by $M^{III}F_6$ octahedra (Fig. 4b). Whatever the description, the cations always form a triangular net, which may induce frustration if M(II) and M(III) are paramagnetic. When M(III) is diamagnetic, as in Na_2NiAlF_7 (Heger, 1973; Tressaud and Dance, 1977; Tressaud et al., 1974), the chains of Ni^{2+} are coupled antiferromagnetically ($T_N = 18$ K), as predicted by Goodenough's rules (1963). In Na_2NiFeF_7, however, ferrimagnetism ($T_C = 90$ K) is observed. Parallel spins of Ni^{2+} lie along [001](Fig. 4a); the sublattice of Fe^{3+} is ferromagnetic and antiparallel to the Ni^{2+} spins (Heger and Viebahn-Hansler, 1972). As indicated by Tressaud et al. (1974; Tressaud and Dance, 1977),

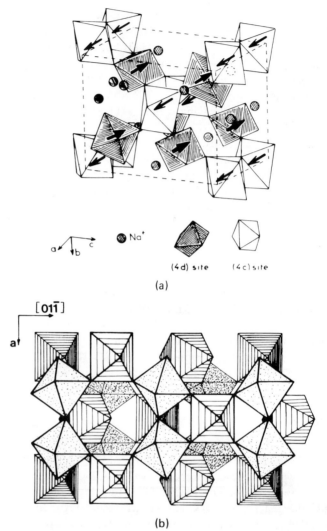

Fig. 4. (a) Perspective view of the crystalline and magnetic structure of Na_2NiFeF_7; Fe(III) octahedra are shaded. From Tressaud and Dance (1977). (b) The (101) projection of Na_2NiFeF_7 showing the HTB layers (Ni^{2+} stippled and Fe^{3+} shaded).

coupling between Ni^{2+} is very weak, and it is the strong antiferromagnetic interaction between Ni^{2+} and Fe^{3+} that obliges Ni^{2+} to adopt a "ferromagnetic" arrangement. The magnetic behavior is then governed by $M^{2+}-M^{3+}$ interactions. The system apparently seems to be unfrustrated, but frustration exists in the $Ni^{2+}-Ni^{2+}$ interactions.

The accidents observed at ~ 40 K both on the $\chi^{-1}(T)$ curves and on the thermal variation of some magnetic peaks indicate a change in the magnetic structure and probably another frustrated arrangement of the spins.

An interesting comparison with the dihydrate $Fe^{II}Fe^{III}F_5 \cdot 2H_2O$ is in progress (Y. Lalligant and G. Ferey, 1985). This compound adopts the same structure, but the M^{2+} and M^{3+} positions (Hall et al., 1977) are opposite to those of Na_2NiFeF_7. The chains are then formed by Fe^{3+} ions, whereas the four fluorine atoms of the $Fe^{II}F_4(H_2O)_2$ octahedron link the chains (Fig. 5). Magnetically, the situations of Na_2NiFeF_7 and $Fe_2F_5 \cdot 2H_2O$ are reversed: the strongest interaction [Fe(III)–Fe(III)] is now in the chains, whereas it was the weakest with Ni^{2+}. The magnetic structures of Na_2NiFeF_7 and $Fe_2F_5 \cdot 2H_2O$ must differ, in spite of the ferrimagnetic behavior of $Fe_2F_5 \cdot 2H_2O$ [$T_C = 48$ K (Jones and Dawson, 1978; Walton and Brown, 1977)], because the Kanamori (1959)–Goodenough (1963) rules predict only antiferromagnetic Fe(III)–Fe(III) interactions when the corresponding octahedra share only corners. Moreover, a previous Mössbauer study (Imbert et al., 1976) showed an accident in the thermal variation of the hyperfine field of Fe^{3+} (Fig. 6a) at $T = 26$ K; neutron diffraction investigations (Y. Lalligant and G. Ferey, 1985) have confirmed that there are two magnetic structures above and below this temperature (Fig. 6b). Their refinement is in progress, but they cannot correspond to a collinear ferrimagnetism as in Na_2NiFeF_7, because it occurs only above 15 kOe with a resulting moment of 1 μ_B (Walton and Brown, 1977). The canted moments will give frustrated structures which

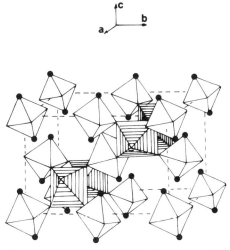

Fig. 5. Perspective view of $Fe_2F_5 \cdot 2H_2O$. Filled circles correspond to oxygen atoms; Fe^{3+} octahedra are shaded.

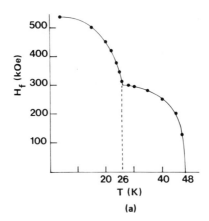

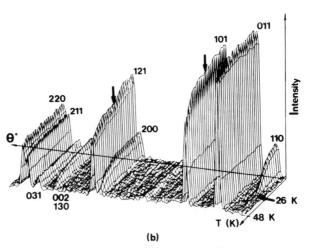

Fig. 6. (a) Thermal variation of the hyperfine field of Fe^{3+} in $Fe_2F_5 \cdot 2H_2O$. (b) Thermal variation of the neutron diffraction spectrum between 4.2 and 48 K. The accident observed at 26 K on some Bragg peaks is noted by arrows.

might be due to the smaller difference in the magnitudes of Fe(III)–Fe(III) and Fe(III)–Fe(II) magnetic interactions, compared with those of $Ni^{2+}-Ni^{2+}$ and $Ni^{2+}-Fe^{3+}$ in Na_2NiFeF_7.

2. MnCrF₅

Monoclinic compound $MnCrF_5$ (space group $C2/c$) (Ferey *et al.*, 1977, 1978; de Kozak, 1971) is commonly described by trans chains along the [001] direction formed by corner-sharing octahedra of Cr^{3+} (Fig. 7), between which

Mn^{2+} ions are inserted. The tilting of $(CrF_5)_n^{2n-}$ chains transforms the octahedra of Mn^{2+} into edge-sharing pentagonal bipyramids, which also run along [001]. They are linked to the chromium chains either by corners or by edges, to ensure the three-dimensional (3D) network.

The metallic triangular sublattice is the same as that observed in HTB FeF_3. The center of the corresponding cavities is now filled by cations, however, and the network is perfectly triangular; each cation therefore has eight neighbors instead of six, as in HTB FeF_3.

This compound is antiferromagnetic ($T_N = 6$ K, $\theta_p = 16$ K), and oddly its magnetic structure (Fig. 7) consists of two orthogonal antiferromagnetic sublattices. The Cr^{3+} spins ($\mu = 2.5\ \mu_B$) lie along [001], whereas the Mn^{2+} moments ($\mu = 3.9\ \mu_B$) are parallel to [010]. Measurements under high magnetic fields (up to 350 kOe) allow one to compare the strength of the different interactions: the magnetization begins to increase linearly with $\vec{H}$ up

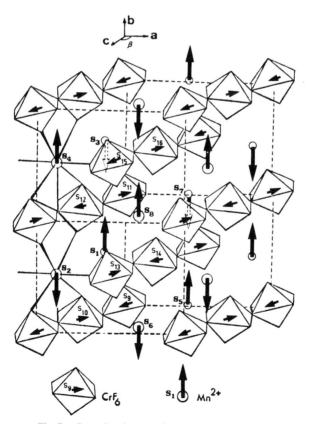

Fig. 7. Crystal and magnetic structure of $MnCrF_5$.

to 5 μ_B per molecule. This value, which corresponds to a ferromagnetic alignment of the Mn^{2+} spins, is obtained for relatively low values of the magnetic field (150 kOe). It supposes a weak coupling of Mn^{2+}. Above 150 kOe, the magnetization increases more slowly. At 350 kOe, it reaches 7.5 μ_B, which is close to the saturation. These measurements confirm that there is practically no interaction between the sublattices of Mn^{2+} and Cr^{3+}. Antiferromagnetic Mn^{2+} interactions are the weakest; they are stronger for the Cr^{3+} ions, which govern the magnetic structure. The Mn^{2+} ions adopt a position that equally satisfies its interactions in each triangle.

3. $Fe^{II}Fe_2^{III}F_8(H_2O)_2$

Compound $Fe^{II}Fe_2^{III}F_8(H_2O)_2$ (Herdtweck, 1983; Leblanc *et al.*, 1984a) is obtained either by hydrothermal synthesis (420°C; 2 kbars) in HF of 49% ion concentration or at high pressure. Its true symmetry is monoclinic (space group $C2/m$), and Fe^{2+}–Fe^{3+} order occurs (Fig. 8a). As in $Fe_2F_5(H_2O)_2$, the surrounding of Fe^{3+} is exclusively fluorinated, whereas $Fe^{II}F_4(H_2O)_2$ octahedra share their four fluorine corners with the corresponding $Fe^{III}F_6$ octahedra. This structure can be better examined using two other projection planes. The ($20\bar{1}$) projection (Fig. 8b) shows that $Fe_3F_8(H_2O)_2$ can be described in terms of HTB layers, which are translated from $\vec{b}/2$, whereas in HTB FeF_3 they are stacked above each other. The Fe^{3+} sublattice forms perovskite layers in the (002) plane (Fig. 8c). This arrangement explains the magnetic properties of this compound.

Susceptibility measurements show that at high temperature the Curie–Weiss law is obeyed. The value of $\theta_p(-240\ K)$ supposes strong antiferromagnetic interactions (Fig. 9a). Below $T_N = 157\ K$, magnetization curves show antiferromagnetism with parasitic ferromagnetism. However, a sharp slope inversion of remanent magnetization (Fig. 9b) and in χ^{-1} curves occurs at 35 K. This peculiarity could be explained only by Mössbauer spectroscopy. As is shown in Fig. 9c, in the paramagnetic state the quadrupolar doublets of Fe^{2+} and Fe^{3+} are well resolved. Between T_N and 35 K, Fe^{3+} orders, but the Zeeman sextet of Fe^{3+} coexists with the doublet of Fe^{2+}. Only below 35 K does the Zeeman sextet of Fe^{2+} appear.

It is clear that there are two temperatures of magnetic ordering in this compound: 157 K for Fe^{3+} and 35 K for Fe^{2+}; therefore, the Fe^{2+}–Fe^{3+} interactions are very weak compared with the Fe^{3+}–Fe^{3+} interactions. Above the last temperature and according to the existence of nonfrustrated perovskite layers of Fe^{3+} in the structure, $Fe_3F_8\cdot2H_2O$ must be considered a 2D magnetic material because Fe^{2+} remains paramagnetic. Preliminary neutron diffraction measurements on single crystals show that the moments lie in the perovskite layers (M. Leblanc *et al.*, unpublished).

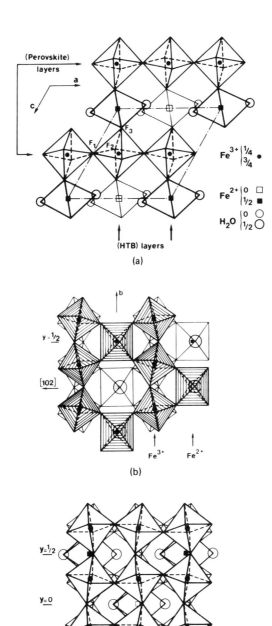

Fig. 8. (a) The (010) projection of $Fe_3F_8 \cdot 2H_2O$; (b) $(20\bar{1})$ projection of $Fe_3F_8 \cdot 2H_2O$ showing HTB layers; (c) (002) projection evidencing perovskite layers.

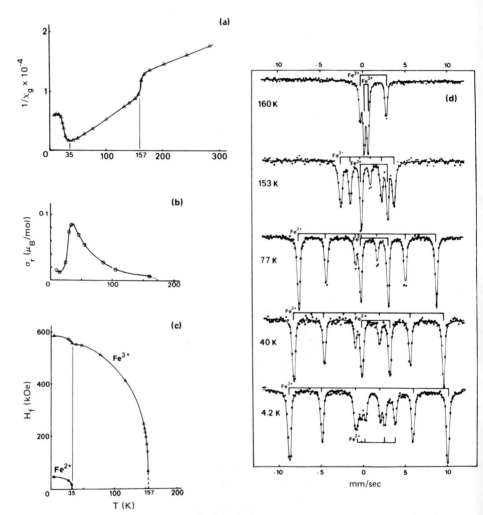

Fig. 9. (a) $\chi^{-1}(T)$ curve of $Fe_3F_8 \cdot 2H_2O$. (b) Thermal variation of the remanent magnetization of $Fe_3F_8 \cdot 2H_2O$. (c) Thermal evolution of the Mössbauer spectra. (d) Hyperfine field versus T for Fe^{3+} and Fe^{2+} in $Fe_3F_8 \cdot 2H_2O$.

The ordering of Fe^{2+} causes the decrease in σ_R and χ^{-1}, enhances the hyperfine magnetic field of Fe^{3+} (Fig. 9d), and probably induces a change in the magnetic structure. All these experimental features are typical of an "idle spin" behavior.

The lack of ordering of Fe^{2+} corresponds to the calculations of Wannier (1950) and Fazekas *et al.* (1974) in the hypothesis of $S = \frac{1}{2}$ Ising spins in antiferromagnetic interactions on a triangular net. Such systems must remain disordered either on a long range or on one of the three sublattices. This second possibility is confirmed (Darcy et al., 1973) in the molecular field approximation. This type of frustration exists when the molecular field of one site is sufficiently weak.

So far, however, it has been impossible to explain the different behavior of Fe^{2+} in $Fe_2F_5 \cdot 2H_2O$ and $Fe_3F_8 \cdot 2H_2O$. In both structures, it has the same nearest neighbors (4 F^- + 2 H_2O) and next nearest neighbors (4 Fe^{3+}). The hyperfine magnetic field has similar low values in $Fe_2F_5 \cdot 2H_2O$ (41 $\pm$ 3 kOe) and in $Fe_3F_8 \cdot 2H_2O$ (47 $\pm$ 3 kOe). Nevertheless, Mössbauer spectroscopy shows that, for the first compound, the sublattices of Fe^{2+} and Fe^{3+} order at the same temperature (48 K) whereas, for the second, Fe^{2+} remains para-magnetic far below the Néel point of the Fe^{3+} sublattice. One might find the reason for the difference after considering the second cation shell around Fe^{2+}.

4. $NH_4Fe^{II}Fe^{III}F_6$

Compound $NH_4Fe^{II}Fe^{III}F_6$ (Ferey et al., 1981, 1984; Tressaud et al., 1970, 1972), first synthesized by Tressaud et al. (1970), derives from the cubic modified pyrochlore structure (Babel, 1972). Its orthorhombic symmetry (space group Pnma) is a consequence of $Fe^{2+}-Fe^{3+}$ ordering in the cell. The structure is then built from crossed corner-sharing trans chains of Fe^{2+} and Fe^{3+} octahedra, which run, respectively, along the [100] and [010] directions (Ferey et al., 1981). These chains link all their F^- ions to give a 3D network (Fig. 10a). The cationic sublattice can then be described as consisting of tetrahedra that share all their vertices (Fig. 10b), and its topology is very frustrating for spins involved in antiferromagnetic interactions. As a matter of fact, each cation belongs to six triangles.

$NH_4Fe_2F_6$ is strictly antiferromagnetic [$T_N = 19$ K (Tressaud et al., 1972)]. Neutron diffraction (Ferey et al., 1984) shows the identity of the nuclear and magnetic cells. Below 19 K the Fe^{2+} spin arrangement is strictly antifer-romagnetic along [010] (Fig. 10c). The Fe^{3+} magnetic sublattice corresponds to a noncollinear antiferromagnetism, with a canting mode $G_xC_yA_z$ according to the notation of Bertaut (1963). The Fe^{2+} and Fe^{3+} magnetic sublattices are quasi-orthogonal ($\alpha = 104°$) with a disposition in which, inside the triangle, a spin adopts a symmetric position toward the others. Moreover, each cationic tetrahedron contains two Fe^{2+} and two Fe^{3+}. This situation confirms the theorical prediction of Villain (1979) in a general survey of insulating spin glasses. He found that, when di- and trivalent ions are arranged in such an ordered way, the system can be described by independent antiferromagnetic chains of Fe^{3+} and M^{2+} if the magnetic interactions J, J',

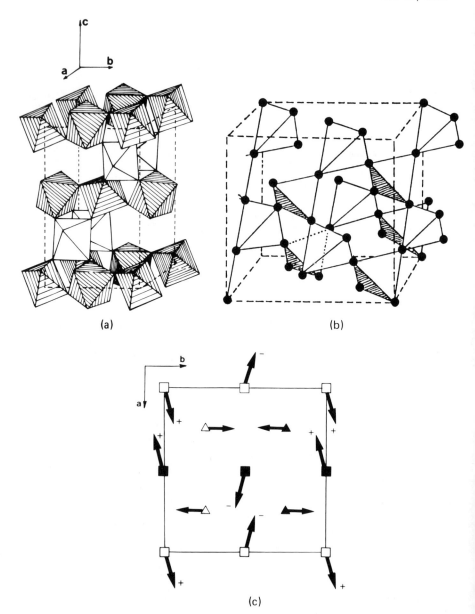

Fig. 10. (a) Perspective view of $NH_4Fe^{II}Fe^{III}F_6$ (Fe^{3+} octahedra are shaded). (b) Tetrahedral metallic sublattice in the pyrochlore structure. (c) The (001) projection of the magnetic structure of $NH_4Fe_2F_6$ [Fe^{3+} represented by squares and Fe^{2+} by triangles; $+$ and $-$ refer to the orientation of the moments toward the (001) plane].

and J'', respectively corresponding to $M^{2+}-M^{2+}$, $M^{3+}-M^{3+}$, and $M^{2+}-M^{3+}$, satisfy the relation

$$|J| + |J'| - 2|J''| > 0.$$

In the case of $NH_4Fe_2F_6$ we can conclude that the $Fe^{2+}-Fe^{3+}$ interactions are very weak and probably responsible for the low T_N.

These four examples allow us to note the following general features of the frustration in ordered triangular nets with antiferromagnetic interactions:

1. The frustration varies with the disparity of the strengths of the interactions.
2. The strongest interaction fixes the relative disposition of the spins.
3. One spin of a triangle adopts a symmetric orientation toward the two others. This conclusion is opposite to the primitive definition of frustration, which assumed that it could not satisfy simultaneously the antiferromagnetic interactions with the two others. Its symmetric position supposes an equal interaction with them.

III. Frustration and Disordering

Cationic ordering was treated as an invariant in the previous section. It is now necessary to examine the influence of disorder on the magnetic properties of fluorides. The disorder can result from a random distribution of $3d$ cations in a crystalline solid, but also in the structure itself (i.e., when the compound is amorphous). The first case is illustrated by $CsMnFeF_6$, and the second by amorphous FeF_3.

A. $CsMnFeF_6$

Compound $CsMnFeF_6$ (Babel et al., 1967; Kurtz, 1977, 1982; Kurtz and Roth, 1977; Kurtz et al., 1976) belongs to the cubic modified pyrochlore structure (space group $Fd3m$); Mn^{2+} and Fe^{3+} statistically occupy the (16c) sites of the structure. Kurtz (1977, 1982) has shown that its magnetic properties could be explained in terms of a spin glass behavior: a cusp of the susceptibility curve appears even for high applied fields and shifts to high temperatures as a function of the field. Below freezing temperature (28 K), a weak magnetization is found, and two small magnetic reflections superimposed on the diffuse intensity have been observed by neutron diffraction (Kurtz and Roth). Some magnetic properties reveal fundamental differences with those of spin glasses, however, (Alba et al., 1982; Beraart et al., 1982; Pappa, 1984; Pappa et al., 1983). The abrupt anomalies of the field-cooled

410   G. Ferey *et al.*

susceptibilities at a characteristic temperature T_g, their large values below T_g, and a broad specific heat maximum near T_g indicate the existence of a transition to a long-range ordered state. The results are presently better analyzed in terms of magnetic clusters. $CsNiFeF_6$ shows approximately the same behavior.

Even if $CsMnFeF_6$ and $CsNiFeF_6$ cannot be considered pure spin glasses [as described by Villain (1979) in the case of a random distribution with $|J| + |J'| - 2|J''| > 0$], they clearly show that cationic disorder is the most important factor in obtaining highly frustrated systems in the crystalline state.

B. AMORPHOUS FeF$_3$

In an amorphous compound, the disorder is complete and implies a wide distribution of the superexchange angles between magnetic carriers. In order to minimize the problems, a compound with only one type of cation was chosen: FeF_3 (Ferey, 1980; Ferey *et al.*, 1979a,b, 1980). It was prepared by quenching FeF_3 vapor on a cold substrate. Its magnetic properties led to a new type of magnetism: speromagnetism, which can be described by an isotropic freezing of the spins (Fig. 11). It occurs at 29 K for FeF_3, whereas the Néel temperature of the nonfrustrated rhomboedral form of FeF_3 (derived from the ReO_3 type) is 363 K. It was shown that the enormous reduction in the spin freezing temperature is not caused by any significant weakening of Fe–F–Fe superexchange interactions, which are of comparable magnitude in both forms (~ 100 K), but is a consequence of their frustration.

Coey and Murphy (1982) proposed for amorphous FeF_3 a structural model that uses exclusively octahedra linked by all their vertices. They form five-,

Fig. 11.   Speromagnetic arrangment of FeF$_3$.

four-, and three-membered rings with respective percentages of 53.5, 43.1, and 3.1 and therefore a great number of odd cycles that describe well the observed frustration, enhanced by the disorder and the distribution of exchange interactions.

The structural disorder is also maximum in vitreous fluorides $PbMnFeF_7$ and Pb_2MnFeF_9, which exhibit spin glass behavior (Miranday et al., 1980).

C. DISORDER ON SQUARE PLATELETS

As mentioned earlier, there are no published examples of disorder on square platelets in fluorine chemistry. The system $Rb_2MnCl_4/Rb_2Cr^{II}Cl_4$ (Munninghoff, 1980; Munninghoff et al., 1980, 1981, 1982), however, provides a good illustration of this type of frustration.

These two compounds crystallize with the K_2NiF_4 structure. Rb_2MnCl_4 is antiferromagnetic ($T_N = 57$ K) with moments along [001]; Rb_2CrCl_4 is a ferromagnet ($T_C = 52.4$ K) (Janke et al., 1983) with moments in the (001) plane. The magnetic study of the $Rb_2Mn_xCr_{1-x}Cl_4$ system shows that for $0.59 > x > 0.41$, spin glass behavior occurs after a decrease in the temperatures of magnetic order on both sides. A similar behavior seems to exist in the system Ba_2CuF_6 ($T_C = 8$ K)/Ba_2NiF_6 ($T_N = 93$ K) (Renaudin and Ferey, 1985).

IV. The Frustration Problem in Fluorides

This brief review has shown that, in spite of the common existence of triangular antiferromagnetic metallic sublattices, the frustrated magnetic character varies with the materials considered: three 120° antiferromagnetic sublattices for HTB FeF_3, idle spin behavior for Fe^{2+} in $Fe_3F_8 \cdot 2H_2O$, orthogonal antiferromagnetic sublattices in $MnCrF_5$ and $NH_4Fe_2F_6$, spin glass behavior in $CsMnFeF_6$, and speromagnetism in amorphous FeF_3.

A preliminary analysis of the results seems to show (Table I) that the frustrated state depends on (1) the number of odd cycles in the structure, and therefore on (2) the number of frustrated and unfrustrated bonds, (3) the number of magnetic neighbors, and (4) the strength of the magnetic interactions. It is characterized mainly by the $|\theta_p|/T_N$ ratio, which increases with the frustrated behavior.

These conclusions concern exclusively structures in which the metallic octahedra share only corners. It would be worthwhile now to study frustration effects in structures that exhibit simultaneously corner and edge sharing.

Other parameters influence frustration, however. Despite the lack of a frustrating factor, β-$LiMnFeF_6$ has idle spin behavior on one of the two sites of

TABLE I

Evolution of the Magnetically Frustrated Character with the Topology of Some Iron Fluorides[a]

Compound	a	b 1	b 2	c 1	c 2	d 1	d 2	T_N (K)	$\frac{|\theta_p|}{T_N}$	$\frac{b+c}{b+c+d}$
HTB FeF$_3$	2	4	—	—	—	2	—	110	6	0.6
Fe$_3$F$_8$·2H$_2$O	2	2	2	0	4	2	0	35	8	0.80
NH$_4$Fe$_2$F$_6$	6	2	4	2	4	0	0	19	10	1.00
CsMnFeF$_6$	6	3	3	3	3	0	0	5	12	1.00
Amorphous FeF$_3$	—	—	—	—	—	—	—	29	17	—

[a] Notation: a = number of odd cycles around each ion, b_1 = frustrated Fe^{3+}—Fe^{3+} bonds around Fe^{3+}, b_2 = frustrated Fe^{3+}—M^{2+} bonds around Fe^{3+}, c_1 = frustrated M^{2+}—M^{2+} bonds around M^{2+}, c_2 = frustrated M^{2+}—Fe^{3+} bonds around M^{2+}, d_1 = nonfrustrated bonds around Fe^{3+}, d_2 = nonfrustrated bonds around M^{2+}. The last column indicates the percentage of frustrated bonds in the structure.

Fe^{3+} (Courbion *et al.*, 1984). Thus, magnetic frustration is still a problem, and many supplementary experiments seem to be necessary to characterize and explain it fully.

Acknowledgments

The authors are very indebted to Dr. Calage and Professor Varret for their constant interest and their help in Mössbauer spectroscopy. The authors also wish to thank Professor J. Rouxel and Dr. P. Chevallier (LA 279 University of Nantes) and Professor A. Hardy and Dr. A. M. Hardy (University of Poitiers) for their aid in the collection of data on crystal structures.

References

Alba, M., Hamman, J., Jacoboni, C., and Pappa, C. (1982). *Phys. Lett. A* **89A** 423.

Babel, D., (1972). *Z. Anorg. Allg. Chem.* **387**, 161.

Babel, D., Pausewang, G., and Viebahn, W. (1967). *Z. Naturforsch., B: Anorg. Chem., Org. Chem., Biochem., Biophys., Biol.* **228**, 1219.

Bertaut, E. F. (1963). *Magnetism* **3**, 149.

Bevaart, L., Tegelarr, P. M. H. L., Van Duyneveldt, A. J., and Steiner, M. (1982). *Phys. Rev. B* **26**, 6150.

Byström, A. (1944). *Ark. Kemi, Miner. Geol.* **1813**, 10.

Coey, J. M. D., Murphy, and P. J. K., (1982). *J. Non-Cryst. Solids* **50**, 125.

Cosier, R., Wise, A., Tressaud, A., Grannec, J., Olazcuaga, R., and Portier, J. (1970). *C. R. Hebd. Seances Acad. Sci., Ser. C* **271**, 142.

Courbion, G., de Pape, R., Teillet, J., Varret, F., and Pannetier, J. (1984). *J. Magn. Magn. Mater.* **42**, 217.

Darcy, L., Wojtowicz, P. J., and Rayl, M. (1973). *Mater. Res. Bull.* **8**, 515.

de Kozak, A. (1971). *Rev. Chim. Miner.* **8**, 328.

Fazekas, P., and Anderson, P. W. (1974). *Philos. Mag.* [8] **30**, 423.

Ferey, G. (1980). *Rev. Phys. Appl.* **15**(6), 68.

Ferey, G. and Lalligant, Y. (1985). *J. Phys. C* (in press).

Ferey, G., Leblanc, M., de Pape, R., Bothorel-Razazzi, M., Passaret, M. (1975). *J. Cryst. Growth* **29**, 209.

Ferey, G., de Pape, R., Poulain, M., Grandjean, D., and Hardy, A. (1977). *Acta Crystallogr., Sect. B* **B33**, 1409.

Ferey, G., de Pape, R., and Boucher, B. (1978). *Acta Crystallogr., Sect. B* **B34**, 1084.

Ferey, G., Leclerc, A. M., de Pape, R., Mariot, J. P., and Varret, F. (1979a). *Solid State Commun.* **29**, 477.

Ferey, G., Varret, F., and Coey, J. M. D., (1979b). *J. Phys. C* **12**, L531.

Ferey, G., Coey, J. M. D., Henry, M., Teillet, J., Varret, F., and Buder, R. (1980). *J. Magn. Magn. Mater* **15-18**, 1371.

Ferey, G., Leblanc, M., and de Pape, R. (1981). *J. Solid State Chem.* **40**, 1.

Ferey, G., Leblanc, M., de Pape, R., Pannetier, J. (1984). *Solid State Commun.* (in press).

Giusepetti, G., and Tadini, C. (1978). *Tschermaks Mineral. Petrogr. Mitt.* **25**, 57.

Goodenough, J. B. (1963). "Magnetism and the Chemical Bond." Wiley (Interscience), New York.

Haegele, R., Verschaeren, W., Babel, D., Dance, J. M., and Tressaud, A. (1978). *J. Solid State Chem.* **24**, 77.

Hall, W., Kim, S., Zubieta, J., Walton, K., and Brown, D. B. (1977). *Inorg. Chem.* **16**, 1884.

Heger, G., and Viebahn-Hansler, R. (1972). *Solid State Commun.* **11**, 1119.

Heger, G. (1973). *Int. J. Magn.* **5**, 119.

Hepworth, M. A., Jack, K. H., Peacock, R. D., and Westland, G. J. (1957). *Acta Crystallogr.* **10**, 63.

Herdtweck, E. (1983). *Z. Anorg. Allg. Chem.* **501**, 131.

Imbert, P., Jehanno, G., and Varret, F. ((1976). *J. Phys. (Orsay, Fr.)* **37**, 969.

Janke, E., Hutchings, M. T., Day, P., and Walker, P. J. (1983). *J. Phys. C* **16**, 5959.

Jones, E. R., and Dawson, R. (1978). *J. Chem. Phys.* **69**, 3289.

Kanamori, J. (1959). *J. Phys. Chem. Solids* **10**, 87.

Knop, O., Cameron, T. S., and Jochem, K. (1982). *J. Solid State Chem.* **43**, 219.

Kurtz, W. (1977). Dissertation, Univ. of Tübingen.

Kurtz, W., and Roth, S. (1977). *Physica, B + C (Amsterdam)* **86–88B**, 715.

Kurtz, W. (1982). *Solid State Commun.* **42**, 871.

Kurtz, W., Geller, R., Dachs, H., and Convert, P. (1976). *Solid State Commun.* **18**, 1479.

Leblanc, M., Ferey, G., Chevallier, P., Calage, Y., and de Pape, R. (1983). *J. Solid State Chem.* **47**, 53.

Leblanc, M., Ferey, G., Calage, Y., and de Pape, R. (1984a). *J. Solid State Chem.* **53**, 360.

Leblanc, M., Ferey, G., and de Pape, R. (1984b). *Mater. Res. Bull.* (in press).

Leblanc, M., Pannetier, J., and Ferey, G. (1985). To be published.

Magneli, A. (1953). *Acta Chem. Scand.* **7**, 315.

Marland, L. G., and Betts, D. D. (1979). *Phys. Rev. Lett.* **43**, 1618.

Miranday, J. P., Renard, J. P., and Varret, F. (1980). *Solid State Commun.* **35**, 41.

Munninghoff, G. (1980). Thesis, Univ. of Marburg.

Munninghoff, G., Kurtz, W., Treutmann, W., Hellner, E., Heger, G., Lehner, N., and Reinen, D. (1981). *Solid State Commun.* **40**, 571.

Munninghoff, G., Hellner, E., Fyne, P. J., Day, P., Hutchings, M. T., and Tasset, F. (1982). *J. Phys. (Orsay, Fr.)* **43** C7, 243.

Munninghoff, G., Treutmann, W., Hellner, E., Heger, G., and Reinen, D. (1980). *J. Solid State Chem.* **34**, 289.

Pappa, C., Hamman, J., and Jacoboni, C. (1983). *J. Magn. Magn. Mater.* **31–34**, 1391.

Pappa, C. (1984). Thesis, Univ. of Saclay.
Passaret, M., Leblanc, M., and de Pape, R. (1974). *High Temp.—High Pressures* **6**, 629.
Renaudin, J., and Ferey, G. (1985). To be published.
Toulouse, G. (1977). *Commun. Phys.* **2**, 115.
Tressaud, A., de Pape, R., and Portier, J. (1970). *C.R. Hebd. Seances Acad. Sci., Ser. C* **270**, 726.
Tressaud, A., Menil, F., Georges, R., Portier, J., and Hagenmuller, P. (1972). *Mater. Res. Bull.* **7**, 1339.
Tressaud, A., Dance, J. M., Portier, J., and Hagenmuller, P. (1974). *Mater. Res. Bull.* **9**, 121.
Tressaud, A., and Dance, J. M. (1977). *Adv. Inorg. Chem. Radiochem.* **20**, 133, and references therein.
Vannimenus, J., and Toulouse, G. (1977). *J. Phys. C* **10**, L537.
Villain, J. (1979). *Z. Phys.* **B 33**, 31.
Walton, E. G., and Brown, D. B. (1977). *Inorg. Chem.* **16**, 2425.
Wannier, G. H. (1950). *Phys. Rev.* **79**, 357.
Wollan, E. O., Child, H. R., Koehler, W. C., and Wilkinson, M. K. (1985). *Phys. Rev.* **112**, 1132.

11

Electronic Conduction in Fluorides

ALAIN TRESSAUD
Laboratoire de Chimie du Solide du CNRS
Université de Bordeaux
Talence, France

I. Introduction

Most fluorides are very good electronic insulators due to large band gaps and empty conduction bands. This behavior is a consequence of the strong electronegativity of fluorine, which results in a low-energy valence band and an electronic localization. A major characteristic of fluorides is therefore a low electronic conductivity with high activation energy.

These properties have been considered for dielectric purposes. Permittivity is lower in fluorides than in oxides, however, and even when it is of the same order of magnitude (at higher temperatures), it is generally associated with high dielectric losses. This feature can be readily attributed to the ionic mobility of F^-.

In addition to numerous series of materials showing a large F^- conductivity, new classes of fluorides possessing a significant electronic conductivity

INORGANIC
SOLID FLUORIDES

415

416   A. Tressaud

have been investigated, including doped semiconductors, one- (1D) and
two-dimensional (2D) metallic fluorides, piezoresistive materials.

II. Semiconducting Fluorides

In metal fluorides, the band gap is often larger than 6 eV; in MgF_2, for
instance, it reaches 12 eV. Semiconducting properties may occur, however,
when the gap contains localized energy levels. Some selected examples,
ranging from conduction mechanisms in insulating FeF_3 to strong semiconducting properties of doped CdF_2, are given in this section.

Insulating thin films of FeF_3 can be prepared either by sublimation of the
trifluoride under a fluorinating atmosphere or by fluorination of thin films of
iron. Vacancies of fluorine atoms and traces of oxygen have been detected
close to the surface. Within the band gap (5.96 eV), an absorption peak is
observed 1.1 eV below the conduction band; it can be correlated to the amount
of fluorine vacancies (Lascaud et al., 1979; Pichon et al., 1982). Current
measurements have been carried out on thin films of FeF_3 under direct electric
fields of about 10^6 V m^{-1}. Between 90 and 200 K, the activation energy is very
small (from 0.008 to 0.1 eV). At higher temperatures the current follows an
exponential law with an activation energy of about 0.8 eV, consistent with the
absorption values; the resistivities range from 10^{17} Ω cm at 240 K to 10^7 Ω cm
at 500 K. An interpretation of this behavior has been proposed by Lascaud,
Pichon et al., on the basis of donor centers associated with fluorine vacancies
and located 1.1 eV below the conduction band. At low temperatures ($T <$
200 K), the conductivity is extremely low and occurs via a tunnel effect
between localized levels close to Fermi level. At higher temperatures
(240 K $< T <$ 500 K) the conduction can be characterized by a mechanism
of thermally activated electrons.

When prepared in an ultravacuum, NiF_2 has a very high resistivity ($\rho \simeq$
10^{12} Ω cm) at 25°C, but when the synthesis occurs in the presence of traces
of oxidizing species, the material can be considered a transparent semiconducting electrode (Salardenne et al., 1982). It may also be mentioned that
NbF_3 is a semiconductor ($\rho \simeq 50$ Ω cm at room temperature) with an
activation energy of 0.17 eV (Pouchard et al., 1971).

The conversion of an insulating material to a highly conducting semiconductor has been widely studied in CdF_2. Pure CdF_2 exhibits an energy gap of
~ 8 eV. When previously doped with trivalent ions (yttrium or rare-earth
ions), CdF_2 can be converted to an n-type semiconductor, provided that
heating is done (1) in cadmium vapor at 500°C (Kingsley and Prener, 1962;
Prener and Kingsley, 1963), (2) in contact with liquid metal (tin, Wood's metal)
at 360°C (Szadkowski et al., 1977), or (3) in a hydrogen atmosphere (200–

500°C) (Wojciechowski and Kaminska, 1979). After conversion, the crystal has a deep blue color. A proper choice of experimental conditions enables one to obtain CdF_2 within a large range of resistivities (down to 0.3 Ω cm) (O'Horo, 1972; Wojciechowski and Kaminska, 1979). Heating in vacuum or in halogens produces a deconversion back to a transparent insulator with resistivity higher than 10^9 Ω cm (Szadkowski et al., 1977).

Prener and Kingsley (1963) proposed the following conduction mechanism. The presence of a trivalent element replacing Cd^{2+} introduces donor levels close to the $4s$ conduction band. When the material is treated with the metallic vapor or liquid, the interstitial F^- ions that had been introduced with M^{3+} migrate toward the surface and combine with cadmium atoms, according to the reaction

$$Cd + 2 F^- \longrightarrow CdF_2 + 2 e^-$$

The released electrons fill the donor levels. A band mechanism takes place by ionization of those levels and excitation of electrons into the conduction band. The role played by the intersitial F^- ions has been confirmed by luminescence of $CdF_2 \cdot MF_3$ doped with Mn^{2+}, and fluorine diffusion has been followed with ^{18}F as tracer (Süptitz et al., 1972; Langer et al., 1974). Slow autodeconversion could be facilitated by the precipitation of cadmium particles in the crystal (Szadkowski et al., 1977).

Semiconducting CdF_2 has been proposed for a couple of applications:

1. As active material in electroluminescent cells; a crystal of insulating CdF_2 containing an activator (MnF_2) for producing electroluminescence is grown onto the surface of a crystal of semiconducting CdF_2 (Prener and Kingsley, 1970)
2. As cathodic material for high-energy primary batteries (Nicholson, 1971)

III. Metallic Fluorides

A. ONE-DIMENSIONAL METALLIC FLUORIDES

In 1D metallic compounds $Hg_{2.9}AsF_6$ and $Hg_{2.9}SbF_6$ prepared by Gillespie and co-workers (Brown et al., 1974; Cutforth et al., 1976), mercury is in an oxidation state intermediate between 0 and 1. The lattice consists of an ionic tetragonal arrangement of octahedral $(AsF_6)^-$ or $(SbF_6)^-$ units. An infinite polymercuric cation is located in between these isolated octahedra, and the chains run along the a and b axes in nonintersecting tunnels (Fig. 1).

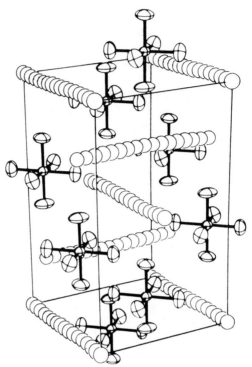

Fig. 1. Structure of $Hg_{2.9}AsF_6$. From Schultz *et al.* (1978). Copyright 1978 American Chemical Society.

In $Hg_{2.9}AsF_6$, the interchain separation (3.09 Å) is close to the Hg–Hg distances in metallic mercury [3.05 Å; $d_{interchain}(Hg_{2.9}SbF_6) = 3.15$ Å]. Within a chain, an even shorter Hg–Hg distance (2.64 Å in both compounds) is incommensurate with the lattice dimensions and suggests a strong overlapping of mercury 6s orbitals. On each mercury atom there is a nonintegral formal charge ($+0.35$) and the conduction band is expected to be partially filled. Neutron diffraction measurements have suggested a stoichiometric Hg_3AsF_6 formula (Schultz *et al.*, 1978; Miro *et al.*, 1978).

Crystals have been grown that are very sensitive to moisture and have a metallic luster with golden faces. Resistivity measurements have shown the conductivity to be identical along the *a* and *b* axes: $\sigma \simeq 10^4\,\Omega^{-1}\,cm^{-1}$ at 300 K. The conductivity ratio σ_a/σ_c is about 100 and 40 for arsenic and antimony compounds, respectively (Table I). The materials show metallic behavior at any temperature below 300 K. They exhibit a large temperature dependence of the conductivity, which increases with decreasing temperature (Cutforth *et al.*, 1977). The superconductivity initially observed in these phases (Chiang *et al.*,

TABLE I

Conductivity in Metallic Fluorides (295 K)

Metallic Fluoride	$\sigma_{\parallel}\ (\Omega^{-1}\,cm^{-1})$ Along Metallic Chains or Layers	$\sigma_{\perp}\ (\Omega^{-1}\,cm^{-1})$
One-dimensional		
$Hg_{2.86}AsF_6$	10^4	2.5×10^2
$Hg_{2.91}SbF_6$	9×10^3	9×10^1
Two-dimensional		
$C_{16}AsF_5$	4×10^5	2.3×10^{-1}
$C_{16}SbF_5$	6.5×10^5	—
Ag_2F	1.1×10^4	10^2

1977; Spal *et al.*, 1977) has been attributed to mercury trapped in the sample (Batalla and Datars, 1983; Schirber *et al.*, 1982).

Two similar compounds, Hg_3NbF_6 and Hg_3TaF_6, have been shown to be superconductors by Datars *et al.* (1983). The critical temperature is 7.0 K and the critical fields extrapolated to $T = 0$ K are 0.17 and 0.13 T for the tantalum and niobium compound, respectively. Because no anisotropy of the critical field is observed, these authors suggest a 3D superconductivity.

B. TWO-DIMENSIONAL METALLIC FLUORIDES

Fluoride Ag_2F has an anti-CdI_2 structure and shows double layers of silver atoms separated by fluorine layers. Following Pauling, this compound can be viewed as presenting a mixed type of ionic–metallic chemical bond. Like metals, crystals of Ag_2F have a positive resistivity coefficient (Andres *et al.*, 1966). The sign of the thermoelectronic force confirms that in Ag_2F free electrons are responsible of the conduction: their number is 3.3×10^{22} cm^{-3} (67×10^{22} cm^{-3} for metallic silver). At 290 K the conductivity is about $10^4\ \Omega^{-1}$ cm^{-1} and increases linearly with decreasing temperatures down to 20 K. The anisotropy of the conductivity is shown in Table I (Kawamura *et al.*, 1974).

The intercalation of acceptor organic molecules (benzonitrile C_6H_5CN, for example) has been achieved between two layers of Ag_2F (Koshkin *et al.*, 1976). An estimate based on a contactless method has shown that the conductivity of the phase exceeds $10^3\ \Omega^{-1}$ cm^{-1}.

We will not deal here with intercalation compounds of graphite because they are investigated elsewhere in this book (Chapter 8). It should be emphasized, however, that the conductivities are sometimes higher than that of metallic copper and are associated with an important anisotropy due to structure (Table I). Because such insertion phases display both ionic and electronic conductions, they have been proposed as electrodes in solid-state galvanic cells (McCarron, 1980).

IV. Piezoresistive Fluorides

Under pressure Pd_2F_6 and other $LiSbF_6$-type fluorides show a reversible electronic transition characterized by a drastic change in resistivity (Langlais *et al.*, 1979). The variation of $\log \rho = f(P)$ measured *in situ* at 25°C is shown in Fig. 2 for Pd_2F_6: between 1 bar and 25 kbars the decrease in $\log \rho$ is ~ 5 $[d(\log \rho)/dP = -0.18 \text{ kbar}^{-1}]$; at higher pressures (40 kbars $< P <$ 80 kbars) the slope decreases down to $d(\log \rho)/dP = -0.015 \text{ kbar}^{-1}$. Simultaneously, the activation energy decreases from 1.34 eV for $P = 1$ bar to 0.07 eV for 60 kbars.

For the $M^{II}M^{IV}F_6$ fluoride series (M^{II}, $M^{IV} = d$ transition metals) identical behavior is observed, with an increase in the phenomenon from $3d$ elements to

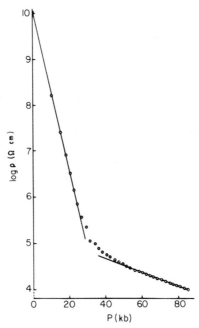

Fig. 2. Pressure dependence of resistivity of Pd_2F_6 at 300 K.

$5d$ elements, that is, for more extended orbitals (Langlais, 1979). Because the structure of these compounds is derived from ReO_3 (see Chapter 3), a similar energy diagram has been proposed. Conduction would be due to a localized electron mechanism via σ^* levels (Goodenough, 1973). In addition, the formation of the $+3$ oxidation state is favored by pressure, according to the reaction $Pd(II) + Pd(IV) \rightleftarrows 2\,Pd(III)$. New levels would thus be generated and a hopping mechanism would arise between $Pd(II)$ and $Pd(III)$ and, in a lower amount, between $Pd(III)$ and $Pd(IV)$.

These piezoresistive fluorides could be used as sensors for a continuous *in situ* control of pressure and for calibration of high-pressure systems (up to 40 kbars). Their advantages, compared with other piezoresistive materials, are that (1) log ρ varies linearly over a large pressure range; (2) single crystals are not needed, because the property is a characteristic of the bulk; and (3) samples can be prepared either as powdered pellets or as thin films (Tressaud *et al.*, 1980).

Insulator $\rightleftarrows$ metal transitions induced under hyperpressures have been claimed for alkaline-earth fluorides CaF_2, SrF_2, and BaF_2. The transition pressures have been evaluated to be intermediate between 220 and 1000 kbars (Vinogradov *et al.*, 1976).

Solid fluorides displaying high ionic conductivity such as $PbSnF_4$ or $Pb_{0.75}Sb_{0.25}F_{2.25}$ (see Chapter 12) also exhibit a linear variation of resistivity with pressure $[d(\log \rho)/dP = 0.04\ \mathrm{kbar}^{-1}]$ in the range 20 kbars $<$ $P <$ 70 kbars. The application range could be extended to higher pressures (Matar *et al.*, 1980; Lucat, 1980).

References

Andres, K., Kuebler, N. A., and Robin, M. A. (1966). *J. Phys. Chem. Solids* **27**, 1747.
Batalla, E., and Datars, W. R. (1983). *Solid State Commun.* **45**, 285.
Brown, I. D., Cutforth, B. D., Davies, C. G., Gillespie, R. J., Ireland, P. R., and Vekris, J. E. (1974). *Can. J. Chem.* **52**, 791.
Chiang, C. K., Spal, R., Denenstein, A. M., Heeger, A. J., Miro, N. D., and McDiarmid, A. G. (1977). *Phys. Rev. Lett.* **39**, 650.
Cutforth, B. D., Datars, W. R., Gillespie, R. J., and Van Schyndel, A. (1976). *Adv. Chem. Ser.* **150**, 1.
Cutforth, B. D., Datars, W. R., Van Schyndel, A., and Gillespie, R. J. (1977). *Solid State Commun.* **21**, 377.
Datars, W. R., Morgan, K. R., and Gillespie, R. J. (1983). *Phys. Rev. B* **28**, 5049.
Goodenough, J. B. (1973). "Transition Metal Oxides." Gauthier-Villars, Paris.
Kawamura, H., Shirotani, I., and Inokuchi, H. (1974). *Chem. Phys. Lett.* **24**, 549.
Kingsley, J. D., and Prener, J. S. (1962). *Phys. Rev. Lett.* **8**, 315.
Koshkin, V. M., Yagubskii, E. B., Mil'ner, A. P., and Zabrodskii, Y. R. (1976). *JETP Lett.* (*Engl. Transl.*) **24**, 110.
Langer, J. M., Langer, T., Pearson, G. L., and Krukowska, F. B. (1974). *Phys. Status Solidi* A **25**, K61.

Langlais, F. (1979). Ph.D. Thesis, Univ. of Bordeaux.

Langlais, F., Demazeau, G., Portier, J., Tressaud, A., and Hagenmuller, P. (1979). *Solid State Commun.* **29,** 473.

Lascaud, M., Lachter, A., Salardenne, J., and Barrière, A. S. (1979). *Thin Solid Films* **59,** 353.

Lucat, C. (1980). Ph.D. Thesis, Univ. of Bordeaux.

McCarron, E. M. (1980). Ph.D. Thesis, Univ. of California, Berkeley.

Matar, S., Reau, J. M., Demazeau, G., Lucat, C., Portier, J., and Hagenmuller, P. (1980). *Solid State Commun.* **35,** 681.

Miro, N. D., McDiarmid, A. G., Heeger, A. J., Garito, A. F., Chiang, C. K., Schultz, A. J., and Williams, J. M. (1978). *J. Inorg. Nucl. Chem.* **40,** 1351.

Nicholson, M. M. (1971). *J. Electrochem. Soc.* **118,** 1047.

O'Horo, M. (1972). Ph.D. Thesis, Pennsylvania State University, University Park.

Pichon, J., Barrière, A. S., and Salardenne, J. (1982). *Phys. Status Solidi* A **69,** 699.

Pouchard, M., Torki, M. R., Demazeau, G., and Hagenmuller, P. (1971). *C. R. Hebd. Seances Acad. Sci.* **273,** 1093.

Prener, J. S., and Kingsley, J. D. (1963). *J. Chem. Phys.* **38,** 667.

Prener, J. S., and Kingsley, J. D. (1970). U.S. Patent 3,503, 812.

Salardenne, J., Salagoïty, J. M., Pichon, J., and Barrière, A. S. (1982). *Thin Solid Films* **88,** 412.

Schirber, J. E., Heeger, A. J., and Nigrey, P. J. (1982). *Phys. Rev. B* **26,** 6291.

Schultz, A. J., Williams, J. M., Miro, N. D., McDiarmid, A. G., and Heeger, A. J. (1978). *Inorg. Chem.* **17,** 646.

Spal, R., Chiang, C. K., Denenstein, A. M., Heeger, A. J., Miro, N. D., and McDiarmid, A. G. (1977). *Solid State Commun.* **39,** 650.

Süptitz, P., Brink, E., and Becker, D. (1972). *Phys. Status Solidi* B **54,** 713.

Szadkowski, A., Lubomirska, W. A., Zareba, A., and Krukowska, F. B. (1977). *Phys. Status Solidi* A **44,** K43.

Tressaud, A., Langlais, F., Demazeau, G., Portier, J., Lucat, C., Rèau, J. M., and Hagenmuller, P. (1980). *Marche Innovation* **43,** 20.

Vinogradov, B. V., Vereshagin, P. F. V., Timofeev, Y. A., and Yakovlev, E. N. (1976). *Pisma Zh. Tekh. Fiz.* **2,** 964.

Wojcieckowski, J., and Kamińska, E. (1979). *J. Phys. D.* **12,** L157.

12

Fast Fluorine Ion Conductors

JEAN-MAURICE RÉAU and JEAN GRANNEC
Laboratoire de Chimie du Solide du CNRS
Université de Bordeaux
Talence, France

I. Introduction

A large number of solids have ionic conductivity comparable to that of molten salts and are referred to as fast ion conductors or solid electrolytes. In recent years these materials have stimulated considerable interest. From the technological viewpoint, solid electrolytes have many potential applications in batteries and other electrochemical devices. The interest in this field is

INORGANIC
SOLID FLUORIDES

reflected in the numerous studies reported in review articles (Hagenmuller and van Gool, 1978; Vashishta et al., 1979; Subbarao, 1980; Bates and Farrington, 1981). Among solid electrolytes, fast fluorine ion conductors have become a research field of growing importance. Indeed, materials with very good electrical performances have been isolated, with particular emphasis on fluorite-type materials, as models of superionic materials due to their relative structural simplicity and availability as large single crystals. Such materials have proved to be useful for thermodynamic measurements, specific electrodes, solid-state batteries, electrochromic devices, gas sensors, etc.

As is well known, the performance of electrolytic oxides with CaF_2-type structure whose conduction is due to O^{2-}, such as those derived from ZrO_2, is good only at high temperatures ($t > 600°C$) (Etsell and Flengas, 1970; Nowick and Park, 1976; Wang et al., 1981) In contrast, isostructural fluorides have a significant anionic conduction at relatively low temperatures (Réau and Portier, 1978). This property is obviously a consequence of the higher mobility of F^- as a result of smaller size, but even more of lower electric charge and less covalent bonding. As a consequence, fluorides can be predicted to be the *best anionic conductors*. Furthermore, these materials are very often electronic insulators, an essential property for their use as electrolytes in electrochemical batteries (Schoonman, 1976; Schoonman and Wolfert, 1981; Portier et al., 1983).

The mechanisms affecting ion transport are almost invariably associated with point defects (Lidiard, 1974). In addition to these thermally produced defects, other point defects can be formed by incorporation of aliovalent impurities in the crystal. Charge compensation requires the generation of extrinsic defects to preserve overall electrical neutrality. For instance, trivalent cations in alkaline-earth fluorides can produce F^- interstitials, whereas monovalent cations produce F^- vacancies. Fluorine ion vacancies can also result from the presence of divalent anions such as oxygen. Ionic diffusion in solids usually proceeds by the motion of these defects within the lattice (Bénière and Catlow, 1983).

Numerous experimental investigations, involving static techniques such as optical spectroscopy and epr as well as dynamic techniques such as electrical conductivity, thermal depolarization (ITC), dielectric relaxation, inelastic relaxation, radioactive tracers, and nuclear magnetic resonance (^{19}F nmr) are used either for selecting or for studying fast ion conductors. The static methods tend to provide information about the nature and site symmetry of the defects, whereas the dynamic ones tend to measure properties such as diffusion constant and activation energy. Finally, the structural and transport properties are investigated by computer simulation methods (Lidiard and Norgett, 1972; Catlow and Mackrodt, 1982).

Optimization criteria for high mobility in fluorides are discussed in the next section of this chapter. Section III proposes a classification of the fast fluorine ion conductors investigated so far. Section IV deals with correlations between electrical properties and crystallographic data determined in particular by neutron diffraction in fluorite-type fluorides. A mechanism for the formation of cubo-octahedral clusters as extended defects in anion-excess fluorides is proposed and discussed in the last section.

II. Optimization Criteria for High Mobility in Fluorides

As already pointed out, the small size of F^- as well as its single charge make it a good candidate as a mobile ion in solids. It is well known that in a solid electrolyte the ionic conductivity may approach that of the liquid (Hladik, 1972). Transitions from a normal to a fast ion conduction regime below melting point may be first or second order (O'Keeffe, 1976; Derrington et al., 1976; Schoonman, 1979). The melting points of fluorides are sufficiently low that high conductivities can be realized at relatively low temperatures.

Not all fluorides are good ionic conductors, however. Certain conditions must be met; these involve the structure and nature of the cationic sublattice associated with fluorine. Fast ion transport results simultaneously from microscopic properties connected with the chemical bonding concept and with the structural features in the material, and from macroscopic properties related to thermodynamic data.

The conductivity of an ion conductor is determined by both the concentration of the mobile ions and the rate at which they hop between equivalent sites:

$$\sigma_i = Nce\mu, \tag{1}$$

where c is the concentration of the mobile ions of charge e and mobility μ in the N crystallographic available sites. The mobility

$$\mu = (e/kT)Z(1 - c)\bar{a}^2 v_0 \exp(\Delta S/k)\exp(-\Delta E/kT) \tag{2}$$

is proportional to the number of empty nearest-neighbor sites $Z(1 - c)$ and to the frequency v_0 of the ion jump attempts across a mean distance $\bar{a}$. The free energy for realizing a jump is $\Delta G = \Delta H - T \Delta S$, where the enthalpy ΔH is equal to the activation energy ΔE of the mobility process (Armstrong et al., 1973; McGeehin and Hooper, 1977; Pouchard and Hagenmuller, 1978). A high ionic conductivity requires a reasonable magnitude for the product $c(1 - c)$ and a small ΔE. As a consequence, the mobile ions can have a high mobility only when they are monovalent and weakly bound to the matrix and when the number of available sites of identical or slightly higher energy is largely in excess of the number of mobile species present.

A study of the F^- conductivity in a large number of fluorides has made it possible to determine several criteria for high mobility in fluorides (Réau and Portier, 1978; Réau et al., 1978d):

1. *Vacancies in the anionic sublattice due to nonstoichiometry.* Fluorides with NaCl, rutile, or ReO_3 structures are not anion deficient, and the F^- ions do not have a large mobility. In contrast, the fluorite, tysonite, and YF_3 structures can lead to nonstoichiometric MF_{2+x} or MF_{3-x} phases in which deviations from stoichiometry result in new vacancies for ionic conduction.

2. *Low coordination number of mobile species.* In cationic conductors, fast ion conduction is favored for mobile species with low coordination numbers (Armstrong et al., 1973). A similar conclusion can probably be applied to anionic conductors: the lower the coordination number of the anions, the higher the mobility. This has been shown for materials such as NaF, CaF_2, and LaF_3, which correspond to respective anionic coordinations of 6,4 and 3. At a given temperature, $\sigma_{NaF} < \sigma_{CaF_2} < \sigma_{LaF_3}$.

Moreover, the cationic coordination numbers in the best fluorine conductors are high; they are, respectively, 8, 9, and 11 in the fluorite, YF_3, and tysonite structures. This notion has to be used cautiously, however.

3. *High cation polarizability.* Analysis of the electrical properties of various fluorides crystallizing with a given structure shows that as a general rule the stronger the polarizability of the cation, the larger is the anionic conductivity. This is illustrated by comparison of the behavior of KF (or RbF) and TlF, CaF_2 (or SrF_2) and β-PbF_2, and YF_3 and BiF_3.

4. *Low melting point and low entropy of melting.* High ionic conductivity appears all the more at lower temperature as the melting point decreases. Good examples are fluorides such as CaF_2, SrF_2, and BaF_2, the melting temperatures of which are much lower than those of zirconia and thoria, but the conductivity of which is consequently much higher at identical temperature.

Furthermore, fast ion conductivity involves an anomaly in the entropy of fusion (Derrington et al., 1975): whereas the entropy change across the melting point is strong in most compounds, it is actually much weaker for materials of high ionic conductivity. A review paper has been published on this problem (Uvarov et al., 1984).

III. Fluorine Ion Conductors

From the considerations developed in the previous section, one could predict that the best F^- conductors would be nonstoichiometric phases involving cations with high polarizability and structure deriving from the fluorite or tysonite type.

A. MATERIALS WITH FLUORITE-TYPE STRUCTURE

1. Nominally Pure and Doped Fluorides

The well-known fluorite-type structure consists of a cubic array of F^- ions with M^{2+} ions occupying half of the centers of the cubes. At low to moderate temperatures the ionic transport in crystals of alkaline-earth fluorides and β-PbF$_2$ is of the type found in normal ionic solids. The prevailing point defects are generally anion–Frenkel pairs (i.e., anion vacancies and interstitial anions with anion transport occurring via the migration of these defects) (Lidiard, 1974). A few hundred degrees below melting point T_m, these fluorides exhibit a broad specific heat anomaly, which passes through a maximum at a temperature T_c (Derrington et al., 1976; Schröter and Nölting, 1980). Above T_c these materials show exceptionally high anion mobility bound to relatively extensive disordering of the anion sublattice.

The concept of sublattice melting has frequently been used to describe the structural transformation accompanying the phase transition to the fast ion state (O'Keeffe, 1977). More recent investigations, however, suggest that this model may be inappropriate and that the level of disordering is not as massive. Experiments on the high-temperature fluorites include thermodynamic (Schröter and Nölting, 1980; Oberschmidt 1981), transport (Carr et al., 1978; Schoonman 1980; Azimi et al., 1984), Brillouin scattering (Hayes, 1980; Catlow et al., 1981), and neutron scattering studies (Clausen et al., 1981; Catlow and Hayes, 1982; Dickens et al., 1982; Bachmann and Schulz, 1983). Theoretical research uses both static defect energy calculations and molecular dynamics computer simulations (Dixon and Gillan, 1980; Catlow, 1980, 1983; Walker et al., 1981). The evidence seems clearly in favor of anion motions by discrete jumps and not by cooperative liquidlike diffusion. The number of mobile defects appears to be relatively small (3–5%). The exact nature and extention of disordering must still be resolved (Chadwick, 1983).

Defect enthalpies for fluorine diffusion in MF$_2$ compounds (M = Ca, Sr, Ba, or Pb) have been the topic of many investigations using the usual techniques of ionic conductivity (e.g., Ure, 1957; Barsis and Taylor, 1968; Bollmann, 1973; Bonne and Schoonman, 1977; Samara, 1979; Jacobs and Ong, 1980). Transport parameters have also been deduced by other methods, for example, nmr (e.g., Figueroa et al., 1978; Wei and Ailion, 1979), dielectric relaxation (e.g., Fontanella et al., 1978b; 1981), and ionic thermal current (e.g., Kitts and Crawford, 1974; Nauta-Leeffers and den Hartog, 1979). Furthermore, computer simulation methods are used, and theoretical values of defect energies have been obtained (Catlow et al., 1977; Corish et al., 1982; Chadwick, 1983).

Table I shows data on defect enthalpies for MF$_2$ (M = Ca, Sr, Ba, or Pb) phases. Whereas the values of h_F, h_{mv}, and h_{mi} obtained by different techniques are generally in agreement, those of h_a are different and seem related to the

TABLE I

Defect Enthalpies for Fluoride Ion Diffusion in MF_2 (M = Ca, Sr, Ba, or Pb)

	Technique	Defect Enthalpy (eV)	Reference
For CaF_2			
h_F (formation of anion–Frenkel pair)	Conductivity	2.7	Jacobs and Ong (1976)
	Theory	2.75	Catlow (1976a); Catlow et al. (1977)
h_{mv} (motion-free vacancy)	Conductivity	0.52	Oberschmidt and Lazarus (1980)
	ITC	0.53	Jacobs and Ong (1980b)
	ITC	0.51	Jacobs et al. (1980)
h_{mi} (motion-free interstitial)	Conductivity	0.79	Jacobs and Ong (1976)
	Conductivity	1.02	Bollmann et al. (1970)
	Conductivity	0.82	Oberschmidt and Lazarus (1980)
	nmr	1.04	Wei and Ailion (1979)
	Theory	0.91	Catlow et al. (1977)
h_a (association of Ln^{3+}–F_i^-)	Conductivity	0.68 (Ln = Y)	Bollmann and Reimann (1973)
	Conductivity	0.65 (Ln = Y)	Jacobs and Ong (1976)
	ITC	0.48 (Ln = La)	Ong and Jacobs (1980)
	Theory	0.66 (Ln = Lu), nn[a] associate	Wapenaar and Catlow (1981)
	Theory	0.45 (Ln = Lu), nnn associate	Wapenaar and Catlow (1981)
For SrF_2			
h_F	Conductivity	2.21	Chadwick et al. (1978)
	Conductivity	2.70	Schoonman and den Hartog (1982)
	nmr	2.58	Chadwick et al. (1978)
	Theory	2.38	Figueroa et al. (1978)
	Theory	2.38	Catlow et al. (1977)
h_{mv}	Conductivity	0.47	Chadwick et al. (1978)
	Conductivity	0.47	Schoonman and den Hartog (1982)

428

	Method	Value	Reference
	nmr	0.59	Chadwick et al. (1978)
	Theory	0.43	Figueroa et al. (1978)
	Theory	0.43	Catlow et al. (1977)
h_{mi}	Conductivity	1.00	Barsis and Taylor (1966)
	Conductivity	1.07	Chadwick et al. (1978)
	Conductivity	0.95	Schoonman and den Hartog (1982)
	nmr	0.82	Chadwick et al. (1978)
	Theory	0.80	Figueroa et al. (1978)
	Theory	0.80	Catlow et al. (1977)
h_a	Conductivity	0.56 (Ln = Yb)	Schoonman and den Hartog (1982)
	ITC	0.44 (Ln = La)	Ong and Jacobs (1980)
	ITC	0.57 (Ln = Lu)	Lenting et al. (1976)
	Theory	0.43 (Ln = Lu), nn associate	Wapenaar and Catlow (1981)
	Theory	0.55 (Ln = Lu), nnn associate	Wapenaar and Catlow (1981)
For BaF$_2$			
h_F	Conductivity	1.91	Figueroa et al. (1978)
	nmr	1.80	Figueroa et al. (1978)
	Theory	1.98	Catlow et al. (1977)
h_{mv}	Conductivity	0.58	Jacobs (1983)
	Conductivity	0.57	Figueroa et al. (1978)
	nmr	0.62	Figueroa et al. (1978)
	Theory	0.46	Catlow et al. (1977)
h_{mi}	Conductivity	0.78	Barsis and Taylor (1966)
	Conductivity	0.76	Figueroa et al. (1978)
	Conductivity	0.72	Jacobs (1983)
	Conductivity	0.71	Wapenaar, et al. (1981); Panhuyzen et al. (1981)
	nmr	0.77	Figueroa et al. (1978)
	Theory	0.72	Catlow et al. (1977)
h_a	Conductivity	0.39 (Ln = La)	Wapenaar et al. (1981)

(cont.)

429

TABLE I (*cont.*)

For BaF$_2$	Technique	Defect Enthalpy (eV)	Reference
	Conductivity	0.33 (Ln = La)	Jacobs (1983)
	ITC	0.46 (Ln = Gd), nn associate	Kitts et al. (1973)
	ITC	0.60 (Ln = Gd), nnn associate	Kitts et al. (1973)
	Theory	0.21 (Ln = Lu), nn associate	Wapenaar and Catlow (1981)
	Theory	0.49 (Ln = Lu), nnn associate	Wapenaar and Catlow (1981)
For β-PbF$_2$			
h_F	Conductivity	1.00	Liang and Joshi (1975)
	Conductivity	0.89	Bonne and Schoonman (1977)
	Conductivity	1.07	Azimi et al. (1984)
	nmr	0.88	Gordon and Strange (1978);
	Theory	1.20	Catlow (1976a); Catlow et al. (1977)
	Theory	1.07	Matar et al. (1983a)
h_{mv}	Conductivity	0.23	Bonne and Schoonman (1977)
	Conductivity	0.33	Mellors and Akridge (1981)
	Conductivity	0.23	Azimi et al. (1984)
	nmr	0.27	Wang et al. (1976)
	nmr	0.29	Chang et al. (1981)
h_{mi}	Conductivity	0.39–0.50	Kennedy and Miles (1976)
	Conductivity	0.50	Bonne and Schoonman (1977)
	Conductivity	0.45	Réau et al. (1975)
	Conductivity	0.54	Ito and Koto (1983)
	Conductivity	0.48	Azimi et al. (1984)
	nmr	0.74	Boyce et al. (1977)
	nmr	0.70	Gordon and Strange (1978)
h_a	Conductivity	0.34 (Ln = La)	Azimi et al. (1984)

[a] nn, Nearest neighbor; nnn, next-nearest neighbor.

nature of Ln^{3+} and the symmetry of the $Ln^{3+}-F_i^-$ association, a tetragonal complex of type I with nearest-neighbor (nn) charge compensation or trigonal complex of type II with next-nearest-neighbor (nnn) charge compensation (Fig. 1).

In fact, a relationship of the type $h_a \sim (r_{Ln})^3$ has been established from ionic conductivity measurements of LnF_3-doped CaF_2 crystals (Bollmann, 1978). Dielectric relaxation studies have confirmed that ion size is the primary factor determining the defect formation (Andeen *et al.*, 1977; Fontanella *et al.*, 1978a). Ionic thermal current experiments have been carried out on CaF_2 (Nauta-Leeffers and den Hartog, 1979), SrF_2 (van Weperen and den Hartog, 1978), and BaF_2 crystals (Laredo *et al.*, 1979, 1980) doped with various trivalent rare-earth and yttrium ions. Three main relaxation peaks have been observed in all these crystals. The relaxation at highest temperature could be

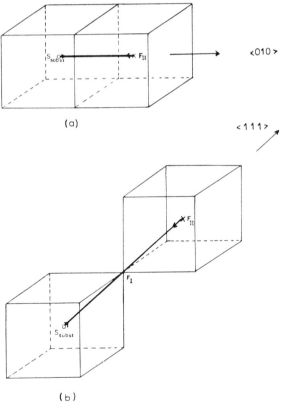

Fig. 1. Structure (a) nearest-neighbor and (b) next-nearest-neighbor dipolar complexes in MF_2-Ln^{3+}.

associated with space charges (den Hartog and Langevoort, 1981). The low-temperature peaks have been identified with increasing temperature as being due to nn and nnn associations.

As far as local compensation is concerned, the predominant dipolar species depends mainly on the nature of the host matrix and size of the contaminating ion. In CaF_2 crystals, for instance, the most frequent defect structure has a tetragonal symmetry, whereas in BaF_2 the most abundant structure is an $Ln^{3+}-F_i^-$ pair in nnn positions coexisting with nn dipoles in lower concentrations. In SrF_2 doped with large Ln^{3+} ions tetragonal complexes are predominant; in SrF_2 doped with smaller Ln^{3+} ions the most frequent defects have trigonal symmetry.

Association energies of nn and nnn clusters $Ln^{3+}-F_i^-$ in CaF_2, SrF_2, and BaF_2 have been obtained by lattice simulation calculations (Wapenaar and Catlow, 1981). The calculations successfully simulate the experimentally observed variations of the dopant–interstitial binding energies with the size of the dopant ion and the lattice parameter of the host (Table II). The stability of nn clusters decreases, whereas the stability of nnn clusters increases as the difference between the ionic radii of the host-lattice cation and of the doping cation increases. The decreasing stability of the nn associations is due mainly to an increase in the F_N-F_N repulsions (F_N = fluorine ion in normal site), as a consequence of the inward shifts of the F_N ions around the Ln^{3+} ions, the size of which decreases (Wapenaar and Catlow, 1981).

The ionic conductivity of nominally pure and doped CdF_2 has been the subject only of preliminary studies (e.g., Tan and Kramp, 1970; Benci et al., 1975; Oberschmidt and Lazarus, 1980). The electrical conduction in pure CdF_2 has been attributed to ionic processes with an activation energy of 0.63 eV. However, CdF_2 crystals that are typical ionic insulators with an energy gap of about 7 eV can be converted to semiconductors if doped with some rare-earth fluorides and heated in Cd vapor (Paracchini and Schianchi, 1982).

TABLE II

Bonding Energies (eV) for Nearest-Neighbor (nn) and Next-Nearest-Neighbor (nnn) Clusters Calculated with Semiempirical Potentials[a]

Lanthanide	CaF_2		SrF_2		BaF_2	
	nn	nnn	nn	nnn	nn	nnn
Lanthanum	0.97	0.29	0.75	0.41	0.49	0.39
Europium	0.82	0.38	0.58	0.49	0.28	0.45
Lutetium	0.66	0.45	0.43	0.55	0.21	0.49

[a] From Wapenaar and Catlow (1981).

2. Anion-Excess Solid Solutions

a. $M_{1-x}M'_xF_{2+x}$ and $M_{1-x}M''_xF_{2+2x}$ (M' = TRIVALENT CATION, M'' = TETRAVALENT CATION) SOLID SOLUTIONS. The comparison of defect enthalpies of MF_2 binary compounds with fluorite-type structure (Table I) shows, in agreement with the optimization criteria (Section II), that the lone pair in Pb^{2+} increases its polarizability and facilitates the diffusion of fluorine.

Another way to raise the conductivity of MF_2 (M = Ca, Sr, Ba, or Pb) is to enhance anionic disordering by the presence of extra F^- anions. Consequently, good fast fluorine ion conductors with fluorite-type (or related) structure are the anion-excess solid solutions $Pb_{1-x}M'_xF_{2+x}$ and $Pb_{1-x}M''_xF_{2+2x}$, where M' and M'' are, respectively, trivalent and tetravalent cations (Réau et al., 1978d; Schoonman and Wapenaar, 1979; Hagenmuller et al., 1981).

It is well known that the alkaline-earth fluorides can dissolve large amounts of trivalent rare-earth fluorides (Ippolitov et al., 1967; Catalano et al., 1969; Fedorov et al., 1974; Joukoff et al., 1976; Sobolev and Fedorov, 1978; Sobolev et al., 1979; Fedorov and Sobolev, 1979; Sobolev and Seiranian, 1981; Sobolev and Tkachenko, 1982) and tetravalent fluorides (Zintle and Udgard, 1938; Catalano and Wrenn, 1975; Wapenaar and Schoonman, 1979; Ratnikova et al., 1980). In the same way, many solid solutions of formulation $Pb_{1-x}M'_xF_{2+x}$ or $Pb_{1-x}M''_xF_{2+2x}$ have been reported in $PbF_2/M'F_3$ systems (Shore and Wanklyn, 1969; Ravez and Dumora, 1969a,b; de Kozak, 1969; Ravez and Vassiliadis, 1970; Grannec and Ravez, 1970; Ravez and Duale, 1970; Ravez et al., 1971; Lucat et al., 1976; Joshi and Liang, 1977; Schoonman and Wapenaar, 1979; Darbon et al., 1981a; Réau et al., 1982; 1983a) or in $PbF_2/M''F_4$ systems (Réau et al., 1978b; Wapenaar and Schoonman, 1979; Darbon et al., 1981b).

The electrical properties of $M_{1-x}Ln_xF_{2+x}$ (M = Ca, Sr, or Ba) solid solutions have been widely studied as a function of temperature and composition. Three regions can be discerned, for instance, in the concentration dependence of the ionic conductivity of $Ba_{1-x}La_xF_{2+x}$ $(0 \le x \le 0.45)$ (Wapenaar et al., 1981): a dilute concentration region $(x \lesssim 10^{-3})$, where only isolated defects are present; an intermediate region $(10^{-3} \lesssim x \lesssim 5.10^{-2})$, where a very steep increase in the conductivity with increasing solute content is shown to be ruled by a superlinear increase in the mobile defect concentration; and finally a concentrated solid-solution region $(x \gtrsim 5.10^{-2})$, in which the conductivity increases less quickly with a slower decrease in activation energy. In the last concentration range, where an exponential increase in the conductivity and a simultaneous linear decrease in activation energy with dopant concentration are observed, the ionic conductivity is characterized by enhancement of the ionic mobility (Wapenaar, 1980). Analogous features have

TABLE III
Electrical Data for Some Fluorine Ion Conductors

Material	$\sigma_{25°C}$ ($\Omega^{-1}\,cm^{-1}$)	$\sigma_{150°C}$ ($\Omega^{-1}\,cm^{-1}$)	ΔE (eV)	Reference
$Ba_{0.894}U_{0.102}Ce_{0.004}F_{2.208}$	$\sim 3.10^{-8}$	$\sim 2.10^{-5}$	0.53	Wapenaar and Schoonman (1979)
$Sr_{0.894}U_{0.102}Ce_{0.004}F_{2.208}$	$\sim 2.10^{-8}$	$\sim 10^{-5}$	0.59	
$Ca_{0.894}U_{0.091}Ce_{0.005}F_{2.187}$	$\sim 3.10^{-10}$	$\sim 8.10^{-7}$	0.78	
$Ba_{0.55}La_{0.45}F_{2.45}$	$\sim 2.10^{-7}$	$\sim 8.10^{-5}$	0.57	Wapenaar et al. (1981)
$Sr_{0.69}La_{0.31}F_{2.31}$	$\sim 7.10^{-8}$	$\sim 3.10^{-5}$	0.67	Fedorov et al. (1982)
$Sr_{0.65}Bi_{0.35}F_{2.35}$	$\sim 8.10^{-7}$	$\sim 2.10^{-4}$	0.56	Réau et al. (1984a)
$Pb_{0.88}In_{0.12}F_{2.12}$	$\sim 2.10^{-4}$	$\sim 8.10^{-3}$	0.34	Réau et al. (1982)
$Pb_{0.75}Sb_{0.25}F_{2.25}$	$\sim 5.10^{-4}$	$\sim 2.10^{-2}$	0.34	Darbon et al. (1981a)
$Pb_{0.75}Bi_{0.25}F_{2.25}$	$\sim 7.10^{-5}$	$\sim 6.10^{-3}$	0.39	Lucat et al. (1976)
$Hg_{0.75}Bi_{0.25}F_{2.25}$	$\sim 10^{-5}$	$\sim 2.10^{-3}$	0.44	Chartier et al. (1982)
$Pb_{0.92}Ge_{0.08}F_{2.16}$	$\sim 6.10^{-5}$	$\sim 5.10^{-3}$	0.40	Granier et al. (1983)
$Pb_{0.90}Zr_{0.10}F_{2.20}$	$\sim 10^{-4}$	$\sim 8.10^{-3}$	0.37	Darbon et al. (1981b)
$Pb_{0.90}Sn_{0.10}F_{2.20}$	$\sim 2.10^{-4}$	$\sim 10^{-2}$	0.37	Granier et al. (1983)
$Pb_{0.875}Th_{0.125}F_{2.25}$	$\sim 4.10^{-4}$	$\sim 10^{-2}$	0.32	Réau et al. (1978b)
$Na_{0.40}Bi_{0.60}F_{2.20}$	$\sim 6.10^{-5}$	$\sim 7.10^{-3}$	0.46	Chartier et al. (1981)
$K_{0.50}Bi_{0.50}F_2$	$\sim 2.10^{-4}$	$\sim 10^{-2}$	0.38	Matar et al. (1980)
$Rb_{0.50}Bi_{0.50}F_2$	$\sim 4.10^{-4}$	$\sim 2.10^{-2}$	0.37	
$Ag_{0.35}Bi_{0.65}F_{2.30}$	$\sim 10^{-5}$	$\sim 5.10^{-3}$	0.47	Grannec et al. (1983)
α'-$PbSnF_4$	$\sim 10^{-3}$	—	0.42	Réau et al. (1978c)
β-$PbSnF_4$	—	$\sim 8.10^{-2}$	0.14	

Compound			E (eV)	Reference
$La_{0.93}Ba_{0.07}F_{2.93}$	—	$\sim 3.10^{-3}$	$\left.\begin{array}{l}0.38 \ (t<150°C)\\ 0.25 \ (t>150°C)\end{array}\right\}$	Roos et al. (1983)
$Ce_{0.95}Ca_{0.05}F_{2.95}$	—	$\sim 6.10^{-3}$	$\left.\begin{array}{l}0.37 \ (t<180°C)\\ 0.25 \ (t>180°C)\end{array}\right\}$	Takahashi et al. (1977)
$Bi_{0.94}K_{0.06}F_{2.88}$	$\sim 7.10^{-4}$	$\sim 2.10^{-2}$	0.32	Shafer et al. (1981)
$Bi_{0.85}Pb_{0.15}F_{2.85}$	$\sim 5.10^{-5}$	$\sim 8.10^{-3}$	0.54	
KSn_2F_5	$\sim 5.10^{-6}$	—	$\left.\begin{array}{l}0.25 \ (t<140°C)\\ 0.55\\ 0.68\\ 0.52 \ (t>140°C)\end{array}\right\}$	Réau et al. (1984b)
$RbSn_2F_5$	$\sim 2.10^{-5}$	—		
$TlSn_2F_5$	$\sim 6.10^{-4}$	$\sim 8.10^{-2}$		
$NH_4Sn_2F_5$	$\sim 5.10^{-4}$	—	$\left.\begin{array}{l}0.22 \ (t<85°C)\\ 0.50\\ 0.30 \ (t>85°C)\end{array}\right\}$	Vilminot et al. (1980)
$TlZrF_5$	$\sim 5.10^{-6}$	$\sim 10^{-4}$		
Tl_2ZrF_6	$\sim 2.10^{-4}$	$\sim 5.10^{-3}$	$\left.\begin{array}{l}0.30\\ 0.32\end{array}\right\}$	Avignant et al. (1980)
Tl_3ZrF_7	$\sim 6.10^{-5}$	$\sim 10^{-3}$		

been observed for the composition dependence of electrical properties of $Ba_{1-x-y}U_xCe_yF_{2+2x+y}$ (Wepenaar and Schoonman, 1979). Some electrical data are given in Table III.

An important property of concentrated solid solutions is the fact that the probability of formation of defect clusters increases with solute content (Cheetham et al., 1971). The importance of the presence of clusters for the conductivity is that the anion defects constituting the clusters do not contribute to the conductivity because of the high association energies of the clusters (Catlow, 1976a). In addition, lattice ion relaxations around clusters influence the height of the activation energy barrier.

The linear decrease in the activation enthalpy with solute content has been accounted for in the enhanced ionic motion model. The mobility of the fluoride interstitials is determined by a Gaussian distribution in the jump enthalpies while the number of mobile interstitials is restricted. In this model the activation enthalpy of the conductivity is given by

$$\Delta H(\sigma) = \Delta H_m - Cx, \tag{3}$$

where C is a proportionality constant, ΔH_m the activation enthalpy for motion in the undisturbed lattice, and x the solute content in mole fraction (Wapenaar, 1980).

The composition dependence of the electrical properties of $Ca_{1-x}Ln_xF_{2+x}$ (Réau et al., 1976; Svantner et al., 1979; 1981) and $Sr_{1-x}La_xF_{2+x}$ (Ivanov-Shitts et al., 1983) has the same features as that of $Ba_{1-x}La_xF_{2+x}$. Yet the break observed at $x \simeq 0.05$ for $Ba_{1-x}La_xF_{2+x}$ appears only at $0.10 < x < 0.15$ for $Ca_{1-x}Ln_xF_{2+x}$ and $Sr_{1-x}La_xF_{2+x}$. On the other hand, the conductivity variation with composition for the $Sr_{1-x}Y_xF_{2+x}$ solid solution, where the size difference is larger for the cations, is very different: a maximum of conductivity associated with a minimum of activation energy is observed at $x \simeq 0.15$ (Réau et al., 1984a). Beyond that value, the decrease in conductivity with increasing x is probably related to a clustering that becomes faster when the rare-earth size is smaller. A comparative study of the electrical properties of $M_{0.90}Ln_{0.10}F_{2.10}$ (M = Sr or Ba) has shown that better electrical performances are obtained when the host lattice is BaF_2 (Fedorov et al., 1982).

It has been shown by ITC experiments on concentrated $Sr_{1-x}La_xF_{2+x}$ and $Ba_{1-x}La_xF_{2+x}$ solid solutions that, in contrast with the peaks associated with nn and nnn dipolar complexes the position of which at a given temperature depends on the concentration of the dipoles, the depolarization peak associated with space charges shifts toward lower temperatures with increasing lanthanum concentrations (den Hartog and Langevoort, 1981). The shifting, which is very quick for $x \lesssim 5.10^{-2}$ but slower beyond, stops when the depolarization peak reaches a position approximately coinciding with that of the nnn peak. This peak has been connected with bulky ion conductivity

(Meuldijk *et al.*, 1982; Meuldijk and den Hartog, 1983). The results are in agreement with the higher conductivity observed with increasing x.

Comparative thermally stimulated polarization and depolarization studies have been carried out on SrF_2 crystals doped with LaF_3 up to molar concentrations of 5.10^{-2} and to temperatures of 500 K (Laredo *et al.*, 1983). Five peaks instead of three have been detected: three low-temperature peaks and two high-temperature ones. The first three relaxations are attributed to reorientations of nn dipoles, L-shaped $La^{3+}-Fi_2^-$ (instead of nnn dipoles), and $2:2:2$ clusters, respectively. The first high-temperature peak, previously related to the relaxation of space charges, is hence a dipolar relaxation; its position is strongly dependent on the thermal history of the sample, and its existence can be interpreted in terms of polarization of dislocations. The second high-temperature peak, in contrast, is associated with space charges accumulated near the electrodes.

High electrical performances are obtained for anion-excess solid solutions when the materials contain cations with *high polarizability*: $Pb_{1-x}In_xF_{2+x}$ (Réau *et al.*, 1982), $Pb_{1-x}Sb_xF_{2+x}$ (Darbon *et al.*, 1981a), $Pb_{1-x}Bi_xF_{2+x}$ (Lucat *et al.*, 1976), $Hg_{1-x}Bi_xF_{2+x}$ (Chartier *et al.*, 1982), $Pb_{1-x}Ge_xF_{2+2x}$ (Granier *et al.*, 1983), $Pb_{1-x}Zr_xF_{2+2x}$ (Darbon *et al.*, 1981b), $Pb_{1-x}Sn_xF_{2+2x}$ (Granier *et al.*, 1983), $Pb_{1-x}Th_xF_{2+2x}$ (Réau *et al.*, 1978b), and $Pb_{1-x}U_xF_{2+2x}$ (Rhandour, 1979) (Table III).

In contrast to $M_{1-x}La_xF_{2+x}$ (M = Sr or Ba) and $Ca_{1-x}Ln_xF_{2+x}$, the lead and mercury solid solutions involving the largest substituting cations show a maximum of conductivity connected with an activation energy minimum for the compositions corresponding approximately to $x_L/2$, x_L being the upper limit of the existence range of solid solutions (Figs. 2 and 3). These maxima of conductivity correspond to compositions involving nearly the same number of extra F^- ions introduced into the matrix of β-PbF_2, independently of the charge of the substituting cation (Table III). Correlations between electrical properties and crystallographic data determined by neutron diffraction for solid solutions show (see Section IV) that a high degree of disordering characterizes these compositions.

Figure 4 gives the variation of ΔE versus ionicity of the M—F bonds f_i(M—F) for various compositions of the $Pb_{1-x}M'_xF_{2+x}$ (M′ = Y, In, Sb, or Bi) and $Sr_{1-x}M'_xF_{2+x}$ (M′ = Y or Bi) within the solid solutions corresponding to increasing disorder with x. Whatever the nature of the trivalent cation, the ionic character of the M—F bond is clearly weaker for lead than for strontium fluorides. This property can be correlated to an activation energy that is lower for lead compounds located below on the left side of Fig. 4 than for strontium fluorides situated above on the right side. These results show the *large influence of the covalent character of the host lattice* on electrical properties of materials with fluorite-type structure. The enhancement of the ionic con-

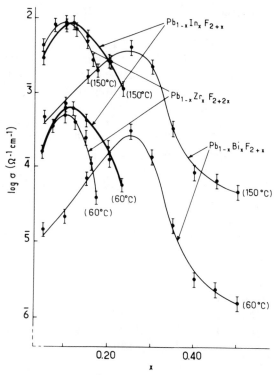

Fig. 2. Composition dependence of the conductivity at $t = 60°C$ and $150°C$ for $Pb_{1-x}In_xF_{2+x}$, $Pb_{1-x}Bi_xF_{2+x}$, and $Pb_{1-x}Zr_xF_{2+2x}$.

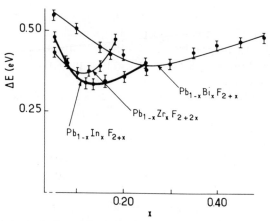

Fig. 3. Variation of the activation energy of $Pb_{1-x}In_xF_{2+x}$, $Pb_{1-x}Bi_xF_{2+x}$, and $Pb_{1-x}Zr_xF_{2+2x}$ with composition.

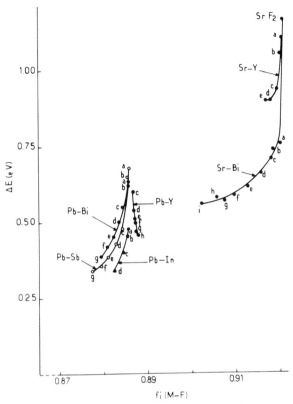

Fig. 4. Variation of the activation energy as a function of the ionicity of the M—F bonds for various compositions of $Sr_{1-x}M'_xF_{2+x}$ (M' = Y or Bi) and $Pb_{1-x}M'_xF_{2+x}$ (M' = Y, In, Sb, or Bi) solid solutions. The values for x are (a) 0.01, (b) 0.02, (c) 0.05, (d) 0.10, (e) 0.15, (f) 0.20, (g) 0.25, (h) 0.30, (i) 0.35.

ductivity observed by substitution of Pb^{2+} by trivalent cations in PbF_2 seems to be due mainly to increasing disorder within the anionic sublattice. For those ternary solid solutions, the role played by the cationic polarizability seems to be a second-order effect. Yet the enhancement of the covalent character of the lattice due to the introduction of trivalent non-close-shell cations with high polarizability improves the electrical performances.

b. $A_{1-x}M'_xF_{1+2x}$ (A = MONOVALENT CATION, M' = TRIVALENT CATION) SOLID SOLUTIONS. These solid solutions have a structure derived from the fluorite type (Thoma *et al.*, 1966). The electrical properties of $Na_{1-x}Y_xF_{1+2x}$ (Pontonnier *et al.*, 1978) are of the same order of magnitude as those of $Ca_{1-x}Y_xF_{2+x}$. In contrast, fast ion conductors are formed when the trivalent ion is Bi^{3+} [Table III: $Na_{1-x}Bi_xF_{1+2x}$ (0.60 ≤ x ≤ 0.70) (Chartier

440   J.-M. Réau and J. Grannec

et al., 1981), $K_{1-x}Bi_xF_{1+2x}$ $(0.50 \leq x \leq 0.70)$, $Rb_{1-x}Bi_xF_{1-2x}$ $(0.50 \leq x \leq 0.60)$ (Matar *et al.*, 1980), $Ag_{1-x}Bi_xF_{1+2x}$ $(0.65 \leq x \leq 0.73)$ (Grannec *et al.*, 1983), and $Tl_{1-x}Bi_xF_{1+2x}$ $(0.50 \leq x \leq 0.60)$ (Matar and Réau, 1982)].

In all solid solutions of $A_{1-x}Bi_xF_{1+2x}$ type, the higher the substitution rate the smaller is the anionic mobility due to the trapping effect of the supernumerary Bi^{3+} cations. Neutron diffraction investigations on $A_{1-x}Bi_xF_{1+2x}$ (A = K or Rb) solid solutions illustrate such a composition dependence (see Section IV). A progressive establishment within $K_{1-x}Bi_xF_{1+2x}$ and $Rb_{1-x}Bi_xF_{1+2x}$ of an order close to that observed in ordered ABi_3F_{10} phases (A = K or Rb) (Matar *et al.*, 1980) has been suggested and has been confirmed by nmr investigations (Réau *et al.*, 1983b). In contrast, the conductivity of $K_{1-x}Bi_xF_{1+2x}$ increases by doping with Th^{4+} because of rising disorder (Burns *et al.*, 1981; Casanho *et al.*, 1983).

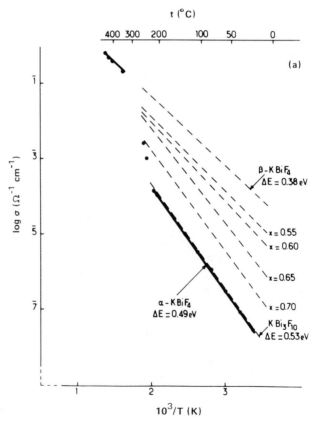

Fig. 5. (a) Variation of $\log \sigma$ versus reciprocal temperature for α-$KBiF_4$, β-$KBiF_4$, KBi_3F_{10}, and $K_{1-x}Bi_xF_{1+2x}$.

The existence of low-temperature forms, α-$KBiF_4$, α- and β-$RbBiF_4$, of the high-temperature phases (called β-$KBiF_4$ and γ-$RbBiF_4$) has been shown by differential thermal analysis (DTA) and microcalorimetry as well as by electrical measurements (Matar *et al.*, 1983b,c). These low-temperature forms are ordered phases deriving, respectively, from the $NaNdF_4$-type structure (Burns, 1965) for α-$KBiF_4$ and β-$RbBiF_4$ and from the $KCeF_4$-type structure (Brunton, 1969) for α-$RbBiF_4$.

The variation of the conductivity with temperature for α-$KBiF_4$ and α-$RbBiF_4$ has been compared with that of $A_{1-x}Bi_xF_{1+2x}$ ($A = K$ or Rb) and ordered ABi_3F_{10} phases (Fig. 5). The electrical properties of α-$KBiF_4$ and α-$RbBiF_4$ are very close to those of KBi_3F_{10} and $RbBi_3F_{10}$, which are isostructural with KY_3F_{10} (Pierce and Hong, 1973). The replacement of alkaline ions ($A = K$ or Rb) by bismuth in $ABiF_4$ involves a weakening of the conductivity for the $A_{1-x}Bi_xF_{1+2x}$ solid solutions. In the same way, a long annealing of the high-temperature varieties of the $ABiF_4$ phases brings about

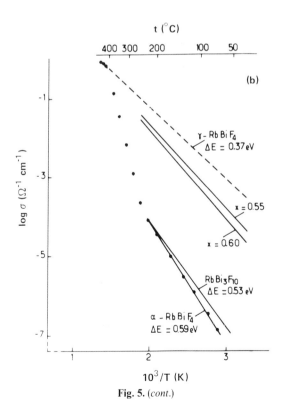

Fig. 5. (*cont.*)

a transformation into α varieties, which are found to be totally ordered. The most striking feature of the Bi^{3+} cation is its capacity to give rise to a stereoactive lone pair. This tendency is materialized in the high-temperature forms of the $ABiF_4$ (A = K or Rb) compounds, which are easily transformed into low-temperature varieties with a distorted environment of Bi^{3+}.

In contrast, the size of Na^+ allows one to consider a certain cationic conductivity in $Na_{1-x}Bi_xF_{1+2x}$. Both ^{19}F and ^{23}Na wide-line nmr studies have shown that, in addition to F^- mobility, a Na^+ mobility appears in these materials from 280 K (Senegas et al., 1983).

3. $Pb_{1-x}Sn_xF_2$ Solid Solution

Substitution of Pb^{2+} by Sn^{2+} in the fluorite-type network of β-PbF_2 is restricted despite the fact that the cations have a very similar ionic radius (Shannon, 1976). For $0 \le x \le 0.25$ $Pb_{1-x}Sn_xF_2$ holds a completely disordered fluorite-type structure (Lucat et al., 1979), but beyond that limit, Sn^{2+} behaves differently from Pb^{2+} because of the greater stereoactivity of the lone pair. That phenomenon leads to a series of cationic networks, which results in the appearance of superstructures. For $x = 0.50$, for instance, an ordered cationic network appears with the sequence PbPbSnSn... perpendicular to the c axis (Perez et al., 1980a; Vilminot et al., 1981b).

Five allotropic forms of $PbSnF_4$ have been detected by DTA and x-ray diffraction as well as by microcalorimetric and dilatometric studies (Réau et al., 1978c; Perez et al., 1980b; Vilminot et al., 1981a). All structures derive from the fluorite type. The study of thermal behavior of $PbSnF_4$ has shown that the sum of entropy changes associated with the different transitions of this material is of the same order of magnitude as the entropy change corresponding to melting (Claudy et al., 1981a).

The presence of cations with high polarizability in $PbSnF_4$ explains the excellent electrical performances (Table III). An electrical transition without discontinuity takes place at $t \simeq 90°C$, which is close to the $\alpha' \rightleftarrows \beta$ crystallographic transition temperature, suggesting that the structures of the monoclinic α' form and tetragonal β form of $PbSnF_4$ are very similar (Réau et al., 1978c; Couturier et al., 1979).

The structural determination of α'-$PbSnF_4$ has corroborated the existence of a cationic network as in the fluorite-type structure but with the cationic ordering PbPbSnSn.... The high value of the F^- mobility is accounted for by the high number of vacancies in the anionic network, F^- ions occupying partially normal and interstitial sites of the fluorite lattice. During the $\alpha' \rightarrow \beta$ transition the decrease in the activation energy is due only to a very weak structural reorganization (Vilminot et al., 1981a). In the same way, the substitution of Pb^{2+} by Sn^{2+} in β-PbF_2 leads to an increase in the conductivity associated with a decrease in the activation energy. This behavior results from the progressive formation with increasing x of F^- vacancies at normal positions by the transfer of a part of these anions from regular positions to interstitial ones close to the tin atoms (Vilminot et. al., 1981b).

4. Anion-Excess Ordered Phases

The $MF_2/M'F_3$ and $MF_2/M''F_4$ systems ($M = Ca$, Sr, Ba, or Pb; $M' =$ trivalent cation; $M'' =$ tetravalent cation) contain extended solid solutions $M_{1-x}M'_xF_{2+x}$ and $M_{1-x}M''_xF_{2+2x}$, the existence ranges of which, however, decrease upon cooling; several fluorite-related anion-excess superstructure phases can be observed (Fedorov et al., 1974; Sobolev et al., 1979; Sobolev and Seiranian, 1981; Sobolev and Tkachenko, 1982). Different superstructure types called t, $rh\alpha'$, $rh\beta$, $rh\gamma$, and $c\beta$ have been detected (Greis, 1978, 1980; Greis and Kieser, 1979; Kieser and Greis, 1980a). They belong to one of the two homologous series T_mF_{2m+5} or T_mF_{2m+6} or eventually to both. For instance, in the system CaF_2/YbF_3, the ordered phases t (Ca_2YbF_7), $rh\alpha$ ($Ca_9Yb_5F_{33}$), $rh\beta$ ($Ca_{8-\delta}Yb_{5+\delta}F_{31+\delta}$) are members of the homologous series ($CaYb$)$_mF_{2m+5}$ ($m = 15$, 14, or 13) and $c\beta$ ($Ca_{17}Yb_{10}F_{64}$) by combination of the phases with $m = 13$ and $m = 14$ (Lechtenböhmer and Greis, 1978). In contrast, tveitite ($rh\gamma$), the only known ordered yttrofluorite mineral, can be described with the most probable formula $Ca_{14}Y_5F_{43}$ as being an $m = 19$ member of the series T_mF_{2m+5} (Greis, 1978; Bevan et al., 1982).

Both superstructure types $rh\beta$, isostructural with $Na_7Zr_6F_{31}$ (Burns et al., 1968), and $rh\gamma$ are characterized by the presence of discrete T_6X_{37} clusters, which consist of octahedral arrangements of six TX_8 square antiprisms sharing corners to enclose a cubo-octahedron of anions with an additional anion at its center. These T_6X_{37} clusters are distributed in an ordered manner within a matrix of virtually unmodified fluorite-type structure.

The fluorite structure can be described by the three-dimensional (3D) arrangements of T_6X_{32} units sharing edges, enclosing thus a seventh anionic cube. The centers of the first six cubes are occupied by cations, whereas the seventh is empty. In anion-excess fluorite-related ordered phases, the introduction of additional anions leads to 45° rotation of the inner faces of the six cation-centered cubes for some of these units. This transformation of six cubes into six antiprisms sharing corners leads to the formation of one T_6X_{36} cluster enclosing an empty cubo-octahedron:

$$T_6X_{32} + 4X \longrightarrow T_6X_{36} \tag{4}$$

The conversion of an elementary cube to a square antiprism is a fundamental first step in the formation of many anion-excess fluorite-related compounds (Aleonard et al., 1978; Bevan et al., 1980; Laval and Frit, 1983a).

The T_6X_{36} cluster is present in fluorides such as KY_3F_{10} (Pierce and Hong, 1973), KTb_3F_{10} (Podberezskaya et al., 1976), and ABi_3F_{10} (A = K, Rb, or et Tl) (Matar and Réau, 1982), which are isostructural with KY_3F_{10}, as well as in α- and γ-KYb_3F_{10} (Labeau et al., 1974), $BaCaLn_2F_{10}$ (Valon et al., 1976), and α-KEr_2F_7 (LeFur et al., 1982). The T_6X_{36} cluster can become a T_6X_{37} cluster when the center of the cubo-octahedron is occupied by an anion, as in tveitite, $Na_7Zr_6F_{31}$, or $PbZr_6F_{22}O_2$ (with a $Zr_6F_{31}O_6$ oxide–fluoride cluster) (Laval and Frit, 1983b). A T_6X_{38} cluster can be detected even with the largest cations (e.g, in CsU_6F_{25}) (Brunton, 1969; Laval and Frit, 1983b); it has been proposed in the $Ba_4Ln_3F_{17}$ phases (Kieser and Greis, 1980b), which would then be the $m = 14$ members of the T_mF_{2m+6} series. It can be suggested that formation of the cubo-octahedron occurs before the onset of long-range ordering, which can be total only when the number of cubo-octahedral motifs becomes sufficient; this is the case, for instance, for the ABi_3F_{10} phases. A mechanism is proposed in Section V for its formation on the basis of segregation of pair defects.

When anion-excess ordered phases contain cations with large size differences, antiprisms are never associated by corners but form 1D T_2X_8 clusters. T_2X_8 is a basic structural unit for an original homologous series, formulated $M_nT_2X_{2n+8}$, of which Pb_3ZrF_{10}, for instance, is an $n = 6$ member (Laval and Frit, 1980; Frit and Laval, 1981).

The investigation of electrical properties of anion-excess ordered materials shows that the performances of ordered phases are always weaker than those of the disordered ones of the same composition (Darbon et al., 1981b. The temperature dependence of conductivity for $A_{1-x}Bi_xF_{1+2x}$ and ABi_3F_{10} (A = K or Rb) phases, given in Fig. 5, demonstrates without doubt that the lowest performances are obtained for ABi_3F_{10}.

B. MATERIALS WITH TYSONITE-TYPE STRUCTURE

1. Nominally Pure and Doped Fluorides

Rare-earth fluorides with the tysonite-type structure have very good electrical performances, although their melting point is very high. The ionic conductivity of nominally pure LaF_3 (or CeF_3) is relatively large even at room temperature, about six orders of magnitude higher than that of nominally pure alkaline-earth fluorides (Nagel and O'Keeffe, 1973).

Many studies of ionic transport in the tysonite-type fluorides including conductivity measurements of LaF_3 crystals (Lee and Sher, 1965; Sher et al., 1966; Solomon et al., 1966), LaF_3 films (Tiller et al., 1973; Lilly et al., 1973), CeF_3 crystals (Nagel and O'Keeffe, 1973), CeF_3 sintered samples (Takahashi et al., 1977), and nmr measurements on LaF_3 (Sher et al., 1966; Goldman and

Shen, 1966; Shen, 1968), LaF_3 powders (Ildstad *et al.*, 1977), or CeF_3 crystals (Lee, 1969) have been reported. It is not easy to summarize the available data for these materials due to the large discrepancies between the results. In fact, there are experimental problems with LaF_3, for instance, due to its reactivity with oxygen of the air and water vapor, and such difficulties increase with temperature. The studies, however, have made it possible to attribute to LaF_3 and activation energy for the F^- motion ($\Delta H_m \simeq 0.45$ eV) below 400 K; nmr measurements have been interpreted in terms of the existence of two distinct fluorine sublattices and an exchange of F^- between inequivalent lattice sites occuring at about 400 K (Goldman and Shen, 1966).

In contrast, the interpretation of transport properties of LaF_3 is difficult due to its complex crystal structure. For a long time the exact structure of LaF_3 of trigonal symmetry ($a = 7.19$ Å, $c = 7.35$ Å) has been uncertain (Mansmann, 1965; Zalkin *et al.*, 1966; Cheetham *et al.*, 1976; Greis and Bevan, 1978). Apparently, LaF_3 can crystallize in at least two different modifications (space groups $P\bar{3}c_1$ and $P6_3/mmc$), the first corresponding to an ordering and a lower symmetry of the anion sublattice ($Z = 6$) and the second corresponding to an anion array that is less ordered and has a higher symmetry ($Z = 2$). This behavior correlates well with the high F^- mobility observed at room temperature.

In the tysonite-type structure ($P\bar{3}c_1$) there are three unequivalent types of F^- with the ratio $12:4:2$ ($F_1/F_2/F_3$). Two of these sites are almost equivalent ($F_2 \simeq F_3$), however, and only two types can actually be considered with the ratio $2:1$ [$F_1/(F_2 + F_3)$]. The F_1-type ions constitute the so-called A sublattice, and the F_2- and F_3-type ions constitute the B sublattice. The crystal structure of LaF_3 contains perpendicular to the c axis hexagonal layers formed by the La^{3+} and one-third of F^- ions ($F_2 + F_3$), the other F^- (F_1) being located between those layers. As a consequence, different fluorine jumps have been investigated in directions parallel and perpendicular to the c axis, the involved anions either belonging or not belonging to two sublattices independently (Jaroszkiewiez and Strange, 1980; Roos, 1983a).

A study of the transport properties of nominally pure LaF_3 single crystals has shown that the conductivity is slightly anisotropic, with $\sigma(\|c) > \sigma(\perp c)$ for temperatures up to ~ 415 K (Schoonman *et al.*, 1980; Roos *et al.*, 1983). Above 415 K the anisotropy disappears, and the extrinsic ionic conductivity of LaF_3 reveals a change of slope. The activation energies are, successively: $\Delta H_1(\perp) = 0.46$ eV and $\Delta H_1(\|) = 0.43$ eV from 300 to 415 K; $\Delta H_2 = 0.26$ eV from 415 to 715 K and $\Delta H_3 = 0.84$ eV above 715 K (Roos *et al.*, 1983).

It is generally accepted that the intrinsic point defects in LaF_3 are thermally generated according to a Schottky mechanism, with a transfer number for F^- close to unity. Hence, the ion transport is governed by the F^- vacancies (Sher *et al.*, 1966). Considering the value of 0.26 eV as the general migration enthalpy of F^- vacancies, a value of 2.3 eV is obtained for the Schottky disorder formation enthalpy ΔH_f (Roos et al., 1983). This value is in satisfactory agreement with the value 2.08 eV previously reported (Chadwick et al., 1979).

2. Anion-Deficient Solid Solutions

A study of the MF_2/LnF_3 systems (M = Ca, Sr, Ba, or Pb) has shown, in addition to anion excess disordered solid solutions and ordered phases deriving from the fluorite type, anion-deficient solid solutions $Ln_{1-y}M_yF_{3-y}$ related to the tysonite type (Tkachenko et al., 1973; Sobolev et al., 1976, 1979; Sobolev and Fedorov, 1978; Fedorov and Sobolev, 1979; Sobolev and Seiranian, 1981; Sobolev and Tkachenko, 1982).

Because the ion transport in LaF_3 (or CeF_3) results from the F^- vacancies, enhancement of the transport properties of LnF_3 by substitution of La^{3+} (or Ce^{3+}) by M^{2+} could be predicted.

The composition dependence of the electrical properties of $La_{1-y}Ba_yF_{3-y}$ $(0 \le y \lesssim 0.10)$ suggests the presence of conductivity maximum for $La_{0.93}Ba_{0.07}F_{2.93}$ (Roos et al., 1983). This maximum could be explained by assuming that the conductivity in the concentrated solutions is ruled by a percolation mechanism (Shante and Kirkpatrick, 1971; Pike et al., 1974). The value of y relative to its maximum value ($y_M = 0.07$) fits with a percolation threshold for an nn → nn conduction path. In the same way, in calcium-stabilized zirconia $Zr_{1-2y}Ca_{2y}O_{2-2y}$ a maximum of conductivity occurs for $2y \simeq 0.13$, and it has also been explained by a percolation mechanism (Nakamura and Wagner, 1980). As in the disordered solid solutions with fluorite-type structure (Greis, 1980), however, the formation for $y > y_M$ of microdomains inside the disordered solid-solution lattice with the tysonite structure could be detected. In fact, ordered $Ln_7Ca_3F_{27}$ phases (Ln = Tb through Yb) with a structure close to tysonite have been found in addition to disordered $Ln_yCa_{1-y}F_{3-y}$ phases (Bevan and Greis, 1978).

In the intermediate domain of the $La_{1-y}Ba_yF_{3-y}$ compounds $(0 < y < 0.07)$ the F^- ions begin with increasing y to exchange more and more easily between both sublattices. The conductivity break flattens out and moves slightly toward lower temperatures; the conductivity anisotropy disappears (Roos et al., 1983). This exchange enhancement has been confirmed by nmr (Aalders et al., 1983).

The $La_{1-y}Ba_yF_{3-y}$ ternary solid solutions show good electrical performances (Table III). The $Ce_{1-y}M_yF_{3-y}$ compounds (M = Ca, Sr, or Ba), for which the maximum of conductivity seems to be attained at $y_M \simeq 0.05$, have a similar behavior (Takahashi et al., 1977).

New solid solutions of trivalent bismuth, the structure of which derives from the tysonite type, have been isolated: $Bi_{1-y}K_yF_{3-2y}$ $(0.02 \le y \le 0.12)$

(Shafer $et\ al.$, 1981) and $Bi_{1-y}Pb_yF_{3-y}$ $(0.075 \leq y \leq 0.175)$ (Réau $et\ al.$, 1984b). The temperature dependence of the conductivity for $Bi_{1-y}K_yF_{3-2y}$ shows a simple Arrhenius-type behavior in the temperature range 25–450°C. In contrast, that of $Bi_{1-y}Pb_yF_{3-y}$ manifestates a break at $t \simeq 150°C$, which is slightly shifted toward lower temperatures with increasing y. This behavior is analogous to that observed for the $Ln_{1-y}M_yF_{3-y}$ (M = Ca, Sr, or Ba) fluorides. Both $Bi_{1-y}K_yF_{3-2y}$ and $Bi_{1-y}Pb_yF_{3-y}$ are excellent ionic conductors (Table III). Their performances result from a highly defective structure and the high polarizability of the Bi^{3+} cation.

C. OTHER MATERIALS

1. β-SnF$_2$

Just like β-PbF$_2$, β-SnF$_2$ exhibits a cation with a high polarizability, but its structure is not of the fluorite type. The SnF_2 lattice contains cyclic Sn_4F_8 tetramers held to each other by weaker Sn—F bonds. The tetramers are puckered eight-membered rings with alternating tin and fluorine atoms strongly bound together. Each tin atom is located at the center of a tetrahedron the face of which is constituted by three fluorine atoms belonging to the same cycle; the fourth apex is the lone pair (McDonald $et\ al.$, 1976; Will and Bargouth, 1980). The good electrical performances of β-SnF$_2$, which are close to those of β-PbF$_2$, result from the high polarizability of Sn^{2+} and from the weak coordination of F^- (Réau $et\ al.$, 1978c). They can be correlated to the weak melting entropy of the material (Claudy $et\ al.$, 1981b).

2. ASn_2F_5

A new family of divalent tin fluorides ASn_2F_5 (A = K, Rb, Tl, or NH$_4$) characterized by a high conductivity has been found (Table III) (Vilminot $et\ al.$, 1980). The low-temperature forms of ASn_2F_5 (A = K, Rb, or Tl) are isostructural and exhibit a layer structure built up by the succession of three kinds of layers perpendicular to the c axis. Both tin atoms have the same geometric environment, that is, a distorted SnF_5E octahedron, E being the lone pair. All fluorine positions around the tin atoms at $z \simeq \frac{2}{3}$ are fully occupied, but those around the tin atoms at $z \simeq 0$ are partially occupied (Vilminot $et\ al.$, 1983).

The temperature dependence of conductivity of KSn_2F_5, $TlSn_2F_5$, and $NH_4Sn_2F_5$ shows breaks observed, respectively, at 155, 85, and 75°C associated with structural modifications. According to the structure of the low-temperature forms of ASn_2F_5 (A = K, Rb, Tl), the conductivity should be very anisotropic and restricted to the layer at $z = 0$. Calculations of the probability density functions and of the atomic potentials confirm this assumption. A ^{19}F-nmr study of $TlSn_2F_5$ is in agreement with a diffusion mechanism of the F^- ions within layers at low temperature. As above the transition temperature, the activation energies deduced from relaxation ($\Delta E = 0.19$ eV) and conductivity measurements ($\Delta E = 0.22$ eV) are very close; participation of all F^- ions in the diffusion process most likely occurs in this temperature domain (Granier et al., 1981).

Two prevailing types of motion have been determined in $NH_4Sn_2F_5$: at low temperature a reorientation of the ammonium ion is observed; above $t = -20°C$ the diffusion of the fluorine atoms is responsible for the relaxation (Battut et al., 1983).

3. $AM''F_5$, $A_2M''F_6$, and $A_3M''F_7$

Studies of the transport properties of compounds with formulas $AM''F_5$, $A_2M''F_6$, and $A_3M''F_7$ (A = K, Rb, Tl; M'' = Zr or Hf) have shown that these materials are good anionic conductors (Avignant et al., 1979, 1980) (Table III). The best performances are obtained for the $Tl_2M''F_6$ phases, the structure of which, however, has so far 'not been determined.

The structure of $AM''F_5$ (M'' = Zr or Hf) consists of sheets of $(M''F_5)^-$ composition that may be described as being formed by edge-sharing and corner-sharing bicapped trigonal prisms ($M''F_8$) (Avignant et al., 1981). The sheets run in a direction perpendicular to the a-axis plane and are bound together by the Tl^+ ions. Cationic polyhedra are joined into a 3D framework and give rise to relatively open tunnels along the b axis. Both fluorine atoms, located inside those tunnels, are most likely responsible for the conductivity, which is probably anisotropic.

The F^- 3D conductivity of the $A_3M''F_7$ (A = K, Rb, or Tl) phases is related to high anionic disorder due to the large number of unoccupied sites (Avignant et al., 1979).

The better performances of thallium compounds can be attributed to the higher polarizability of Tl^+.

4. $ABiF_4$

The $ABiF_4$ (A = Na or Ag) phases are isostructural with $NaNdF_4$ (Burns, 1965) and involve a partially ordered structure. The conductivity of $NaBiF_4$ and $AgBiF_4$ is therefore much lower than those observed for $Na_{0.44}Bi_{0.60}F_{2.20}$ and $Ag_{0.35}Bi_{0.65}F_{2.30}$, the structures of which derive from the fluorite type (Chartier et al., 1981; Grannec et al., 1983).

In contrast, no cationic conductivity has been detected in the ordered $NaBiF_4$ phase between 20 and 200°C; this result can most likely be extrapolated to $AgBiF_4$ (Senegas et al., 1983).

5. Oxide–Fluorides

Fast fluorine ion conduction has also been investigated in oxide–fluorides. A $Pb_{1-x}Bi_xO_xF_{2-x}$ solid solution ($0 \leq x \leq 0.80$) has been isolated in the $PbF_2/BiOF$ system. For $x \leq 0.67$ the structure is of the fluorite type, and for $0.67 < x \leq 0.80$ it shows a fluorite-derived rhombohedral distortion (Matar et al., 1982). The presence of oxygen atoms in normal anionic positions of the fluorite lattice and the shift of some of the fluorine atoms from normal positions into interstitial ones have been shown by neutron diffraction (Soubeyroux et al., 1984). The transport properties of $Pb_{1-x}Bi_xO_xF_{2-x}$ for $x > 0.50$ are similar to those of $Pb_{0.75}Bi_{0.25}F_{2.25}$. Furthermore, the introduction of oxygen in β-PbF_2 gives the oxide–fluoride a higher thermal stability.

$Tl_2Zr_3OF_{12}$, which belongs to the family of oxide–fluorides with formula $A_2M''_3OF_{12}$ (A = Rb, Tl, or NH; M'' = Zr or Hf) exhibits predominant fluorine conduction at low temperature; at temperature higher than 450 K, there is probably thallium contribution to the conductivity (Avignant et al., 1982). The structure of this compound consists of double layers of edge-sharing and corner-sharing distorted square antiprisms ZrX_8 (X = F or O). It has been related to that of the fluorite, and the existence of a new cluster T_6X_{34} intermediate between T_6X_{32} and $T_6X_{36(or\ 37)}$ entities has been shown (Mansouri and Avignant, 1984). Increasing fluorine conduction has been observed at rising x within the $Tl_2Zr_{3-x}M'_xOF_{12-x}$ (M' = Y or In) solid solutions.

IV. Determination of Conduction Mechanisms by Neutron Diffraction in Fluorite-Type Fluorides

The various selection criteria discussed in Section II have allowed investigators to isolate many highly conducting fluorides with a fluorite-type structure. It has been necessary to perform further investigations, however, in order to acquire an understanding of the long-range conduction mechanisms in that class of materials. Both nmr and neutron diffraction studies have proved to be very helpful tools for many studies. Consistent results that account for the dependence of ionic conductivity on substitution rates within solid-solution domains of composition $M_{1-x}M'_xF_{2+x}$, $M_{1-x}M''_xF_{2+2x}$, or $A_{1-x}M'_xF_{1+2x}$ have been obtained. They result, in particular, from correlations between electrical properties and precise crystallographic data obtained by neutron diffraction. Two of the most representative examples of such studies, the solid solutions $Pb_{1-x}Th_xF_{2+2x}$ ($0 \leq x \leq 0.25$) (Soubeyroux et al., 1981) and $A_{1-x}Bi_xF_{1+2x}$ ($0.50 \leq x \leq x_L$) (A = K, $x_L = 0.70$; A = Rb, $x_L = 0.60$) (Soubeyroux et al., 1982), are described in the following sections.

A.　$Pb_{1-x}Th_xF_{2+2x}$ $(0 \leq x \leq 0.25)$ SOLID SOLUTION

Figure 6 shows a fluorite unit cell with anions at normal sites (called F_I) and the cations occupying only half of the octacoordinated sites. The excess F^- ions resulting from the replacement of M^{2+} cations by M^{4+} cations are located in interstitial sites (they are called here F_{II}).

By means of Fourier difference maps, the nuclear densities of interstitial fluorine atoms have been determined. Three types of interstitials have been distinguished: F'_{II} shifted along the [110] directions from the $(\frac{1}{2},\frac{1}{2},\frac{1}{2})$ crystallographic position and the others F''_{II} and F'''_{II} shifted along the [111] direction. Coordinates of interstitials after refinement are, respectively: F'_{II}, $(\frac{1}{2},u,u)$ with $u = 0.40$; F''_{II}, (v_1,v_1,v_1) with $v_1 = 0.40$; and F'''_{II}, (v_2,v_2,v_2) with $v_2 = 0.30$. The F'''_{II} interstitials can be regarded as F_I ions at normal positions slightly relaxed from their original position. The presence of interstitial fluorine atoms F_{II} is a general feature of fluorite-type compounds containing an excess of anions. Let us quote UO_{2+x} (Willis, 1964), $Ca_{1-x}Y_xF_{2+x}$ (Cheetham et al., 1970, 1971; Laval and Frit, 1983a), and $Pb_{1-x}Bi_xF_{2+x}$ (Lucat et al., 1980).

Figure 7 gives the composition dependence of the total number of interstitial fluorine atoms $(n_{F_{II}} = n_{F'_{II}} + n_{F''_{II}} + n_{F'''_{II}})$ and that of the vacancies $(n_{V_{F_I}} = 2 - n_{F_I})$. For $x < 0.10$ a strong simultaneous increase in $n_{F_{II}}$ and $n_{V_{F_I}}$ is observed. For higher values of x, however, $n_{F_{II}}$ increases linearly, whereas $n_{V_{F_I}}$ becomes nearly constant ($\simeq 0.50$). This behavior is actually similar to that of the $Pb_{1-x}Bi_xF_{2+x}$ solid solution. Nevertheless, the increase in $n_{F_{II}}$ and $n_{V_{F_I}}$ for $Pb_{1-x}Th_xF_{2+2x}$ is two times more rapid. Experimental values fit with the substitution model $Pb^{2+} + 2 F_I^- = Th^{4+} + 4 F_{II}^- + 2 V_{F_I}$ for the boundary composition, which can be written $Pb_{0.75}Th_{0.25}F_{1.50}F_{II}V_{F_{I_{0.50}}}$. An analogous

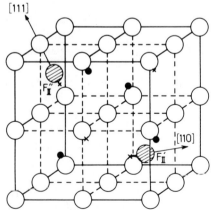

Fig. 6.　Fluorite-type structure ($\bigcirc$, F_1: $\bullet$, Ca; x, $V_{F_{II}}$).

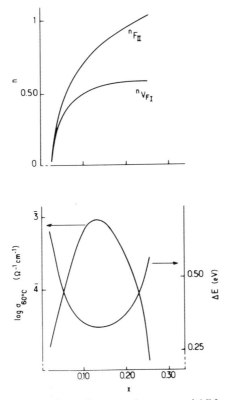

Fig. 7. Composition dependence of $n_{V_{F_I}}$, $n_{F_{II}}$, $\log \sigma_{60^\circ C}$, and ΔE for $Pb_{1-x}Th_xF_{2+2x}$.

model had been established for the upper limit of the $Pb_{1-x}Bi_xF_{2+x}$ solid solution $Pb^{2+} + F_I^- = Bi^{3+} + 2 F_{II}^- + V_{F_I}$ (Lucat *et al.*, 1980).

$Pb_{0.75}Th_{0.25}F_{1.50}F_{II}V_{F_{I0.50}}$ has a structure close to that of Pb_3ZrF_{10} (Laval and Frit, 1980), the elementary composition of which is $Pb_3ZrF_{16}F_{II_4}V_{F_{I_2}}$, in good agreement with the substitution mechanism proposed for $Pb_{0.75}Th_{0.25}F_{2.50}$. The formation of defects due to the replacement of one M^{2+} by one M'^{3+} or M''^{4+} in the fluorite-type solid solutions would be related, therefore, to the establishment of a local ordering, which becomes progressively a long-range one for compositions close to the boundaries.

The elementary pathway of the mobile ions is visualized by the distribution map of interstitial fluorides. Figure 8 gives the diffusion pathways of the composition $Pb_{0.77}Th_{0.23}F_{2.46}$ of the $Pb_{1-x}Th_xF_{2+2x}$ solid solution. This path is similar to that proposed for β-PbF_2 at high temperature (Koto *et al.*, 1980, 1981). Moreover, this result shows clearly that there is no direct exchange between normal sites along the edges.

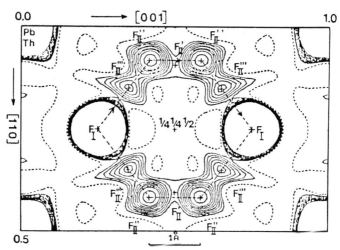

Fig. 8.   Envelope of crystallographic sites of fluorine atoms detected by neutron diffraction for $Pb_{0.77}Th_{0.23}F_{2.46}$.

Figure 7 gives the variation with x of $n_{V_{F_I}}$, $n_{F_{II}}$, log $\sigma_{60°C}$, and ΔE. For $x < 0.125$ a large increase in the number of vacancies and interstitial fluorine atoms occurs corresponding to increasing disorder. The result is a decrease in activation energy and an enhancement of conductivity. For $x > 0.125$ the number of vacancies has reached an upper limit, whereas the number of interstitial F^- ions increases regularly. The consequence is a conductivity drop, whereas activation energy increases. The conductivity decrease can be explained by progressive ordering between vacancies and interstitials, first locally and then at long range, when the upper limit composition is reached. A maximum of disorder appears, therefore, for a $Pb_{0.875}Th_{0.125}F_{2.25}$ composition, which presents the best electrical performances (Soubeyroux *et al.*, 1981).

Such conduction mechanisms can be applied to other fluorite-type solid solutions of composition $M_{1-x}M'_xF_{2+x}$ and $M_{1-x}M''_xF_{2+2x}$ as long as the evolution of the number of vacancies and interstitial anions when x increases is analogous to that observed for $Pb_{1-x}Th_xF_{2+2x}$. That is the case, for instance, for $Pb_{1-x}Bi_xF_{2+x}$ (Lucat *et al.*, 1980) or $Pb_{1-x}Zr_xF_{2+2x}$ (Laval *et al.*, 1984).

B. $A_{1-x}Bi_xF_{1+2x}$ (A = K or Rb) SOLID SOLUTIONS

The $ABiF_4/BiF_3$ systems include, in addition to $A_{1-x}Bi_xF_{1+2x}$ ($0 \leq x \leq x_L$) solid solutions (A = K, $x_L = 0.70$; A = Rb, $x_L = 0.60$), two single

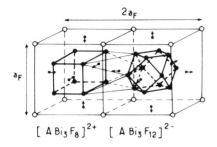

$[\,A\,Bi_3\,F_8\,]^{2+}$ $[\,A\,Bi_3\,F_{12}\,]^{2-}$

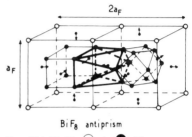

Bi F$_8$ antiprism

Fig. 9.   Crystal structure of ABi_3F_{10} (A = K or Rb) (●, Bi; ○, A; ●, F).

ABi_3F_{10} phases at $x = 0.75$. Whereas the solid-solution domain is crystallo-graphically disordered, the ABi_3F_{10} phases isostructural with KY_3F_{10} (Pierce and Hong, 1973) are characterized by a 1:3 order between A and bismuth, and the repetition in the lattice of two types of motifs, $|ABi_3F_8|^{2+}$ (fluorite type) and $|ABi_3F_{12}|^{2-}$, in which fluorine atoms form a cubo-octahedral cluster. The coordination polyhedron of bismuth in these phases is an antiprism $|BiF_8|$ formed of four F''_{II} (32f relaxed fluorine atoms from normal positions in the fluorite cell) and four F'_{II} (48i interstitial fluorine atoms) (Fig. 9).

The temperature dependence of the conductivity for several compositions of the $A_{1-x}Bi_xF_{1+2x}$ solid solutions and for the ordered phases ABi_3F_{10} is reported in Fig. 5. When x increases, at a given temperature the conductivity decreases, and one observes an increase in the activation energy. The lowest performances are obtained for ABi_3F_{10}.

Neutron diffraction investigations have been carried out in order to explain such a variation of the electrical properties within $A_{1-x}Bi_xF_{1+2x}$ (Soubey-roux et al., 1982). Interstitial positions in $A_{1-x}Bi_xF_{1+2x}$ are found only in F'_{II} sites ($\frac{1}{2},u,u$; $u = 0.34$). They are very close to the interstitial positions constituting the cubo-octahedral cluster in KY_3F_{10}.

Figure 10 gives the composition dependence of the number of fluorine atoms at normal and interstitial positions as well as that of the number of vacancies in normal sites ($n_{V_{F_I}} = 2 - n_{F_I}$). High-temperature forms of $ABiF_4$

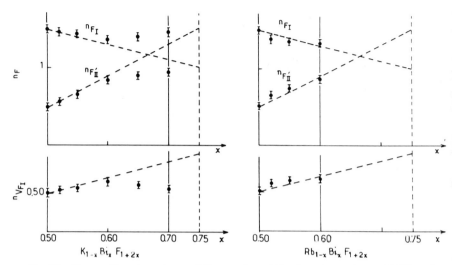

Fig. 10. Composition dependence of the number of fluorine atoms at normal (F_I) and interstitial (F'_{II}) sites and of the number of vacancies (V_{F_I}) at normal sites for $A_{1-x}Bi_xF_{1+2x}$ (A = K or Rb).

phases may be written $A_{0.50}Bi_{0.50}F_{I_{1.50}}F'_{II_{0.50}}V_{F_{I_{0.50}}}$. When x increases within the solid solution, $n_{F_{II}}$ increases whereas n_{F_I} decreases linearly. By extrapolating up to $x = 0.75$, $n_{F_{II}} = 1.50$ and $n_{F_I} = 1$. The theoritical formulation of ABi_3F_{10} would then be $A_{0.25}Bi_{0.75}F_IF_{II_{1.50}}V_{F_I}$. This is in agreement with the structure proposed for this phase. As far as the compositions at $x = 0.65$ and $x = 0.70$ of the $K_{1-x}Bi_xF_{1+2x}$ solid solution are concerned, the experimental points deviate from the dashed line. This deviation is due to the approximation made by using a simple fluorite-type unit cell for these two compositions, the diffraction spectra of which contain a superstructure line that has not been involved in the calculations. The presence of this superstructure suggests, in fact, the early formation of precursor clusters of cubo-octahedral type. This hypothesis is supported by an nmr study of the $K_{1-x}Bi_xF_{1+2x}$ solid solution (Réau et al., 1983b). Therefore, it is not surprising to find lower values of $n_{F'_{II}}$ with respect to those primarily expected. These results suggest a mechanism of substitution when bismuth replaces an alkaline ion in $ABiF_4$:

$$A^+ + 2 F_I^- = Bi^{3+} + 4 F_{II}^- + 2 V_{F_I}$$

β-$KBiF_4$ and γ-$RbBiF_4$ show, as does $Pb_{0.875}Th_{0.125}F_{2.25}$ (see the previous section), a maximum of disordering.

When x increases in $A_{1-x}Bi_xF_{1+2x}$, $n_{F_{II}}$ and $n_{V_{F_I}}$ increase linearly, and it is possible to predict the formation of isolated BiF_8 antiprisms. This results in the onset of an ordering between A and bismuth atoms, on one hand, and elementary cubes F_8 and cubooctahedra F_{12}, on the other hand. Local ordering becomes progressively long range order at rising x; it is well established for ABi_3F_{10}. This hypothesis is confirmed by the nmr study of the $A_{1-x}Bi_xF_{1+2x}$ solid solutions.

The increase with x in the number of vacancies $n_{V_{F_I}}$ should lead to an enhanced conductivity; however, such vacancies are trapped within the cubo-octahedra F_{12}, and their contribution to long-range conductivity may be neglected.

The progressive onset of anionic cubo-octahedra in the $A_{1-x}M'_xF_{1+2x}$ solid solutions seems to be a general feature. In fact, in the disordered fluorite-type solid solutions $Na_{1-x}Y_xF_{1+2x}$ ($0.50 \leq x \leq 0.64$) the crystallographic investigations undertaken to determine the local fluorine arrangement agree with the assumption that the anion excess is trapped near the yttrium atoms to form thermally stable fluorine cubo-octahedra (Pontonnier et al., 1983).

V. Theoretical Approach to Complex Clusters in Anion-Excess Fluorides

In previous sections we have shown that the transport properties of anion-excess solid solutions are related to the exact nature of the disorder in these materials. The role played by trivalent dopants, for instance, is to increase the relative concentration of interstitial anions that are associated with them in nn or nnn monomers. As x increases, these clusters begin to aggregate in the form of dimers, trimers, and even more extended defect units (Catlow et al., 1983).

Neutron diffraction studies of some solid solutions have shown interstitial fluorine atoms shifted from the cube-center sites along the [110] and [111] directions. Cluster models have been used to account for the distribution of normal and interstitial fluorine atoms versus composition (Cheetham et al., 1971). The 2:2:2 cluster consists of a planar dimer formed from two $Ln^{3+}-$ F_i^- pairs and stabilized by the occurrence of two anion vacancies and two anion interstitials due to relaxation of two lattice F^- ions in the [111] directions. The dimerization of these defects leads to a 4:3:2 cluster that contains four $Ln^{3+}-F_i^-$ monomers and three anion vacancies (Fig. 11): one in the center of the defect and two due to displacement of two lattice F^- ions into interstitial sites along [111] directions, as in the 2:2:2 cluster.

Other configurations have been proposed, such as the 2:2:2 + 1 cluster, which has been characterized by computer simulation (Catlow et al., 1983). It is a 2:2:2 cluster that has captured an additional F^- interstitial. It appears, generally, that the bonding energies of the clusters are strongly dependent on the ion size (Corish et al., 1982). Unfortunately, the $F'_{II}-F'_{II}$ distances of these 2:2:2 and 4:3:2 clusters and those deriving from these models are too short.

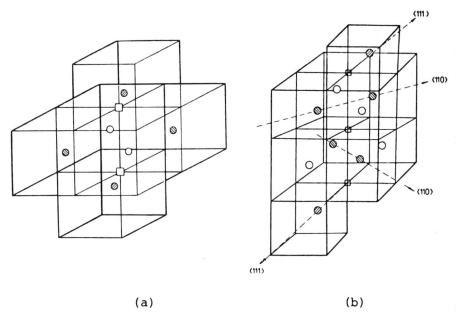

(a)   (b)

Fig. 11. (a) The 2:2:2 cluster with substitutional cations; (b) the 4:3:2 cluster with substitutional cations ($\square$, anion vacancy; $\otimes$, [110] interstitial; $\oslash$, [111] relaxed lattice ion; $\bigcirc$, trivalent substitutional).

A new polyhedral cluster built up on the basis of a square antiprism and labeled 4:4:3 provides a short-range order model perfectly suitable for the whole range of compositions observed in the $Ca_{1-x}Y_xF_{2+x}$ solid solution and avoids F'_{II}–F'_{II} distances that are too short in spite of locally closer packing of the anions (Laval and Frit, 1983a). Such 4:4:3 clusters could be precursors for the largest clusters (i.e., the ordered microdomains observed for samples annealed for a long time) (Greis, 1978).

Knowledge of the nature of the clusters within anion-excess solid solutions and of transformations with increasing x of small clusters into *superclusters* is imperative for understanding transport properties (Archer et al., 1983). The example of $A_{1-x}Bi_xF_{1+2x}$ solid solutions is significant and suggests stronger bonding for superclusters than for the smaller ones.

A. CUBO-OCTAHEDRAL CLUSTER

The existence of cubo-octahedral motifs in ordered phases such as KY_3F_{10} (Pierce and Hong, 1973), $Na_7Zr_6F_{31}$ (Burns et al., 1968), and tveitite $Ca_{14}Y_5F_{43}$ (Bevan et al., 1982) has led to computer simulation studies of their stability. Table IV shows the results of the relaxation for KY_3F_{10}, KBi_3F_{10},

TABLE IV

AM_3F_{10} Phases

Atomic positions in AM_3F_{10} phases before and after relaxation

		After Relaxation		
Atom	Before Relaxation	KY_3F_{10}	KBi_3F_{10}	$RbBi_3F_{10}$
M (K, Rb) (8c)	$(\frac{1}{4},\frac{1}{4},\frac{1}{4})$	$(\frac{1}{4},\frac{1}{4},\frac{1}{4})$	$(\frac{1}{4},\frac{1}{4},\frac{1}{4})$	$(\frac{1}{4},\frac{1}{4},\frac{1}{4})$
	$(\frac{3}{4},\frac{3}{4},\frac{3}{4})$	$(\frac{3}{4},\frac{3}{4},\frac{3}{4})$	$(\frac{3}{4},\frac{3}{4},\frac{3}{4})$	$(\frac{3}{4},\frac{3}{4},\frac{3}{4})$
N (Y, Bi) (24e)	$(x,0,0); x = 0.2401$	$x = 0.2388$	$x = 0.2402$	$x = 0.2403$
F_I (32f)	$(v,v,v); v = 0.1081$	$v = 0.1095$	$v = 0.1110$	$v = 0.1095$
F_{II}' (48i)	$(\frac{1}{2},u,u); u = 0.3353$	$u = 0.3395$	$u = 0.3399$	$u = 0.3408$

Calculated physical constants for the AM_3F_{10} phases

Material	Cohesive Energy (eV)[a] U_L	Elastic Constants $(10^{11}$ dynes/cm)			Dielectric Constants		Bulk-Lattice Strains
		C_{11}	C_{12}	C_{44}	ϵ_s	ϵ_∞	
KY_3F_{10}	-41.57	14.54	6.30	5.66	8.705	3.53	0.005
KBi_3F_{10}	-39.31	11.69	5.44	4.55	11.690	3.48	-0.009
$RbBi_3F_{10}$	-39.16	11.28	5.09	4.26	10.714	3.48	-0.002

[a] Constant-volume energies calculated for an elementary motif.

and $RbBi_3F_{10}$ phases (Matar et al., 1984). The following observations have been made. (1) The starting atomic positions are in good agreement with those obtained after relaxation of the structure. (2) The average value of the F_{II}' (48i) coordinates (0.50,0.34,0.34) is very close to those of interstitial anions in $A_{1-x}Bi_xF_{1+2x}$ (A = K or Rb) solid solutions; this suggests, in fact, formation of cubo-octahedral *precursors* in the apparently disordered solid-solution domain.

Table IV also gives the calculated physical constants for the three phases. The values of U_L obtained for the ordered AM_3F_{10} phases are clearly lower than those corresponding to disordered MF_2 fluorides (M = Ca, Sr, or Ba): $U_L(CaF_2) = -26.76$ eV, $U_L(SrF_2) = -25.33$ eV, $U_L(BaF_2) = -23.81$ eV (Axe et al., 1965). These results emphasize the stability of the AM_3F_{10} phases.

B. STABILIZATION OF CUBO-OCTAHEDRAL CLUSTERS IN ANION-EXCESS SOLID SOLUTIONS

The cubo-octahedral cluster has proved to be very stable as a building block in ordered anion-excess fluorite phases. Therefore, computer simulation studies have been carried out to stabilize that complex defect within CaF_2

itself. The knowledge of short-range and shell-model parameters for the CaF_2/U^{4+} fluoride has oriented the investigations on this system (Matar et al., 1984).

The response of a perfect lattice to the introduction of a defect can be visualized through relaxations of the ions in order to adopt a new, stable configuration with a minimum of energy. These relaxations affect most directly the defect itself and a small region surrounding it. They decrease fairly rapidly for distances away from the defect (see Chapter 5).

Point defects and nn and nnn pair defects have been considered first. The nn pair is energetically favored over the nnn pair (Table V). In contrast, the angular dimers formed by the association of two pairs are more stable than the elementary pairs, and the nn–nn dimer is favored over the nn–nnn dimer. The structure of the nn–nn defect is shown in Fig. 12a before and after relaxation (which is visualized by arrows).

Substitution of Ca^{2+} by U^{4+} involves the introduction of two F^- anions in interstitial sites. The relaxation of such a configuration leads to an attraction of the interstitials by the substitutional cation and also to a shift of two F_I^- from normal positions (on both sides of the plane formed by the dimer) into interstitial positions. Such behavior corresponds to the substitution mechanisms proposed in the neutron diffraction study of $Pb_{1-x}Th_xF_{2+2x}$ (Soubeyroux et al., 1981). This relaxed configuration has been confirmed by quasi-elastic neutron scattering experiments on $Ba_{0.9}U_{0.1}F_{2.2}$ (Andersen et al., 1983; Ouwerkerk et al., 1983).

After relaxation of the angular dimer, the positions of fluorine atoms in the immediate neighborhood of U_{Ca}^{4+} are nearly equivalent (Table V). One may notice that other surrounding anions at normal positions are only slightly affected by relaxation phenomena.

These results have suggested a new type of defect, the $1:3:1$ trimer (nn–nn–nn) involving, respectively, one U_{Ca}^{4+}, three equivalent F_{II}', and one V_{F_I} (Fig. 12b). This configuration is electrically neutral, hence favorable. This $1:3:1$ trimer has been simulated by inserting interstitials into ideal lattice positions $(\frac{1}{2},0,0)$ and by creating a vacancy at normal position $(\frac{1}{4},\frac{1}{4},\frac{1}{4})$. Its relaxation yields a stable configuration very close to that of the relaxed angular dimer (Table V). The calculations show that the $1:3:1$ trimer is stable with regard to all other defects so far studied. Therefore, the formation of this new type of defect could follow immediately that of the angular dimer (nn–nn).

The cubo-octahedral cluster has been simulated within CaF_2 itself in the following way. Formally, in CaF_2 the anionic sublattice is modified by creating 8 vacancies at normal sites and intercalating 12 interstitials in the same manner as in KY_3F_{10}. The relaxation of such a configuration leads to satisfactory convergence. However, the bonding energy of a cluster in which

TABLE VA

Formation Energies of Isolated Defects (eV) at
Constant Volume

Defect	Formation Energy (eV)
Anion vacancy	5.76
Anion interstitial	−3.12
Anion–Frenkel pair	2.64
Cation vacancy	23.60
U_{Ca}^{4+}	−44.19

TABLE VB

Defect Pair Formation and Bonding Energies per Substitutional Cation in
CaF_2/U^{4+} at Constant Volume

Defect	Formation Energy (eV)	Bonding Energy (eV)
nn pair, [110]	−48.69	−1.38
nnn pair, [111]	−48.09	−0.78
Angular dimer nn–nn	−53.77	−3.34
Angular dimer nn–nnn	−52.45	−2.02
1 : 3 : 1 trimer	−53.61	−6.04

TABLE VC

Bond Distances and Angles in the Angular Dimer (nn–nn)

Distance or Angle	Before Relaxation	After Relaxation
$d_{(U_{Ca}-2F'_{II})}$ (Å)	2.725	2.329
$d_{(U_{Ca}-F_I)}$ (Å)	2.360	2.323
$F'_{II}-U-F'_{II}$ angle (deg)	90	72
$F'_{II}-U-F_I$ angle (deg)	54.73	68.6
$d_{(U_{Ca}-V_{F_I})}$ (Å)	—	2.36

TABLE VD

Bond Distances and Angles in the 1 : 3 : 1 Trimer

Distance or Angle	Before Relaxation	After Relaxation
$d_{(U_{Ca}-3F'_{II})}$ (Å)	2.725	2.319
$d_{(U_{Ca}-V_{F_I})}$ (Å)	2.360	2.360
$F'_{II}-U_{Ca}-F'_{II}$ angle (deg)	90	71.7

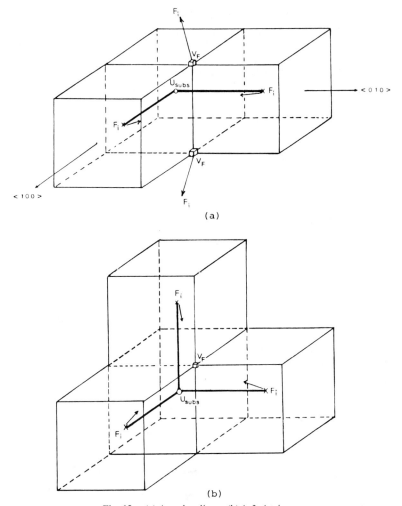

Fig. 12. (a) Angular dimer; (b) 1:3:1 trimer.

the cationic sublattice would not be modified is positive; the F_{12} cluster as such is not stable. In contrast, the relaxation of the effective $|Ca_3UF_{12}|^{2-}$ cluster (U^{4+} replacing Ca^{2+} at the fluorite-type cube corners) yields a valid minimization of the energy as well as a stable configuration. The bonding energy of the defect per substitution cation is found to be -12.65 eV. The large difference between the bonding energies of the cubo-octahedron, on the one hand (-12.65 eV), and the 1:3:1 trimer (-6.04 eV), on the other hand, favors the former. The cubo-octahedral cluster has been successfully simulated as well in $SrF_2(U^{4+})$ and $BaF_2(U^{4+})$.

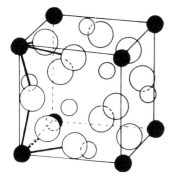 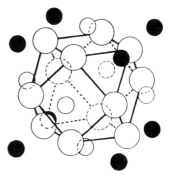

Fig. 13. Formation of the $|Ca_3UF_{12}|^{2-}$ cluster by association of $1:3:1$ trimers two by two ($\bigcirc$, F'_{II}; $\bigcirc$, Ca; $\bullet$, U_{Ca}).

An increasing number of substitutional cations in the CaF_2 lattice could lead to an increasing number of $1:3:1$ trimers and ultimately to their clustering. This most likely occurs through formation of the cubo-octahedron by associating eight $|U_{Ca}F'_{II_3}V_{F_I}|$ defects (i.e., eight $1:3:1$ trimers by sharing F'_{II} interstitials two by two). This result is illustrated in Fig. 13 and supported by crystallographic data.

Such studies are developing quickly to account for the conductivity mechanisms of F^- in appropriate fluorides.

References

Aalders, A. F., Polman, A., Arts, A. F. M., and De Wijn, H. W. (1983). *Solid State Ionics* **9–10**, 539.

Aleonard, S., Le Fur, Y., Pontonnier, L., Gorius, M. F., and Roux, M. T. (1978). *Ann. Chim. (Paris)* **3**, 417.

Andeen, C., Link, D., and Fontanella, J. (1977). *Phys. Rev.* **16**, 3762.

Andersen, N. H., Clausen, K., and Kjems, J. K. (1983). *Solid State Ionics* **9–10**, 543.

Archer, J. A., Chadwick, A. V., Jack, I. R., and Zequiri, B. (1983). *Solid State Ionics* **9–10**, 505.

Armstrong, R. D., Bulmer, R. S., and Dickinson, T. (1973). *J. Solid State Chem.* **8**, 219.

Avignant, D., Mansouri, I., Sabatier, R., and Cousseins, J.-C. (1979). *Ann. Chim. Fr.* **8**, 585.

Avignant, D., Mansouri, I., and Cousseins, J.-C. (1980). *C. R. Hebd. Seances Acad. Sci., Ser. C* **291**, 203.

Avignant, D., Mansouri, I., Chevalier, R., and Cousseins, J.-C. (1981). *J. Solid State Chem.* **38**, 121.

Avignant, D., Mansouri, I., Cousseins, J. C., Alizon, J., Battut, J. P., Dupuis, J., and Robert, M. (1982). *Mater. Res. Bull.* **17**, 1982.

Axe, J. D., Gaglianello, J. W., and Scardefield, J. (1965). *Phys. Rev.* **139**, 4A, 1211.

Azimi, A., Carr, V. M., Chadwick, A. V., Kirkwood, F. G., and Saghafian, R. (1984). *J. Phys. Chem. Solids* **45**, 23.

Bachmann, R., and Schulz, H. (1983). *Solid State Ionics* **9–10**, 521.

Barsis, E., and Taylor, A. (1966). *J. Chem. Phys.* **45**, 1154.

Barsis, E., and Taylor, A. (1968). *J. Chem. Phys.* **48**, 4362.

Bates, J. B., and Farrington, G. C., eds. (1981). "Fast Ionic Transport in Solids." North-Holland Publ., Amsterdam.

Battut, J. P., Dupuis, J., Robert, H., and Granier, W. (1983). *Solid State Ionics* **8**, 77.

Benci, S., Manfredi, M., Paracchini, C., and Schianchi, G. (1975). *Solid State Commun.* **17**, 779.

Bénière, F., and Catlow, C. R. A., eds. (1983). "Mass Transport in Solids, Series B, Physics," Vol. 97. Plenum, New York.

Bevan, D. J. M., and Greis, O. (1978). *Rev. Chim. Miner.* **15**, 346.

Bevan, D. J. M., Greis, O., and Strähle, J. (1980). *Acta Crystallogr., Sect. A* **A36**, 889.

Bevan, D. J. M., Strähle, J., and Greis, O. (1982). *J. Solid State Chem.* **44**, 75.

Bollmann, W. (1973). *Phys. Status Solidi A* **18**, 313.

Bollmann, W. (1978). *Phys. Status Solidi A* **45**, 187.

Bollmann, W., and Reimann, R. (1973). *Phys. Status Solidi A* **16**, 187.

Bollmann, W., Görlich, P., Hauck, W., and Mothes, H. (1970). *Phys. Status Solidi A* **2**, 157.

Bonne, R. W., and Schoonman, J. (1977). *J. Electrochem. Soc.* **124**, 28.

Boyce, J. B., Mikkelsen, J. C., and O'Keeffe, M. 1(977). *Solid State Commun.* **21**, 955.

Brunton, G. (1969). *Acta Crystallogr., Sect. B* **B25**, 600.

Burns, G., Dacol, F. H., Shafer, M. W., and Mahan, G. D. (1981). *Solid State Ionics* **5**, 645.

Burns, J. H. (1965). *Inorg. Chem.* **4**, 881.

Burns, J. H., Ellison, R. D., and Levy, M. A. (1968). *Acta Crystallogr., Sect. B* **B24**, 230.

Carr, V. M., Chadwick, A. V., and Saghafian, R. (1978). *J. Phys. C* **11**, L637.

Casanho, A., Guggenheim, H., and Walstedt, R. E. (1983). *Phys. Rev. B* **27**, 6587.

Catalano, E., and Wrenn, W. (1975). *J. Cryst. Growth* **30**, 54.

Catalano, E., Bedford, R. G., and Silveira, V. G. (1969). *J. Phys. Chem. Solids* **30**, 1613.

Catlow, C. R. A. (1976a). *J. Phys. C* **9**, 1845.

Catlow, C. R. A. (1976b). *J. Phys. C* **9**, 1859.

Catlow, C. R. A. (1980). *J. Phys.* **41**, C6-53.

Catlow, C. R. A. (1983). *Solid State Ionics* **8**, 89.

Catlow, C. R. A., and Hayes, W. (1982). *J. Phys. C* **15**, L9.

Catlow, C. R. A., and Mackrodt, W. C., eds. (1982). "Computer Simulation of Solids." Springer-Verlag, Berlin and New York.

Catlow, C. R. A., Norgett, M. J., and Ross, J. A. (1977). *J. Phys. C* **10**, 1627.

Catlow, C. R. A., Comins, J. D., Germano, F. A., Harley, R. T., Hayes, W., and Owen, I. B. (1981). *J. Phys. C* **14**, 329.

Catlow, C. R. A., Chadwick, A. V., and Corish, J. (1983). *J. Solid State Chem.* **48**, 65.

Chadwick, A. V. (1983). *Solid State Ionics* **8**, 209.

Chadwick, A. V., Gordin, R. E., Kirkwood, F. G., and Strange, J. H. (1978). *Int. Meet. Solid Electrolytes*, 2nd, Abstract 2.B.

Chadwick, A. V., Hope, D. S., Jaroszkiewicz, G., and Strange, J. H. (1979). *In* "Fast Ion Transport in Solids" (P. Vashista, J. N. Mundy, and G. K. Shenoy, eds.), p. 683. North-Holland Publ., Amsterdam.

Chang Hsu, Engelsberg, M., and Lowei, J. (1981). *Solid State Ionics* **5**, 609.

Chartier, C., Grannec, J., Réau, J. M., Portier, J., and Hagenmuller, P. (1981). *Mater. Res. Bull.* **16**, 1159.

Chartier, C., Grannec, J., Réau, J. M., and Hagenmuller, P. (1982). *Mater. Res. Bull.* **17**, 1283.

Cheetham, A. K., Fender, B. E. F., Steele, D., Taylor, R. I., and Willis, B. T. M. (1970). *Solid State Commun.* **8**, 171.

Cheetham, A. K., Fender, B. E. F., and Cooper, H. J. (1971). *J. Phys. C* **4**, 3107.

Cheetham, A. K., Fender, B. E. F., Fuess, H., and Wright, A. F. (1976). *Acta Crystallogr., Sect. B* **B32**, 94.

Claudy, P., Letoffe, J. M., Perez, G., Vilminot, S., Granier, W., and Cot, L. (1981a). *J. Fluorine Chem.* **17**, 145.

Claudy, P., Letoffe, J. M., Vilminot, S., Granier, W., Al Ozaibi, Z., and Cot, L. (1981b). *J. Fluorine Chem.* **18**, 203.

Clausen, K., Hayes, W. Hutchings, M. T., Kjems, J. K., Schnabel, P., and Smith, C. (1981). *Solid State Ionics* **5**, 589.

Corish, J., Catlow, C. R. A., Jacobs, P. W. M., and Ong, S. H. (1982). *Phys. Rev. B* **25**, 6425.

Couturier, G., Danto, Y., Pistre, J., Salardenne, J., Lucat, C., Réau, J. M., Portier, J., and Vilminot, S. (1979). *In* "Fast Ion Transport in Solids" (P. Vashista, J. N. Mundy, and G. K. Shenoy, eds.), p. 687. North-Holland Publ., Amsterdam.

Darbon, P., Réau, J. M., and Hagenmuller, P. (1981a). *Mater. Res. Bull.* **16**, 273.

Darbon, P., Réau, J. M., Hagenmuller, P., Depierrefixe, C., Laval, J. P., and Frit, B. (1981b). *Mater. Res. Bull.* **16**, 389.

De Kozak, A. (1969). *C. R. Hebd. Seances Acad. Sci.* **268**, 2184.

Den Hartog, H. W., and Langevoort, J. C. (1981). *Phys. Rev. B* **24**, p. 3547.

Derrington, C. E., Lindner, A., and O'Keeffe, M. (1975). *J. Solid State Chem.* **15**, 171.

Derrington, C. E., Navrotsky, A., and O'Keeffe, M. (1976). *Solid State Commun.* **18**, 47.

Dickens, M. H., Hayes, W., Hutchings, H. T., and Smith, C. (1982). *J. Phys. C* **15**, 4043.

Dixon, M., and Gillan, M. J. (1980). *J. Phys.* **41**, C6-24.

Etsell, T. H., and Flengas, S. N. (1970). *Chem. Rev.* **70**, 339.

Fedorov, P. P., and Sobolev, B. P. (1979). *J. Less-Common Met.* **63**, 31.

Fedorov, P. P., Izotova, O. E., Alexandrov, V. B., and Sobolev, B. P. (1974). *J. Solid State Chem.* **9**, 368.

Fedorov, P. P., Turkina, T. M., Sobolev, B. P., Mariani, E., and Svantner, M. (1982). *Solid State Ionics* **6**, 331.

Figueroa, D. R., Chadwick, A. V., and Strange, J. M. (1978). *J. Phys. C* **11**, 55.

Fontanella, J., Andeen, C., and Schuele, D. (1978a). *Phys. Rev.* **17**, 3429.

Fontanella, J., Jones, D. L., and Andeen, C. (1978b). *Phys. Rev. B* **18**, 4454.

Fontanella, J., Wintersgill, M. C., Welcher, P. J., Chadwick, A. V., and Andeen, C. G. (1981). *Solid State Ionics* **5**, 585.

Frit, B., and Laval, J. P. (1981). *J. Solid State Chem.* **39**, 85.

Goldman, M., and Shen, L. (1966). *Phys. Rev.* **144**, 321.

Gordon, R. E., and Strange, J. H. (1978). *J. Phys. C* **11**, 3213.

Granier, W., Bernier, P., Dohri, M., Alizon, J., and Robert, H. (1981). *J. Phys.* **42**, L301.

Granier, W., Gendry, M., El Mansouri, A., Vilminot, S., and Cot, L. (1983). *Z. Anorg. Allg. Chem.* **506**, 59.

Grannec, J., and Ravez, J. (1970). *Bull. Soc. Chim. Fr.* **5**, 1753.

Grannec, J., Chartier, C., Réau, J. M., and Hagenmuller, P. (1983). *Solid State Ionics* **8**, 73.

Greis, O. (1978). *Rev. Chim. Miner.* **15**, 481.

Greis, O. (1980). *In* "The Rare Earths in Modern Science and Technology" (G. J. McCarthy, J. J. Rhyne and M. B. Silber, eds.), Vol. 2, p. 167. Plenum, New York.

Greis, O., and Bevan, D. J. M. (1978). *J. Solid State Chem.* **24**, 113.

Greis, O., and Kieser, M. (1979). *Rev. Chim. Miner.* **16**, 520.

Hagenmuller, P., and van Gool, W., eds. (1978). "Solid Electrolytes." Academic Press, New York.

Hagenmuller, P., Réau, J. M., Lucat, C., Matar, S., and Villeneuve, G. (1981). *Solid State Ionics* **3/4**, 341.

Hayes, W. (1980). *J. Phys.* **41**, C6-7.

464   J.-M. Réau and J. Grannec

Hladik, J., ed. (1972). "Physics of Electrolytes," Vol 1. Academic Press, New York.
Ildstad, E., Svare, I., and Fjeldly, T. A. (1977). *Phy. Status Solidi A* **43**, K65.
Ippolitov, E. G., *et al.* (1967). *Inorg. Mater.* (*Engl. Transl.*) **3**, 59.
Ito, Y., and Koto, K. (1983). *Solid State Ionics* **9–10**, 527.
Ivanov-Shitts, A. K., Sorokin, N. I., Fedorov, P. P., and Sobolev, B. P. (1983). *Solid State Phys.* **25**, 1748.
Jacobs, P. W. M. (1983). *In* "Mass Transport in Solids, Series B" (F. Benière and C. R. A. Catlow, eds.), Physics, Vol. 97, p. 81, Plenum.
Jacobs, P. W. M., and Ong, S. H. (1976). *J. Phys.* **37**, C7-331.
Jacobs, P. W. M., and Ong, S. H. (1980a). *J. Phys. Chem. Solids* **41**, 431.
Jacobs, P. W. M., and Ong, S. H. (1980b). *Cryst. Lattice Defects* **8**, 177.
Jacobs, P. W. M., Ong, S. H., Chadwick, A. V., and Carr, V. M. (1980). *J. Solid State Chem.* **33**, 159.
Jaroszkiewicz, G. A., and Strange, J. H. (1980). *J. Phys.* **41**, C6-246.
Joshi, A. V., and Liang, C. C. (1977). *J. Electrochem. Soc.* **124**, 1253.
Joukoff, B., Primot, J., and Tallot, C. (1976). *Mater. Res. Bull.* **11**, 1201.
Kennedy, J. H., and Miles, R. C. (1976). *J. Electrochem. Soc.* **123**, 47.
Kieser, M., and Greis, O. (1980a). *J. Less-Common Met.* **71**, 63.
Kieser, M., and Greis, O. (1980b). *Z. Anorg. Allg. Chem.* **469**, 164.
Kitts, E. L., and Crawford, J. H. (1974). *Phys. Rev. B* **9**, 5264.
Kitts, E. L., Ikeya, M., and Crawford, J. H. (1973). *Phys. Rev. B* **8**, 5840.
Koto, K., Schulz, H., and Huggins, R. A. (1980). *Solid State Ionics* **1**, 355.
Koto, K., Schulz, H., and Huggins, R. A. (1981). *Solid State Ionics* **3–4**, 381.
Labeau, M., Aleonard, S., Vedrine, A., Boutonnet, R., and Cousseins, J. C. (1974). *Mater. Res. Bull.* **9**, 615.
Laredo, E., Puma, M., and Figueroa, D. R. (1979). *Phys. Rev. B* **19**, 2224.
Laredo, E., Figueroa, D. R., and Puma, M. (1980). *J. Phys.* **41**, C6-451.
Laredo, E., Puma, M., Suarez, N., and Figueroa, D. (1983). *Solid State Ionics* **9–10**, 497.
Laval, J. P., and Frit, B. (1980). *Mater. Res. Bull.* **15**, 45.
Laval, J. P., and Frit, B. (1983a). *J. Solid State Chem.* **49**, 237.
Laval, J. P., and Frit, B. (1983b). *Rev. Chim. Miner.* **20**, 368.
Laval, J. P., Depierrefixe, C., Frit, B., and Roult, G. (1984). *J. Solid State Chem.* **54**, 260.
Lechtenböhmer, C., and Greis, O. (1978). *J. Less-Common Met.* **61**, 177.
Lee, K. (1969). *Solid State Commun.* **7**, 367.
Lee, K., and Sher, A. (1965). *Phys. Rev. Lett.* **14**, 1027.
Le Fur, Y., Aleonard, S., Gorius, M. F., and Roux, M. Th. (1982). *Acta Crystallogr., Sect. B* **B38**(5), 1431.
Lenting, B. P. M., Numan, A. J., Bijvank, E. J., and den Hartog, H. W. (1976). *Phys. Rev. B* **14**, 1811.
Liang, C. C., and Joshi, A. V. (1975). *J. Electrochem. Soc.* **122**, 466.
Lidiard, A. B. (1974). *In* "Crystals with the Fluorite Structure" (W. Hayes, ed.). Oxford Univ. Press (Clarendon), London and New York.
Lidiard, A. B., and Norgett, M. J. (1972). *In* "Computational Solid State Physics" (F. Herman and T. R. Koehler, eds.). Plenum, New York.
Lilly, A. C., Laroy, B. C., Tiller, C. O., and Whiting, B. (1973). *J. Electrochem. Soc.* **120**, 1673.
Lucat, C., Campet, G., Claverie, J., Portier, J., Réau, J. M., and Hagenmuller, P. (1976). *Mater. Res. Bull.* **11**, 167.
Lucat, C., Rhandour, A., Cot, L., and Réau, J. M. (1979). *Solid State Commun.* **32**, 167.
Lucat, C., Portier, J., Réau, J. M., Hagenmuller, P., and Soubeyroux, J. L. (1980). *J. Solid State Chem.* **32**, 279.
McDonald, R. C., Ho-Kuen Hau, H., and Eriks, K. (1976). *Inorg. Chem.* **15**, 762.
McGeehin, P., and Hooper, A. (1977). *J. Mater. Sci.* **12**, 1.

Mansmann, M. (1965). *Z. Kristallogr.* **122**, 399.

Mansouri, I., and Avignant, D. (1984). *J. Solid State Chem.* **51**, 91.

Matar, S. F., and Réau, J. M. (1982). *C. R. Hebd. Seances Acad. Sci.* **294**, 649.

Matar, S. F., Réau, J. M., Lucat, C., Grannec, J., and Hagenmuller, P. (1980). *Mater. Res. Bull* **15**, 1295.

Matar, S. F., Réau, J. M., and Hagenmuller, P. (1982). *J. Fluorine Chem.* **20**, 529.

Matar, S. F., Catlow, C. R. A., and Réau, J. M. (1983a). *Solid State Ionics* **9–10**, 511.

Matar, S. F., Réau, J. M., Rabardel, L., Grannec, J., and Hagenmuller, P. (1983b). *Mater. Res. Bull.* **18**, 1485.

Matar, S. F., Réau, J. M., Grannec, J., and Rabardel, L. (1983c). *J. Solid State Chem.* **50**, 1.

Matar, S. F., Réau, J. M., Catlow, C. R. A., and Hagenmuller, P. (1984). *J. Solid State Chem.* **52**, 114.

Mellors, G. W., and Akridge, J. R. (1981). *Solid State Ionics* **5**, 605.

Meuldijk, J., and den Hartog, H. W. (1983). *Phys. Rev.* **27**, 6376.

Meuldijk, J., Mulder, H. H., and den Hartog, H. W. (1982). *Phys. Rev. B* **25**, 5204.

Nagel, L. E., and O'Keeffe, M. (1973). *In* "Fast Ion Transport in Solids" (W. Van Gool, ed.), p. 165. North-Holland Publ., Amsterdam.

Nakamura, A., and Wagner, J. B. (1980). *J. Electrochem. Soc.* **127**, 2325.

Nauta-Leeffers, Z. C., and den Hartog, H. W. (1979). *Phys. Rev. B* **19**, 4162.

Nowick, A. S., and Park, D. S. (1976). *In* "Super Ionic Conductors" (G. D. Mahan and W. L. Roth, eds.), p. 395. Plenum, New York.

Oberschmidt, J. (1981). *Phys. Rev. B* **23**, 5038.

Oberschmidt, J., and Lazarus, D. (1980). *Phys. Rev. B* **21**, 5823.

O'Keeffe, M. (1976). *In* "Super Ionic Conductors" (G. D. Mahan and W. L. Roth, eds.), p. 101. Plenum, New York.

O'Keeffe, M. (1977). *Comments Solid State Phys.* **7**, 163.

Ong, S. H., and Jacobs, P. W. M. (1980). *J. Solid State Chem.* **32**, 193.

Ouwerkerk, M., Kelder, E. M., Schoonman, J., and Van Miltenburg, J. C. (1983). *Solid State Ionics* **9–10**, 531.

Panhuyzen, R. A., Arts, A. F. M., Wapenaar, K. E. D., and Schoonman, J. (1981). *Solid State Ionics* **5**, 641.

Paracchini, C., and Schianchi, G. (1982). *J. Phys. C* **15**, L627.

Perez, G., Granier, W., and Vilminot, S. (1980a). *C. R. Hebd. Seances Acad. Sci. Ser. C* **290**, 337.

Perez, G., Vilminot, S., Granier, W., Cot, L., Lucat, C., Réau, J. M., Portier, J., and Hagenmuller, P. (1980b). *Mater. Res. Bull.* **15**, 587.

Pierce, J. W., and Hong, H. Y. P. (1973). *Proc. Rare Earth Res. Conf., 10th, 1973* A2, p. 527.

Pike, G. E., Camp, W. J., Seager, C. H., and McVay, G. L. (1974). *Phys. Rev. B* **10**, 4909.

Podberezskaya, N. V., Potapova, O. G., Borisov, S. U., and Gatilov, Yu. V. (1976). *J. Struct. Chem.* **17**, 948.

Pontonnier, L., Vicat, J., Hammou, A., and Filhol, A. (1978). *Int. Meet. Solid Electrolytes, 2nd* Abstract 6–13.

Pontonnier, L., Patrat, G., Aleonard, S., Capponi, J. J., Brunel, M., and De Bergevin, F. (1983). *Solid State Ionics* **9–10**, 549.

Portier, J., Réau, J. M., Matar, S., Soubeyroux, J. L., and Hagenmuller, P. (1983). *Solid State Ionics* **11**, 83.

Pouchard, M., and Hagenmuller, P. (1978). *In* "Solid Electrolytes" (P. Hagenmuller and W. van Gool, eds.), p. 191. Academic Press, New York.

Ratnikova, I. D., Korenev, Iu., M., and Novoselova, A. V. (1980). *J. Inorg. Chem.* **25**, 816.

Ravez, J., and Duale, M. (1970). *C. R. Hebd. Seances Acad. Sci.* **270**, 56.

Ravez, J., and Dumora, D. (1969a). *C. R. Hebd. Seances Acad. Sci.* **269**, 235.

Ravez, J., and Dumora, D. (1969b). *C. R. Hebd. Seances Acad. Sci.* **269**, 331.

Ravez, J., and Vassiliadis, M. (1970). *C. R. Hebd. Seances Acad. Sci.* **270**, 219.

Ravez, J., Darriet, M., von der Mühll, R., and Hagenmuller, P. (1971). *J. Solid State Chem.* **3**, 234.

Réau, J. M., and Portier, J. (1978). *In* "Solid Electrolytes" (P. Hagenmuller and W. van Gool, eds.), p. 313. Academic Press, New York.

Réau, J. M., Claverie, J., Campet, G., Deportes, Ch., Ravaine, D., Souquet, J. C., and Hammou, A. (1975). *C. R. Hebd. Seances Acad. Sci.* **280**, 325.

Réau, J. M., Lucat, C., Campet, G., Portier, J., and Hammou, A. (1976). *J. Solid State Chem.* **17**, 123.

Réau, J. M., Rhandour, A., Lucat, C., Granier, W., Vilminot, S., and Cot, L. (1978a). *Mater. Res. Bull.* **13**, 435.

Réau, J. M., Rhandour, A., Lucat, C., Portier, J., and Hagenmuller, P. (1978b). *Mater. Res. Bull.* **13**, 827.

Réau, J. M., Lucat, C., Portier, J., Hagenmuller, P., Cot, L., and Vilminot, S. (1978c). *Mater. Res. Bull.* **13**, 877.

Réau, J. M., Portier, J., Levasseur, A., Villeneuve, G., and Pouchard, M. (1978d). *Mater. Res. Bull.* **13**, 1415.

Réau, J. M., Matar, S., Kacim, S., Champarnaud-Mesjard, J. C., and Frit, B. (1982). *Solid State Ionics* **7**, 165.

Réau, J. M., Fedorov, P. P., Rabardel, L., Matar, S. F., and Hagenmuller, P. (1983a). *Mater. Res. Bull.* **18**, 1235.

Réau, J. M., Matar, S., Villeneuve, G., and Soubeyroux, J. L. (1983b), *Solid State Ionics* **9–10**, 563.

Réau, J. M., Rhandour, A., Matar, S. F., and Hagenmuller, P. (1984a). *J. Solid State Chem.* **55**, 7.

Réau, J. M., Rhandour, A., Matar, S., and Hagenmuller, P. (1984b). *Mater. Res. Bull.* (in press).

Rhandour, A. (1979). Thesis, Univ. of Bordeaux.

Roos, A. (1983a). Thesis, Utrecht Univ., The Netherlands.

Roos, A. (1983b). *Mater. Res. Bull.* **18**, 405.

Roos, A., Aalders, A. F., Schoonman, J., Arts, A. F. M., and De Wijn, H. W. (1983). *Solid State Ionics* **9–10**, 571.

Samara, G. A. (1979). *J. Phys. Chem. Solids* **40**, 509.

Schoonman, J. (1976). *J. Electrochem. Soc.* **123**, 1772.

Schoonman, J. (1979). *In* "Fast Ion Transport in Solids" (P. Vashishta, J. N. Mundy, and G. K. Shenoy, eds.), p. 631. North-Holland Publ., Amsterdam.

Schoonman, J. (1980). *Solid State Ionics* **1**, 121.

Schoonman, J., and den Hartog, H. W. (1982). *Solid State Ionics* **7**, 9.

Schoonman, J., and Wapenaar, K. E. D. (1979). *J. Electrochem. Soc.* **126**, 1385.

Schoonman, J., and Wolfert, A. (1981). *Solid State Ionics* **314**, 373.

Schoonman, J., Overslinzen, G., and Wapenaar, K. E. D. (1980). *Solid State Ionics* **1**, 211.

Schröter, W., and Nölting, J. (1980). *J. Phys. Orsay, Fr.* **41**, C6–20.

Senegas, J., Chartier, C., and Grannec, J. (1983). *J. Solid State Chem.* **49**, 99.

Shafer, M. W., Chandrashekhar, G. V., and Figat, R. A. (1981). *Solid State Ionics* **5**, 633.

Shannon, R. D. (1976). *Acta Crystallogr., Sect. A* **A32**, 751.

Shante, V. K., and Kirkpatrick, S. (1971). *Adv. Phys.* **20**, 325.

Shen, L. (1968). *Phys. Rev.* **177**, 259.

Sher, A., Solomon, R., Lee, K., and Muller, M. W. (1966). *Phys. Rev.* **144**, 593.

Shore, R. G., and Wanklyn, B. M. (1969). *J. Am. Ceram. Soc.* **52**, 79.

Sobolev, B. P., and Fedorov, P. P. (1978). *J. Less-Common Met.* **60**, 33.

Sobolev, B. P., and Seiranian, K. B. (1981). *J. Solid State Chem.* **39**, 337.

Sobolev, B. P., and Tkachenko, N. L. (1982). *J. Less-Common Met.* **85**, 155.

Sobolev, B. P. Fedorov, P. P., Seiranian, K. B., and Tkachenko, N. L. (1976). *J. Solid State Chem.* **17**, 201.

Sobolev, B. P., Seiranian, K. B., Garashina, L. S., and Fedorov, P. P. (1979). *J. Solid State Chem.* **28**, 51.

Solomon, R., Sher, A., and Muller, M. W. (1966). *J. Appl. Phys.* **37**, 247.

Soubeyroux, J. L., Réau, J. M., Mater, S., Hagenmuller, P., and Lucat, C. (1981). *Solid State Ionics* **2**, 215.

Soubeyroux, J. L., Réau, J. M., Matar, S., Villeneuve, G., and Hagenmuller, P. (1982). *Solid State Ionics* **6**, 103.

Soubeyroux, J. L., Matar, S. F., Réau, J. M., and Hagenmuller, P. (1984). *Solid State Ionics* **14**, 337.

Subbarao, E. C., ed. (1980). "Solid Electrolytes and their Applications." Plenum, New York.

Svantner, M., Mariani, E., Fedorov, P. P., and Sobolev, B. P. (1979). *Krist. Tech.* **14**, 365.

Svantner, M., Mariani, E., Fedorov, P. P., and Sobolev, B. P. (1981). *Cryst. Res. Technol.* **16**, 617.

Takahashi, T., Iwahara, H., and Ishikawa, T. (1977). *J. Electrochem. Soc.* **124**, 280.

Tan, Y. T., and Kramp, D. (1970). *J. Chem. Phys.* **53**, 3691.

Thoma, R. E., Insley, H., and Hebert, G. M. (1966). *Inorg. Chem.* **5**, 1222.

Tiller, C. O., Lilly, A. C., and Laroy, B. C. (1973). *Phys. Rev. B* **8**, 4787.

Tkachenko, N. L., Garashina, L. S., Izotova, O. E., Aleksandrov, V. B., and Sobolev, B. P. (1973). *J. Solid State Chem.* **8**, 213.

Ure, R. W. (1957). *J. Chem. Phys.* **26**, 1363.

Uvarov, N. F., Hairetdinov, E. F., and Boldyrev, V. V. (1984). *J. Solid State Chem.* **51**, 59.

Valon, P., Cousseins, J. C., Vedrine, A., Gacon, J. C., Boulon, G., and Fong, F. G. (1976). *Mater. Res. Bull.* **11**, 43.

van Weperen, W., and den Hartog, H. W. (1978). *Phys. Rev.* **18**, 2857.

Vashishta, P., Mundy, J. N., and Shenoy, G. K., eds. (1979). "Fast Ion Transport in Solids." North-Holland Publ., Amsterdam.

Vilminot, S., Perez, G., Granier, W., and Cot, L. (1980). *Rev. Chim. Miner.* **17**, 397.

Vilminot, S., Perez, G., Granier, W., and Cot, L. (1981a). *Solid State Ionics* **2**, 87.

Vilminot, S., Perez, G., Granier, W., and Cot, L. (1981b). *Solid State Ionics* **2**, 91.

Vilminot, S., Bachmann, R., and Schulz, M. (1983). *Solid State Ionics* **9–10**, 559.

Walker, A. B., Dixon, M., and Gillan, M. J. (1981). *Solid State Ionics* **5**, 601.

Wang, D. Y., Park, D. S., Griffith, J., and Nowick, A. S. (1981). *Solid State Ionics* **2**, 95.

Wapenaar, K. E. D. (1980). *J. Phys. (Orsay, Fr.)* **41**, C6–220.

Wapenaar, K. E. D., and Catlow, C. R. A. (1981). *Solid State Ionics* **2**, 245.

Wapenaar, K. E. D., and Schoonman, J. (1979). *J. Electrochem. Soc.* **126**, 667.

Wapenaar, K. E. D., and Schoonman, J. (1981). *Solid State Ionics* **5**, 637.

Wapenaar, K. E. D., Van Koesveld, J. L., and Schoonman, J. (1981). *Solid State Ionics* **2**, 145.

Wei, S. H. N., and Ailion, D. C. (1979). *Phys. Rev. B.* **19**, 4470.

Will, G., and Bargouth, M. O. (1980). *Z. Kristallogr.* **153**, 89.

Willis, B. T. M. (1964). *Proc. Br. Ceram. Soc.* **1**, 9.

Zalkin, A., Templeton, D. H., and Hopkins, T. E. (1966). *Inorg. Chem.* **5**, 1466.

Zintle, E., and Udgard, A. (1938). *Z. Anorg. Allg. Chem.* **240**, 150.

13

Nonlinear Properties of Fluorides

J. RAVEZ
Laboratoire de Chimie du Solide du CNRS
Université de Bordeaux
Talence, France

I. Introduction

The nonlinear properties considered here concern mainly insulating materials (piezoelectricity, nonlinear optics, pyroelectricity, ferroelectricity, and ferroelasticity). These materials are of great interest because they offer a potential basis for many applications [i.e., electromechanical transducers, infrared detectors, memories and displays, and electro-optica modulators (Ravez and Micheron, 1979)]. Although their properties have been well known for a long time [e.g., see "Pyroelectricity: A 2300-Year History" (Lang, 1974), the use of inorganic solid fluorides is much more recent in such research fields. $(NH_4)_2BeF_4$, the first ferroelectric, was not reported until 1957 (Pepinsky and Jona, 1957). In this chapter the properties of such materials are described. They are classified by topic and structural type. An explanation of the origin of the nonlinear properties is given in most cases.

INORGANIC
SOLID FLUORIDES

469

II. Piezoelectric and Nonlinear Optical Materials

Only the noncentrosymmetric phases have piezoelectric and nonlinear optical properties. Table I lists most of the materials that potentially have these properties.

Very few piezoelectric and second harmonic generation (SHG) measurements have been performed on fluorides. A strong piezoelectric activity has been detected for some of them, however. In contrast, the nonlinear dielectric susceptibilities that characterize the SHG are weak.

The following values are measured on crystals; they are compared with those of α quartz or $LiNbO_3$ as references:

1. Electromechanical coupling factor: $k_{31}(BaMgF_4) = 0.17$; $k_{24}(BaMnF_4) = 0.26$; $k_{26}(\alpha$ quartz$) = 0.14$; $k_{24}(LiNbO_3) = 0.60$
2. Piezoelectric strain coefficients (pC N^{-1}): $d_{33}(BaMgF_4) = 8.1$; $d_{31}[(NH_4)_2BeF_4] = 12.5$; $d_{11}(\alpha$ quartz$) - 2.3$; $d_{15}(LiNbO_3) = 74$
3. Piezoelectric stress coefficients (C m^{-2}): $e_{33}(BaMgF_4) = 0.61$; $e_{11}(\alpha$ quartz$) = 0.17$; $e_{15}(LiNbO_3) = 3.8$
4. Nonlinear dielectric susceptibilities (10^{-12} V^{-1} m): $d_{32}(BaMgF_4) = 0.07$; $d_{32}(BaZnF_4) = 0.13$; $d_{11}(\alpha$ quartz$) = 0.50$; $|d_{33}(LiNbO_3)| = 33$ (Landolt-Börnstein, 1979)

III. Pyroelectric and Ferroelectric Materials

Pyro- and ferroelectric materials are noncentrosymmetric and, in addition, are polar. The main characteristics are the Curie temperature T_C, the pyroelectric coefficient p, and the spontaneous polarization P_s. The most important families are the following:

1. A theoretical study taking into account the analogy between ferroelectric HCl and HF led to the conclusion that the latter is chemically the simplest ferroelectric (Merkel and Blumen, 1976).
2. A dielectric study has shown that cubic PbF_2 exhibits a relatively large decrease in the permittivity ϵ'_r with increasing temperature and obeys a Curie–Weiss law (Samara, 1976). It could be suggested that this anomalous temperature dependence and the relatively large value of ϵ'_r are associated with a ferroelectric mode at very low temperature. This is the first example of a soft ferroelectric mode in a crystal having the relatively simple fluorite structure.
3. The $A^{II}M^{II}F_4$ phases with the $BaMnF_4$ structure cyrstallize with the polar $2mm$ point group at room temperature. They have been shown to be

TABLE I

Fluoride Compounds and Families Noncentrosymmetric at Room Temperature

Fluoride	Space Group	Nonpolar	Polar	Fluoride	Space Group	Nonpolar	Polar
NH_4F, ND_4F	$P6_3mc$		×	$A^I MnM^{III}F_6$ (NH_4MnFeF_6 type)	$Pb2n$		×
N_2H_5F	$P2_12_12$	×		Li_2MoF_6	$P4_22_12$	×	
MF_2 (M = Ge or Sn)	$P2_12_12_1$	×		Na_2NiF_6	$P6_3mc$		×
SbF_3	$Ama2$		×	$A_2^I M^{IV}F_6$ (K_2MnF_6 type)	$P6_3mc$		×
TeF_4	$P2_12_12_1$	×		Rb_2AmF_6	$Cmc2_1$		×
MF_5 (M = Pa or U)	$I\bar{4}2d$	×		$LiScSiF_6$	$P6_322$	×	
$ABeF_3$ (A = Rb or NH_4)	$P2_12_12_1$	×		$Ba_2M^{II}F_6$ (Ba_2ZnF_6 type)	$I422$	×	
$AgNiF_3$	$P4_22_12$	×		$BaCd_2F_6$	$C222_1$	×	
NH_4SnF_3	$C2$		×	$Li_3M^{III}F_6$ (Li_3AlF_6 type)	$Pna2_1$		×
A_xFeF_3 (A = Rb, Cs, or Tl)	$Ima2$		×	K_2LiAlF_6	$P3m1$		×
$A^{II}M^{III}F_4$ ($BaMnF_4$ type)	$A2_1am$		×	$A_3^{II}(M^{III}F_6)_2$ [$Sr_3(FeF_6)_2$ type]	$P4$		×
$BaCrF_4$	$I4cm$ or $I\bar{4}c2$	×		$A^I Eu_xF_7$(A = Rb or Cs)	$P6_322$	×	
$BaCuF_4$	$Cmc2_1$		×	$KM_2^{III}F_7$ (M = Yb or Ln)	$P2$		×
$A_2^I BeF_4$ (A = K, Rb, or Cs)	$Pn2_1a$		×	β-KEr_2F_7	$Pna2_1$		×
$A^I LiBeF_4$ (A = K, Rb, or Tl)	$P6_3$		×	$Na_2LiBe_2F_7$	$P\bar{4}2_1m$	×	
$(NH_4)_2BeF_4$	$Pn2_1a$		×	$A_2^I M^{III}M^{III}F_7$ (Na_2MgAlF_7 type)	$Imm2$		×
$Li(NH_4)BeF_4$	$Pc2_1n$		×	Li_3PaF_8	$P4_22_12$	×	
$Li(N_2H_5)BeF_4$	$Pna2_1$		×	Li_2CaUF_8	$I\bar{4}m2$	×	
$MnAlF_5$	$Ama2$		×	Li_4ZrF_8	$P4_22_12$	×	
$CaM^{III}F_5$ ($CaFeF_5$ type)	$C222_1$	×		$Sr_2Fe_2F_9$	$I4$		×
$A^{II}M^{III}F_5$ ($SrAlF_5$ type)	$P4$		×	K_5ThF_9	$Cmc2_1$	×	
$BaGaF_5$	$P2_12_12_1$	×		$CsYb_3F_{10}$	Pc		×
Na_2SbF_5	$P2_12_12_1$	×		β-KEr_3F_{10}	Cm		×
$RbBe_2F_5$	$P1$		×	$RbTh_3F_{13}$	$P2_1ma$		×
$A_2^I M^{III}F_5$ (K_2EuF_5 type)	$Pna2_1$		×	γ-$Na_5Fe_3F_{14}$	$P\bar{4}2_11$	×	
$A^I M^V F_6$ ($KNbF_6$ type)	$P\bar{4}c2$	×		$Rb_5Zr_4F_{21}$	$P2_1$		×

ferroelectric at room temperature. The spontaneous polarization P_s varies from 6.7 ($BaNiF_4$) to 9.7 μC cm^{-2} ($BaZnF_4$) at 300 K (Eibschütz et al., 1969). The polarity reversal has been explained by atomic displacements in the (001) planes where the a axis is the polar direction (Keve et al., 1969). These compounds are two-dimensional (2D) ferroelectrics. Unfortunately, the ferroelectric–paraelectric transition has not been observed, because T_C lies above the melting point (Di Domenico et al., 1969). The strong pyroelectric effect associated at room temperature with the very small value of ϵ'_r makes these crystals suitable for pyroelectric devices.

In addition, studies on $BaMnF_4$ crystals have shown two dielectric anomalies at low temperature (26 and 250 K). The stable phase below 250 K is incommensurate. Below 26 K antiferromagnetic ordering occurs, the spins being parallel to the b axis. In fact, a weak ferromagnetism appears due to spin canting resulting from a magnetoelectric coupling to the spontaneous polarization (Scott, 1978).

4. $(NH_4)_2BeF_4$ shows two transitions at 177 and 183 K. The sequence is the following:

$$\text{Improper ferroelectric } (mm2) \xrightarrow{T_C = 177 \text{ K}} \text{incommensurate} \xrightarrow{183 \text{ K}} \text{paraelectric } (mmm)$$

The low value of P_s at 158 K (0.20 μC cm^{-2}) is in agreement with a 3D ferroelectric behavior. Ferroelectric domains have been observed for $T < T_C$. Polarity is reversed here by rotation of the three discrete tetrahedra. The situation is similar in $(ND_4)_2BeF_4$, where T_1 and T_C are slightly lower (Pepinsky and Jona, 1957; Uesu et al., 1981).

5. The atomic arrangement in polar $SrAlF_5$ is related to that of an identical structure but with opposite polarity because of atomic shifts smaller than 0.5 Å. Such displacements reverse the spontaneous polarization. Calorimetric, piezoelectric, nonlinear optical, and dielectric measurements have shown that compound to be ferroelectric with $T_C = 715$ K. The other members of the same $A^{II}M^{III}F_5$ family also exhibit a transition between 500 and 715 K. The T_C decreases as the size of M^{3+} increases [i.e., $T_C(SrAlF_5) = 715$ K; $T_C(SrGaF_5) = 605$ K] (Ravez et al., 1981).

6. $(NH_4)_2PF_7$ shows two transitions at 172 and 228 K. A strong dielectric anomaly takes place at 172 K. The phase with the lowest temperature ($T < 172$ K) is antiferroelectric (Vedam et al., 1959).

7. Calorimetric and dielectric measurements have shown the weberite $Na_2M^{II}M^{III}F_7$ compounds to have a transition above 700 K. In addition, the crystals are piezoelectric and generate second harmonics at room temperature. The polar ($mm2$)–non polar (mmm) transition is potentially a ferroelectric–paraelectric one (Tressaud et al., 1981).

IV. Ferroelastic Materials

A ferroelastic material has two or more stable states in the absence of mechanical stress or electric field, and it can be reproducibly transformed from one to another of these states by the application of a mechanical stress. In the ferroelastic state the crystal symmetry is reduced to a subgroup of a higher symmetry class (prototype) by a small distortion. The main characteristics are the ferroelastic-prototype Curie temperature and the spontaneous strain e_s. In the following examples, ferroelasticity has been either confirmed experimentally or predicted crystallographically for fluorides:

The orthorhombic β-SnF_2 (222)–tetragonal δ-SnF_2 (422) transition at $T_C = 339$ K would be a pure ferroelastin one, according to Denes (1980).

Many $A^IM^{II}F_3$ compounds with a perovskite-type structure undergo a ferroelastic-type phase transition from tetragonal (4/mmm) to cubic ($m3m$) symmetry corresponding to an alternate rotation of the MF_6 octahedra along the cubic c axis. This first-order improper transition occurs at low temperature (e.g., $KMnF_3$ and $RbCaF_3$, respectively, at 184 and 194 K) (Ridou et al., 1980).

An orthorhombic (mmm)–tetragonal (4/mmm) transition has been observed in the $A^IM^{III}F_4$ compounds based on the $TlAlF_4$ structure. A uniaxial compressive stress of 22 MN m^{-2} applied along the room-temperature [001] direction interchanges the a and c axes (Abrahams and Bernstein, 1972). In addition, 90° domains have been observed under a polarizing microscope in $CsVF_4$ crystals. These two experiments confirm ferroelasticity. The spontaneous room-temperature strain is relatively low [e.g., $e_s(RbFeF_4) = 2.10^{-4}$, $e_s(CsFeF_4) = 3.10^{-4}$ at 300 K]. The structure in the distorted phase can be represented schematically by tilted MF_6 octahedra. The transition may possibly be related to the condensation of soft phonons at particular zone-boundary points of the tetragonal high-temperature phase (Hidaka et al., 1979).

K_2NiF_4 exhibits 2D antiferromagnetic ordering from ~ 200 K down to Néel temperature $T_N = 97$ K. The orthorhombic (mmm)–tetragonal (4/mmm) transition at T_N is predicted to be ferroelastic (Abrahams, 1971). The application of a stress along the [100] or [010] direction is thus expected to reverse the sense of each spin at $z = \frac{1}{2}$ relative to those at $z = 0$. This would imply a ferromagnetic–ferroelastic coupling.

The $A^{II}M^{IV}F_6$ compounds present a hexagonal ($R\bar{3}$) $LiSbF_6$-type–cubic ($m3m$) ReO_3-type transition (Reinen and Steffens, 1978). The transition is potentially ferroelastic, the corresponding Curie temperature varying with the size and the electronic properties of the A^{II} and M^{IV} ions. Examples are the increase from $T_C(MnZnF_6) = 143$ to $T_C(NiZrF_6) = 390$ K and from $T_C(MnZrF_6) = 143$ to $T_C(MnSnF_6) = 530$ K.

474   J. Ravez

The $A_3^I M^{III} F_6$ fluorides with an $(NH_4)_3 FeF_6$-related structure exhibit one or more transitions. The highest temperature transition is a noncubic ferroelastic–cubic prototype ($m3m$). In addition, dielectric and thermocurrent measurements on $(NH_4)_3 MF_6$ (M = Al or Fe) have shown that the low-temperature phase is characterized by a remanent polarization (Lorient et al., 1981).

V. Discussion

The number of fluorides with nonlinear properties is relatively low compared with that of the corresponding oxides. Two reasons may be given for the difference of behavior:

1. Research on such fluorides is relatively recent. In addition, nonlinear properties have been studied principally by physicists, and the difficulty of synthesizing fluorides has led them to the study of other compounds.

2. Compounds with nonlinear properties present a crystallographic distortion of a prototypical centrosymmetric unit cell. When fluorides and oxides crystallize with the same structural type (e.g., perovskite, pyrochlore, tungsten bronze types), the greater ionicity of the M—F bonds than of the M—O bonds gives less distortion for fluorides. That is why such fluorides are either centrosymmetric or have a distortion only at low temperature. Nevertheless, fluorides offer a specific advantage over oxides due to their good transparency in the visible region and their electronic insulating properties.

References

Abrahams, S. C. (1971). *Mater. Res. Bull.* **6**, 881.
Abrahams, S. C., and Bernstein, J. L. (1972). *Mater. Res. Bull.* **7**, 715.
Denes, G. (1980). *Mater. Res. Bull.* **15**, 807.
Di Domenico, M., Eibschütz, M., Guggenheim, H. J., and Camlibel, I. (1969). *Solid State Commun.* **7**, 1119.
Eibschütz, M., Guggenheim, H. J., Wemple, S. H., Camlibel, I., and Di Domenico, M. (1969). *Phys. Lett.* 29, 409.
Hidaka, M., Wood, I. G., and Garrard, B. J. (1979). *Phys. Status Solidi* 56, 349.
Keve, E. T., Abrahams, S. C., and Bernstein, J. L. (1969). *J. Chem. Phys.* **51**, 4928.
Landolt–Börnstein (1979). "Elastic, Piezoelectric and Related Constants of Crystals," Part 3, p. 11. Springer-Verlag, Berlin and New York.
Lang, S. B. (1974). *Ferroelectrics* 7, 231.
Lorient, M., von der Mühll, R., Tressaud, A., and Ravez, J. (1981). *Solid State Commun.* **40**, 847.
Merkel, C., and Blumen, A. (1976). *Solid State Commun.* **20**, 755.
Pepinsky, R., and Jona, F. (1957). *Phys. Rev.* **105**, 344.

Ravez, J., and Micheron, F. (1979). *Actual. Chim.* **1**, 9.
Ravez, J., Abrahams, S. C., Chaminade, J. P., Simon, A., Grannec, J., and Hagenmuller, P. (1981). *Ferroelectrics* **38**, 773.
Reinen, D., and Steffens, F. (1978). *Z. Anorg. Allg. Chem.* **441**, 63.
Ridou, C., Rousseau, M., and Nouet, J. (1980). *Ferroelectrics* **26**, 685.
Samara, G. A. (1976). *Phys. Rev.* **13**, 4529.
Scott, J. F. (1978). *Ferroelectrics* **20**, 69.
Tressaud, A., Lozano, L., and Ravez, J. (1981). *J. Fluorine Chem.* **19**, 61.
Uesu, Y., Kobayashi, J., Ogawa, J., and Strukov, B. A. (1981). *Ferroelectrics* **36**, 355.
Vedam, K., Pepinsky, R., Lajzerowicz, J., Okaya, Y., and Stemple, N. (1959). *Bull. Am. Phys. Soc.* [2] **4**, 63.

14

Optical Properties of Fluorides

C. FOUASSIER

Laboratoire de Chimie du Solide du CNRS
Université de Bordeaux
Talence, France

I. Introduction

In the field of optical properties, fluorides have several specific features: small refractive index, wide transmission wavelength range, weak probability of intraconfigurational transitions, etc. Interest in the optical applications of these materials also arises from two characteristics: (1) the low melting points compared with those of oxides generally facilitate crystal growth from the

INORGANIC
SOLID FLUORIDES ·

liquid state; (2) most crystals have a high chemical stability and can be maintained in air without deterioration of their optical properties.

This chapter deals only with crystalline fluorides; the optical properties of fluoride glasses are described in Chapter 7.

II. Index of Refraction

Owing to the weak polarizability of the fluoride ion, fluorides show the lowest refractive indices. Typical values in the visible range lie between 1.35 and 1.5, the lowest being those for sodium phases [e.g., Na_2SiF_6, 1.310; NaF, 1.336 ($\lambda = 589$ nm)].

III. Transmission Properties

Due to the wide transparency wavelength range of fluorides, various optical components are made with these materials, the most common being the nonhygroscopic alkaline-earth fluorides.

A. ABSORPTION EDGE IN THE ULTRAVIOLET RANGE

When the cations have a noble gas configuration, fluorides are transparent in the uv range down to short wavelengths (Fig. 1). For some (LiF, MgF_2, and $LiYF_4$) the absorption edge lies below 120 nm (Tomiki and Miyata, 1969; Rehn et al., 1977). The valence bands are composed of the outer p states of fluorine. Owing to the strong electronegativity of this element, transfer of an electron to a cationic state requires high energy. The energy of the first exciton

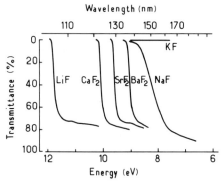

Fig. 1. Transmittance of fluoride crystals in the ultraviolet (thickness $\simeq$ 0.2 cm). After Tomiki and Miyata (1969).

TABLE I

Energy of the First Exciton Peak for Alkali and Alkaline-Earth
Fluorides at Room Temperature

Fluoride	Energy (eV)	Fluoride	Energy (eV)
LiF	12.8^a		
NaF	$10.5^{a,b}$	MgF_2	11.7^c
KF	$9.6^{a,b}$	CaF_2	11.1^c
RbF	9.4^b	SrF_2	10.4^a
CsF	9.1^b	BaF_2	10.0^a

[a] Li (1980).
[b] Tomiki and Miyata (1969).
[c] Stephan *et al.* (1969).

associated with interband transitions is given in Table I for the alkali and
alkaline-earth fluorides. It increases with decreasing size of the cation due to
the covalent contribution of the M—F bond.

B. ABSORPTION EDGE IN THE INFRARED RANGE

On the long-wavelength side, the transparent region is limited by multiphonon absorption, so transmission is higher than in oxides but lower than in
other halides. Fluorides of heavy-metal cations such as BaF_2 show high
transmittance up to ~ 10 μm.

IV. Absorption and Emission Properties of Transition Element, Rare-Earth, and Mercury-like-Configuration Ions

Fluorides are attractive host lattices for the investigation of spectroscopic
properties of transition element and rare-earth ions.

Optical transitions can be observed over a wide wavelength range, so that a
large number of data for calculating crystal field parameters of $3d^n$ and $4f^n$
configurations can be collected. Transitions to configurations lying at high
energy have been observed only in fluorides; this is the case for $3d \to 4s$
transitions present above 6 eV in the spectra of divalent $3d$ ions incorporated
into LiF or $KMgF_3$ (McClure, 1976).

Owing to the insulating properties of fluorides, crystals containing a
transition element in an intermediate oxidation state remain transparent at
high concentration. They are therefore particularly suitable for the investigation of transitions resulting from interactions between ions (exciton–
exciton and exciton–magnon transitions).

Moreover, many high oxidation states of the transition metals are stabilized almost uniquely by the fluoride ligand.

A. TRANSITION ELEMENTS

1. Absorption Spectra

The absorption spectra of transition elements in discrete MF_6 groups have been compiled by Allen and Warren for $3d$ (1971) as well as $4d$ and $5d$ (1974) ions. Fluorine is the element inducing the smallest nephelauxetic effect (decrease in the interelectronic repulsions resulting from the involvement of cationic orbitals in molecular orbital formation with ligand orbital combinations). Some values of the nephelauxetic ratio, $\beta = B_{\text{fluoride}}/B_{\text{free ion}}$ (B is the Racah parameter), are listed in Table II. The decrease in β within each series with the number of d electrons reflects the increasing tendency toward covalency. The increase in metal–ligand mixing with the oxidation state can also be noted.

Also included in Table II are the crystal field parameters Dq. The crystal field is lower than in oxides but higher than in other halides. It follows the usual trends, $Dq(3d) < Dq(4d) < Dq(5d)$, and for a given element $Dq(M^{2+}) < Dq(M^{3+}) < Dq(M^{4+})$, etc.

The absorption spectra at room temperature consist of broad bands due to phonon-assisted transitions. The intensities are low (typical values of the oscillator strengths for spin-allowed transitions lie between 10^{-5} and 10^{-4}) and decrease when the temperature is lowered. At low temperature a vibrational structure generally appears (Ferguson et al., 1964; Simo et al., 1969). Interionic coupling in fluorides in which the transition ions are bridged by common anions (e.g., rutile- or perovskite-type fluorides) enhances the intensity of some spin-forbidden transitions by a factor greater than 10 (Lohr and McClure, 1968).

Intense charge-transfer transitions (from the highest filled, predominantly anionic orbitals to the empty, predominantly cationic orbitals) are observed for the trivalent and tetravalent ions in the uv range below 330 nm.

2. Emission Spectra

At 300 K ions with $3d^3$ (Cr^{3+} and Mn^{4+}) and $3d^5$ (Mn^{2+} and Fe^{3+}) configurations show luminescent properties in the visible range. Table III gives the characteristics of the emission.

The low-temperature emission spectra of the d^3 ions possess a very large number of lines (Ferguson et al., 1971). The electric dipole transitions resulting from the mixing of the even d states with odd internal vibrations (t_{1u}, t_{2u}) and

TABLE II

Nephelauxetic Ratios β_{35} (Deduced from Spin-Allowed Bands) and Crystal Field Splitting Parameter Dq for Hexafluoro Complexes of Transition Element Ions[a]

Oxidation State	V	Cr	Mn	Fe	Co	Ni	Cu
β_{35}							
+3	0.78	0.80	0.80	0.77	0.70	0.61	0.53
+4	—	0.57	0.55	—	0.53	0.41	—
$Dq\,(\text{cm}^{-1})$							
+3	1610	1520	1740	1300	1410	1620	1410
+4	—	2200	2220	—	2030	2010	—

	Mo	Tc	Ru	Rh	Pd	Ag
β_{35}						
+3	0.90	—	0.80	0.64	—	0.61
+4	—	0.75	0.68	0.54	0.43	—
$Dq\,(\text{cm}^{-1})$						
+3	2350	—	2200	2230	—	1840
+4	—	2840	2500	2050	2600	—

[a] Allen and Warren (1975).

TABLE III

Characteristics of Emission of d^3 and d^5 Ions

Ion	Nature of Transition	Emission Wavelength (nm)	Color
Cr^{3+}	$^2E_g \longrightarrow {}^4A_{2g}$	660	Red
Mn^{4+}		620–630	Red
Mn^{2+}	$^4T_{1g} \longrightarrow {}^6A_{1g}$	Octahedral site, 580–590	Orange
		Cubic site, 490–530	Green
Fe^{3+}		Octahedral site, 730	Red

the magnetic dipole transition are followed by vibronic progressions essentially formed with the internal vibrational modes of the MF_6 group (Chodos et al., 1976; Greenough and Paulusz, 1979).

For the four ions mentioned the quenching temperatures are generally well above 300 K. The transitions arc parity forbidden and spin forbidden; as a consequence of the weak mixing of the cationic and fluorine orbitals, the lifetimes are very long [e.g., 0.1 sec for Mn^{2+} in CaF_2 (Garlick and Sayer, 1962)].

In a few fluorides such as AlF_3, due to a very weak crystal field the $^4T_{2g}$ state of Cr^{3+} lies below ^2E_g, giving rise to a broad luminescence band at ~ 770 nm with a decay time of ~ 0.5 msec.

The luminescence of manganese is of particular interest for applications. Manganese-doped fluorides (MgF_2 and $KMgF_3$) characterized by a long persistence are used for radar screens. The electroluminescent properties of crystals (Langer et al., 1979) or thin films (Williams, 1981) of CdF_2: Mn have been extensively studied. Although the generation of carriers usually occurs in the luminescent layer, in the SiO/ZnF_2:Mn/SiO thin-film cells carriers are created in the SiO layer and injected into ZnF_2: Mn; such devices may operate with either ac or dc voltages.

At liquid-nitrogen temperature, Co^{2+} and Ni^{2+} show an emission in the near infrared (ir) (1.6–2 μm), which allows one to obtain a tunable continuous coherent radiation using MgF_2 (Moulton and Mooradian, 1979) or $KZnF_3$ (Künzel et al., 1981) as host lattice. The $3d$ ion occupies a centrosymmetric site. The spin-allowed zero-phonon d–d transitions are accompanied by broad phonon sidebands, which allow the tunability.

B. RARE-EARTH ELEMENTS

The visible absorption spectra of trivalent rare-earth ions consist of forbidden dipole electric f–f transitions, which gain intensity from admixture of the nearest opposite-parity configurations (such as $4f^{n-1}5d^1$) into the $4f^n$

states by the odd crystal field components. The probability of these transitions is therefore lower than in oxides because the $5d$ orbitals lie at higher energy and the crystal field is weaker.

Owing to their high transmission properties and the high symmetry of the rare-earth sites, such fluorides as $LiYF_4$ (point group S_4) and KY_3F_{10} (C_{4v}) are particularly suitable for theoretical investigations of crystal field effects.

The main particularities of the luminescence of fluorides with respect to oxides arise from the following properties:

1. *Weaker crystal field.* The $4f$ orbitals being internal, the influence of the crystal field on their energy levels is small (the splitting of the J levels is of the order of a few hundred cm^{-1}). This small perturbation, however, plays an important role in the concentration quenching of the luminescence. The overlapping of absorption and emission lines, which causes nonradiative losses by cross-relaxation, is reduced in weak crystal field materials. The rare-earth interactions are particularly decreased when the rare-earth polyhedra are isolated from each other. $K_5NdLi_2F_{10}$, which fulfills both conditions, is one of the stoichiometric materials showing the weakest concentration quenching (quenching ratio of 1.7 compared with 2.3 for NdP_5O_{14}) (Lempicki *et al.*, 1979).

The $5d$ orbitals being external, the influence of the host lattice on the energy of the $4f^{n-1}5d^1$ levels is much more pronounced. As a consequence of the weaker nephelauxetic effect and less intense crystal field, the $5d \rightarrow 4f$ emission of Ce^{3+} and Eu^{2+} in fluorides is markedly displaced to short wavelengths with respect to oxides. For Eu^{2+}, in some host lattices ($BaAlF_5$, BaY_2F_8, and $BaSiF_6$, divalent europium being substituted for barium), the shift to high energies is so pronounced that the bottom of the $5d$ band lies above the first excited level of the $4f^7$ configuration, $^6P_{7/2}$; consequently, the emission spectrum consists of intense lines originating from this level instead of the usual $d \rightarrow f$ band (Hewes and Hoffman, 1971; Fouassier *et al.*, 1976). In these low crystal field host lattices, the Sm^{2+} $5d$ orbitals lie above the $^5D_0(4f^6)$ level, likewise giving rise to a line spectrum (Gros and Gaume-Mahn, 1980).

2. *Weaker nonradiative probabilities.* The probability of a nonradiative transition from a level to the next lowest level decreases exponentially with the number of phonons emitted. The weak energies of phonons in fluorides compared with those in oxides and the smaller crystal field splittings reduce the nonradiative decay rates, so that the rates of multiphonon emission are 10 times smaller in LaF_3 (phonons of highest energy, 350 cm^{-1}) than in Y_2O_3 (550 cm^{-1}) (Riseberg, 1976). The quantum efficiency of the visible emission of ions having closely spaced levels, such as Ho^{3+} or Er^{3+}, is therefore much higher in fluorides than in oxides (Weber, 1967).

The existence of various levels decaying radiatively with high efficiency is favorable for the phenomenon of up-conversion, for example, for ir-to-visible conversion. In materials containing ytterbium and erbium, the energy of two ir photons ($\lambda = 0.97$ μm) absorbed by ytterbium can be transferred to an erbium ion, which is promoted to the green-emitting $^4S_{3/2}$, $^2H_{11/2}$ states ($\lambda = 0.52$–0.54 μm) (Auzel, 1966; Sommerdijk, 1971; Caro and Porcher, 1979). Transfer of the energy of three ir photons to a thulium ion allows excitation of the blue-emitting $^1G_4$ level ($\lambda = 0.47$ μm). The best conversion efficiencies have been obtained with fluorides YF_3 and $NaYF_4$ (Auzel and Pecile, 1973; Bril et al., 1975).

3. *Long lifetime of the excited states.* As a result of the low probability of f–f transitions, the trivalent rare-earth ions are characterized in fluorides by long lifetimes.

The high optical quality of fluoride crystals make them particularly attractive for laser applications. The most commonly used host lattice is $LiYF_4$. The shortest and longest wavelengths so far reported for solid lasers have been obtained with this fluoride doped, respectively, with cerium ($\lambda = 325$ nm) (Ehrlich et al., 1978) and holmium ($\lambda = 3.91$ μm) (Esterowitz et al., 1979) as activator. Lanthanide fluorides will perhaps allow laser emission at much shorter wavelength because Nd^{3+}, Er^{3+}, and Tm^{3+} show an efficient $5d \rightarrow 4f$ emission in the vacuum uv at ~ 170 nm when incorporated in fluorides such as LaF_3, YF_3, and $LiYF_4$ (Yang and De Luca, 1976). Using a molecular H_2 laser as a pumping source, one could construct a tunable laser system.

Fluoride materials are of interest for laser fusion because, owing to their low nonlinear index, they can sustain high peak powers without causing beam distortion. The neodymium emission in $LiYF_4$ matches the peak emission wavelength of neodymium phosphate or fluorophosphate glasses used as amplifiers (Le Goff et al., 1978).

Rare-earth-doped alkaline-earth fluorides are used as scintillators for the detection of x rays and charged particles. Incorporation of lanthanide fluorides LnF_3 in electroluminescent ZnS thin films enables one to obtain various colors (Kahng, 1968; Benoît et al., 1981; Ohnishi et al., 1983).

C. IONS WITH ns^2 CONFIGURATION

Ions with ns^2 configuration such as Tl^+ (Mayer et al., 1975) and Pb^{2+} (Arkhangelskaya et al., 1979) show the usual three uv $s \rightarrow p$ absorption bands displaced to high energies with respect to oxides or other halides as a consequence of the weak nephelauxetic effect. Excitation into these bands gives rise to one or two emission bands in the uv range below 300 nm.

V. Color Centers

The nature of color centers formed in alkali and alkaline-earth fluorides has been extensively investigated.

A. Alkali Fluorides

Mostly F centers (electrons trapped at negative-ion vacancies) are formed when alkali fluorides are irradiated by radiations of high energy (x-rays or electron beam) or warmed in alkaline metal vapor. The F centers are stable to ~ 400 K. The absorption band lies at shorter wavelength than in other halides (Buchenauer and Fitcher, 1968).

The probability of radiative emission increases with decreasing size of the anion, so that in alkali fluorides, with the exception of LiF, the F centers have luminescent properties (Bartram and Stoneham, 1975; Leung and Song, 1980). Thermal quenching of the emission occurs between 100 and 150 K. At room temperature, F aggregate centers have luminescent properties. Tunable laser emission has been obtained in the red or near ir (Table IV).

B. ALKALINE-EARTH FLUORIDES

Irradiation of alkaline-earth fluorides at low temperature produces F and V_k (trapped hole) centers. The thermal stability of F centers is much lower than in alkali fluorides. Most F centers are destroyed by recombination with V_k centers between 85 and 140 K.

The F centers in MF_2 fluorides do not show luminescence. When doped with lanthanum, cerium, gadolinium, terbium, lutetium, or yttrium, the MF_2 crystals have photochromic properties. They can be colored either by heating in an alkaline-earth vapor or by x or γ irradiation at 300 K. The "four-band spectrum" has been ascribed to a rare-earth–fluorine vacancy complex that has trapped two electrons to give rise to a neutral center (Staebler and

TABLE IV

Wavelength Range of the Emission of F Aggregate Centers

Host Lattice	Color Center	λ_{em} (μm)	Reference
LiF	F_2	0.665–0.715	Kulinski et al. (1980)
LiF	F_2^+	0.83–1.15	Basiev et al. (1979)
LiF	F_2^-	1.09–1.23	Basiev et al. (1982)
NaF	F_2^+	0.95–1.35	Khulugurov et al. (1978)

Schnatterly, 1971; Anderson and Sabisky, 1971). Under uv irradiation, the center is photoionized, and a new absorption spectrum corresponding to the ionized form appears (Staebler and Kiss, 1969; Alig, 1971). The change in color so produced is reversible. Optical erasure can be carried out by exposing the crystals to visible light. The possibility of using rare-earth-doped MF_2 crystals for information storage has been extensively investigated (Duncan and Staebler, 1977).

Thermoluminescent fluoride crystals ($CaF_2:Mn$, $LiF:Mg$) are used as dosimeters.

References

Alig, R. C. (1971). *Phys. Rev. B* **3**, 536.
Allen, G. C., and Warren, K. D. (1971). *Struct. Bonding (Berlin)* **9**, 49.
Allen, G. C., and Warren, K. D. (1974). *Struct. Bonding. (Berlin)* **19**, 101.
Allen, G. C., and Warren, K. D. (1975). *Coord. Chem. Rev.* **16**, 227.
Anderson, C. H., and Sabisky, E. S. (1971). *Phys. Rev. B* **3**, 527.
Arkhangelskaya, V. A., Lushchik, N. E., Reiterov, V. M., Soovik, Kh. A., and Trofimova, L. M. (1979). *Opt. Spectrosc. (Engl. Transl.)* **47**, 393.
Auzel, F. (1966). *C. R. Hebd. Seances Acad. Sci., Ser. B* **262**, 1016.
Auzel, F., and Pecile, D. (1973). *J. Lumin.* **8**, 32.
Bartram, R. H., and Stoneham, A. M. (1975). *Solid State Commun.* **17**, 1593.
Basiev, T. T., Mirov, S. B., and Prokhorov, A. M. (1979). *Sov. Phys.—Dokl. (Engl. Transl.)* **24**, 384.
Basiev, T. T., Voron'ko, Y. K., Mirov, S. B., Osiko, V. V., Prokhorov, A. M., Soskin, M. S., and Taranenko, V. B. (1982). *Kvantovaya Elektron* **9**, (*Moscow*) 1741.
Benoit, J., Benalloul, P., and Blanzat, B. (1981). *J. Lumin.* **23**, 175.
Bril, A., Sommerdijk, J. L., and De Jager, A. W. (1975). *J. Electrochem. Soc.* **122**, 660.
Buchenauer, C. J., and Fitchen, D. B. (1968). *Phys. Rev.* **167**, 846.
Caro, P., and Porcher, P. (1979). *J. Lumin.* **18/19**, 257.
Chodos, S. L., Black, A. M., and Flint, G. D. (1976). *J. Chem. Phys.* **65**, 4816.
Duncan, R. C., and Staebler, D. L. (1977). *Top. Appl. Phys.* **20**, 133.
Ehrlich, D. J., Moulton, P. F., and Osgood, R. M. (1978). *Opt. Lett.* **4**, 184.
Esterowitz, L., Eckardt, R. C., and Allen, R. E. (1979). *Appl. Phys. Lett.* **35**, 236.
Ferguson, J., Guggenheim, H. J., and Wood, D. L. (1964). *J. Chem. Phys.* **40**, 822.
Ferguson, J., Guggenheim, H. J., and Wood, D. L. (1971). *J. Chem. Phys.* **54**, 504.
Fouassier, C., Latourrette, B., Portier, J., and Hagenmuller, P. (1976). *Mater. Res. Bull.* **11**, 933.
Garlick, G. F., and Sayer, M. (1962). *J. Electrochem. Soc.* **109**, 678.
Greenough, P., and Paulusz, A. G. (1979). *J. Chem. Phys.* **70**, 1967
Gros, A., and Gaume-Mahn, F. (1980). *C. R. Hebd. Seances Acad. Sci.* **291**, 141.
Hewes, R. A., and Hoffman, M. V. (1971). *J. Lumin.* **3**, 261.
Khang, D. (1968). *Appl. Phys. Lett.* **13**, 210.
Khulugurov, V. M., and Lobanov, B. D. (1978). *Sov. Tech. Phys. Lett. (Engl. Transl.)* **4**, 472.
Kulinski, T., Kaczmarek, F., Ludwiczak, M., and Blaszczak, Z. (1980). *Opt. Commun.* **35**, 120.
Künzel, W., Knierim, W., and Dürr, U. (1981). *Opt. Commun.* **36**, 383.
Langer, T., Krukowska-Fulde, B., and Langer, J. M. (1979). *Appl. Phys. Lett.* **34**, 216..
Le Goff, D., Bettinger, A., and Labadens, A. (1978). *Opt. Commun.* **26**, 108.

Lempicki, A., McCollum, B. C., and Chinn, S. R. (1979). *IEEE Quantum Electron.* **QE-15**, 896.
Leung, C. H., and Song, K. S. (1980). *Solid State Commun.* **33**, 907.
Li, H. H. (1980). *J. Thermophys.* **1**, 97.
Lohr, L. L., and McClure, D. S. (1968). *J. Chem. Phys.* **49**, 3516.
McClure, D. S. (1976). *J. Lumin.* **12/13**, 67.
Mayer, U., Schmid, D., and Seidel, H. (1975). *Phys. Status Solid: B* **70**, 269.
Moulton, P. F., and Mooradian, A. (1979). *Appl. Phys. Lett.* **35**, 838.
Onishi, H., Sakuma, N., and Ieyasu, K. (1983). *J. Electrochem. Soc.* **130**, 2115.
Rehn, V., Burdick, D. L., and Jones, V. O. (1977). *NBS. Spec. Publ. (U.S.).* **509**.
Riseberg, L. A. (1976). *Prog. Opt.* **14**, 130.
Simo, C., Banks, E., and Holt, S. (1969). *Inorg. Chem.* **8**, 1446.
Sommerdijk, J. L. (1971). *J. Lumin.* **4**, 441.
Staebler, D. L., and Kiss, Z. J. (1969). *Appl. Phys. Lett.* **14**, 93.
Staebler, D. L., and Schnatterly, S. E. (1971). *Phys. Rev. B* **3**, 516.
Stephan, G., Le Calvez, Y., Lemonier, J. C., and Robin, S. (1969). *J. Phys. Chem. Solids* **30**, 601.
Tomiki, T., and Miyata, T. (1969). *J. Phys. Soc. Jpn.* **27**, 658.
Weber, M. J. (1967). *Phys. Rev.* **157**, 262.
Williams, F. (1981). *J. Lumin.* **23**, 1.
Yang, K. H., and De Luca, J. A. (1976). *Appl. Phys. Lett.* **29**, 499.

15

Fluorides for Electrochromic Devices

TETSU OI and KATSUKI MIYAUCHI
Central Reserach Laboratory
Hitachi, Ltd.
Tokyo, Japan

JEAN-MICHEL DANCE
Laboratoire de Chimie du Solide du CNRS
Université de Bordeaux
Talence, France

I. Introduction

Electrochromic devices are expected to become important nonemissive display devices in which a thin-film transition metal oxide, typically WO_3, is used as the coloring material. The principle of operation requires double injection of an ion and an electron into the coloring material and can be expressed as

$$WO_3 + x(e^- + M^+) = M_xWO_3$$

Colorless   Blue

INORGANIC
SOLID FLUORIDES

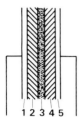

Fig. 1. Schematic layer configuration of an all-solid-state electrochromic cell with nonpolarizable interfaces: 1, transparent electrode (electronic conductor); 2, electrochromic material (electronic and ionic); 3, solid electrolyte (ionic); 4, mixed conductor (electronic and ionic); 5, counterelectrode (electronic). From Zeller (1976).

1 2 3 4 5

where M^+ denotes a lithium ion or proton, and $x \approx 0.1$. In order to supply the M^+ ion and stop the electron, an electrolyte is required. In the case of the all-solid-state device shown in Fig. 1 (Zeller, 1976), a thin-film solid electrolyte is utilized.

The electrochromic display device thus constructed features better clarity and wider viewing angle than a liquid crystal display device. It requires lower power consumption and offers better visibility in a lighter atmosphere than emissive display devices such as light-emitting diodes.

II. Use of Fluorides as Thin-Film Ionic Conductors

As thin-film solid electrolytes for such electrochromic display devices, fluorides appear so far to be the most successful materials. Ease of thin-film preparation, relative working stability, and flexibility of the methods, which leads to high ionic conductivities, seem to be the main reasons for such success.

Fluorides can be used as protonic, anionic, or lithium ion conductors.

A. PROTON CONDUCTORS

Such fluoride thin films as LiF and MgF_2 are proton conductors under moderate partial pressure of H_2O due to some hydrolysis. Following preliminary works of Deb and Witzke (1975) and of Stocker *et al.* (1979), Deneuville *et al.* (1980) extensively studied the influence of evaporation conditions and the role of the water molecule in MgF_2 thin films. Yoshimura *et al.* (1982), moreover, found that gas bubbles, as a decomposition product of H_2O, could be observed in a device after coloration or bleaching. They also interpreted the coloring–bleaching characteristics of the WO_3/MgF_2 system in terms of electronic conduction in such materials (Yoshimura *et al.*, 1983a).

B. ANION CONDUCTORS

In order to construct an all-solid-state device using an iridium oxide thin-film coloring material, which is believed to color by simultaneous injection of an anion and an electron, Rice and Bridenbaugh (1981) utilized evaporated polycrystalline PbF_2 and $PbSnF_4$ thin films. These materials had been found to be fluorine conducting materials by Réau et al. (1978) (see Chapter 12). A response time as short as 0.1 sec was obtained, and the device could be tested for up to 300 cycles of coloring and bleaching.

C. LITHIUM ION CONDUCTORS

Stable lithium ion conducting thin films applicable to electrochromic display devices were found using some double fluorides (Oi and Miyauchi, 1981). Their composition is typically expressed as $mLiF \cdot nAlF_3$, where $n/(m + n)$ ranges from 0.05 to 0.5, as shown in Fig. 2. These thin films are amorphous and have a lithium ion conductivity of $\sim 10^{-4} \, \Omega^{-1} \, cm^{-1}$ at room temperature.

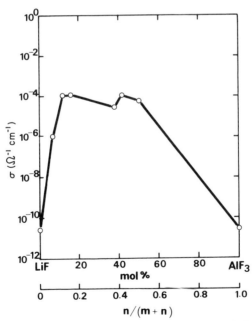

Fig. 2. Relationship between ac ionic conductivity σ and film chemical composition given either in mole percent AlF_3 or as an $n/(m + n)$ ratio ($T = 25°C$; dry N_2). Adapted from Oi and Miyauchi (1981).

The feasibility of using $LiAlF_4$ (i.e., $LiF \cdot AlF_3$) thin film in an electrochromic display device was first tested with a $WO_3/LiAlF_4/LiIn$ sandwich (Oi et al., 1982). A $WO_3/LiAlF_4/2Li_2O \cdot 4FeO \cdot 7WO_3$ layer material leads to a coloration time of 0.5 sec and 10^6 cycles of coloration and bleaching (Uehara et al., 1983).

Furthermore, the idea of using fluorides as cationic conductors in depositing amorphrous thin films has been extended to such other fluoride systems as $NaF \cdot MF_3$ (M = Al or Cr) (Dance and Oi, 1983a) and $LiF \cdot NiF_2$ (Dance and Oi, 1983b). The effect of adding small amounts of di- or trivalent fluorides to LiF has also been investigated (Oi, 1984). An MgF_2 thin film doped with metallic lithium, moreover, has proved to be a good lithium ion conductor (Yoshimura et al., 1983b).

References

Dance, J. M., and Oi, T. (1983a). Mater. Res. Bull. 18, 263–268.
Dance, J. M., and Oi, T. (1983b). Thin Solid Films 104, L71–L73.
Deb, S. K., and Witzke, H. (1975). Proc. Int. Electron. Device Meet., 1975 pp. 393–397.
Deneuville, A., Gerard, P., and Billat, R. (1980). Thin Solid Films 70, 203–223.
Oi, T. (1984). Mater. Res. Bull. 19, 451–457.
Oi, T., and Miyauchi, K. (1981). Mater. Res. Bull. 16, 1281–1289.
Oi, T., Miyauchi, K., and Uehara, K. (1982). J. Appl. Phys. 53, 1823.
Réau, J. M., Lucat, C., Portier, J., Hagenmuller, P., Cot, L., and Vilminot, S. (1978). Mater. Res. Bull. 13, 877–882.
Rice, C. E., and Bridenbaugh, P. M. (1981). Appl. Phys. Lett. 38, 59–61.
Stocker, H. J., Singh, S., Van Uitert, L. G., and Zydzik, G. J. (1979). J. Appl. Phys. 50, 2993–2994.
Uehara, K., Oi, T., and Kudo, T. (1983). Abstr. 44th Jpn. Appl. Phys. Meet., 1983.
Yoshimura, T., Watanabe, M., Kiyota, K., and Tanaka, M. (1982). Jpn. J. Appl. Phys. 21, 128–132.
Yoshimura, T., Watanabe, M., Koike, Y., Kiyota, K., and Tanaka, M. (1983a). Thin Solid Films 101, 141–151.
Yoshimura, T., Watanabe, M., and Koike, Y. (1983b). Jpn. J. Appl. Phys. 22, 157–160.
Zeller, H. R. (1976). In "Nonemissive Electrooptic Displays" (A. R. Kmetz and F. K. von Willisen, eds.), pp. 149–154. Plenum, New York.

16

Nuclear Magnetic Resonance of Fluorides

GÉRARD VILLENEUVE
Laboratoire de Chimie du Solide du CNRS
and École Nationale Supérieure de Chimie et de Physique de Bordeaux
Talence, France

I. Introduction

Many comprehensive textbooks are devoted to nmr with applications to solid-state physics and chemistry (Abragam, 1961; Slichter, 1980), and several hundred articles related to ^{19}F nmr in solid inorganic fluorides have been published. The purpose of this chapter is neither to develop general ideas about nmr nor to draw up an exhaustive catalog of the available literature. Its main purpose is to point out to specialists in the physics and chemistry of fluorides what nmr can bring to the knowledge of the properties of these materials. A few significant examples are given; these were selected because they seem to illustrate the power of the technique as well as its limitations.

A. BASIC PRINCIPLES OF NUCLEAR MAGNETIC RESONANCE

Since its discovery by Bloch *et al.* (1946) and Purcell *et al.* (1946), nmr has proved to be a powerful tool for investigating certain aspects of the local properties of solids. It gives valuable information about local structures, dynamic properties as well as electron transfers.

Nuclear magnetic resonance spectroscopy is a technique based on the nuclear Zeeman effect: all nuclei with odd mass number have a spin I that is an odd integral multiple of $\frac{1}{2}$. Nuclei with an even isotope number may either be spinless (if the nuclear charge is even) or possess an integral spin (if the nuclear charge is odd).

A magnetic field H_0 removes the degeneracy of the $2I + 1$ nuclear levels of an isolated nucleus, the energies of which become

$$E_m = -g_n \beta_n m_I H_0 \tag{1}$$

with respect to the energy in the absence of a magnetic field. In this expression β_n is the nuclear magneton (5.05×10^{-24} erg/G), g_n the nuclear g factor, the value of which is specific for a given nucleus, and m_I the nuclear spin quantum number; $g_n \beta_n / \hbar$ is often called the gyromagnetic ratio γ of the nucleus.

The interlevel spacing is $g_n \beta_n H_0$; according to the selection rule $\Delta m_I = \pm 1$, one readily obtains the fundamental relation

$$\omega_0 = \gamma H_0, \tag{2}$$

where ω_0 is the angular frequency of the oscillating magnetic field $H_1 \cos \omega_0 t$ that is needed to induce allowed transitions between two consecutive spin levels.

In real systems, that is, when the nuclei can no longer be considered isolated, other interactions are superimposed on the Zeeman effect, mainly nuclear dipole–dipole, electric quadrupole, and nucleus–electron interactions. They modify the linewidth, alter the shape, and shift the position of the absorption line.

Taking into account the interactions just mentioned, we can write the total Hamiltonian in the following short form:

$$\hat{\mathscr{H}} = \hat{\mathscr{H}}_Z + \hat{\mathscr{H}}_D + \hat{\mathscr{H}}_Q + \hat{\mathscr{H}}_{n-e}. \tag{3}$$

Here, $\hat{\mathscr{H}}_Z$ is the Zeeman interaction (it gives roughly the position of the absorption line); $\hat{\mathscr{H}}_D$ is the $n-n$ dipolar interaction, generally ~ 1000 times lower than the Zeeman interaction in solids [it brings an important contribution to the linewidth (1–10 G)]; $\hat{\mathscr{H}}_Q$ is the quadrupolar interaction, present only when the nuclear spin is greater than $\frac{1}{2}$; and $\hat{\mathscr{H}}_{n-e}$ describes the magnetic interactions between the investigated nucleus and its electronic environment (it shifts the position of the absorption line).

Each interaction includes, in addition to the static part, a time-fluctuating component, which is responsible for the nuclear relaxation, that is, the facility with which the different components of the magnetization return to the thermal equilibrium destroyed by H_1.

Three relaxation times are used in this chapter: the spin–spin relaxation time T_2 describes the dephasing of the transverse magnetization after an H_1 pulse; the spin–lattice relaxation time T_1 is the time characterizing the recovery of the longitudinal magnetization equilibrium; the spin-lattice relaxation time in the rotating frame $T_{1\rho}$ is the relaxation time when the magnetization is locked along H_1.

B. PROPERTIES OF THE ^{19}F NUCLEUS

The ^{19}F nucleus has an $I = \frac{1}{2}$ spin, with a gyromagnetic ratio γ of 25,165 radians sec^{-1} G^{-1} (i.e., the absorption occurs at a field of ~ 4993 G for a frequency of 20 MHz). Its natural abundance is 100% and its sensitivity very high (0.83 at constant field and for an equal number of nuclei compared with that of a proton, assumed to be unity). The spectra are generally obtained without many difficulties; ^{19}F is thus a good candidate for nmr studies. Owing to the fact that its spin is $\frac{1}{2}$, quadrupolar interaction does not occur and the Hamiltonian reduces to only Zeeman, dipolar, and nucleus–electron contributions.

C. INTERACTIONS INVOLVED

1. Nuclear Diople–Dipole Interaction

The nuclear dipole–dipole interaction is common to all nuclear spins in a solid. In the absence of other interactions, the dipolar fields are the main source of resonance linewidth (of the order of a few gauss).

The expression of the dipole–dipole interaction between two nuclear spins is

$$\hat{\mathcal{H}}_{12} = \frac{\hbar^2 \gamma_1 \gamma_2}{r_{12}^3} \left[\hat{I}_1 \hat{I}_2 - \frac{3(\hat{I}_1 \hat{r}_{12})(\hat{I}_2 \hat{r}_{12})}{r_{12}^2} \right], \tag{4}$$

where γ_1 and γ_2 are the gyromagnetic ratios of nuclei 1 and 2, $\hat{I}_1$ and $\hat{I}_2$ the spin operators, and r_{12} is the internuclear vector.

In a real solid the total hamiltonian is obtained by summation over all pairs of nuclei:

$$\hat{\mathcal{H}}_D = \sum_{i<j} \hat{\mathcal{H}}_{ij}. \tag{5}$$

Except for a few cases, for which well-resolved spectra can be obtained (two or three lines when two or three nuclei are coupled together and isolated from each other), the dipolar interactions give only a broad resonance with little or no resolved structure. Despite these difficulties, valuable information can often be obtained through measurement of the second moment of the absorption line.

If $f(H)$ represents the normalized line shape as a function of the field value H, the first moment of the line gives the center of the resonance (i.e., the average magnetic field):

$$M_1 = H_{av} = \int_0^\infty H f(H) \, dH. \tag{6}$$

The second moment is the mean square width measured from the center of the resonance line:

$$M_2 = \int_0^\infty (H - H_{av})^2 f(H) \, dH. \tag{7}$$

It is evaluated numerically from the experimental absorption line.

Furthermore, the dipolar contribution to the second moment can be calculated theoretically, if the atomic positions are known, from the classical Van Vleck formula (1948). If a crystal contains N identical resonant nuclei in the unit cell, the second moment is

$$M_2 = \frac{3}{4} g_n^2 \beta_n^2 I(I+1) \frac{1}{N} \sum_{j,k} \frac{(1 - 3\cos^2 \theta_{jk})^2}{r_{jk}^6} \tag{8}$$

where r_{jk} is the distance between two nuclei j and k, and θ_{jk} the angle between r_{jk} and the field direction. The sum runs over each nucleus j in the unit cell and all its neighbors k within the crystal.

The crystal may contain different nuclei. They all contribute to the second moment, but their contribution is smaller:

$$M_2 = \frac{1}{3}\frac{1}{N}\sum_{j,f} g_f \beta_n I_f(I_f + 1)\frac{[1 - 3(\cos^2 \theta_{jf})]^2}{r_{jf}^6}. \tag{9}$$

Now the sum runs over all nonresonant nuclei labeled f, and the total second moment is the sum of Eqs. (8) and (9).

In a polycrystalline sample $[3(\cos^2 \theta) - 1]^2$ is replaced by its average value over all orientations (i.e., $\frac{4}{5}$). The final expression for the second moment of a powder thus becomes

$$M_2 = \frac{3}{5}g_n^2\beta_n^2\frac{1}{N}I(I + 1)\sum_{j,k}\frac{1}{r_{jk}^6} + \frac{4}{15}\frac{1}{N}\sum_{j,f}g_f^2\beta_n^2 I_f(I_f + 1)\frac{1}{r_{jf}^6}. \tag{10}$$

Hence, it appears that the second moment, even for polycrystalline samples, is strongly dependent on the internuclear distances and as a consequence on the atomic positions.

2. Nucleus–Electron Interactions

The observed resonance field differs more or less from the theoretical value for a free nucleus. This is called the "screening effect," the "chemical shift," or the "magnetic shift," and it arises from the local magnetic field created by the electronic environment of the nuclei. This additional field, positive or negative with respect to the applied field H_0, has a strength proportional to H_0. Whatever its origin, the effective magnetic field seen by a nucleus becomes

$$H_{eff} = (1 + \sigma)H_0, \tag{11}$$

where σ is called the chemical shift with respect to the resonance of the bare nucleus:

$$\sigma = \Delta H / H_0. \tag{12}$$

II. Structural Investigations

A. APPLICATION OF THE METHOD OF SECOND MOMENTS IN A RIGID LATTICE

It is shown in Section I that the analysis of the second moment of the absorption line could, in principle, give information about atomic positions insofar as it is very sensitive to the interatomic distances, but x-ray diffraction usually is a more powerful technique for obtaining accurate results. Nevertheless, this technique sometimes fails, as in the case of oxide–fluorides, due to the vicinity of F^- and O^{2-} diffusion factors. Only anionic positions can be determined without knowledge of the distribution of each individual species (F^-, O^{2-}) within the anionic sublattice. In such a situation, nmr, from ^{19}F second-moment analysis, can often support classical diffraction technique on the basis of the fact that the ^{16}O nucleus (natural abundance $> 99.96\%$) is spinless. Hence, a simple comparison between the experimental ^{19}F second moment and the theoretical one calculated from Van Vleck's relation for each structural hypothesis gives an unambiguous answer with respect to F^- localization inside the anionic sublattice.

A striking example is the oxide–fluoride ScOF, which crystallizes with the baddeleyite (ZrO_2) structure (Holmberg, 1966). An attempt to distinguish the positions of O^{2-} and F^- on the basis of x-ray diffraction alone was a failure. Two independent sites are occupied by the anions, and three hypotheses can be set up in order to calculate the theoretical second moment (Vlasse *et al.*, 1979):

1. Oxygen in site 1, fluorine in site 2: $M_2 = 14.65 \text{ G}^2$
2. Oxygen in site 2, fluorine in site 1: $M_2 = 10.69 \text{ G}^2$
3. Random distribution between both sites: $M_2 = 12.10 \text{ G}^2$

The experimental value observed, $M_2 = 10.2 \pm 1.0 \text{ G}^2$, showed clearly the existence of an anionic ordering following the second hypothesis (Fig. 1).

In principle, this procedure could be extended to other oxide–fluorides, but if the number of possibilities is too high, in particular if several sites are involved, different hypotheses may lead to close values of the theoretical (Van Vleck, 1948) second moment, and it becomes impossible to decide among them

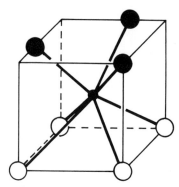

Fig. 1.   Ordering between O^{2-} and F^- in scandium oxyfluoride. Key: ●, Sc; ○, O; ●, F. After Vlasse *et al.* (1969). Copyright 1969 Pergamon Press Ltd.

from nmr considerations alone. In this case the first step consists in eliminating the hypotheses whose second moment deviates too much from the experimental second moment; then the final choice is the consequence of crystal chemistry considerations (Villeneuve *et al.*, 1979).

The second-moment method is based on the relation between the nuclear dipole–dipole interaction and the crystal structure. It must be pointed out that it is limited to diamagnetic materials. As a matter of fact, the interactions between electronic and nuclear spins bring another contribution to the second moment, which becomes frequency dependent. Even for diamagnetic materials, the anisotropy of the chemical shift introduces a small contribution to the linewidth proportional to the frequency. Hence, it is desirable to work at various frequencies in order to estimate this effect.

B. CHEMICAL SHIFT STUDIES

The chemical shift due to electronic currents induced by the applied magnetic field depends on the electronic distribution over the atom. It is sensitive to local environment, and it may be isotropic or anisotropic.

In principle, each fluorine located in a given position in the lattice could give rise to an independent absorption line with a shift characteristic of the nature of its surrounding, because coordination number, distortion, and covalency modify slightly the position of the line with respect to that of a "perfect" F^- (338 ppm toward high fields compared with the resonance of bare ^{19}F).

Nevertheless, the resolution of these lines is very difficult to observe in a solid by conventional methods (continuous wave as well as Fourier transform experiments) because the nuclear dipole–dipole interaction leads to significant broadening of the line (Section I,C), and as a consequence the width of the individual lines is much larger than their separation. Hence, except very few favorable cases (when the lines are far apart), one can observe only the envelope of an unresolved spectrum.

In order to overcome this difficulty new techniques have been developed:

1. It is now possible to work at high frequencies (frequencies up to 400 MHz are available in commercial multinuclear spectrometers) associated with superconducting magnets working at high (fixed) fields. Examples are given in Table I, where the corresponding proton frequency resonance is listed for the sake of comparison.

The advantage of high-frequency/high-field techniques for determining the chemical shift is easy to understand insofar as the dipolar linewidth is field independent, whereas the absolute shift σH_0 is proportional to the applied field. Thus, the higher the applied field, the higher is the separation between two adjacent lines and the better the resolution.

TABLE I

Conditions of ^{19}F Resonance for Some Nuclear Magnetic
Resonance Spectrometers

nmr Frequency for ^{1}H (MHz)	Magnetic Field (G)	nmr Frequency for ^{19}F (MHz)
10.000	2,3488[a]	9.4077
90.000	21,139[a]	84.669
200.000	46,975[b]	188.154
400.000	93,950[b]	376.308

[a] The source of the magnetic field is an electromagnet.
[b] The source of the magnetic field is a superconducting magnet.

Figure 2 illustrates as an example the case of two lines, both having a dipolar width of 5 G (i.e., 20 kHz as expressed in frequency units) with a relative chemical shift difference $\Delta\sigma$ of 100 ppm. The simulated spectra obtained in the hypothesis of Gaussian line shapes shows clearly the increase in resolution with increasing frequency.

2. In spite of this improvement, the use of very high frequencies is not a panacea, because the anisotropic part of the chemical shift, which is another cause of broadening in polycrystalline or vitreous samples, increases with frequency. In order to achieve high resolution in solids one must remove the dipolar interactions as well as the anisotropic chemical shift. For nuclei such as ^{19}F there are two ways of accomplishing this: magic angle spinning and

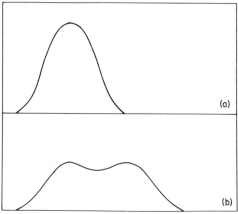

Fig. 2. Example of improvement of resolution by using high frequencies. (a) Spectrum of two lines at 10 MHz; (b) the same at 400 MHz.

multipulse sequences. The first method was developed by Andrew and Newing (1962), who showed that when a sample rotates around an axis that makes an angle $\theta = 54.7°$ with the magnetic field [i.e., when $3(\cos^2 \theta) - 1 = 0$, θ being called the "magic angle"] the dipole–dipole interaction and the anisotropic part of the chemical shift average to zero, provided that the frequency of the rotation is larger than the linewidth as expressed in frequency units. At present, the upper limit of the rotation frequency is 8–10 kHz, limiting the efficiency of the method to lines with a width of less than about 2 to 2.5 G. The second method makes use of appropriate multipulse sequences, which time average the dipole–dipole interactions and the anisotropic chemical shift (Waugh et al., 1968).

1. Solid Hexafluorides

Nuclear magnetic resonance studies on solid hexafluorides have been carried out by conventional methods (Blinc et al., 1963, 1965; Michel et al., 1970). In UF_6 the complex shape of the line obtained at low temperature is analyzed as an overlapping of two components, the intensities of which are in the ratio 2:1 (Fig. 3). It appears that UF_6 octahedra are no longer regular as in the liquid state, but are axially distorted, the two apical and four equatorial fluorine atoms having different chemical shifts. Table II gives as an example

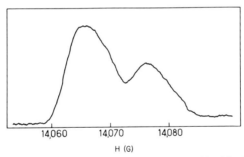

Fig. 3. Absorption spectrum of UF_6 recorded at 54.6 MHz. After Virlet (1971).

TABLE II

Chemical Shifts of Fluorine in UF_6

	Regular Octahedron	Short F—U Bonds (Equatorial Fluorine)	Long U—F Bonds (Apical Fluorine)
Shift with respect to bare ^{19}F nucleus (ppm)	-540	-824	-43

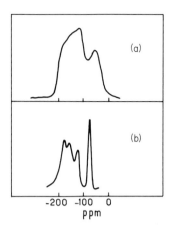

-200 -100   0
ppm

Fig. 4. Spectra obtained for α-SnF$_2$ with b axis perpendicular to the field. (a) Conventional spectrum; (b) high-resolution spectrum. After Le Floch-Durant *et al.* (1982).

the chemical shift of fluorine in regular UF$_6$ (liquid) and its modification when the F—U bonds are either elongated or shortened.

Obviously, this example is rather favorable to this type of study, the large difference between chemical shift ($\Delta\sigma \simeq 800$ ppm) allowing a good resolution even at moderate fields.

2. α-SnF$_2$

The low-temperature α phase of SnF$_2$ crystallizes in the monoclinic system, and there are four crystallographically unequivalent fluorine atoms (McDonald *et al.*, 1976). The unit cell contains four Sn$_4$F$_8$ tetramers and thus 32 different fluorine atoms. If these 32 fluorine atoms were not related by symmetry elements of the space group C2/c, 32 lines would be observed, but the number of magnetically unequivalent fluorine atoms is reduced to 8 and even to 4 when the magnetic field lies in the plane perpendicular to the b axis of the monoclinic cell.

Conventional nmr experiments performed at 90 MHz by Fourier transform technique give a poor resolution. Multipulse sequences used to remove the F–F dipolar interactions by Le Floch-Durant *et al.* (1982) improved the resolution, as shown in Fig. 4, and it is possible to assign each line of the resolved spectrum to each unequivalent fluorine atom on the basis of a single-crystal rotation pattern.

C. DYNAMIC STRUCTURES

1. Second-Moment Reduction

Nuclear magnetic resonance provides valuable information about the dynamic state of reorienting entities from the analysis of the second moment of the absorption line versus temperature.

In the "rigid-lattice regime," the second moment of the line is given by Van Vleck's equation (see Section I,C), but even in the solid state the hypothesis assuming that the atoms (or ions) are fixed in well-defined positions of the lattice is only a rough approximation. In addition to the classical lattice vibrations, reorientation motions may occur in apparently rigid structures.

When a group reorients itself at a frequency large enough compared with the linewidth, one can calculate the reduced second moment of the narrowed line by averaging the $[3(\cos^2 \theta) - 1]^2$ term in Eq. (8). Its value depends on the reorientation mechanism: (1) reorientation around a single symmetry axis of the concerned entity, (2) spherical rotation, or (3) reorientation around all symmetry axes at random.

Comparison between experimental and theoretical reduced second moments reveals the dynamic structure of the group at a given temperature.

The most striking results have been obtained on alkali hexafluorophosphates MPF_6 by Miller and Gutowsky (1963); these salts crystallize in an NaCl-type lattice at room temperature, with the cations and phosphorus atoms occupying the face-centered lattice points. The second moments of the ^{19}F absorption at room temperature are given in Table III. They do not change up to 135°C.

The calculated value for the intragroup contribution of a rigidly fixed PF_6^- ion is 12.2 G^2 (i.e., 10 times more than the observed value, indicating that a dynamic state is involved in the line narrowing. If the rotation occurs around a single symmetry axis of the octahedra (two-, three-, or fourfold), the second moment would be reduced by the factor $\frac{1}{4}[3(\cos^2 \gamma_{ij}) - 1]^2$, where γ_{ij} is the angle between the internuclear vector r_{ij} and the rotation axis.

Miller and Gutowsky obtained 2.48, 2.55, and 2.27 G^2 for intragroup contributions, respectively, for C_2, C_3, and C_4 rotations; in any event the calculated second moments are too large compared with the experimental value, even without taking into account the intergroup interactions.

A more disordered reorientation is the spherical rotation. In this model the PF_6 group is assimilated to a sphere that reorients itself at random around the entire continuous range of axes. As a consequence the intragroup interactions average out to zero, and only the intergroup interactions remain. In this hypothesis the intergroup terms $[3(\cos^2 \theta_{ij}) - 1]/r_{ij}$ are equivalent to the

TABLE III

Second Moments of the ^{19}F Absorption Line Observed in Alkali Hexafluorophosphates[a]

	$NaPF_6$	KPF_6	$RbPF_6$	$CsPF_6$
Second moment (G^2)	1.32 ± 0.03	1.04 ± 0.03	1.07 ± 0.02	0.85 ± 0.02

[a] After Miller and Gutowski (1963).

TABLE IV

Second Moments of ^{19}F Absorption in Alkali Hexafluorophophates
Calculated for Spherical Rotation of $PF_6{}^-$ Groups

	$NaPF_6$	KPF_6	$RbPF_6$	$CsPF_6$
Second moment (G^2)	1.276	1.071	0.971	0.809

rigid-lattice values obtained by putting each nucleus at the center of the sphere (McCall and Douglas, 1960).

The calculated second moments on the basis of the spherical rotation model are given in Table IV. These values are close to the experimental values, but the differences are larger than the standard deviations of Table III.

A reorientation model restricted to rotations about all symmetry axes of the octahedron is then anticipated. It is a "six-position model" with the net effect that each fluorine atom spends one-sixth of the time in each corner of the octahedron. In this case, the intragroup contribution is also averaged to zero, and the intergroup term $[3(\cos^2 \theta_{ij}) - 1]/r_{ij}$ is calculated, giving good agreement with the experimental value for $NaPF_6$ (1.340 G^2). For the other alkali hexafluorophosphates it has been suggested that the P—F bonds are canted from the edges of the unit cell.

Nuclear magnetic resonance also provides information about the temperature dependence of the PF_6 reorientations. The thermal variation of the second moment is given in Fig. 5. One can see that the reorientation occurs more and more easily from sodium to cesium salt (the line is still narrow at 60 K in $CsPF_6$, whereas the broadening begins to occur at 150 K in $NaPF_6$).

The activation energy for PF_6 reorientation can be obtained from the thermal variation of the second moment in the narrowing regime (Fig. 5).

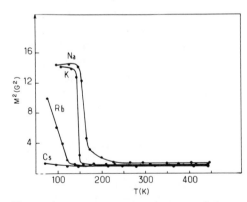

Fig. 5. Fluorine-19 second-moment versus temperature variation as observed for alkali fluorophosphates. After Miller and Gutowsky (1963).

Several approximations have been derived (Section III); one obtains a value of 0.23 ± 0.04 eV for $NaPF_6$.

2. Relaxation Studies

Fluorine-19 spin–lattice relaxation can be a useful tool for examining the possible reorientation of fluorine. When it is due to fluctuation of dipolar interactions between a pair of nuclear spins, the relaxation rate can be expressed as

$$T_1^{-1} = \tfrac{3}{2}\gamma^4\hbar^2 I(I + 1)[J^{(1)}(\omega_0) + J^{(2)}(2\omega_0)], \qquad (13)$$

where γ is the nuclear gyromagnetic ratio, and ω_0 the Larmor angular frequency. The spectral density functions $J^{\alpha}(\omega)$ are the Fourier transforms of the time-correlation functions of the fluctuating part of the dipole coupling tensor. The values of spectral densities can be calculated from the reorientation mechanism. It must be pointed out at this stage that the same equation governs the spin–lattice relaxation rate whatever the nature of the motion (reorientation or diffusion), as will be seen in the next section. The different behaviors come only from the calculated spectral densities $J^{\alpha}(\omega)$.

An example of the use of spin–lattice relaxation rate is the study of the reorientation of the square-pyramidal ion SbF_5^{2-} in K_2SbF_5 (Miyajima *et al.*, 1980). The orthorhombic unit cell (*Cmcm*) contains nearly square pyramidal SbF_5^{2-} units, each having its pseudo-C_4 axis parallel to the crystallographic b axis.

The thermal evolution of the spin–lattice relaxation rate has been investigated in order to elucidate the nature of the reorientation. In Fig. 6 the experimental results obtained by Miyajima *et al.* (1980) at 10.0 and 26.6 MHz show a T_1^{-1} maximum (38.5 sec^{-1}) at 375 K and 10.0 MHz. For $T > 220$ K, T_1^{-1} is proportional to ω_0^{-2}, whereas in the low-temperature region it is proportional to $\omega_0^{-1/2}$. The authors have explained the high-temperature behavior by a model of anisotropic reorientation of the pyramidal SbF_5^{2-} group around its pseudo-C_4 axis. They calculated, therefore, the spectral densities due to this kind of motion, taking into account the intraionic as well as interionic interactions. They showed that a maximum relaxation rate occurs when $\omega_0\tau_c = 0.51$, and assuming an Arrhenius-type activation process,

$$\tau_c = \tau_0 \exp(E_a/kT), \qquad (14)$$

where τ_c is the correlation time of the reorientation mechanism and E_a its activation energy, they were able to calculate the spin–lattice relaxation rate

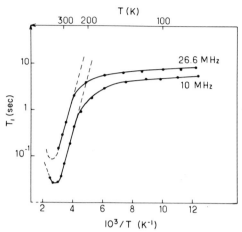

Fig. 6. Fluorine spin–lattice relaxation time of powdered K_2SbF_5. After Miyajima *et al.* (1980).

as indicated by dashed lines in Fig. 6. They used following parameters:

$$\tau_0 = 4.2 \times 10^{13} \quad \text{sec},$$

$$E_a = 0.31 \quad \text{eV}.$$

A rather good agreement is obtained at $T > 220$ K, corroborating the uniaxial (pseudo-C_4) reorientation of SbF_5^{2-}.

The behavior of the spin–lattice relaxation rate observed for K_2SbF_5 below 220 K frequently occurs at low temperature. It cannot be interpreted in terms of molecular reorientation because the extrapolated value due to this mechanism is much lower than the experimental one. A possible predominant mechanism governing the spin–lattice relaxation in this region is often attributed to spin diffusion in the presence of paramagnetic impurities with a concentration as low as a few parts per million. This mechanism predicts an $\omega_0^{-1/2}$ law for T_1^{-1} (De Gennes, 1958); this is actually confirmed by the experimental data.

III. Diffusion of the Fluoride Ion

This topic has been intensively developed in previous review papers including applications to F^- diffusion (see, for instance, Whittingham and Silbernagel, 1978; Berthier, 1979; Ailion, 1981; Brinkman, 1983). Therefore, we restrict this section to the main features of nmr in the presence of diffusive motion, and we point out some practical applications of the technique.

A. ABSORPTION SPECTRA

When the nuclei diffuse with a jump frequency higher than the dipole–dipole interaction (i.e., when $M^{1/2}\tau_c \lesssim 1$), the absorption line is no longer described by the rigid-lattice situation. In the case of a translational motion the effective second moment averages to zero, whereas for simple reorientation motion it takes a finite value (see Section II,C). It appears that the square root of the second moment acts as a time reference, and in most cases motional narrowing occurs when $\tau_c \lesssim 10^{-4}$ sec. In the narrowing regime the second moment at a temperature T depends on the jump frequency at this temperature. Several expressions have been derived (Gutowksy and Pake, 1950; Abragam, 1961; Whittingam *et al.*, 1972), but they give the same qualitative behavior.

B. RELAXATION

When the relaxation is due only to dipolar interactions, it is expressed for a pair of identical spins by the relations

$$T_2^{-1} = \tfrac{3}{8}\gamma^4\hbar^2 I(I+1)[J^{(0)}(0) + 10J^{(1)}(\omega_0) + J^{(2)}(2\omega_0)], \tag{15}$$

$$T_1^{-1} = \tfrac{3}{2}\gamma^4\hbar^2 I(I+1)[J^{(1)}(\omega_0) + J_2(2\omega_0), \tag{16}$$

$$T_{1\rho}^{-1} = \tfrac{3}{8}\gamma^4\hbar^2 I(I+1)[J^{(0)}(12\omega_1) + 10J^{(1)}(\omega_0) + J^{(2)}(2\omega_0)]. \tag{17}$$

The spectral densities $J^{(\alpha)}(\omega)$ can be calculated in principle if the mechanism of the motion is known. Nevertheless, the more sophisticated the model, the more difficult the calculations, which must often be limited to simple cases.

1. Bloembergen, Purcell, and Pound Model

The Bloembergen, Purcell, and Pound (BPP) model (1948) is an extension to solids of the relaxation rates established for viscous liquids. It assumes an isotropic motion of atoms, neglecting the lattice site problems, and it considers only a single diffusion process characterized by a single correlation time thermally activated:

$$\tau_c = \tau_0 \exp(E_a/kT). \tag{18}$$

The calculated relaxation rates versus $1/T$ are shown in Fig. 7. Relaxation rate T_2^{-1} shows a plateau at low temperature corresponding to the rigid-lattice regime when $\tau_c > M_2^{1/2}$; then it decreases with increasing temperature, as does the linewidth. Relaxation time T_1^{-1} includes two symmetric parts with respect to a maximum, which occurs when $\tau_c = 0.62/\omega_0$. Similar behavior is observed for $T_{1\rho}^{-1}$, with a maximum when $\tau_c = \tfrac{1}{2}\omega_1$.

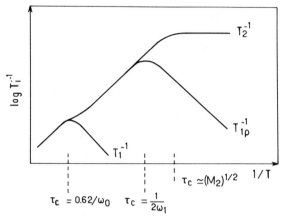

Fig. 7. Schematic behavior of T_1^{-1}, $T_{1\rho}^{-1}$, and T_2^{-1} in the BPP model.

Hence, it appears that the different relaxation rates act as time (or frequency) references for observing ionic motions:

$$M^{1/2} \quad \text{for} \quad T_2^{-1}, \qquad M^{1/2} \simeq 10^4 - 10^5 \quad \text{sec}^{-1},$$

$$\omega_0 \quad \text{for} \quad T_1^{-1}, \qquad \omega_0/2\pi \simeq 10^6 - 10^8 \quad \text{sec}^{-1},$$

$$\omega_1 \quad \text{for} \quad T_{1\rho}^{-1}, \qquad \omega_1/2\pi \simeq 10^5 - 10^6 \quad \text{sec}^{-1}.$$

When $\omega_0 \tau_c \gg 1$ (slow motion with respect to ω_0), T_1^{-1} is proportional to $\omega_0^{-2} \tau_c^{-1}$:

$$T_1^{-1} \propto \omega_0^{-2} \tau_0^{-1} \exp(-E_a/kT). \tag{19}$$

When $\omega_0 \tau_c \ll 1$ (fast motion with respect to ω_0), T_1^{-1} is proportional to τ_c and is frequency independent:

$$T_1^{-1} \propto \tau_0 \exp(E_a/kT). \tag{20}$$

The BPP model is simple and easily applied, even in relatively complex systems.

2. Torrey Model

The Torrey model (1953, 1954) considers the random walk of the atoms from a lattice site to a near-neighboring one. It completely neglects the correlation effects; that is, it assumes that all lattice positions are available for the mobile species. Spectral densities are calculated for cubic lattices, and it has been shown that the values of the maximum relaxation rates depend on the nature of the lattice [simple cubic (s.c.), body-centered cubic (b.c.c.), or face-centered cubic (f.c.c.)].

3. Correlated Jump Models

Correlated jump models (Einsenstadt and Redfield, 1963; Wolf, 1977; Bustard, 1980) take into account correlation effects: the ion can jump only into a site that is empty. The number and distribution of vacancies are thus important parameters governing the jump probability. Wolf calculated T_1^{-1} and $T_{1\rho}^{-1}$ supposing the number of vacancies to be very low; his model has been extended to any concentration of vacancies, but all calculations have been performed only for cubic Bravais lattices.

C. EXAMPLES

Extensive nmr studies have been carried out on fluoride ionic conductors to determine the jump parameters (activation energy and preexponential factor), to elucidate the jumping mechanisms, and to test new models (see Chapter 12).

1. Materials with Fluorite-Type Structure

The compound β-PbF$_2$ is one of the most thoroughly investigated materials with fluorite-type structure due to intrinsic vacancies in the lattice. Continuous-wave experiments show that the absorption line begins to narrow at room temperature (Fig. 8). Linewidth and second-moment analysis in the narrowing regime give an activation energy of 0.63 eV, quite comparable to that obtained from conductivity measurements (Schoonman *et al.*, 1975). Relaxation rates have been determined by Boyce *et al.* (1977). Below 250°C the experimental results agree well with the BPP model. Above this temperature some discrepancies appear (Fig. 9): (1) T_2^{-1} and $T_{1\rho}^{-1}$ begin to increase and then decrease again; (2) T_2^{-1}, $T_{1\rho}^{-1}$, and T_1^{-1} are not equal in the "extreme narrowing regime"; (3) T_1^{-1} increases above 450°C.

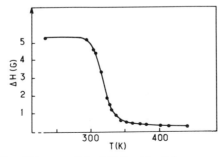

Fig. 8. Linewidth versus T for β-PbF$_2$. After Schoonman *et al.* (1975).

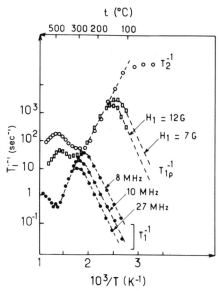

Fig. 9. Relaxation rates T_2^{-1}, T_1^{-1}, and $T_{1\rho}^{-1}$ versus $1000/T$ for different values of H_1 and ω_0. After Boyce *et al.* (1977). Copyright 1977 Pergamon Press Ltd.

A tentative explanation of this apparently anomalous behavior has been proposed by Boyce *et al.* They assume that two kinds of motions occur in β-PbF_2. The faster could be due to vacancy diffusion within the normal F^- sublattice. A slower exchange between F^- ions involving normal and interstitial positions could be superimposed on the first motion; at high temperature the first mechanism would be too fast to govern the relaxations. This assertion could explain the behavior of T_2^{-1}, which would never increase with T if it were due only to conventional nuclear dipole–dipole inter-actions.

Nevertheless, Hogg *et al.*, (1977) seem to have supplied a better answer to this problem by doping PbF_2 with small amounts of paramagnetic impurities (Mn^{2+}). They detected the same T_2^{-1} anomaly as Boyce *et al.* in nominally pure PbF_2, and they showed that it shifted toward low temperature at rising impurity rate. Hence, they were able to explain the anomalies in relaxation results by the presence of paramagnetic impurities with concentrations as low as 10 ppm or even less.

For a long time, one of the most frequently discussed questions concerning the jump mechanism in fluorite-type materials was the location of the crystallographic sites (normal, interstitial, or both) involved in the F^- jumps. Figueroa *et al.* (1978), using an encounter model that takes into account correlated motions (Wolf, 1977) and assuming a vacancy diffusion process, obtained a diffusion coefficient from nmr relaxation data that was in rather good agreement with that calculated from ionic conductivity and the Nernst–Einstein relation.

This hypothesis might actually be true, at least in the case of BaF_2, which is a poor ionic conductor with small disorder, so that Wolf's model based on a small number of vacancies is applicable. Nevertheless, the problem of strongly disordered materials is quite different. In the $Rb_{1-x}Bi_xF_{1+2x}$ fluorite-type solid solution, experimental $T_{1\rho}^{-1}$ values suggest a random walk process in an fcc lattice, thus involving both normal and interstitial sites of the structure (Villeneuve *et al.*, 1985). This analysis confirms the conclusion of neutron diffraction investigations performed on the same materials (Soubeyroux *et al.*, 1982; see also Chapter 12).

2. Materials with Tysonite Structure

Tysonite LaF_3 crystallizes in trigonal symmetry in which there are three distinct fluorine positions, F_1, F_2, and F_3, in the ratio $12:4:2$ (Maximov and Shulz, quoted by Roos *et al.*, 1983). Earlier nmr experiments revealed an equivalence of the dynamic behavior on sublattices 2 and 3 (Goldman and Shen, 1966), so that we can consider only two sublattices, F_1 and F_2, with occupancy rates in the ratio $2:1$.

Goldman and Shen, from an analysis of the T_2^{-1} spin–spin relaxation rate (Fig. 10) and cross-relaxation experiments, derived a model based on a prevailing F^- motion on one F^- sublattice, with slow exchange between the two sublattices. They determined the ratio F fast/F slow $= 2$ and concluded that the motion of F_1^- was the fastest. The anomalous behavior of T_2^{-1} was related to the exchange between F_1^- and F_2^- sublattices. It must be observed that T_2 experiments have been performed on LaF_3 single crystals doped with $1\% \ Pr^{3+}$ in order to lower the spin–lattice relation time and to make the measurements easier. Hence, it may be asked why the anomalous behavior of

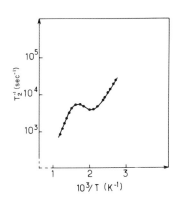

Fig. 10.   Temperature dependence of the relaxation rate T^{-1} of the mobile spins in LaF_3, with H_0 parallel to the c axis. After Goldman and Shen (1966).

T_2^{-1} above 500 K is not due to the influence of dilute paramagnetic impurities. Aalders et al. (1983) have reinvestigated the spin–spin relaxation rate on pure LaF_3; they claim that the fast motion occurs within the F_2 sublattice. This example shows that the mechanism of ionic conduction in tysonite is far from being resolved.

3. Nuclear Magnetic Resonance: A Powerful Technique for Elucidating the Nature of Mobile Species in Ionic Solids

The originality of nmr comes from its selectivity; each nucleus is characterized by its own gyromagnetic ratio, and it can be studied separately, whereas conductivity measurements reflect only the bulk properties of a material without identifying the ion responsible for its conduction properties.

In substituted $CaF_{2-x}H_x$ fluorides, thermal evolution of the ^1H- and ^{19}F-nmr linewidth in the same range of temperature showed unambiguously that the ionic conductivity was due to H^- mobility (Villeneuve, 1979). In thallium zirconium fluoride $TlZrF_5$, the conductivity results from the diffusion of Tl^+ cations, despite their size (Alizon et al., 1983). Cationic motion also occurs in Li_3ThF_7 (Avignant et al., 1983).

In fluoride glasses containing lithium, nmr has proved that both Li^+ and F^- diffuse in the lattice (Senegas et al., 1984).

IV. Chemical Bonding in Fluorides Investigated by Nuclear Magnetic Resonance

In diamagnetic materials, electronic currents give rise to a shift of the absorption line with respect to that of the bare nucleus (Section II,B). The presence of neighbors with single electrons is another source of shift for the absorption line. The electronic spin creates at the nuclei of nonmagnetic ions a local field produced by electron–nucleus dipole–dipole and hyperfine interactions. Hence, the nuclei experience an effective field

$$H_{eff} = H_0 + H_D + H_{HF},$$

where H_0, H_D, and H_{HF} are, respectively, the applied, dipolar, and hyperfine fields.

A. ELECTRON–NUCLEUS DIPOLE–DIPOLE INTERACTIONS

The dipolar field at the nucleus is given by the relation (Shulman and Knox, 1960)

$$H_D = \sum_i \frac{\langle \mu_i \rangle}{r_i^3} [3(\cos^2 \theta_i) - 1]. \tag{21}$$

The summation runs over all magnetic ions; r_i is the distance between the ith magnetic ion and the resonant nucleus, θ_i the angle between the axis ion nucleus and the applied field, and $\langle \mu_i \rangle$ the thermodynamic average of the magnetic moment of the ions with spin S_i.

In the paramagnetic region the dipolar field can be expressed as

$$H_D = \frac{H_0 \chi_m}{N} \sum_i \frac{3(\cos^2 \theta_i) - 1}{r_i^3}, \tag{22}$$

where χ_m is the molar susceptibility, and N Avogadro's number. Thus, the dipolar field is proportional to the applied field. It can be calculated if the structure has been determined, and for a single crystal its value also depends on the orientation.

It is important to observe for a polycrystalline sample that the line broadens but that its centroid is not shifted because the term $3(\cos^2 \theta) - 1$ averages to zero.

B. HYPERFINE INTERACTIONS

Their origins lie in the interactions between the magnetic moment of the nucleus and the moment of the single-electron spins in the orbitals of the resonant atom. When that atom has a closed shell structure, the observation of a hyperfine interaction is a direct evidence of an electron transfer toward the orbitals of the magnetic ion.

In this eventuality the hyperfine field is given by the following relation (assuming the absence of spin–orbit coupling):

$$H_{HF} = \frac{-\chi_m H_0}{N g_n \beta_n g_e \beta_e} \left[A_s + \sum_p A_p (3 \cos^2 \theta_p) - 1 \right], \tag{23}$$

where A_s is the "contact" interaction requiring a finite spin density at the nucleus. A nonzero value of A_s involves some covalency between the $2s$ orbitals of the fluoride ion and the paramagnetic ion since at the nucleus the only nonvanishing wave functions are those describing s orbitals. Variable A_p is the interaction between unpaired spins in $2p$ orbitals and their nucleus. Since there is no electron density at the nucleus, the interaction is dipolar; thus it possesses an anisotropic character.

The hyperfine field is proportional to H_0, so that the effective field seen by the nucleus can be written as

$$H_{eff} = H_0(1 + \sigma), \tag{24}$$

where σ represents the relative magnetic shift:

$$\sigma = \frac{H_{eff} - H_0}{H_0}. \tag{25}$$

It can be separated into two parts:

$$\sigma_{iso} = \frac{-\chi_m}{N g_n \beta_n g_e \beta_e} A_s, \tag{26}$$

$$\sigma_{aniso} = \frac{\chi_m}{N} \left(\sum_i \frac{3(\cos^2 \theta_i) - 1}{r_i} - \frac{1}{g_n \beta_n g_e \beta_e} \sum_p A_p [3(\cos^2 \theta_p) - 1] \right). \tag{27}$$

The position of the line in a single crystal and its shape for a powder sample give σ_{iso} and σ_{aniso}; from their plot versus molar susceptibility with temperature variation as an implicit variable, one can obtain A_s and A_p. In order to be more precise, one must distinguish two hyperfine interactions involving p orbitals: the A_σ interaction arises from unpaired spins in the fluorine p orbitals, which lie along the cation–F^- axis; the A_π interaction arises from electrons in the p orbitals perpendicular to this direction. The analytic expression of the 2s and 2p wave functions of the fluorine atom are well known, and it is possible to calculate the hyperfine interaction between one electron in a 2s or 2p shell and the ^{19}F nucleus (Shulman and Sugano, 1963):

$$a_{2s} = 1.503 \quad cm^1, \tag{28}$$

$$a_{2p} = 0.0429 \quad cm^{-1}. \tag{29}$$

A comparison between experimental A and quantum mechanical values a_{2s} and a_{2p} gives the fraction of unpaired electrons in the fluorine orbitals. As in isolated F^- ions the number of spin-up atoms is equal to that of the spin-down atoms; there is no polarization, and the hyperfine field at the nucleus vanishes. If a hyperfine interaction is observed, it comes from an electron transfer to the orbitals of the magnetic cation; this transfer is "up" or "down" according to the Pauli principle and Hund's rule, and as a result a spin polarization occurs. Thus, the fraction of unpaired electrons corresponds to the fraction of electrons transferred to the magnetic ions.

It appears that nmr of ^{19}F in paramagnetic systems is a suitable method for the quantitative study of covalency as well as superexchange.

C. EXAMPLES

Since the pioneering work of Shulman and Jaccarino (1957) on ^{19}F nmr in paramagnetic MnF_2, many studies have been carried out to determine the spin densities f_s and f_p in the 2s and 2p orbitals of fluorine (Turov and Petrov, 1972). Symmetry considerations led to a separation of f_p into two parts; f_σ from σ bonds and f_π from π bonds.

An interesting comparison has been made of the percentage of unpaired 2s and 2p electrons between MnF_2 and NiF_2 (Shulman, 1961), which crystallize

TABLE V

Spin Densities in Perovskites $KMnF_3$, $KNiF_3$, and K_2NaCrF_6

Spin Density	$KMnF_3$	$KNiF_3$	K_2NaCrF_6
f_s (%)	0.52 ± 0.02	0.50 ± 0.05	-0.02 ± 0.01
$f_\sigma - f_\pi$ (%)	0.2 ± 0.1	4.95 ± 0.6	-4.90 ± 0.8

[a] After Shulman and Knox (1960).

with the rutile structure. In this case the measurements can determine the spin density f_s in $2s$ orbitals, but only the difference $f_\sigma - f_\pi$ in occupancy of the p_σ orbitals as compared with the p_π orbitals (Tinkham, 1956). Experimental f_s values are quasi identical for MnF_2 ($f_s \simeq 0.5\%$ in both cases), but $f_\sigma - f_\pi$ varies dramatically from MnF_2 ($0.2 \pm 0.3\%$) to NiF_2 ($4.1 \pm 0.4\%$). In MnF_2, the Mn^{2+} ion (d^5) is capable of forming σ bonds as well as π bonds with the F^- p orbitals; in NiF_2, the $3d^8$ configuration of Ni^{2+} in the approximately octahedral field means that the two unpaired electrons are in $d_{x^2-y^2}$ and d_{z^2} orbitals, which can form only σ bonds F^-. As a consequence, experiments on NiF_2 give f_σ, whereas $f_\sigma - f_\pi$ is obtained from MnF_2. It appears that the small value of $f_\sigma - f_\pi$ in MnF_2 arises from the cancellation of f_σ by f_π, the surprising conclusion being that p_π bonding, when allowed, is almost as important as p_σ bonding.

The comparison between σ and π bonds has also been made in fluorides with perovskite structure (Shulman and Knox, 1960). These authors studied the ^{19}F-nmr shift in single crystals of $KMnF_3$, $KNiF_3$, and K_2NaCrF_6. The Mn^{2+} ion gives rise to both σ and π bonds with $2p$ F^- orbitals, whereas Ni^{2+} forms only σ bonds, and Cr^{3+} only π bonds (we consider, of course, only the "magnetic" bonds"). Table V gives the spin densities obtained from the three fluorides from ^{19}F-nmr experiments.

Spin density f_s comes from interactions between the $2s$ orbital of F^- and e_g orbitals of the magnetic cation, so its value of close to zero for K_2NaCrF_6 is not surprising. The values of $f_\sigma - f_\pi$ confirm the unexpectedly large interactions.

References

Aalders, A. F., Polman, A., Arts, A. F. M., and De Wijn, H. W. (1983). *Solid State Ionics* **9–10**, 539–542.

Abragam, A. (1961). "The Principles of Nuclear Magnetism." Oxford Univ. Press (Clarendon), London and New York.

Ailion, D. C. (1981). *In* "Nuclear and Electron Resonance Spectroscopy Applied to Material Science" (E. N. Kaufmann and G. K. Shenoy, eds.), pp. 55–68. Elsevier North-Holland, New York.

Alizon, J., Battut, J. P., Dupuis, J., Robert, H., Mansouri, I., and Avignant, D. (1983). *Solid State Commun.* **47**, 969–972.

Andrew, E. R., and Newing, R. (1958). *Proc. Phys. Soc., London* **72**, 959–963.

Avignant, D., Metin, J., Djurado, D., Cousseins, J. C., Battut, J. P., Dupuis, J., and Hamdi, M. (1983). *Stud. Inorg. Chem.* **3**, 271–274.

Berthier, C. (1979). *In* "Fast ion Transport in Solids Electrodes and Electrolytes" (P. Vashishta, J. N. Mundy, and J. K. Shenoy, eds.), pp. 171–176. Elsevier North-Holland, New York.

Blinc, R., Marinkovic, V., Pirkmajer, E., Zupancic, I., and Maricic, S. (1963). *J. Chem. Phys.* **38**, 2474–2477.

Blinc, R., Pirkmajer, E., Zupancic, I., and Rigny, P. (1965). *J. Chem. Phys.* **43**, 3417–3418.

Bloch, F., Hansen, W. W., and Packard, M. (1946). *Phys. Rev.* **69**, 127.

Blombergen, N., Purcell, E. M., and Pound, R. V. (1948). *Phys. Rev.* **73**, 679–705.

Blombergen, N., Purcell, E. M., and Pound, R. V. (1948). *Phys. Rev.* **73**, 679–705.

Boyce, J. B., Mikkelsen, J. C., Jr., and O'Keefe, M. 1977). *Solid State Commun.* **21**, 955–958.

Brinkman, D. (1983). *In* "Progress in Solid Electrolytes" (T. A. Wheat, A. Ahmad and A. K. Kuriakose, eds.), pp. 1–26.

Bustard, L. D. (1980). *Phys. Rev. B* **22**, 1–11.

De Gennes, P. G. (1958). *J. Phys. Chem. Solids* **7**, 345.

Eisenstadt, M., and Redfield, A. G. (1963). *Phys. Rev.* **132**, 635–643.

Figueroa, D. R., Chadwick, A. V., and Strange, J. H. (1978). *J. Phys. C* **11**, 55–73.

Goldman, M., and Shen, L. (1966). *Phys. Rev.* **144**, 321–331.

Gutowsky, H. S., and Pake, G. E. (1950). *J. Chem. Phys.* **18**, 162–170.

Hogg, S. P., Vernon, S. P., and Jacarino, V. (1977). *Phys. Rev. Lett.* **39**, 481–484.

Holmberg, B. (1966). *Acta Chem. Scand.* **20**, 1082–1088.

Le Floch-Durand, M., Haeberlen, U., and Müller, C. (1982). *J. Phys. (Orsay, tr.)* **43**, 107–112.

McCall, D. W., and Douglas, D. C. (1960). *J. Chem. Phys.* **33**, 777–778.

McDonald, R. C., Ho Kuen Hau, H., and Eriks, K. (1976). *Inorg. Chem.* **15**, 762–765.

Michel, J., Drifford, M., and Rigny, P. (1970). *J. Chim. Phys. Phys.-Chim. Biol.* **67**, 31.

Miller, G. R., and Gutowsky, H. S. (1963). *J. Chem. Phys.* **39**, 1983–1994.

Miyajima, S., Nakamura, N., and Chihara, H. (1980). *J. Phys. Soc. Jpn.* **49**, 1867–1873.

Purcell, E. M., Torrey, H. C., and Pound, R. V. (1946). *Phys. Rev.* **69**, 37–38.

Roos, A., Aalders, A. F., Schoonman, J., Arts, A. F. M., and de Wijn, H. W. (1983). *Solid State Ionics* **10–11**, 571–574.

Schoonman, J., Ebert, L. B., Hsieh, C-H., and Huggins, R. A. (1975). *J. Appl. Phys.* **46**, 2873–2876.

Senegas, J., Reau, J. M., Aomi, T., and Poulain, M. (1984). Private communication (to be published in *J. Solid State Chem.*).

Shulman, R. G. (1961). *Phys. Rev.* **121**, 125–143.

Shulman, R. G., and Jaccarino, V. (1957). *Phys. Rev.* **108**, 1219–1231.

Shulman, R. G., and Knox, K. (1960). *Phys. Rev.* **119**, 94–101.

Shulman, R. G., and Sugano, S. (1963). *Phys. Rev.* **130**, 506–516.

Slichter, C. P. (1980). *In* "Springer Series in Solid State Sciences" (M. Cardona, P. Fulde, and H. J. Queisser, eds.). Springer-Verlag, Berlin and New York.

Soubeyroux, J. L., Reau, J. M., Matar, S., Villeneuve, G., and Hagenmuller, P. (1982). *Solid State Ionics* **6**, 103–111.

Tinkham, M. (1956). *Proc. R. Soc. London, Ser. A* **236**, 535–548.

Torrey, H. C. (1953). *Phys. Rev.* **92**, 962–969.

Torrey, H. C. (1954). *Phys. Rev.* **96**, 690.

Turov, E. A., and Petrov, M. P. (1972). "Nuclear Magnetic Resonance in Ferro- and Antiferromagnets." Wiley, New York.

Van Vleck, J. H. (1948). *Phys. Rev.* **74**, 1168–1183.

Villeneuve, G. (1979). *Mater. Res. Bull.* **14**, 1231–1234.
Villeneuve, G., Echegut, P., and Chaminade, J. P. (1979). *Mater. Res. Bull.* **14**, 691–695.
Villeneuve, G., Asai, T. Mater, S., and Relau, J. M. (1985). *Phys. Status Solidi*, in press.
Virlet, J. (1971). Thesis, Univ. of Paris XI–Orsay.
Vlasse, M., Saux, M., Echegut, P., and Villeneuve, G. (1979). *Mater. Res. Bull.* **14**, 807–812.
Waugh, J. S., Huber, L. M., and Haeberlen, U. (1968). *Phys. Rev. Lett.* **20**, 180–182.
Wittingham, M. S., and Silbernagel, B. G. (1978). *In* "Solid Electrolytes" (P. Hagenmuller and W. Van Gool, eds.), pp. 93–107. Academic Press, New York.
Whittingham, M. S., Connel., P. S., and Huggins, J. (1972). *J. Solid State Chem.* **5**, 321–327.
Wolf, D. (1977). *J. Phys. C* **10**, 3345.

17

Mössbauer Spectroscopy
of Fluoride Compounds

P. B. FABRITCHNYI

Department of Chemistry
Moscow State University
Moscow, USSR

I. Introduction

Progress in the application of Mössbauer spectroscopy to the fields of physics and chemistry of solid fluorides is due to the feasibility of acquiring various kinds of information about electronic shells of the resonant atom by analysis of different interactions of the Mössbauer nucleus with its environment. These may be observed in the spectrum via the isomer shift and the quadrupole and magnetic splittings of nuclear levels. Because the informativeness of the spectra depends largely on the nature of both the material and the isotope used, studies of fluorides are generally performed using very sensitive and experimentally easy ^{57}Fe resonance in magnetically ordered iron compounds involving simultaneously all the hyperfine parameters.

INORGANIC
SOLID FLUORIDES

II. Iron-57 Mössbauer Spectroscopy of Fluorides

Mössbauer resonance of ^{57}Fe has been used for the determination of valence and spin states of iron atoms, as well as the characterization of crystallographic sites and polyhedra distortions, in studies of phase transitions, space orientation of the magnetic moments, spin canting, etc. The investigation of two perovskite-type fluorides, $KFeF_3$ and $RbFeF_3$, performed by Davidson et al. (1973) and Someya and Yto (1981) provides a good example of the extremely varied information that can be obtained by means of the Mössbauer effect. In more recent years, along with such traditional applications, ^{57}Fe Mössbauer systematic studies have taken two principal directions: low-dimensional magnetic interactions and properties of amorphous ferric fluorides.

A. LOW-DIMENSIONAL ANTIFERROMAGNETS

Space isolation of magnetic cation sheets in layer-type fluorides results in their more or less pronounced two-dimensional (2D) behavior. Theoretical studies predicted specific features of magnetism in such crystals (see Chapter 9), and these predictions were largely checked by Mössbauer measurements. In fact, Mössbauer resonance makes possible an easy determination of the 3D ordering temperature T_N and of the temperature variation of magnetization by measurement of the hyperfine field $H_{hf} = f(T)$. The experiments performed in the vicinity of T_N provide the value of the critical exponent β and enable one to characterize the relative importance of 2D and 3D magnetic correlations at $T > T_N$. The results obtained for a large number of fluorides exhibiting 2D magnetic properties have been compiled by Tressaud and Dance (1982).

In layered fluorides with general formula $AFeF_4$ ($A = K$, Cs, Rb, or NH_4), Mössbauer experiments have also shown significant quadrupole interactions for Fe^{3+} ions, their origin being interpreted in terms of the polarizable point charge model by Teillet et al. (1982).

More recently, ^{57}Fe resonance has been intensively applied to investigations of chain-structure fluorides with quasi-1D behavior. The results obtained up to 1981 for $AFeF_5$ compounds ($A = K_2$, Cs_2, Rb_2, N_2H_6, or Ca) can be found in the survey mentioned (Tressaud and Dance, 1982); subsequent studies concerned refined experiments on K_2FeF_5 single crystals (Cooper et al., 1982; Boersma et al., 1982) and a new 1D antiferromagnet, β-$Rb_2FeF_5 \cdot H_2O$ (Fourquet et al., 1982). In such fluorides Mössbauer measurements have shown the existence of extremely low 3D ordering temperature, anomalously reduced values of the saturation H_{hf} at ^{57}Fe nuclei (zero-point spin reduction), and variations of T_N with magnitude and direction of the applied magnetic field. All these results provide a Mössbauer resonance confirmation of the validity of the spin-wave theory predictions.

B. AMORPHOUS FERRIC FLUORIDES

A Mössbauer contribution to the investigation of amorphous materials is especially promising, because conventional diffraction methods are insufficiently informative in this case. Experiments on amorphous FeF_3 and $AFeF_4$ compounds (A = Na, K, or Rb) carried out by Henry et al. (1980) and Eibschütz et al. (1981) demonstrate the capacity of ^{57}Fe resonance not only to display the amorphization but also to provide an experimental basis for checking different models of disordered fluorides. A discussion (Czjzek, 1982) and some calculations (Lines, 1982) have stimulated new efforts to establish the extent to which the distortions of anionic octahedra are random in amorphous ionic materials. Moreover, the study of partly crystalline FeF_3 thin films (Lachter et al., 1980) has given rise to the suggestion that a substantial number of the Fe^{3+} ions are in tetrahedral coordination, accounting for texture changes during crystallization processes.

III. Other Mössbauer Isotopes in the Investigation of Fluorides

The number of studies carried out on fluorides of elements other than iron is rather limited due to lower sensibility or difficulties in making resonance observations for most Mössbauer nuclei. Generally, since the discovery of a new resonant nucleus, the simplest fluorides have been studied among other inorganic compounds only in order to establish the correlations of isomer shifts and quadrupole splittings on the electronegativity of ligands. More specific studies (Ballard et al., 1978; Birchall et al., 1981) have dealt with tin (^{119}Sn) or antimony (^{121}Sb) compounds. The data obtained concern the analysis of mixed-valence states and the determination of the quadrupole coupling constant, providing information about the p_z character of a lone pair and the bond hybridization of the metallic atom in various environments.

Although all problems concerning ^{57}Fe resonance are at present still far from being solved, in the near future one can expect, a relative increase in the investigations on other nuclei.

IV. Diamagnetic Mössbauer Probe Ions

Doping with impurities makes it possible not only to extend investigations to materials not possessing any resonant isotope, but also to acquire information that cannot be obtained when the Mössbauer element is a constituent component. The main advantage of diamagnetic probe ions (e.g., $^{119}Sn^{4+}$, $^{121}Sb^{5+}$, and $^{125}Te^{6+}$) introduced as impurities in a magnetically ordered lattice is the exceptional sensibility of such ions to their cationic environment. Because the hyperfine field transferred at the nucleus of the nonmagnetic ion results from exchange interactions with its magnetic neighbors, it enables one to detect the impurity insertion in the material and to investigate different processes occuring in the impurity environment.

Such studies, so far performed mainly on magnetic oxides, become more sophisticated in the case of fluoride compounds because the high volatility of tin, antimony, and especially tellurium fluorides makes the doping processes difficult. A possible method of doping is based on the fluorination of an appropriate oxide in which probe ions were previously introduced by coprecipitation. In this way ferric trifluoride doped with $^{119}Sn^{4+}$ was obtained (Fabritchnyi et al., 1973). The spectra showed an important decrease in the H_{hf} value at $^{119}Sn^{4+}$ nuclei and parallel diminution of the isomer shift, in agreement with the expected weakening of covalent charge transfer of polarized electrons via Fe–F–Sn chains as compared with Fe–O–Sn chains. A more recent study of FeF_3 doped with both $^{119}Sn^{4+}$ and Mg^{2+} (Fefilatiev, 1979) showed the appearance of different impurity pairs and nonmagnetic clusters, revealing an increasing tendency toward local compensation of heterovalent impurity charges with enhancement of the ionicity of chemical bonds.

Finally, it should be mentioned that the stabilization of $^{119}Sn^{2+(4+)}$ probe ions on the surface of a ferromagnetic oxide (Fabritchnyi et al., 1981) provides a new possibility for investigating impurity behavior in mixed oxyfluoride layers.

References

Ballard, J. G., Birchall, T., Gillespie, R. J., Maharajh, E., Tyrrer, D., and Vikris, J. E. (1978). *Can. J. Chem.* **56**, 2417–2421.

Birchall, T., Dénès, G., Ruelenbauer, K., and Pannetier, J. (1981). *J. Chem. Soc., Dalton Trans.* pp. 1831–1836.

Boersma, F., Cooper, D. M., de Jonge, W. J. M., Dickson, D. P. E., Johnson, C. E., and Tinus A. M. C. (1982). *J. Phys. C* **15**, 4141–4145.

Cooper, D. M., Dickson, D. P. E., and Johnson, C. E. (1982). *J. Phys. C* **15**, 1025–1033.

Czjzek, G. (1982). *Phys. Rev. B* **25**, 4908–4910.

Davidson, C. R., Eibschütz, M., and Guggenheim, H. J. (1973). *Phys. Rev. B* **8**, 1864–1880.

Eibschütz, M., Lines, M. E., Van Uitert, L. G., Guggenheim, H. J., and Zydzik, G. J. (1981). *Phys. Rev. B* **24**, 2343–2348.

Fabritchnyi, P. B., Dance, J. M., Ménil, F., Portier, J., and Hagenmuller, P. (1973). *Solid State Commun.* **13**, 655–658.

Fabritchnyi, P. B., Protskii, A. N., Gorkov, V. P., Tran Minh Duc, Demazeau, G., and Hagenmuller, P. (1981). *Sov. Phys.—JETP (Engl. Transl.)* **54**, 608–611.

Fefilatiev, L. P. (1979). Ph.D. Thesis, Univ. of Moscow (abstract).

Fourquet, J. L., de Pape, R., Teillet, J., Varret, F., and Paraefthymiou, G. C. (1982). *J. Magn. Magn. Mater* **7**, 209–214.

Henry, M., Varret, F., Teillet, J., Ferey, G., Massenet, O., and Coey, J. M. D. (1980). *J. Phys. (Orsay, Fr.)* **41**, C1/279–280.

Lachter, A., Gianduzzo, J. C., Barrière, A. S., Fournès, L., and Ménil, F. (1980). *Phys. Status Solidi A* **61**, 619–630.

Lines, M. E. (1982). *J. Phys. Chem. Solids* **43**, 723–730.

Someya, Y., and Yto, A. (1981). *J. Phys. Soc. Jpn.* **50**, 1891–1897.

Teillet, J., Calage, Y., and Varret, F. (1982). *J. Phys. Chem. Solids* **43**, 863–869.

Tressaud, A., and Dance, J. M. (1982). *Struct. Bonding (Berlin)* **52**, 87–146.

18

Local and Cooperative Effects in the Electron Paramagnetic Resonance Spectra of Transition Metal Fluorides

DIRK REINEN
Fachbereich Chemie
Universität Marburg
Marburg, Federal Republic of Germany

JEAN-MICHEL DANCE
Laboratoire de Chimie du Solide du CNRS
Université de Bordeaux
Talence, France

I. Introduction

Electron paramagnetic resonance spectroscopy (epr) is one of the most efficient probe techniques for determining the local properties of paramagnetic centers in an extended solid. The site symmetry and the geometry of the

INORGANIC
SOLID FLUORIDES

ligand coordination are directly accessible. In particular, transitions from static to dynamically averaged geometries can be studied by temperature-dependent measurements. In cases in which the ground-state wave functions contain admixtures of angular momentum, detailed knowledge of the transition ion–ligand bond can also be obtained.

If the concentration of the paramagnetic ions in the host lattice is high, electronic interactions between the centers may occur, which usually change the epr spectra drastically. The new epr parameters now reflect the properties of all d^n cations in the unit cell. It is shown later that in those cases an analysis of the epr spectra may often yield information not only on the local but also on the cooperative properties, such as the strength of the exchange interaction or the geometric arrangement of the transition metal polyhedra in the unit cell. It is also possible to study with the epr technique phase transitions of bulk transition metal compounds that imply changes in the elastic properties (dynamic-to-static transitions) or in the electronic configuration (high-spin/low-spin equilibria).

This chapter is not intended to be comprehensive but to show by a selection of typical examples the utility of epr spectroscopy for the chemist. For the sake of simplicity the background theory is not outlined; equations are given only for rather simple model cases for which powder and single-crystal data are available. The epr spectra of *local* d^n centers are discussed only briefly, although many examples of transition metal impurity centers in various fluoride host lattices have been reported in the literature [see the books of Abragam and Bleaney (1970) and Altschuler and Kosyrew (1974)]. The main emphasis here is given to a discussion of exchange-narrowed epr spectra and the information that can be deduced from them. We consider the influence of exchange interactions on the g tensor and on the epr line width as well, taking representative examples mainly from Cu^{2+}, Ni^{2+}, and Mn^{2+} chemistry.

II. Electron Paramagnetic Resonance Spectra of Local Centers

The investigation of solids doped with transition metal ions serves two purposes. On one hand, the d^n cation can be used as a probe for the symmetry of the host site; on the other, it is possible to study the bonding properties and geometry of the specific paramagnetic center in the environment provided by the host lattice. In order to determine the local properties, usually a diamagnetic dilution that avoids interactions between different transition metal ions is needed.

Information of this type can be derived from the g tensor, the hyperfine structure, and the ligand hyperfine structure in the epr spectra. The presence of ligand hyperfine splittings resulting from the interaction between unpaired electrons in the antibonding MOs of the transition metal polyhedron and the nuclear spins of the ligands gives the most direct evidence for covalent bond contributions. It allows one to calculate the spin density of the d electrons, which is transferred to the ligands as a consequence of delocalized MOs (Abragam and Bleaney, 1970; Sugano et al., 1970). In classical experiments on Ni^{2+} and Mn^{2+} in an octahedral F^- coordination of the perovskite structure, it could be demonstrated that $\sim 12\%$ (1.5%) of the e_g electrons are transferred onto the ligand $p(s)$ orbitals via σ bonding. As expected for the ionic F^- ligand the extent of delocalization in the σ antibonding MOs is rather small. It is surprising, however, that the transferred spin density in the π antibonding t_{2g} MOs (Mn^{2+} and Cr^{3+}) is not much smaller than the σ contribution in the chosen host lattice, indicating significant π contributions to the transition metal–ligand bonds. Slightly smaller numbers are obtained for the σ transfer in Cu^{2+}-doped ZnF_2 (Swalen et al., 1970).

Information on the site symmetry as well as on the bonding properties can be deduced from the g tensor if larger orbital contributions are admixed to the otherwise pure spin ground state. In particular, d^1, d^9, and low-spin d^7 configurations are of interest, because the "one-electron" or "one-hole" wave functions are fairly simple, and the physics is not encumbered by complicated algebra. Taking Cu^{2+} in octahedral coordination as a representative example, we obtain a ^2E ground state, which is σ antibonding and hence subject to a strong Jahn–Teller effect. The twofold orbital degeneracy is lifted by tetragonal or orthorhombic distortions, leaving a Kramers' doublet as the new energetically stabilized ground state. The following g values ($g_0 = 2.0023$) are induced by this state (Bleaney et al., 1955):

$$g_x = g_0 + 2u_x(\cos\varphi + \sqrt{3}\sin\varphi)^2,$$

$$g_y = g_0 + 2u_y(\cos\varphi - \sqrt{3}\sin\varphi)^2, \tag{1}$$

$$g_z = g_0 + 8u_z\cos^2\varphi.$$

The parameters u_i are orbital contributions due to interactions between the ground state and excited levels resulting from the octahedral T_{2g} state (A_g and B_{1g}, B_{2g}, B_{3g} in D_{2h}; Fig. 1) via spin–orbit coupling:

$$u_i = k_i\zeta_0/E_i \qquad (i \equiv x, y, z; \quad E_{x(y,z)} = E[A_g \to B_{1g(2g,3g)}]). \tag{2}$$

Only linear u contributions have been included in Eq. (1). The ground-state wave function (without the admixtures from the excited octahedral T_{2g} level) is

$$\psi_g = \sin\varphi d_{z^2} + \cos\varphi d_{x^2-y^2}. \tag{3}$$

For $\varphi = 0°$ (60°, 120°), a $d_{x^2-y^2}$ ground state corresponding to a tetragonal elongation is obtained:

$$g_z = g_\parallel = g_0 + 8u_\parallel, \qquad g_x = g_y = g_\perp = g_0 + 2u_\perp, \tag{1a}$$

whereas $\varphi = 90°$ (150°, 30°) is equivalent to a d_{z^2} ground state and a tetragonal

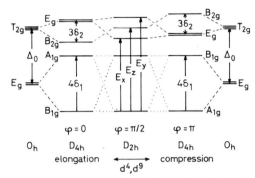

Fig. 1. Energy level diagram of Cu^{2+} in (distorted) octahedral coordination.

compression:

$$g_z = g_{||} = g_0, \qquad g_x = g_y = g_\perp = g_0 + 6u_\perp \tag{1b}$$

Thus, a tetragonal elongation and compression (or any orthorhombic distortion, expressed by φ) can be easily recognized in an epr powder spectrum (Fig. 2a–c). The covalency parameter k can be calculated from the experimental u_i values, if the energies of the ligand field transitions are known [Eq. (2)]; k represents the reduction of the orbital momentum by the delocalization of the d orbitals toward the ligands and also includes the reduction of the spin–orbit coupling constant ζ_0 ($\simeq 830$ cm^{-1} for the free Cu^{2+} ion) by similar effects. The magnitude of the experimental k parameters (quadratic u contributions to g included) for CuF_6^{4-} polyhedra (0.71 ± 0.02) (Reinen and Friebel, 1979) is in the range expected in comparison to the information from the ligand hyperfine structure if the corresponding theoretical equations are applied.

The hyperfine splitting that results from the interaction between the electrons and the nuclear spin of the paramagnetic ion is described by the parameters A_i. For Cu^{2+} one obtains (Bleaney et al., 1955)

$$\varphi = 0°: \quad A_{||/p} = (-\kappa - \tfrac{4}{7})\alpha^2 + 8u_{||} + \tfrac{6}{7}u_\perp,$$
$$A_{\perp/p} = (-\kappa + \tfrac{2}{7})\alpha^2 + \tfrac{11}{7}u_\perp,$$
$$\varphi = 90°: \quad A_{||/p} = (-\kappa + \tfrac{4}{7})\alpha^2 - \tfrac{6}{7}u_\perp, \tag{4}$$
$$A_{\perp/p} = (-\kappa - \tfrac{2}{7})\alpha^2 + \tfrac{45}{7}u_\perp,$$

where α is the MO coefficient of $d_{x^2-y^2}$ in the ground state. Parameter p is a radial one (0.036 cm^{-1}) and κ ($\sim \tfrac{1}{3}$) represents the admixture of configurations with unpaired s electrons. The hyperfine splitting yields no basic additional information to that deduced from the g values.

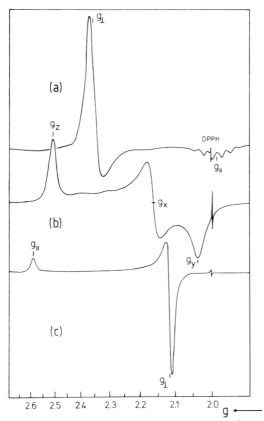

Fig. 2. Electron paramagnetic resonance powder spectra of Cu^{2+} in F^- coordination. (a) Cu^{2+} in Ba_2ZnF_6: compressed octahedra ($g_{||} = 1.99_0$, $g_\perp = 2.36_1$, $k \simeq 0.71$; 77 K). (b) $Ba_2CuCaGa_2F_{14}$: elongation with strong orthorhombic component ($\varphi \approx 10°$; $g_x = 2.16_5$, $g_y = 2.03_8$, $g_z = 2.50_8$, $k \simeq 0.71_5$; 298 K) (Friebel, Holler, and Babel, 1983). (c) $CuPbF_6$: elongated octahedra ($g_{||} = 2.60_5$, $g_\perp = 2.11_7$, $k \simeq 0.71$; 298 K); Cu^{2+}-doped $ZnZrF_6$ has an analogous g tensor.

We now discuss whether the epr spectra reflect the symmetry of the host-lattice site, which is occupied by the transition metal ion, or a symmetry that is induced by the specific d^n configuration of the paramagnetic center. Cations that possess an orbitally nondegenerate ground state in the environment imposed by the host structure [e.g., d^8, d^5, or d^3 (Ni^{2+}, Mn^{2+}, or Cr^{3+}) in octahedral coordination] in general adopt the host site symmetry if they substitute a diamagnetic cation of the same charge and of comparable size. In the case of cations with an orbitally degenerate ground state, however, Jahn–Teller splittings generally occur, which are due to vibronic coupling effects and which may reduce the symmetry of the transition metal polyhedron with respect to the host site symmetry. Octahedrally coordinated Cu^{2+} with a σ antibonding ^2E_g ground state is again chosen as a representative example. The incorporation of Cu^{2+} in an octahedral site usually leads to an isotropic epr signal at high temperatures, indicative of motional narrowing due to a dynamic Jahn–Teller effect:

$$g_i = \tfrac{1}{3}(g_x + g_y + g_z) = g_0 + 4u. \tag{5}$$

At low temperatures the underlying distorted polyhedra are nearly always frozen out; usually these are elongated octahedra in the case of Cu^{2+} [Eq. (1a)]. This geometry is energetically slightly favored with respect to the compressed coordination. In only very few reported cases is the isotropic g tensor [Eq. (5)] transformed into two low-temperature g tensors, which exhibit cubic anisotropy and represent a different variant of the dynamic Jahn–Teller effect (Coffman, 1968):

$$g([100]\circlearrowleft) = g_0 + 4u \pm 2u,$$

$$g([110]\circlearrowleft) = g_0 + 4u \pm u, \tag{6}$$

$$g([111]) = g_0 + 4u.$$

If Cu^{2+} is doped into a site with a lower symmetry in which the orbital degeneracy is already lifted, pseudo-Jahn–Teller effects can stabilize the Cu^{2+} polyhedron by increasing the extent of the distortion and by further reducing the symmetry. An example is Cu^{2+}-doped ZnF_2 (Swalen et al., 1970). Although the ZnF_6 octahedra are tetragonally compressed, the g tensor is distinctly orthorhombic and the d_{z^2} ground state contains an appreciable $d_{x^2-y^2}$ admixture. This result reflects the preference of Cu^{2+} for an elongated rather than a compressed octahedron. Similar phenomena are observed for Cu^{2+} in the tetragonally compressed ZnF_6 sites of K_2ZnF_4 (Reinen and Krause, 1981). Dynamic effects are also involved in this host structure, however. If the host site compression occurs along the molecular z direction ($\varphi = 90°$), the geometric possibilities for realizing elongated octahedra are restricted to the x and y axes. A dynamic averaging process over those two conformations [$\varphi = 90° \pm \alpha$; $30° > \alpha > 0°$; Eq. (1)] will then lead, at least at higher temperatures, to the following g tensor (Reinen and Freibel, 1979):

$$g_{\|(dyn)} = g_0 + 2u(1 + 2\cos^2\alpha), \qquad g_{\perp(dyn)} = g_0 + 8u\sin^2\alpha. \tag{7}$$

In contrast to the equilibration process, which induces the isotropic g value of Eq. (5), this case represents "planar dynamics" and yields a compressed octahedron in time average. Although the g values [Eq. (7); i.e., with $\alpha \simeq 30°$] resemble those of a compressed octahedron [Eq. (1b)] with the same sequence

$g_\perp > g_{||}$, the orbital contributions are quite different and allow one to distinguish between the two alternatives.

The epr spectra of Ni^+, doped into LiF and NaF (Hayes and Wilkens, 1964) and Ag^{2+} (Allen et al., 1972; Friebel, 1974; Friebel and Reinen, 1975), have similar features, although not many epr studies have been performed with these cations. The d^7 configurated Ni^{3+} is usually low spin in F^- coordination and also possesses an ^2E_g ground state in octahedral coordination. The g tensor obeys Eq. (1b), which implies a d_{z^2} ground state, and as for the d^9 configuration, there is a Jahn–Teller-induced tetragonal elongation (Reinen et al., 1974).

Apparently the epr spectra of d^n cations with orbitally degenerate ground states are determined not only by the geometric properties of the host polyhedron, but also by the electronic effects connected with the d^n configuration of the chosen transition ion. The latter influence is particularly remarkable if the ground state is strongly σ antibonding. This was demonstrated for Cu^{2+} and other cations with an ^2E_g ground state.

Although most of the epr studies on fluoride compounds deal with transition ions in (distorted) octahedral coordination, some results are reported for the coordination numbers 4 and 8. In $SrCuF_4$ and $CaCuF_4$ (von Schnering et al., 1971), Cu^{2+} has a square-planar environment. The epr data are consistent with the expected $d_{x^2-y^2}$ ground state (see Section III). The epr spectra of Ni^{2+} in CaF_2 (Zaripov et al., 1968) and CdF_2 (Gehlhoff and Ulrici, 1974) indicate a strong Jahn–Teller distortion of the cubic host site, which splits the orbital triplet ground state $^3T_{1g}$. A trigonal elongation of the cube along one [111] direction and possibly a rearrangement of the remaining six ligands toward an octahedron are proposed for the distortion mechanism.

If the concentration of the d^n cation in the host lattice increases continuously, dipolar and exchange interactions between different paramagnetic centers may arise and eventually change the epr spectra. In the next section, we demonstrate that geometric information about the cooperative and frequently also the local properties can be derived from the g tensor in those cases, sometimes quite easily.

III. Exchange Interactions and Cooperative Properties

If two transition metal polyhedra in a solid matrix are coupled by sufficiently strong exchange interactions, cooperative effects are observed in the epr spectrum. Let us consider two identical molecular g tensors of orthorhombic symmetry that are oriented with respect to each other as depicted in Fig. 3a.

Whereas the g_x components have parallel orientations, g_y and g_z are geometrically connected by the canting angle 2γ. For an isotropic exchange integral J that is much larger than the magnetic anisotropy energy $[J > 4|g_z - g_y|\beta H]$, the following cooperative g values are expected (see also Chao, 1973):

$$g_x^c = g_x, \qquad g_y^c = \cos^2\gamma g_y + \sin^2\gamma g_z, \qquad g_z^c = \cos^2\gamma g_z + \sin^2\gamma g_y. \qquad (8)$$

These equations are generalized expressions of the cases $2\gamma = 90°$,

$$g_{||}^c = g_x, \qquad g_\perp^c = \tfrac{1}{2}(g_y + g_z), \qquad (9)$$

and $2\gamma = 180°$,

$$g_x^c = g_x, \qquad g_y^c = g_y, \qquad g_z^c = g_z, \qquad (10)$$

which have been treated by Abragam and Bleaney (1970). Similar expressions [Eq. (8)], but quadratic in g, have also been reported (Abe and Ono, 1956; Hathaway and Billing, 1970); they are not relevant to the exchange problem, however. Some examples from Cu^{2+} chemistry may illustrate the effects of J on the g tensor. Compounds $CuM^{IV}F_6$ have unit cells with two different orientations of tetragonally elongated CuF_6 octahedra. Whereas the epr spectrum of $CuPbF_6$ exhibits the molecular g tensor expected for a tetragonal elongation [Eq. (1a)], the spectrum of $CuZrF_6$ is clearly exchange narrowed (Fig. 4a; Friebel et al., 1983). The canting angle is near $2\gamma = 90°$, and the g

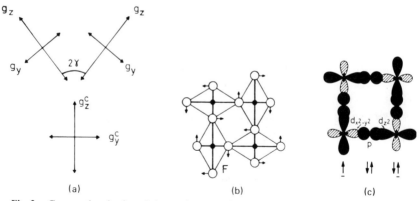

(a)   (b)   (c)

Fig. 3. Cooperative elastic and electronic properties. (a) Coupling of molecular g tensors with canting angle 2γ $[g_x = g_x^c$ perpendicular to coupling plane, Eq. (8)]. (b) Antiferrodistortive order of (corner-connected) elongated octahedra (possibly with an orthorhombic component superimposed). Arrows indicate possible ligand positions with $2\gamma > 90°$ $[K_2CuF_4: 2\gamma = 90°$; Ba_2CuF_6: $2\gamma = 116°]$. (c) Orbital order pattern for Cu^{2+} $(d_{z^2}^2 d_{x^2-y^2}^1)$ correlated with (b) $(g_x = g_y$ and $2\gamma = 90°)$.

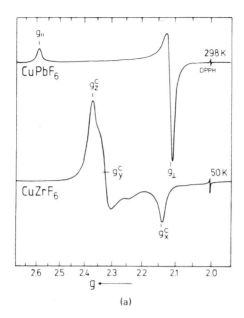

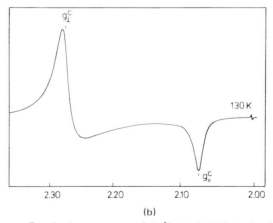

Fig. 4. Exchange effects in the epr spectra of Cu^{2+}. (a) $CuPbF_6$ (molecular g tensor) and $CuZrF_6$ (cooperative g values: $g_{||}^c = 2.13_4$; $g_\perp^c = 2.32_4, 2.37_0$). (b) $CaCuF_4$ (cooperative g values: $g_{||}^c = 2.07_5$, $g_\perp^c = 2.27_5$). (c, p. 534) Ba_2CuF_6 [exchange-coupled g tensor at 77 K (Table I); long-range magnetic order at 4.2 K: $H_x^i = -1160$ Oe, $H_y^i = -330$ Oe, $H_z^i = 1470$ Oe].

values correspond approximately to

$$g_{||}^c = g_\perp = g_0 + 2u_\perp, \qquad g_\perp^c = \tfrac{1}{2}(g_{||} + g_\perp) = g_0 + 4u_{||} + u_\perp. \qquad (9a)$$

Apparently the exchange J is smaller than $4|g_{||} - g_\perp|\beta H$ under X- and

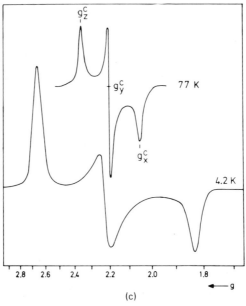

(c)

Fig. 4. (*cont.*)

Q-band conditions for $CuPbF_6$ and larger for $CuZrF_6$, presumably due to a more favorable exchange pathway in the latter compound. The order pattern of the molecular $d_{x^2-y^2}$ orbitals, which underlies Eq. (9a), is usually called "antiferrodistortive" (Fig. 3b). Single-crystal and powder epr data for K_2CuF_4 (K_2NiF_4 structure type) are also consistent with the antiferrodistortive order pattern (Reinen and Krause, 1981). The same two-sublattice structure is realized for the CuF_4 square planes in the unit cell of $CaCuF_4$ and $SrCuF_4$ (Dance, 1981). The g tensors are exchange narrowed (Fig. 4b), and, by means of Eq. (9a), the molecular g values for $CaCuF_4$, for example, are calculated to be $g_{\parallel} = 2.47_5$, $g_{\perp} = 2.07_5$, in agreement with Eq. (1a). Completely analogous results are obtained for the isostructural $SrAgF_4$ (Friebel and Reinen, 1975). In contrast, $BaCuF_6$ has a different structure with elongated octahedra in ferrodistortive order (Dance, 1981). The orthorhombic g tensor is in accord with the local distortion component of the CuF_6 entities, which is induced by the strongly anisotropic connection pattern of the polyhedra in the lattice. Ba_2CuF_6, (space group *Cmca*) which has a layer structure similar to that of K_2CuF_4, is a particularly interesting example with respect to the cooperative order pattern. The epr spectrum is exchange narrowed and strongly orthorhombic (Fig. 4c). Assuming an isotropic covalency factor k (in agreement with the experimental evidence from other Cu^{2+} compounds) and making use of the energies of the ligand field transitions; one can calculate the unknown

TABLE I

Structure, Electron Paramagnetic Resonance, and Ligand Field Data
for Ba_2CuF_6 (298 K)

a_x	1.87^a	$1.85^{b,c}$	$_aA_g \longrightarrow {_b}A_g \simeq$	$9,800 \text{ cm}^{-1}$
a_y	1.86	1.93_5	$\longrightarrow B_{1g}$	$10,000 \text{ cm}^{-1}$
a_z	2.32	2.32_5	$\longrightarrow B_{2g} \simeq$	$12,000 \text{ cm}^{-1}$
φ'	1.5°	10°	$\longrightarrow B_{3g}$	$13,300 \text{ cm}^{-1}$
2γ	101°	116°		
g_x	$2.09_0{}^d$	$2.05_5{}^{c,d}$	g_x^c	2.05_4
g_y	2.08_5	2.10_8	g_y^c	2.20_4
g_z	2.45_5	2.45_2	g_z^c	2.35_3

[a] Cu—F bond lengths (angstroms) according to v. Schnering (1973).
[b] Cu—F bond lengths (angstroms) according to Reinen and Weitzel (1977).
[c] This set yields cooperative g_i^c parameters [Eqs. (1) and (8)] very close to those observed.
[d] Molecular g values calculated from the respective bond length data with $k = 0.71$ and the reported ligand field transitions.

parameters k, φ, and γ as well as the molecular g tensor from Eqs. (1) and (8). Equation (8) is valid, however, only if the canting angle 2γ has no out-of-plane component (Fig. 3b). This assumption is likely to be true for layer structures of the type discussed. The obtained φ value can be compared with the φ' parameter, derived from the structural data by using the following expressions, in which the Δi are the deviations from the average Cu—F bond length (Reinen and Friebel, 1979):

$$\Delta x(y) = \frac{\rho}{\sqrt{3}}\cos(2\varphi'(\mp)120^0), \qquad \Delta z = \frac{\rho}{\sqrt{3}}\cos 2\varphi', \qquad (11)$$

where $\rho = (\Sigma\, 2\,\Delta i^2)^{1/2}$ $(i = x, y, z)$.

The data set calculated from the spectroscopic results (Friebel, 1974) differed considerably from that derived from single-crystal x-ray results (von Schnering, 1973; Table I). A subsequent refinement of the x-ray data by a powder neutron diffraction analysis (Reinen and Weitzel, 1977) yielded improved structural parameters very close to those derived from the epr spectra.[†] Obviously, one can deduce quantitative structural information about local symmetry as well as geometric order patterns of the polyhedra in the unit cell from the g tensor in favorable cases, if exchange narrowing

[†] The data proposed by Friebel (1974) were based on the equations of Abe and Ono (1956). A new calculation [Eqs. (1) and (8) with the inclusion of quadratic u contributions] led to exactly the same parameter set as that resulting from the neutron diffracion analysis (Table I).

determines the epr spectra. The direct observation of the cooperative and molecular g tensor can be predicted for sufficiently small exchange integrals if the microwave frequency is varied. For $v \gg (7.5\bar{g}/\delta g) J$ [v (GHz), J (cm^{-1})], the molecular g tensor should be observed, whereas for frequencies significantly smaller than the critical value, exchange narrowing can be expected. So far only very few experimental examples of this behavior are known. A frequency dependence of the g tensor has been observed for Cu^{2+}-doped pyridine–N-oxide complexes (van Kalkeren et al., 1983), which is due to this phenomenon.

Whereas at higher temperatures pair interactions of the kind just described are observed, magnetic long-range order is expected below the critical temperature T_c. For elongated octahedra the $d_{x^2-y^2}$ orbitals of Cu^{2+} are singly occupied. The orbital order pattern induced by antiferrodistortive ordering (Fig. 3b) will hence lead to ferromagnetism (Khomskii and Kugel, 1973), as illustrated in Fig. 3c. Similarly, the elastic order of "ferrodistortive" symmetry ($2\gamma = 180°$, long Cu^{2+}–ligand spacings parallel to each other) is correlated with antiferromagnetism. Whereas in the latter case the epr signal vanishes below T_c ($S = 0$ ground state) drastic charges of the epr spectrum may be observed for an antiferrodistortive pattern, which are due predominantly to the magnetic anisotropy field connected with the long-range spin–spin order in the specific structure. The transition from the exchange-narrowed epr spectra (pair interactions) to those characteristic of "short-range" and finally of "long-range" order is continuous and starts even far above T_c for K$_2$CuF$_4$ (Geick and Strobel, 1983; Reinen and Krause, 1981) and Ba$_2$CuF$_6$ (Friebel et al., 1976). The internal field H^i shifts the signal drastically (Fig. 4c). The briefly sketched phenomena of ferromagnetic resonance in transition metal fluorides and chlorides and also of antiferromagnetic resonance are reviewed elsewhere (Geick and Strobel, 1983).

We have discussed so far the two limiting cases of a transition metal ion that is isomorphously substituted into a host site in very low concentration, as well as the stoichiometric transition metal compound. We now consider the information that can be deduced from the epr spectra of intermediate compositions. The mixed crystals Ba$_2$Zn$_{1-v}$Cu$_v$F$_6$, for example (Friebel et al., 1976), show epr spectra corresponding to compressed CuF$_6$ polyhedra (host-lattice effect) at low v values (Figs. 2a and 5a). With increasing Cu^{2+} concentration the g values [Eq. (1b)] move toward those of Eq. (7) or (9). This

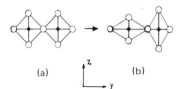

(a)   (b)

Fig. 5. Changes of the CuF$_6$ geometry in mixed Ba$_2$Z[n$_{1-v}$Cu$_v$F$_6$ crystals with increasing v. (a) Pair of octahedra compressed along $x - v \lesssim 0.3$. (b) Change toward octahedra elongated along $z(y) - v \simeq 0.4$.

change is consistently interpreted by the preference of Cu^{2+} for an elongated rather than a compressed octahedron. Because of the host site compression along x the molecular in-plane axes are the easy directions for such an elongation. The elongation is energetically favorable in particular if neighboring Cu^{2+} polyhedra choose alternative directions for these geometric changes (Fig. 5b). The g tensor cannot distinguish, however, whether the long axis of each polyhedron fluctuates between y and z [planar dynamics, Eq. (7)] or whether pairs of exchange-coupled Cu^{2+} polyhedra with (statically) perpendicular oriented long Cu—F bond lengths are already present [Eq. (9)]. At very high Cu^{2+} concentrations ($v > 0.6$) we definitely have exchange-narrowed epr spectra and the long-range order pattern depicted in Fig. 3b. Similar results are obtained for $K_2Zn_{1-x}Cu_xF_4$ mixed crystals (Reinen and Krause, 1981).

The g tensor is an order parameter and can also be used to characterize phase transitions. For example, $CuZrF_6$ is triclinic at low temperature and exhibits an orthorhombic g tensor (Fig. 4a). At 353 K the space group becomes trigonal and the epr spectrum isotropic [Eq. (5)]. The phase transition is of the order–disorder type and is caused by the change from a static to a dynamic Jahn–Teller distortion. It is also noteworthy from the experimental stand point that spectra of transition metal fluorides, even with high concentrations of paramagnetic ions, are usually sufficiently resolved. Only if the dimensionality of the interactions becomes larger than 2 in the case of corner-connected CuF_6 polyhedra (e.g., $K_2CuF_4 \rightarrow KCuF_3$) or if the CuF_6 polyhedra are face connected with short Cu–Cu spacings, are spectra observed that are too broad for an analysis. In any case, the presence of exchange interactions induces characteristic changes in the line shape of the signals, and hyperfine structures are no longer resolved.

We may conclude that the analysis of the g tensor of Cu^{2+} compounds as representative examples has yielded a wealth of information on the local geometry of the transition metal polyhedra and the cooperative ordering of the polyhedra with respect to one another. Although at low concentrations of the paramagnetic centers the influence of the host-lattice site symmetry is rather strong, the structure is determined by the electronic properties of the transition metal ion at high concentrations. In Section IV we discuss the linewidth effects induced by magnetic interactions between d^n cations (Poole and Farach, 1979). Although line-shape measurements can be very informative concerning these interactions, not many epr studies on (undiluted) transition metal fluorides have been reported.

High-spin/low-spin transitions and equilibria can also be followed by epr spectroscopy. For example, the F^- ligand is strong enough to force Ni^{3+} (d^7) into the low-spin configuration in ordered perovskites of the $A_2A'NiF_6$ type (A, A' = alkaline ion), although the separation between the doublet ground state and the first excited high-spin state is rather small ($500-1000$ cm^{-1}) (Reinen et al., 1974; Reinen and Friebel, 1979). Below 298 K the cubic lattices undergo second-order phase transitions to tetragonal unit cells, which contain elongated NiF_6 octahedra in ferrodistortive order (Grannec et al., 1976). The low-temperature g tensor obeys Eq. (1b) in agreement with this geometry but contains additional higher-order orbital contributions, which are induced by interactions with excited quartet states (Lacroix et al., 1964). With increasing temperature the anisotropic epr signals of Cs_2KNiF_6 and Rb_2NaNiF_6 become broad and isotropic and shift toward larger average g values. Obviously, the Jahn–Teller distortion of the NiF_6 polyhedra is dynamic above the phase transition, but quartet states are also populated (Reinen et al., 1974; Reinen and Friebel, 1979). From the 4.2 K spectra of Cs_2NaNiF_6 with a hexagonal structure it could be deduced that half of the Ni^{3+} is high-spin, the other half low-spin configurated.

IV. Linewidth and Dimensionality of Magnetic Interactions

Electron paramagnetic resonance spectra of magnetically condensed systems usually show exchange-narrowed lines. Very informative parameters in this case are the linewidth and the line shape (for insulators, we are restricted to Gaussian, Lorentzian, or intermediate line shapes). From the variation of the linewidth with temperature and with the angle θ between the magnetic field direction and the internuclear vector, information concerning the strength of the magnetic interactions and their dimensionality can be deduced. Since the late 1960s, many epr linewidth studies have been devoted to the determination of spin correlation in one- (1D), two- (2D), and three-dimensional (3D) magnetic systems. Most of them are extensions of the pioneering publications of Richards and co-workers (Dietz et al., 1971; Richards and Salamon, 1974).

A. THEORY

The simplest interaction in a magnetic system is of the dipole–dipole type. Its influence on the linewidth can be separated into the secular term of the angular form $(3\cos^2\theta - 1)^L$ and the nonsecular terms following an $A\sin^2\theta\cos^2\theta + B\sin^4\theta$ law. The sum of the two terms leads (for $L = 1$, $A = B = 2$) to a $1 + \cos^2\theta$ dependence. If the exchange field is larger than the observation field, the nonsecular terms are not averaged out, and the variation is of the mentioned $1 + \cos^2\theta$ type. In the reverse case the nonsecular terms vanish and the typical $(3\cos^2\theta - 1)^L$ variation is observed with a minimum linewidth at the "magic angle" $\theta = 54°43'$. In the first approximation L is $\frac{4}{3}$ for 1D systems and 2 for 2D systems (Kokoszka, 1975; Richards, 1975). More sophisticated models take into account a third term due to a slow decay of the spin correlation (Cheung and Soos, 1978; Ferrieu and Pomerantz, 1981). Concerning the variation of the linewidth with temperature, most theories use a $[(T/T_N) - 1]^p$ dependence; p differs considerably with the model used (Huber, 1972, 1973).

B. EXPERIMENTAL RESULTS

Unfortunately, only few experiments have been performed on fluorides. The only study of 1D magnetic fluorides is that on $CsNi^{II}F_3$ (Barjhoux, 1979). Linear chains of face-sharing NiF_6 octahedra constitute the structure. One-dimensional ferromagnetism is observed down to 2.6 K, whereas the chains order antiferromagnetically at even lower temperatures. The reported V-band spectra at 300 K are very broad due to the presence of dipole–dipole interactions and anisotropy. The variation of the linewidth with θ is accounted for by the assumption that the anisotropy A is of the same order of magnitude as the exchange interaction J (Fig. 6).

Most of the 2D compounds studied have the K_2NiF_4 structure, in which layers of corner-sharing NiF_6 octahedra are separated by alkali ions. The extensively investigated K_2MnF_4 was at the origin of the relatively recent theories of spin dynamics (Richards, 1975). The dependence of the linewidth

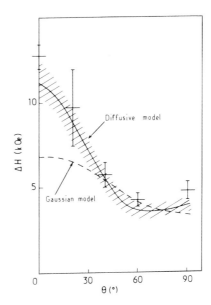

Fig. 6. Angular dependence of the peak-to-peak semilinewidth of $CsNiF_3$ as a function of θ. The lines are theoretical curves corresponding to different models. The dashed zone accounts for the uncertainty of $J = 10 \pm 0.5$ K and $A = 8.5 \pm 0.5$ K.

ΔH on θ follows a $(3\cos^2\theta - 1)^2$ + constant law (Fig. 7). As the temperature is lowered toward the antiferromagnetic ordering temperature $T_N = 45$ K, ΔH initially decreases, passes through a minimum, and then increases rapidly near T_N, in agreement with the theory. In the case of K_2CuF_4 the variation of ΔH with θ is more complex and in disagreement even with a $1 + \cos^2\theta$ dependence. The observed $(1 + \cos^2\theta) - (1 - 3\cos^2\beta)$ dependence is attributed to anisotropic exchange interactions (β is the angle between the external field and the c axis of the unit cell) (Yamada and Ikebe, 1972). $BaMnF_4$, with a staggered layer structure, shows similar features. The $(3\cos^2\theta - 1)^2$ + constant law is followed down to 40 K. Below this temperature a $\cos^2\theta$ dependence due to anisotropy is reported. In the case of the double-layer compound $K_3Mn_2F_7$ a $1 + \cos^2\theta$ law is observed, which indicates the presence of 2D as well as 3D correlations (Van Uijen and de Wijn, 1981).

Another interesting study is that of percolation phenomena in semidiluted systems such as $K_2Mg_{1-x}Mn_xF_4$ and $Rb_2Mg_{1-x}Mn_xF_4$. In the first mixed-crystal series a $(3\cos^2\theta - 1)^2$ law is followed (minimum linewidth at $54°43'$), indicating predominant 2D interactions. If one plots the ratio $I(H_0)/I(H)$ versus $[H - H_0/\Delta H_{1/2}]^2$ at $\theta = 0°$ and $\theta = 54°43'$ [$I(H)$ is the intensity at field H; H_0: center field], striking effects in dependence on x are observed (Fig. 8). With decreasing Mn^{2+} concentration the classical 2D magnetism changes to a 1D magnetism (around the percolation threshold at $x = 0.59$) and finally to the magnetism of isolated Mn^{2+} clusters (Gaussian curve). A similar result is obtained for $Rb_2Mg_{1-x}Mn_xF_4$. Particular attention has been paid to the temperature dependence of the linewidth near the percolation threshold. Whereas the linewidth is dominated by a $(3\cos^2\theta - 1)^2$ dependence at high

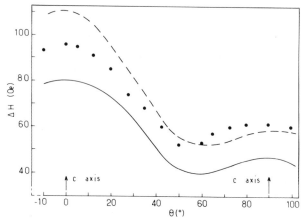

Fig. 7. Angular dependence of the linewidth for K_2MnF_4 at 298 K. The two curves correspond to different models.

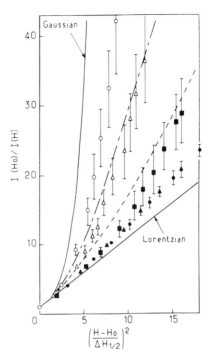

Fig. 8. Line shapes for $K_2Mn_xMg_{1-x}F_4$ mixed crystals at $\theta = 0$. The theoretical curves and for a 1D system (——·——) for $x = 0.59$ as a 1D system with interchain coupling (- - -) are shown; full curves correspond to Gaussian and Lorentzian line shapes. The values of x are as follows: ○, 0.35; △, 0.52; ■, 0.56; ●, 0.86; ▲, 1.00.

temperatures, antiferromagnetic correlation effects govern the linewidth below 30 K. At 1.6 K a nine-line hyperfine structure appears, suggesting the critical slowdown of the spin dynamics (at the scale of epr). It is believed to be due to isolated Mn^{2+} ions (just at the percolation threshold), which interact with their next-nearest neighbors locked in clusters (Walsh *et al.*, 1979).

V. Summary

Electron paramagnetic spectra of d^n cations in extended solids are valuable sources of information with respect to transition metal–ligand bonding and structural properties, particularly if significant orbital contributions in the ground-state wave function are present. Whereas at low concentrations of the paramagnetic ions the epr data reflect predominantly the host site symmetry, the electronic properties of the d^n cation may have a strong influence on the structure at higher concentrations. If exchange effects are present, not only the local geometry of the transition metal polyhedron, but also the cooperative patterns of the polyhedra in the unit cell can frequently be derived from the epr spectra. Various examples of transition metal fluorides, mainly with d^9-configurated cations, have been presented. Also, a short account of the possibility of following high-spin/low-spin equilibria by epr spectroscopy has been given. Finally, the influence of magnetic interactions on the epr line width has been considered. In particular, the dependence on the dimensionality of the magnetic system and on the concentration of the d^n cation in mixed crystals has been discussed.

References

Abe, H., and Ono, K. (1956). *J. Phys. Soc. Jpn.* **11**, 947.

Abragam, A., and Bleaney, B. (1970). "Electron Paramagnetic Resonance of Transition Ions." Oxford Univ. Press (Clarendon), London and New York.

Allen, G. C., Meeking, R. F., Hoppe, R., and Müller, B. (1972). *J. Chem. Soc., Chem. Commun.* p. 291.

Altschuler, S. A., and Kosyrew, B. M. (1974). "Electron Paramagnetic Resonance in Compounds of Transition Elements." Wiley, New York.

Barjhoux, Y., Boucher, J. P., and Cibert, J. (1979). *J. Appl. Phys.* **50**(3), 1764.

Bleaney, B., Bowers, K. D., and Pryce, M. H. L. (1955). *Proc. R. Soc. London, Ser. A* **228**, 166.

Chao, C. (1973). *J. Magn. Reson.* **10**, 1.

Cheung, T. T. P., and Soos, Z. G. (1978). *J. Chem. Phys.* **69**(8), 3845.

Coffman, R. E. (1968). *J. Chem. Phys.* **48**, 609.

Dance, J. M. (1981). *Mater. Res. Bull.* **16**, 599.

Dietz, R. E., Merritt, F. R., Dingle, R., Hone, D., Silbernagel, B. G., and Richards, P. M. (1971). *Phys. Rev. Lett.* **26**(19), 1186.

Ferrieu, F., and Pomerantz, M. (1981). *Solid State Commun.* **39**, 707.

Friebel, C. (1974). *Solid State Commun.* **15**, 639.

Friebel, C., and Reinen, D. (1975). *Z. Anorg. Allg. Chem.* **413**, 51.

Friebel, C., Propach, V., and Reinen, D. (1976). *Z. Naturforsch., B: Anorg. Chem., Org. Chem.* **31B**, 1574.

Friebel, C., Pebler, J., Steffens, F., Weber, M., and Reinen, D. (1983). *J. Solid State Chem.* **46**, 253.

Gehlhoff, W., and Ulrici, W. (1974). *Phys. Status Solidi B* **65**, K93.

Geick, R., and Strobel, K. (1983). *Rev. Infrared Millimeter Waves* **1**, 249.

Grannec, J., Sorbe, P., Chevalier, B., Etourneau, J., and Portier, J. (1976). *C. R. Hebd. Seances Acad. Sci., Ser. C* **282**, 915 (1976).

Hathaway, B. J., and Billing, D. E. (1970). *Coord. Chem. Rev.* **5**, 143.

Hayes, W., and Wilkens, J. (1964). *Proc. Soc. London, Ser. A* **281**, 340.

Huber, D. L. (1972). *Phys. Rev. B* **6**, 3180.

Huber, D. L. (1973). *Phys. Rev. A* **43**, 311.

Khomskii, D. J., and Kugel, J. (1973). *Solid State Commun.* **13**, 763.

Kokozka, G. F. (1975). *Nato Adv. Study Inst., Ser., B* **7**, 171.

Lacroix, R., Hoechli, U., and Müller, K. A. (1964). *Helv. Phys. Acta* **37**, 627.

Poole, C. P., and Farach, H. A. (1979). *Bull. Magn. Reson.* **1**, 4.

Reinen, D., Friebel, C., and Propach, V. (1974). *Z. Anorg. Allg. Chem.* **408**, 187.

Reinen, D., and Friebel, C. (1979). *Struct. Bonding (Berlin)* **37**, 1.

Reinen, D., and Krause, S. (1981). *Inorg. Chem.* **20**, 2750.

Reinen, D., and Weitzel, H. (1977). *Z. Naturforsch., B: Anorg. Chem., Org. Chem.* **32B**, 476.

Richards, P. M. (1975). *Nato Adv. Study Inst., Ser. B* **7**, 147.

Richards, P. M., and Salamon, M. B. (1974). *Phys. Rev. B* **9**(1), 32.

Sugano, S., Tanabe, Y., and Kamimura, H. (1970). "Multiplets of Transition Metal Ions in Crystals." Academic Press, New York.

Swalen, J. D., Johnson, B., and Gladney, H. M. (1970). *J. Chem. Phys.* **52**, 4078.

van Kalkeren, G., Keijzers, C. P., Srinivasan, R., de Boer, D., and Wood, J. S. (1983). *Mol. Phys.* **48**, 1.

van Uijen, C. M. J., and de Wijn, H. W. (1981). *Phys. Rev. B* **24**(*g*), 5368.

von Schnering, H. G. (1973). *Z. Anorg. Allg. Chem.* **400**, 201.

von Schnering, H. G., Kolloch, B., and Kolodziejczyk, A. (1971). *Angew. Chem.* **83**, 440.

Walsh, W. M., Jr., Birgeneau, R. J., Rupp, L. W., Jr, and Guggenheim, H. J. (1979). *Phys. Rev. B* **20**(11), 4645.

Yamada, I., and Ikebe, M. (1972). *J. Phys. Soc. Jpn.* **33**(5), 1334.

Zaripov, M. M., Kropotov, V. S., Livanova, L. D., and Stepanov, V. G. (1968). *Sov. Phys.–Solid State* (*Engl. Transl.*) **9**, 2344.

19

Fluoride Chemistry
and Biomineralization

CHARLES A. BAUD

Institute of Morphology
University Medical Center
Geneva, Switzerland

I. Introduction

Soluble inorganic fluorides distributed throughout the body water of living organisms may be involved in the processes of biomineralization. With increasing fluoride concentrations in the environment, fluoride content rises in cells and extracellular fluids and therefore in deposited minerals. Several cell

INORGANIC
SOLID FLUORIDES

ever, have the capacity to concentrate fluoride far above the ion found in the environment; these include marine microorganisms (Baas-Becking, 1959), bacteria (Jenkins et al., 1969), and specialized vertebrate cells (Suga et al., 1981). Accumulated fluoride can be released at relatively high local concentrations and be available for binding to inorganic deposits (Jenkins and Edgar, 1977). This chapter is an attempt to document the role played by fluorine in the formation of minerals by organisms.

II. Fluorine and Biological Apatites

In many biological systems, the mineral that forms in hard tissues is of an apatitic nature, and hydroxylapatite is generally considered the prototype for such minerals. In the presence of fluorine, fluoride ions are incorporated into the crystalline lattice of the mineral, and the incorporation brings about changes in its chemical and physical properties.

A. MINERAL ACCRETION

The degree of mineralization of bone tissue (DMBT), as measured by means of a combined microradiographic and microdensitometric method, expressed as grams of mineral per unit tissue volume, varies widely among species, from bone to bone, and even from one site to another in a single bone. Fluorine was found to cause an increase in DMBT. In industrial fluorosis, DMBT is significantly higher than in control subjects of the same age group; a direct correlation is observed between the increase in DMBT and that of fluoride content, at least until the latter reaches 0.8% (Bang et al., 1978). A similar increase in DMBT is experimentally induced in male mice by fluoride treatment (Gössi, 1978).

Such an increased bone tissue density may be due, as indicated by Legeros et al. (1982), to enhanced apatite formation. The presence of fluorine has been shown to promote the formation of apatite under conditions in which it would normally not form without fluorine, particularly in conditions of undersaturation with respect to hydroxylapatite.

In the same way, fluorine exerts a significant effect on the remineralization of partly demineralized tissues. Even at a concentration of 0.1 mM in topical agents or calcifying solutions, fluorine accelerates the remineralization rate of softened dental enamel by a factor of 4 or 5 (Koulourides, 1968) and increases the total amount of apatite deposited (Feagin, 1977).

B. CRYSTALLINITY

From a physical viewpoint, biological apatite crystallites can be characterized by their three-dimensional regular periodicity over longer or shorter distances. In the infrared absorption spectra, the resolution of the absorption bands belonging to the P–O vibration modes of the PO_4 groups between 700 and 550 cm^{-1} illustrates the regularity of the environment of the PO_4 groups and gives a "crystallinity index," which can be used to evaluate the crystalline state.

In compact bone tissue of mice, Tochon-Danguy (1978) showed a good linear correlation between level of fluorine incorporation into bone mineral and crystallinity index, a finding confirmed by Legeros et al. (1982). In cases of industrial fluorosis, the crystallinity index of the bone mineral substance was found to increase as a function of increasing fluoride content (Bang et al., 1978). Likewise, in compact bone tissue of osteoporotic patients treated with fluoride, the crystallinity index of bone mineral was increased by the treatment, as shown by Baud et al. (1983). These observations suggest greater crystal perfection and/or crystallite size as an effect of fluoride action.

Crystal size and/or perfection can also be evaluated by measuring the width of the x-ray diffraction peaks. The first studies providing evidence of changes in the morphological appearance of apatite crystals with fluoride uptake *in vivo* (Zipkin et al., 1962; Posner et al., 1963; Baud and Moghissi-Buchs, 1965) showed that the x-ray diffraction lines became progressively narrower as the percentage of fluorine increased. These changes in x-ray diffraction lines may indicate either an increase in crystal size or a decrease in crystal defects: small-angle x-ray diffraction analysis showed that these changes can be accounted for entirely by an increase in crystallite size (Eanes et al., 1965). High-resolution electron microscopy, however, showed that fluorine was also effective in compensating for crystal structural disorder caused by carbonate inclusion in the apatite lattice (Featherstone and Nelson, 1980; Featherstone et al., 1981).

For patients with osteoporosis, fluoride therapy induces larger crystal formation (Bernstein and Cohen, 1967). These changes in crystal size correspond almost exclusively to enlargement in width and thickness, without alteration in crystal length (Baud et al., 1983).

The increase in size of biological apatite crystallites with increasing fluoride content might originate from an increase in the rate of crystal growth induced by fluorine. It has been suggested that this acceleration of crystal growth is in part related to fluoride enhancement of the rate of hydrolysis of nonapatitic calcium phosphate intermediates, which may accumulate during the crystal growth process (Meyer and Nancollas, 1972; Amjad and Nancollas, 1979;

Eanes, 1980). An alternate possibility is related to the influence that fluoride has on the magnitude of the driving force for precipitation, that is, on the supersaturation of the solution with respect to the fluoride-containing phase that precipitates (Varughese and Moreno, 1981; Margolis et al., 1982).

C. CRYSTAL STRUCTURE

Hydroxyl ions of biological apatites can be partly replaced by fluoride ions, leading to the formation of fluorhydroxylapatite, the existence of which is proved by measuring the parameters of the unit cell in the crystal lattice by high-resolution x-ray diffraction. The a parameter diminishes progressively in proportion to the increase in fluoride content, in agreement with Vegard's law, which establishes that in crystals the dimensions of the unit cell varies linearly, at least on a first approximation, with the concentration of the substituents. This substitution of fluoride for hydroxyl in vivo in bone apatite has been extensively demonstrated in humans with endemic (Baud and Bang, 1971) or industrial (Bang et al., 1978) fluorosis, in osteoporotic patients treated with fluoride (Baud et al., 1978), in fluorotic bovines (Baud and Slatkine, 1962; Baud and Bang, 1972), and in experimental mice (Baud and Bang, 1973).

This substitution of fluoride into the bone apatite structure appears to occur primarily during crystal nucleation and growth (Eanes and Reddi, 1979), so that fluoride is not homogeneously distributed but deposited mainly in bone tissue formed during periods of fluoride ingestion (Baud and Bang, 1970a; Bang, 1978).

Topical application of fluoride ions on calcified tissue surfaces, especially at low concentrations and when the pH is lowered, causes the formation of fluorapatite (Baud et al., 1967). Whether the fluorapatite formation occurs from dissolution followed by reprecipitation, or from direct fluoride exchange in the lattice of preexisting crystals, is not certain. The shortening of the a axis, however, was observed in samples that had been treated for only a few minutes (Baud and Bang, 1970b). It is difficult to envision exchange of any depth into the crystallites in such a short period of time, and dissolution followed by reprecipitation therefore would seem to be the more likely mechanism (Grøn, 1977). High fluoride concentrations favor calcium fluoride formation in the same way, involving dissolution and reprecipitation.

D. MINERAL DISSOLUTION

It is quite clear that the mineral substance deposited in bone and other calcified tissues when fluorine is present is both qualitatively and quantitatively different from that deposited in its absence. The presence of fluorine causes not only heteroionic substitutions in the crystalline lattice of the apatite phase, but also important alterations of its physicochemical properties. These modifications affect the demineralization processes of the calcified tissues by mechanisms in which several factors are involved (Brown, 1974).

Among these factors, some are thermodynamic, in particular the heteroionic substitutions in the crystalline lattice; they affect mineral solubility. The decrease observed in the solubility of fluoride-substituted apatites is related to their greater structural stability, the position of the fluoride ions being slightly different from that of the hydroxyl ions and more stable (Young and Elliot, 1966).

Others are kinetic factors, which affect the dissolution rate. The degree of mineralization of bone tissue directly influences the volume of water spaces and therefore the accessibility of the mineral substance to demineralizing agents and ion diffusion (Fenwick, 1974; Baud et al., 1977). Likewise, the increase in crystal size results in a diminution of the surface of attack and a slackening of the dissolution (Neuman and Neuman, 1958). The suppression of lattice defects eliminates the sites of preferential dissolution (Arends, 1973, 1977).

The resistance of high-fluoride bone tissue to dissolution processes is due not only to the substitution of fluoride for hydroxyl in the crystalline lattice, but also to a greater degree of mineralization and a higher crystallinity.

E. CHEMICAL COMPOSITION

Fluoride enrichment of bone mineral can also be associated with other alterations in the chemical composition of bone mineral. The calcium phosphorus ratio of bone mineral increases with an increase in fluoride, progressively approaching the stoichiometric value for apatite (Gedalia et al., 1965; Bang, 1978). These changes are the result of an increase in calcium, the phosphorus content being unchanged.

The magnesium concentration in bone mineral tends to increase with an increase in fluoride (McCann and Bullock, 1957; Zipkin et al., 1960; Singer et al., 1974), suggesting a magnesium–fluoride interrelationship, but the chemical form of magnesium in biological apatite has not been established.

Quantitative evaluations of carbonate content in bone mineral in relation to fluoride level have given puzzling results: a decrease in carbonate with an increase in fluoride (McCann and Bullock, 1957) or quite the contrary (Singer et al., 1974).

III. Fluorine and Nonapatitic Mineralization

A. FLUORITE

Mineral fluorite is synthesized in the hydrosphere by two groups of organisms. The statoliths of a wide range of mysid crustaceans (Mysidacea), which occupy marine or freshwater environments, contain fluorite. Fluorite is also one of the mineralization products of the gizzard plates of some tectibranch gastropods of the genus *Scaphander* (Lowenstam and McConnell, 1968). These biogenic crystals are acicular in shape, whereas inorganically formed precipitates are distinctly different (Lowenstam, 1981).

The amorphous precursor of this mineral, also not known to be formed inorganically in the hydrosphere, is found in the spicules of certain common marine nudibranch gastropods, where it occurs together with amorphous calcium carbonate monohydrocalcite (Lowenstam, 1981).

B. POTASSIUM FLUOROSILICATE

Fluorine constitutes about 10% of the dry weight of the marine sponge *Halichondria moorei*; fluorine occurs as potassium fluorosilicate, most of which must be present in the solid state (Gregson *et al.*, 1979).

C. INFLUENCE OF FLUORINE ON THE CRYSTAL TYPE OF CALCIUM CARBONATE

Fluoride anions are not exchanged with carbonate in calcium carbonate. The presence of fluoride ions, however, favors calcite formation, whereas the presence of sulfate ions favors aragonite formation (Kitano *et al.*, 1976), but the mechanism of these processes is obscure.

References

Amjad, F., and Nancollas, G. H. (1979). *Caries Res.* **12**, 250–258.

Arends, J. (1973). *Caries Res.* **7**, 261–268.

Arends, J. (1977). *Caries Res.* **11**, 186–188.

Baas-Becking, L. G. M. (1959). *N. Z. Oceanogr. Inst. Mem.* **3**, 48–64.

Bang, S. (1978). *In* "Fluoride and Bone" (B. Courvoisier, A. Donath, and C. A. Baud, eds.), pp. 56–61. Huber, Bern.

Bang, S., Baud, C. A., Boivin, G., Demeurisse, C., Gössi, M., Tochon-Danguy, H. J., and Very, J. M. (1978). *In* "Fluoride and Bone" (B. Courvoisier, A. Donath, and C. A. Baud, eds.), pp. 168–175. Huber, Bern.

Baud, C. A., and Bang, S. (1970a). In "Fluoride in Medicine" (T. L. Vischer, ed.), pp. 27–36. Huber, Bern.

Baud, C. A., and Bang, S. (1970b). Caries Res. 4, 1–13.

Baud, C. A., and Bang, S. (1971). IADR Abstr. 49, 72.

Baud, C. A., and Bang, S. (1972). Proc. Int. Conf. X-Ray Opt. Microanal., 6th,1971 pp. 835–840.

Baud, C. A., and Bang, S. (1973). J. Dent. Res. 52, 589.

Baud, C. A., and Moghissi-Buchs, M. (1965). C. R. Hebd. Seances Acad. Sci. 260, 1793–1794.

Baud, C. A., and Slatkine, S. (1962). C. R. Hebd. Seances Acad. Sci. 255, 1801–1802.

Baud, C. A., Bang, S., and Baud, J. P. (1967). C. R. Hebd. Seances Acad. Sci., Ser. D 265, 1334–1336.

Baud, C. A., Bang, S., and Very, J. M. (1977). J. Biol. Buccale 5, 195–202.

Baud, C. A., Lagier, R., Bang, S., Boivin, G., Gössi, M., and Tochon-Danguy, H. J. (1978). In "Fluoride and Bone" (B. Courvoisier, A. Donath, and C. A. Baud, eds.), pp. 290–292. Huber, Bern.

Baud, C. A., Courvoisier, B., Tochon-Danguy, H. J., and Very, J. M. (1983). Calcif. Tissue Int. 35, A18.

Bernstein, D. S., and Cohen, P. (1967). J. Clin. Endocrinol. Metab. 27, 197–210.

Brown, W. E. (1974). J. Dent. Res. 53, 204–216.

Eanes, E. D. (1980). J. Dent. Res. 59, 144–150.

Eanes, E. D., and Reddi, A. H. (1979). Metab. Bone Dis. Relat. Res. 2, 3–10.

Eanes, E. D., Zipkin, I., Harper, R. A., and Posner, A. S. (1965). Arch. Oral Biol. 10, 161–173.

Feagin, F. F. (1977). Caries Res. 11, Suppl., 79–83.

Featherstone, J. D. B., and Nelson, D. G. A. (1980). Aust. J. Chem. 33, 2363–2368.

Featherstone, J. D. B., Nelson, D. G. A., and McLean, J. D. (1981). Caries Res. 15, 278–288.

Fenwick, J. C. (1974). Can. J. Zool. 52, 755–764.

Gedalia, I., Menczel, J., Antebi, S., Zuckermann, H., and Pinchevski, Z. (1965). Proc. Soc. Exp. Biol. Med. 119, 694–697.

Gössi, M. (1978). In "Fluoride and Bone" (B. Courvoisier, A. Donath, and C. A. Baud, eds.), pp. 62–67. Huber, Bern.

Gregson, R. P., Baldo, B. A., Thomas, P. G., Quinn, R. J., Bergquist, P. R., Stephens, J. F., and Horne, A. R. (1979). Science 206, 1108–1109.

Grøn, P. (1977). Caries Res. 11, Suppl., 172–204.

Jenkins, G. N., and Edgar, W. M. (1977). Caries Res. 11, 226–242.

Jenkins, G. N., Edgar, W. M., and Ferguson, D. B. (1969). Arch. Oral Biol. 14, 105–119.

Kitano, Y., Kanamori, N., and Yoshioka, S. (1976). In "The Mechanisms of Mineralization in the Invertebrates and Plants" (N. Watabe and K. M. Wilbur, eds.), pp. 191–202. Univ. of South Carolina Press, Columbia.

Koulourides, T. (1968). In "The Art and Science of Dental Caries Research" (R. S. Harris, ed.), pp. 355–378. Academic Press, New York.

LeGeros, R. Z., Singer, L., Ophaug, R. H., Quirolgico, G., Thein, A., and LeGeros, J. P. (1982). In "Osteoporosis" (J. Menczel, G. C. Robin, M. Makin, and R. Steinberg, eds.), pp. 327–341. Wiley, New York.

Lowenstam, H. A. (1981). Science 211, 1126–1131.

Lowenstam, H. A., and McConnell, D. (1968). Science 162, 1496–1498.

McCann, H. G., and Bullock, F. A. (1957). J. Dent. Res. 36, 391–398.

Margolis, H. C., Varughese, K., and Moreno, E. C. (1982). Calcif. Tissue Int. 34, S33–S40.

Meyer, J. L., and Nancollas, G. H. (1972). J. Dent. Res. 51, 1443–1450.

Neuman W. F., and Neuman, M. W. (1958). "The Chemical Dynamics of Bone Mineral." Univ. of Chicago Press, Chicago, Illinois.

Posner, A. S., Eanes, E. D., Harper, R. A., and Zipkin, I. (1963). Arch. Oral Biol. 8, 549–570.

552   C. A. Baud

Singer, L., Armstrong, W. D., Zipkin, I., and Frazier, P. D. (1974). *Clin. Orthop. Relat. Res.* **99,** 303–312.
Suga, S., Wada, K., and Ogawa, M. (1981). *Jpn. J. Ichthyol.* **28,** 304–312.
Tochon-Danguy, H. J. (1978). *In* "Fluoride and Bone" (B. Courvoisier, A. Donath, and C. A. Baud, eds.), pp. 73–76. Huber, Bern.
Varughese, K., and Moreno, E. C. (1981). *Calcif. Tissue Int.* **33,** 431–439.
Young, R. A., and Elliot, J. C. (1966). *Arch. Oral Biol.* **11,** 699–707.
Zipkin, I., McClure, F. J., and Lee, W. A. (1960). *Arch. Oral Biol.* **2,** 190–195.
Zipkin, I., Posner, A. S., and Eanes, E. D. (1962). *Biochim. Biophys. Acta* **59,** 255–258.

20

Fluorine Chemistry and Energy

JOSIK PORTIER

Laboratoire de Chimie du Solide du CNRS
Université de Bordeaux
Talence, France

I. Introduction

The main feature of the 1973 economic crisis was the oil price increase. A secondary phenomenon was a heightened awareness of the energy problem in a world that until then had been skeptical about the pessimistic prophecies of

INORGANIC
SOLID FLUORIDES

experts. The scientific community also felt more concern, either from conviction or because of the financial pressure produced by new research programs with energy objectives.

Fluorine chemists were also led to reevaluate the role of their research in the field of energy. The fact that fluorine is, so far, absolutely necessary in the nuclear fuel cycle partially justified their studies. However, it became important to develop other energy aspects of this field of chemistry.

In the first part of this chapter the fundamental properties of fluorine and fluoro compounds are examined in the context of energy. The second part concerns the application of fluorides to energy conversion, storage, and savings.

II. Fundamental Properties of Fluorine and Fluoro Compounds in the Energy Context

A. THERMODYNAMIC PROPERTIES

It is difficult to consult a book concerning fluorine chemistry, whether organic or inorganic, devoted to biology or geology, without finding this preliminary statement: fluorine is among the most electronegative elements. This chapter is not an exception; in the energy field, this is also the main feature.

Fluorine tends to accept an extra electron in its electronic cloud and adopts the octet structure of its neighbor neon ($1s^2$, $2s^2$, $2p^6$). The resulting bond is particularly strong. The consequence is a high thermal stability for fluoro compounds. This fact is important for energy applications. Thermally stable materials are especially useful for heat transfer and storage. In this respect fluorides are of great interest.

The electronegativity of fluorine also gives rise to reactions ($x M + y F_2 \rightarrow M_x F_{2y}$) with high free energy of formation $|\Delta G|$. In the conversion of chemical energy to electrical energy, in batteries, for instance, this property might induce high voltages.

The unique charge of fluorine ions leads to low lattice energies. The consequence is a low melting and boiling point for most fluorides. This property together with their stability in the liquid or gaseous state might have numerous applications (storage, cooling fluids, etc.) Moreover, it is the moderate boiling point of UF_6 and the stability of its vapor that are used in the isotopic separation of uranium. Other compounds having such properties do not exist; UCl_6, for instance, which is also volatile, decomposes in the vapor state.

The electronegativity of fluorine gives to the C—F bond considerable energy (C—F $= -107$ kcal mol^{-1} in comparison with C—H $= -87$ kcal mol^{-1}). As a consequence the organofluoro compounds possess a good thermal stability but also a high resistance to ultraviolet (uv) light. This feature is important for solar energy uses.

B. OPTICAL PROPERTIES

In the solid state fluorides have a broad band gap. This property is also a consequence of fluorine electronegativity. Indeed, transparency to high-energy radiation is limited in the salts by possible electron transfer from the anion to the cation. Such a transfer is difficult in the case of fluorides. For instance, alkali or alkaline-earth fluorides have a band gap wider than 10 eV. This property may be of interest for materials used in solar energy, where transparency to the uv region of the solar spectrum is required. It is obvious, however, that fluorides are generally not convenient for photovoltaic conversion because of transparency to sunlight. The semiconductors used (e.g., Si and Cu_2S/CdS) have a smaller band gap (1 eV $< E_g <$ 2.5 eV).

Another noteworthy optical property is the low refraction index due to the weak polarization of fluorine (see Chapter 7). Whereas oxides have a refraction index higher than 1.5, fluorides containing light elements possess indexes close to 1.3. This property is used for solar energy conversion and laser fusion.

C. ELECTRICAL PROPERTIES

Solid fluorides are, in general, electronic insulators; this property is obviously connected to halogen electronegativity. Some fluorides, however, have a high ionic conductivity, which can be explained by the small size of F$^-$ (see Chapter 5, 11, and 12). Both properties can be used for electricity storage.

D. THERMAL PROPERTIES

In addition to thermal stability, the most important properties to be considered are specific heat, transformation or fusion enthalpy, and thermal conductivity.

The thermal conductivity of fluorides is generally as low as that of insulators. In contrast, fluorides possess heat capacities and heats of fusion per unit weight or per unit volume that are appreciably higher than those of other salts. This property is due to the low mass and small radius of the fluorine atom. If one compares, as an example, sodium fluoride and sodium chloride, one observes that the molar fusion enthalpies are of the same order of magnitude, whereas the enthalpies per unit weight or per unit volume are much larger for the fluoride, as shown in the following tabulation:

	T_F (K)	$\Delta H_{f\,mol}$ (kJ)	$\Delta H_f/g$ (kJ)	H_f/cm^3 (kJ)
NaCl	1074	29.5 kJ	0.5 kJ	0.8 kJ
NaF	1266	33.6 kJ	0.8 kJ	1.6 kJ

III. Energy Conversion

A. NUCLEAR ENERGY

The role of fluorine in isotopic separation is described in Chapter 21. However, some less classical applications of fluorides to nuclear techniques must be emphasized.

1. Fused Salt Reactors

A move to replace the conventional pressurized water reactors employing uranium by molten salt breeders using thorium is developing. Such breeders would have numerous advantages:

1. They use thorium instead of uranium. It is estimated that deposits of thorium throughout the world are four times as large as those of uranium.
2. They do not yield plutonium as the end product, but rather uranium-233. If it is easy to manufacture atomic bombs with plutonium, it is quite impossible with uranium-233, the handling of which is extremely difficult because of strong γ-ray emission, making the element useless for nuclear weapons.
3. They are safer than conventional reactors because they are easier to stop.

A fused salt breeder is quite simple in structure. A cylindrical vessel is fed with fused lithium, beryllium, and thorium fluorides at 500°C. It is bombarded by a fast-proton beam, and thorium in the liquid state changes into fissionable uranium-233, discharging heat and neutrons that are used for rebreeding uranium-233.

Fluorides could be selected for their good thermal stability, which makes them less corrosive of the structure materials (Society of Molten-Salt Thermal Technology, 1980).

2. Fusion Reactors

The realization of fusion would solve the energy problem of the industrial world. This is why so much research is devoted to this area. Various fluorides can be considered in this program (Steiner, 1976).

The fusion reaction $^2D_1 + {}^3T_1 \rightarrow {}^4He\,(3.2\ \mathrm{MeV}) + n(14.1\ \mathrm{MeV})$ occurs in a plasma according to the Lawson criterion: at a temperature T the product of the plasma density with confinement duration t must reach a minimal value (e.g., $T = 10^8$ K, $n = 10^4$ nuclei per cubic centimeter, $t = 1$ sec). Two types of confinement are used. The first is realized with magnetic fields. The second is called the laser–pellet concept: A tritium–deuterium target is illuminated with a laser beam; the target is heated and ionized; the implosion compresses the central part of the target.

The laser power must be enormous. In the scope of the Shiva program (Lawrence Livermore Laboratory), the lasers provide 20–30 TW during 10^{10} sec. Such power is obtained with optical pieces of large dimension. Consequently, single crystals cannot be used as in classical solid lasers. Neodynium-doped glasses have been selected. Among the materials studied, it seems that fluoride glasses (fluoroberyllates and fluorophosphates) are the most promising because of their low refractive index (Weber, 1976).

Other components such as molten LiF or Li_2BeF_4 have been considered for fusion reactors in the breeding part or for cooling.

B. SOLAR ENERGY

1. Photovoltaic Conversion

As mentioned before, the band gap of fluorides is not adapted to photovoltaic conversion. It has been proposed, however, that fluorine can improve the materials used (Ovshinsky and Madan, 1978).

Commercially available photovoltaic devices are made with silicon single crystals, expensive materials, because of the crystal-growing operation. Much less expensive amorphous silicon has been proposed. Unfortunately, the conversion yield is very low; the diffusion length of the electron hole pair is short because of the defects. The yield is improved by the presence of hydrogen in silicon. The Si—H bond, however, is not stable under uv radiation; "Si—H" doped with fluorine, "Si—H—F", would be stable under irradiation with a yield of 10%.

2. Photothermal Conversion

The simplest and cheapest conversion of solar energy is conversion in the form of heat. Some studies concern the use of fluoro compounds for solar collectors. The cover of a collector is generally glass; however, a part of the sunbeam is reflected. It is possible to limit this reflection by deposition on the glass of a thin film of material of lower refractive index. Some experiments were carried out with MgF_2; a significant increase of 3.4% in the energy transmitted to the collector was observed (Hsieh and Coldewey, 1974).

It may be worthwhile replacing the glass by polymers. Indeed, those materials have certain advantages:

1. *Low refraction indices.* This results in low losses by reflection. This property is explained by the fact that the elements forming polymers (carbon, fluorine, oxygen, nitrogen, etc.) are lighter that those found in glasses (silicon, sodium, calcium, etc.). This property is enhanced in the case of fluoro polymers by the weak polarizability of fluorine:

Material	Refractive Index
Glasses	$1.51 < n_0 < 1.52$
Polycarbonate	1.59
Polyvinyl fluoride	1.45
Polytetrafluoroethylene	1.34

2. *Weak absorption.* Polymers can be used in very thin films. The absorption due to fundamental extinction is consequently reduced.

A disadvantage of polymers is the gradation involved in the UV part of sunlight. In this respect, fluoro polymers are more stable than chloro polymers because of the strength of the C—F bond (Zerlaut, 1976).

3. Conversion of Solar Energy to Mechanical Energy

Freons are used in engines designed for pumping underground water. Liquid Freon is evaporated by the heat resulting from a solar collector. The compression work is used to put the pump in action. Then the pumped water cools the Freon, which is again condensed. This system, however, has a low efficiency due to the similar temperatures of the cold and hot sources ($\Delta T \simeq 30°C$). It is nevertheless useful in areas where dryness is severe (Vaillant, 1978).

IV. Energy Storage

Energy storage is a very important issue. This is obvious when one is using discontinuous sources of energy (e.g., solar). In addition, good energy management requires efficient storage. Indeed, if the production of energy is, as a rule, nonstop (e.g., nuclear) the consumption can be time or season dependent.

A. THERMAL STORAGE

Heat content and latent heat are used independently or together. The desired properties are a high specific heat, a high heat of fusion, thermal stability, weak corrosion for metals and oxides, and low cost. Various materials are considered according to the temperature (Dumon, 1977); many of them are fluorides.

1. Low Temperature

The storage of heat at low temperature ($t < 100°C$) is especially required in the case of solar energy using planar collectors. The materials used are various hydrates ($Na_2SO_4 \cdot 10H_2O$, $Na_2SO_3 \cdot 10H_2O$, etc.) that correspond to the previously mentioned criteria. A fluoride, $KF \cdot 4H_2O$, has been proposed. The advantage of that compound is its low melting point (18.5°C), close to room temperature; thus, no substantial isolation is needed (Schröder, 1977).

2. Moderate Temperature

Organic fluids can be stored at $\sim 300°C$. High-temperature stable oils are needed. The stability of fluoro compounds could solve this problem.

3. High Temperature

It is in the field of high temperature that fluorides seem to show the greatest potential. Indeed, molten fluorides are thermally more stable and much less corrosive to metals than other salts, particularly chlorides. In addition, as previously shown, fluorides have a very high specific energy. For instance, a 1-kWhr storage capacity has a volume of 40 liters in the case of a lead battery, compared with 1.3 liters for a heat storage cell using light fluorides. A heat storage system using eutectic fluoride mixtures associated with a Stirling engine was proposed for space utilization (Schröder, 1975).

B. ELECTROCHEMICAL STORAGE

Electrochemical storage is important not only for balancing electrical production and consumption (peak saving) but also for the propulsion of vehicles. In both cases electrochemical systems with specific energy ($Whr\,kg^{-1}$) as high as possible are required. That criterion is more important for mobile storage, however, than for static storage.

The maximum amount of energy that could be obtained in a chemical reaction (temperature and pressure being constant) is given by Gibbs energy. Furthermore, the minimum weight of a battery is that of the reactants if the weights of the electrolyte and container are neglected. Consequently, the theoretical specific energy is

$$-\Delta G \Big/ \sum M = nFE \Big/ \sum M,$$

where n is the number of electrochemical equivalents, F the Faraday constant, E the voltage, and $\sum M$ the sum of the molecular masses of the reactants.

High ΔG and consequently high voltages are obtained when one selects for the negative electrode elements with a weak electronegativity (left side of the periodic table) and for the positive electrode elements with a strong electronegativity (right side of the periodic table). In both cases, in order to obtain a low mass, one must select the elements in the first rows of the periodic table. It is obvious that the best battery that can be designed should have lithium at the cathode and fluorine at the anode.

Such a device has been produced. A lithium primary battery was designed by Matsushita Company on the basis of research of fluorine chemists using intercalation of fluorine into graphite (see Chapter 8) (Watanabe, 1973). The following reaction is used:

$$n\,Li + (CF)_n \longrightarrow n\,LiF + n\,C$$

The specific energy is very high (300 Whr kg^{-1}). It can provide energy for a watch for 20 years. Unfortunately, this system is not reversible; as a consequence, it cannot be used for massive storage of energy.

An interesting property of fluorides is the high ionic conductivity they may have (Reau and Portier, 1978). The performances are of the same order of magnitude as those of β-alumina, which is often considered the most promising solid electrolyte. Consequently, it was tempting to design solid batteries using F^- conductors. Several systems have been designed in the form of ceramics or thin films, as shown in the following tabulation:

Cell	ocv (v)
$Pb/\beta\text{-}PbF_2/BiF_3$	0.353
$Pb/\beta\text{-}PbF_2/AgF$	1.25
$Pb/\alpha'\text{-}PbSnF_4/C_8AsF_6$	1.90

A significant voltage was obtained for the last cell using a graphite salt. Unfortunately, it exhibits a high internal resistance, limiting the current by the formation of an ionic insulating film on the surface of the electrolyte due to the oxidizing power of C_8AsF_6. Most likely, more valuable batteries could be built in which one electrode were made of graphite intercalated with an alkali metal and the other an oxidizing graphite compound. The remaining problem is to find a solid electrolyte that is compatible with such an electrode couple.

V. Energy Savings

Until 1973 the economic efficiency of a country was measured by its gross national product. Energy was almost free; so energy consumption was neglected. Since that time this point of view has had to be reconsidered. Indeed, importation of oil often became disastrous for national economies, especially because the production level had to be maintained to avoid unemployment. Consequently, a more or less elaborate program of energy savings has been developed in many countries. This effort is not uniform, however. To produce 1 U.S. dollar of gross national product, 15 MJ are necessary in Sweden, 18 in France, 23 in Japan, 39 in the United States, and 56 in the Soviet Union (Claasen, 1976).

Energy savings can be realized in various and numerous domains. In the fluorine industry itself it is possible to save energy. The preparation of fluorine by electrolysis requires high overvoltages due to the formation on the anodes of insulating compounds. A better knowledge of the mechanism of anodic oxidation should provide new solutions.

Many uses of fluoro compounds in energy savings could be quoted, but only three examples are given here: the reduction of friction, the valorization of low-temperature calories, and the limitation of corrosion.

A. REDUCTION OF FRICTION

Twenty percent of the primary energy used in the world is transformed into mechanical energy. Obviously, this conversion is limited by Carnot's yield. For instance, 10–40% of the energy released by the combustion of fuel is used as mechanical energy. Unfortunately, it is often used in an inefficient manner and it quickly returns to the form of heat due to friction. As a significant example, the energy yield of a car is only 3%.

Some of the loss is due to the friction of mechanical parts. Teflon can be used to advantage in the manufacture of these parts. This fluoro polymer has a very low friction coefficient (about five times lower than that of steel). It is used to make dry bearings and pinions. In certain European cars the distribution chains are replaced by Teflon belts. The possibility of manufacturing engines that run without oil is being studied, the lubrication being provided by Teflon.

Another interesting compound is graphite fluoride $(CF)_n$. Because of its layer structure, it has a very low friction coefficient, even lower than those of Teflon, graphite, and molybdenum sulfide. It is used as a lubricant (Play and Godet, 1975). Solid lubricants based on alkali and alkaline-earth fluorides have also been proposed (Sliney, 1982) (see Chapter 12).

B. VALORIZATION OF LOW-TEMPERATURE CALORIES

An important part of fuel and electricity is used for the production of low-temperature calories, in particular for heating houses. It is possible to save energy in this domain. Some energy-saving systems make use of fluoro compounds.

The conversion of thermal energy into mechanical energy and then into electricity depends on the Carnot's yield:

$$W/Q = (T_H - T_C)/T_H,$$

where W/Q is the ratio of the obtained work to the heat taken out of the hot source T_H. Of course, the yield is lower than unity. Consequently, it is absurd to transform electricity into heat by the Joule effect to heat a house, for example. If electricity is used to produce work W, however, it is possible with a heat pump to extract calories from the cold source and inject them into the hot source:

$$R' = Q'/W = T'_C/(T'_H - T'_C).$$

In this case the "yield" exceeds unity. In a heat pump a gas is liquefied by an electrical compressor releasing heat to the hot source. Then, the liquid is evaporated by absorbing heat from the cold source.

Various fluids can be used; generally, heat pumps work with Freons. In addition to their low toxicity, Freons are thermally stable and have great versatility. Their boiling points are vary considerably with composition, allowing their utilization for various temperatures of hot and cold sources ($92.8 \,°C$ for CCl_2/CCl_2F to $-82.2 \,°C$ for CHF_3).

More complicated cycles could be used in which the work given by the electrical compressor is replaced by a third source at high temperature (combustion of fuel, sun, etc.). High-temperature stable fluids are required. Some fluo aromatic compounds have been proposed (Herker, 1977).

C. CORROSION PREVENTION

The oxidation of metals by atmospheric agents corresponds to an enormous loss of energy. The corrosion of 1 kg of steel costs 30 MJ of primary energy.

A new process for the passivation of metallic surfaces using fluoro compounds has been discovered. The chromium salts used for the treatment of sheet iron are replaced by potassium monofluorophosphate. The nature of the passivation mechanism is not yet entirely clear. One knows, however, that the lifetime of metallic parts is extended by 25% with respect to the chromium process (Cot, 1978).

References

Cot, L. (1978). U.S. Patent 789,658.
Claasen, R. S. (1976). *In* "Critical Materials Problems" (C. Stein, ed.), 5. Academic Press, New York.
Dumon, R. (1977). "Energie Solaire et Stockage d'Energie." Masson, Paris.
Herker, F. E. (1977). *J. Fluorine Chem.* **9,** 113.
Hsieh, C. K., and Coldewey, R. W. (1974). *Sol. Energy* **16,** 63.
Ovshinsky, S. R., and Madan, A. (1978). *Nature (London)* **267,** 482.
Play, D., and Godet, M. (1975). *Proc. Colloq. Int. CWRS* **233,** 441.
Réau, J. M., and Portier J. (1978). *In* "Solid Electrolytes" (P. Hagenmuller and W. Van Gool, eds.), p. 313. Academic Press, New York.
Schröder, J. (1975). *J. Eng. Ind.* **97,** 893.
Schröder, J. (1977). *In* "Energy Storage" (H.P. Silverman and J. B. Berkowitz, eds.), p. 206. Electrochem. Soc., Princeton, New Jersey.
Society of Molten-Salt Thermal Technology. (1980). "Fundamentals of Molten-Salt Thermal Technology." Foundation for Industrial & Economic Research, Tokyo.
Steiner, D. (1976). *In* "Critical Materials Problems" (C. Stein, ed.), p. 39. Academic Press, New York.
Vaillant, R. J. (1978). "Utilisations et promesses de l'énergie solaire." Eyrolles, Paris.
Watanabe, N. (1980). *Solid State Ionics* **1,** 87.
Weber, M. J. (1976). *In* "Critical Materials Problems" (C. Stein, ed.), p. 261. Academic Press, New York.
Welch, M. J. (1976). *In* "Critical Materials Problems in Energy Production" (C. Stein), p. 26. Academic Press, New York.
Zerlaunt, G. A. (1976). *In* "Critical Materials Problems in Energy Production" (C. Stein, ed.), p. 389. Academic Press, New York.

21

Industrial Uses of Inorganic Fluorides

BERNARD COCHET-MUCHY
Atochem
Centre de Recherches de Lyon
Pierre Bénite, France

JOSIK PORTIER
Laboratoire de Chimie du Solide du CNRS
Université de Bordeaux
Talence, France

I. Inorganic Fluorides in Industry: Commodities and Specialities

A. HISTORICAL DEVELOPMENT OF THE USE OF FLUORIDES

The use of natural inorganic fluorides is not new. We find traces of their use from Greek and Roman times, when large crystals of fluorspar were used to make dishes, vases, and other ornamental items. In antiquity, the Chinese also discovered in fluorspar an interesting material for cutting jewels and other adornments.

Of Latin origin, the word *fluorine* etymologically refers to another, also very old, use. The Latin verb *fluere* means *flowing* and refers to the property some fluorides show when melted, that is, the property of decanting easily as a liquid. The idea of using this property in metallurgy to produce a flux seems to date back to 1556, when it was described by Agricola (1950). It was not until much later, however, that a truly industrial demand for fluorspar for metallurgical purposes arose. Its origin is connected with the introduction of the Bessemer and Siemens–Martin processes into iron and steel metallurgy.

The development of the first natural and then synthetic fluorine compounds of aluminum for the manufacture of nonferrous metals resulted from the simultaneous discovery by Hall in the United States and by Heroult in France of a process of electrolytic preparation of aluminum by reducing its oxide, which is dissolved in a bath of molten cryolite.

As a result of first Scheele's work in the laboratory (1771) and then Fremy's work (1856), the preparation of hydrogen fluoride reached the industrial stage, but only in 1930, with the creation of two production units, one in the United States and one in France. This date marks an important step in the industrial development of fluorides, because the manufacture of this acid from fluorspar made the rapid development of another chemistry possible: that of fluorine and its derivatives. The original properties of fluorine compounds has given them an important role in products, and fields as varied as refrigerants, plastics, petrochemical industry, aerosol propellants, surfactants, space industry, health service, electrical industries, metallurgy, and nuclear industry.

As far as the last field is concerned let us remember the decisive emphasis given in the 1950s to the development of fluorides by the work carried out in the United States under the Manhattan Project. Linked originally to an essentially military aim, because it was a way of preparing large amounts of uranium hexafluoride necessary for isotopic separation plants, this work showed clearly the privileged role and the indispensable character of fluorides, particularly inorganic fluorides, in the nuclear fuel cycle they helped to promote.

Except for some developments in Germany in the 1940s (Neumark, 1947), it is this period that marks significantly the transition to the truly industrial stage from the work of Moissan (in 1886) on the electrolytic preparation of elemental fluorine.

Finally, the large amount of research related to the development of nuclear energy for military and peaceful purposes and carried out by the industrialized countries has contributed to a better knowledge of the access roads and properties of numerous fluorides, leading to many advances and new uses. It can probably be said that the atomic era is the era of fluorine.

To summarize, if we do not include the metallurgical uses, the production of considerable tonnages of inorganic fluorides for industrial use is relatively recent. This explains the distinction between the term *commodities*, which is applicable to a limited number of fluorides that are mined or produced on a large scale, and the term *specialities*, which is reserved for products whose demand is limited by the specificity of their use.

B. MAIN SOURCES OF FLUORIDES AS RAW MATERIALS IN THE INDUSTRIAL PRODUCTION OR USE OF INORGANIC FLUORIDES

The earth's crust contains between 0.06 and 0.07% in weight of fluorine, which places this element in the thirteenth position on the scale of abundance. Numerous minerals contain some fluorine associated with other elements in various amounts and with various degrees of complexity. The only usable sources in the industrial stage, however, are the ores whose nature and fluorine content allow profitable exploitation at an economic level. In fact, there are three basic raw materials:

1. *Natural cryolite* $AlF_3 \cdot 3NaF$ (54.3% F). The richest of the ores, largely mined when industrial aluminum production began. It is practically given up now; a deposit at Ivigtut in Greenland was exploited until the late 1960s.
2. *Fluorspar* or *fluorite* CaF_2 (48.7% F). The most largely used. Its average content in exploited deposits is nearly 80%, but it can also be exploited with a lower content.
3. *Fluorapatite* $CaF_2 \cdot 3Ca_3(PO_4)_2$ (3.8% F). Raw material in which fluorine appears as a by-product of the phosphorus and fertilizer industries. Its potential reserves are large, and it likely represents the ore of the future.

1. Fluorspar

a. RESOURCES AND NEEDS. Widely distributed in all parts of the world, fluorspar is presently the main basic material.

The assessment of the economic and subeconomic reserves is closely linked to the technoeconomic criteria applied to the estimates: CaF_2 content, nature of the associated minerals, type of deposit, distance between the mining and consumption sites, and market prices.

In 1978 the economic fluorspar reserves were estimated, on a worldwide scale, to be more than 43 million metric tons expressed in fluorine contained in the ores with more than 35% CaF_2. An additional 36 million metric tons of reserves are subeconomic in present economic circumstances.

Geographically, fluorspar reserves are relatively well distributed, as shown in Table I. It is advisable to point out the privileged place South Africa holds, for it alone controls almost 30% of the world reserves. Mexico also has an excellent position on a worldwide level (13% of the reserves) and on the American continent.

TABLE I

World Fluorine Reserves Contained in Fluorspar, 1978[a]

Location	Reserves	Others[b]	Total
America	10,200	17,100	27,300
North			
Canada	800	450	1,250
Mexico	5,400	2,150	7,550
United States	2,500	13,000	15,500
South			
Argentina	700	700	1,400
Brazil	800	800	1,600
Europe	10,250	9,000	19,250
Western			
France	1,300	500	1,800
Italy	1,000	3,350	4,350
Spain	1,450	900	2,350
United Kingdom	3,200	1,850	5,050
Eastern			
Soviet Union	2,200	1,200	3,400
Other	1,100	1,200	2,300
Africa	15,900	4,000	19,900
Kenya	1,200	550	1,750
Namibia	900	900	1,800
South Africa	12,700	1,650	14,350
Other	1,100	900	2,000
Asia	6,350	5,700	12,050
China	1,700	450	2,150
India	1,300	300	1,600
Mongolia	1,600	750	2,350
Thailand	1,550	4,100	5,650
Other	200	100	300
Oceania	200	200	400
	42,900	36,000	78,900

[a] Thousands of metric tons. From U.S. Department of Interior (1980).

[b] Known but subeconomic reserves.

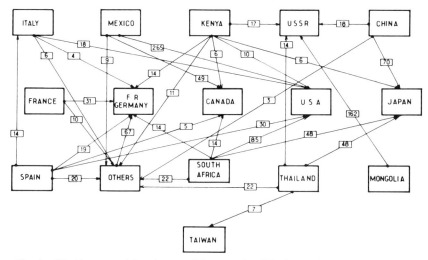

Fig. 1.   World commercial exchanges of fluorspar in 1978 (thousand metric tons of fluorine contained in fluorspar). From Morse (1980).

The imbalance between localization of the resources and need leads to an exchange movement, the volume of which represents about 50–60% of production. The principal places of importation are North America (United States, Canada), ~40% of the exchange, 65% coming from Mexico; western Europe, ~15%; the Soviet Union and socialist countries, ~20%; and Japan, ~15%.

Figure 1 illustrates the principal exchange movements between the producing and consuming countries in 1978; the transactions, when expressed in fluorine content, corresponded to about 1.2 million metric tons. One may say that the world economy of fluorspar is dominated largely by North America and particularly by the United States as far as consumption or control of the resources is concerned.

Table II gives an idea of the needs of the main consumers for the year 1978. These needs have been satisfied by the producing countries in proportions ranging from a few percent to more than 20%, as shown in Table III for the year 1978. A part of the produced tonnage is used for storage.

b. GRADES.   Fluorspar ores can be classified into three main groups on the basis of their CaF_2 content and their final utilization. The first consists of *metallurgical-grade fluorspar* (or metspar), the most common of which contains at least 60% CaF_2.[†] The second group, *ceramic-grade fluorspar*,

---

[†] In the United States one often speaks of CaF_2 effective content for metspar. This means a value obtained by subtracting $2\frac{1}{2}$ times the silica content from the CaF_2 content.

TABLE II

Fluorspar Demand, 1978[a]

Country	Reported Consumption
United States	1100
Mexico	360
Japan	380
South Africa	50
Western Europe	1110
	3000

[a] Thousands of metric tons. From U.S. Department of Interior (1981) and Chemical Daily Co., Ltd. (1980).

TABLE III

Breakdown of World Production of Fluorspar, 1978[a,b]

	% of Total World Production		
Location	Continent	Region	Country
America	25		
North		23	
United States			2.5
Mexico			20.5
South		2	2.0
Europe	38		
Western		23	
France			6.5
Italy			3.5
Spain			6.5
United Kingdom			4.0
Other			2.5
Eastern		15	
Soviet Union			11.0
Other			4.0
Africa	12.5		
Kenya			2.2
South Africa			8.5
Other			1.8
Asia	24.5		
China			8.5
Mongolia			9.8
Thailand			5.0
Other			1.2
	100.0		100.0

[a] From U.S. Department of Interior (1982).

[b] World Production in 1978 was estimated to be 4660×10^3 metric tons of fluorspar and 2100×10^3 metric tons of fluorine.

containing 85–97% CaF_2, consists of two types: grade 1, 95–97% CaF_2; and grade 2, <93% CaF_2. The third group, *chemical-grade fluorspar for acid* (acid-grade fluorspar, or acidspar) contains at least 97% CaF_2.

The main impurity of fluorspar is generally between 1 and 1.5% silica. Sulfur (0.03–0.2%), calcite (1%), iron oxide, lead, and zinc are also found.

c. TREATMENT OF THE ORE.   With a few exceptions the crude ore is seldom pure enough to be used as is or after simple physical preparation: hand sorting, crushing, grinding, and screening.

Depending on the content and purity ascertained, the ore is subjected to various enrichment and purification treatments, at various degrees of complexity; these are almost always of a physical nature. They can range from a simple manual screening to flotation, if it is necessary to go beyond a first enrichment by conventional methods employing, for example, heavy medium cone, cyclone, and drum separators. At the flotation stage it is possible to recover certain impurities, barite, galena, and sphalerite, in order to use them independently.

To meet the demands of certain users, particularly those of U.S. steel manufacturers, various agglomeration and compacting processes have been developed.Their final purpose is to provide a dense and nonfriable material in the form of balls, briquets, or pellets. The technique makes use of a mixture made from fine-powder fluorspar (generally the fines that are the result of the enrichment treatments) diluted by other products (e.g., magnesium, iron, aluminum oxides, or hydrated lime).and thickened with a binder (molasses, silicates, tall oil, or lignin). The shaping is ensured by the use of classical equipment: roll compactor and briquet fabricating press.

A considerable portion of the production is delivered in the form of briquets, especially for metallurgy. The distribution of production among the various qualities depends on the importance of the respective uses. As shown in Table IV, it varies from one country to another and also with time.

TABLE IV

Uses of Fluorspar in Various Countries

End Use	United States (%)	South Africa (%)	Japan (%)	Western Europe (%)
Metallurgy	40	72	70	
Ceramics and other	7	18	} 30	} 40
Chemistry	53	10		60
	100	100	100	100

TABLE V

Prices of Various Qualities of Fluorspar by Country of Origin, First Quarter of 1982[a]

Fluorspar	Country of Origin				
	United States	Mexico	Europe	South Africa	China
Metspar, $>70\%$ CaF_2	121	118–123	100	122	—
Ceramic grade, 88–90% CaF_2	110	—	—	—	—
Ceramic grade, 95–96% CaF_2	182	—	—	—	—
Acidspar, $>97\%$ CaF_2	193–198	149–154	130–160	169–171[b]	100[b]

[a] U.S. dollars per metric ton, free on board (FOB) (border or port). From *Industrial Minerals* (June 1982), *Engineering and Mining Journal* (April 1982).
[b] Cost, insurance, and freight (CIF) U.S. ports.

d. PRICE.   Prices evidently depend on several factors: quality, tonnage, origin, presentation, packaging (bulk, in bags, or in containers), and the supply-and-demand ratio. Tables V and VI give approximate prices.

2. Natural Cryolite

As pointed out earlier, natural cryolite is practically vanishing. Mining of the only known deposit in Greenland at Ivigtut was stopped after the constitution of a considerable store. Indispensable to the aluminum industry, where it was first used, natural cryolite has progressively been replaced by synthetic products made from hydrofluoric and fluosilicic acids.

3. Phosphate Rock Deposits

Phosphate rock deposits, practically unexploited at present as a primary source of fluorine for economic reasons, constitute enormous reserves. They

TABLE VI

Prices of Fluorspar Imported by Japan, 1981[a]

Value[b]	Country of Origin				
	North Korea	China	Thailand	Kenya	South Africa
Thousands of yen per metric ton	12.4	21.1	18.9	24.4	31.6
U.S. dollars per metric ton	56.3	95.7	85.7	110.7	143.4

[a] From Japan Exports and Imports, Japan Tariff Association (1982).
[b] CIF Japan.

are estimated (Morse, 1980) to contain ~ 345 million metric tons of fluorine, of which it would be easy to recover $\sim 20\%$, that is, either about 70–72 million metric tons (almost double of the reserves of fluorspar presently known, i.e., 43 million metric tons; see Table I).

Numerous processes for the recovery of fluorine from natural phosphates have been investigated by the producing countries, the United States and Europe in particular, to be sure of their supplies of raw materials in case of a shortage of fluorspar. To our knowledge, there is no industrial plant that has been set up for this single purpose. There are some localized units, however, although there are only a limited number of them, mainly in the United States. These ensure the recovery and reuse of fluosilicic acid, a by-product of the phosphoric acid and fertilizer industries (superphosphates).

Statistics show a production of fluosilicic acid from phosphate rock of 50,000 to 60,000 metric tons per year, or 40,000–50,000 metric tons per year fluorine, that is, less than 10% of U.S. needs.

In conclusion, it can be pointed out that the balance between needs and resources will be maintained at medium term (~ 10 years) without major difficulty, as shown by following tabulation:

Fluorine Reserves	Metric Tons
In the form of fluorspar	43,000,000
In the form of natural phosphates	72,000,000
	115,000,000
Prediction of consumption until the year 2000	68,000,000

C. FLUORIDES IN THE CHEMICAL INDUSTRY

1. Hydrofluoric Acid

Figure 2 illustrates the importance of hydrofluoric acid, the true center of the inorganic and organic fluorine industry. Although this chapter is devoted to solid inorganic fluorides, this acid should be mentioned here.

Hydrofluoric acid is one of the main end products of fluorspar. An average of 2.2 metric tons of fluorspar are consumed per metric ton of hydrogen fluoride produced. It is the reagent that opens the door to the majority of inorganic fluorides manufactured in large quantities.

a. MADE FROM FLUORSPAR. Called hydrofluoric acid when it is in aqueous solution and hydrogen fluoride when it is anhydrous, its preparation requires the reaction of sulfuric acid on fluorspar according to the equation

$$CaF_2 + H_2SO_4 \longrightarrow CaSO_4 + 2\,HF\uparrow$$

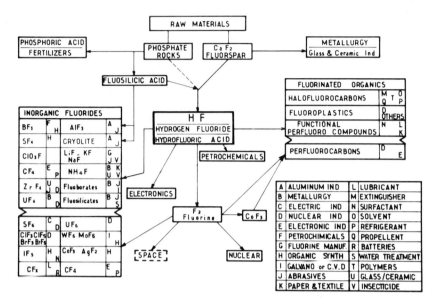

Fig. 2. Hydrogen fluoride: a basic raw material in fluorine chemistry. C.V.D., Chemical vapor deposition.

Today this reaction is carried out by a continuous process in large-capacity units that are entirely automatic and perfectly salubrious.

Various processes (Gall, 1980) using the basic reaction exist all over the world. They require some closely and some not so closely connected techniques, the variations of which are often very similar and relate to (1) eventual preliminary concentration and purification of fluorspar and its drying if required by the nature of the raw material; (2) spar–sulfuric acid mixing (3) thermal treatment leading to the production of gaseous acid, generally carried out in a rotary kiln, with external heating, at 200 to 300°C, and (4) condensation and purification of the acid contained in the output gas.

The reaction gives a sterile product with a low content of residual HF–CaSO$_4$ and hydrofluoric acid contained in a gaseous flux made up of, in addition to water vapor and sulfuric acid, volatile impurities such as SO$_2$, SiF$_4$, and H$_2$S due to the presence of silica and sulfur. So to obtain an acid as pure as possible, it is important to use a high-content and high-purity raw material such as acid-grade fluorspar. A wise combination of dust removal, demisting, condensation, washing, and distillation techniques allows a high-purity anhydrous acid to be obtained, whatever the process.

Diluted acid can be obtained either during the stages of concentration leading to the anhydrous product or by dilution.

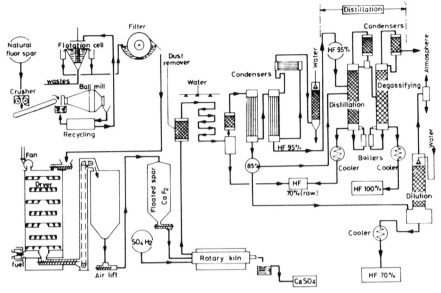

Fig. 3.   Flow sheet of a conventional hydrogen fluoride production unit.

Figure 3 represents the flow sheet of a production unit. This fabrication is characterized by a few tricky stages: (1) correct proportioning and mixing of the reactives, on which the yield depends; (2) corrosion, partly controlled by the choice of suitable materials; and (3) elimination of the sterile product under acceptable ecological conditions if it cannot be reused.

   b. MADE FROM FLUOSILICIC ACID.   Although these processes still do not have any industrial reality, it is useful to remember that hydrofluoric acid can be made from fluosilicic acid as a result of the recovery of fluorine from various sources, including the aluminum, phosphate, fertilizer, and fluoride industries.

   Fluosilicic acid is obtained by the dust removal and washing of the residual gases containing fluorine in SiF_4 and HF forms using a solution of recycled fluosilicic acid. The operation, which is carried out in wet cyclones and then in scrubbers generally equipped with nozzles, gives a solution of 20 to 25% of H_2SiF_6. The following reactions were observed:

$$4\,HF + SiO_2 \longrightarrow SiF_4 + 2\,H_2O$$
$$SiF_4 + 2\,HF \longrightarrow H_2SiF_6$$
$$2\,SiF_4 + 2\,H_2O \longrightarrow H_2SiF_6 + 2\,HF + SiO_2$$

Among the proposed processes for reusing the solution of H_2SiF_6 thus obtained in the form of hydrofluoric acid, three groups of techniques have been considered:

1. Reaction of ammonia on the solution according to

$$H_2SiF_6 + 6\,NH_3 + 3\,H_2O \longrightarrow 6\,NH_4F + SiO_2$$

2. Acidification of the H_2SiF_6 (Stauffer Chemical Company, 1963a,b, 1966), using concentrated sulfuric acid, which allows the SiF_4 recycled at absorption to be displaced:

$$H_2SiF_6 + H_2SO_4 \longrightarrow 2\,HF(aq) + SiF_4(g)$$

The $H_2SO_4/HF/H_2O$ mixture is separated using various techniques (distillation and stripping).

3. Concentration of fluosilicic acid and thermal decomposition; this is the Buss process (Buss A.G., 1962).

The gaseous $HF/SiF_4/H_2O$ mixture is treated in a column with a selective organic solvent (polyethers or polyols), which separate hydrogen fluoride in the form of a complex. The vapors that contain H_2O and SiF_4 are retransformed into H_2SiF_6 by absorption and are then recycled.

c. PRODUCTION AND PRICE OF HYDROFLUORIC ACID. Economically, it is difficult to determine from the market data only how much hydrofluoric acid is manufactured. Indeed, the manufacturers themselves consume mostly what they produce, thus reducing the amount of trade movement.

Need, however, is narrowly linked with the development of the aluminum and uranium industries and the demand for fluorides. The figures in Table VII, which are valid for the year 1980, give a rough estimate of the tonnages for the

TABLE VII

Hydrogen Fluoride Production and Consumption, 1980[a]

Country	100% HF Produced (1000 metric tons)	Fluoro-carbons (%)	Aluminum (%)	Other Petro-chemical, Nuclear (%)
United States	340–350	42	33	25
Mexico	80 (90% exported)			
Eastern Europe	300	45	40	15
Japan	95–100	35	45	20

[a] From U.S. and Japanese reviews; communications from private companies.

most important production zones and their distribution among consumption areas. The mean price level in the United States in 1980 was fixed at $1300/metric ton of anhydrous hydrofluoric acid delivered to the utilization sites.

2. Aluminum Fluoride and Cryolite

About 40% of the hydrofluoric acid produced is used in the preparation of aluminum fluoride and synthetic cryolite, of which the aluminum industry is a large consumer. The quantity generally used to produce 1 metric ton of aluminum fluoride is about 2.9 metric tons ore. This industry, therefore, indirectly consumes considerable amounts of fluorspar.

a. ALUMINUM FLUORIDE.

The production of aluminum fluoride heads that of all inorganic fluorides. Two types of processes are used: the wet process and the dry process, the industrialization of which is more recent.

i. WET PROCESS. Traditionally, the manufacture of aluminum fluoride involved the following stages:

1. Attack of a suspension of hydrated alumina by an aqueous solution of hydrofluoric acid from which the fluosilicic acid had been previously removed by sodium carbonate if necessary:

$$Al(OH)_3 + 3\,HF(aq) \longrightarrow AlF_3 \cdot 3H_2O(l)$$

2. Precipitation of the trihydrate from the oversatured solution, separation of the solid, and partial recycling of the solution.

3. Drying and calcination of the trihydrate into an anhydrous product. This process, which may seem simple, is in fact complicated due to the crystallization of $AlF_3 \cdot 3H_2O$, which can be delayed or can be begun during the attack; filtration of the trihydrate, which can be difficult as a result of granulometry variations or the presence of silica; and too rapid drying due to the presence of water and temperature, if too high, can lead to pyrohydrolysis and partial decomposition in alumina:

$$2\,AlF_3 + H_2O \longrightarrow Al_2O_3 + 6\,HF$$

Insufficient drying, however, has the disadvantage of leading to fluorine loss when it is introduced into the electrolysis cells due to hydrolysis of the molten fluorides.

When the preparation of aluminum fluoride from hydrofluoric acid by the wet process gradually evolved into more modern processes, such as the dry process, new units for the production of aluminum fluoride from fluosilicic acid, such as that of the Aluminum Corporation of America at Fort Meade, Florida, appeared in the 1970s. The availability of larger and larger amounts of fluosilicic acid, a by-product of the phosphate industry, the link between the limited fluosilicate market and ecological considerations, and the progressive exhaustion of fluorspar reserves explain this evolution.

A new process has been described (Morlock *et al.*, 1980). The principal stages of the fabrication, represented in Fig. 4, are the following:

1. Fluosilicic acid trapping from the concentration, under vacuum at boiling point, of acid with 30% P_2O_5 into acid with 52 to 54% P_2O_5 content. It is necessary to limit as much as possible the carrying away of small particles of P_2O_5, which would interfere with the electrolysis.

2. Precipitation of silica. This is carried out by a batch process by adding aluminum hydroxide to a solution of fluosilicic acid at 90 to 100°C, according to

$$H_2SiF_6 + 2\,Al(OH)_3 \longrightarrow 2\,AlF_3 + SiO_2 + 4\,H_2O$$

The capacity of the silica to filtrate is narrowly linked to the operation conditions, for example, concentration of the original acid, pH, temperature at the end of the reaction, and stirring.

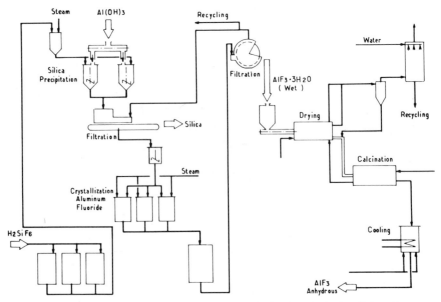

Fig. 4.   Wet process for aluminum fluoride production from hydrofluosilicic acid.

3. Filtration of the silica. This is a delicate part of the process. It is necessary to work quickly to avoid aluminum fluoride losses in the cake, and it is necessary to wash the silica methodically. The choice of materials and the type of filter are also important in order to avoid corrosion and undesirable deposits. This impure silica can be recycled at the ore attack, which allows higher recovery rate of the fluorine contained in the phosphate.

4. Crystallization and filtration of aluminum fluoride. The first stage is carried out by a batch process near boiling point (95°C) with slow stirring:

$$AlF_3(l) + 3 H_2O \longrightarrow AlF_3 \cdot 3H_2O(s)$$

After precipitation the trihydrate is separated from the mother liquor by filtration at a well-controlled pH in order to achieve easy filtration. The mother liquor is used for washing silica.

5. Drying and dehydration of trihydrated aluminum fluoride. This operation is similar to that previously described. It requires the same precautions to avoid partial decomposition.

ii. DRY PROCESS. In this type of process the reaction is no longer carried out in an aqueous medium but between a gaseous and solid phase. The reactor is supplied, by means of a continuous and countercurrent process, with two fluxes. One consists of warm gases rich in hydrofluoric acid which is produced *in situ*. The other consists of dried, dehydrated alumina in the form of a powder with suitable granulometry. The corresponding chemical reaction is

$$Al_2O_3(s) + 6 HF(g) \longrightarrow 2 AlF_3(s) + 3 H_2O(g)$$

Figure 5 diagramatically represents the principle of the unit exploited at Salindres in France by Pechiney-Aluminium. This manufacturing unit is one of the biggest in the world (capacity, 50,000 metric tons aluminum fluoride per year). It consists of two main parts: (1) a rotary kiln with internal heating for the production of hydrofluoric acid (it is possible to use this type of heating because silicon tetrafluoride, the volatile impurities containing sulfur, and the combustion products CO, CO_2, H_2O, etc., do not combine with alumina); and (2) fluidized bed reactor working at high temperature (550–600°C), linked to other devices allowing drying of the solids or gas–solid separation (cyclones, electrofilters, etc.), heating or cooling of the gases, and washing and neutralizing of the gaseous or solid effluents.

Such a unit (Seigneurin, 1969) allows us to manufacture a product that contains from 90 to 92% of aluminum fluoride. More than 95% of fluorine is fixed, and the conversion output of the spar to acid reaches 98–99%.

This process has many advantages. It ensures the methodical exhaustion of the gases. It uses the hydrofluoric acid coming from the reaction of sulfuric acid on the fluorspar, just as it is (i.e., without isolation, purification, or

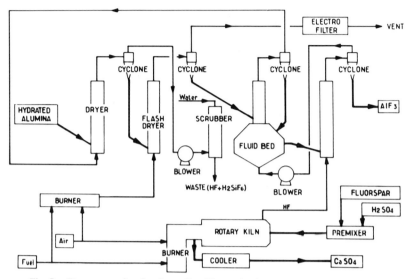

Fig. 5. Dry process for aluminum fluoride production from hydrogen fluoride.

condensation). It gives much running flexibility. Finally, its heat balance is excellent thanks to the recovery of energy that has been carried out during the various stages of fabrication.

TONNAGE AND PRICES. With reference to the amount of aluminum produced on a worldwide scale, ~16 million metric tons in 1980, it is possible to get an idea of the yearly demand for aluminum fluoride. Even by admitting a low consumption of 30 kg of aluminum fluoride per metric ton of produced metal, these needs are considerable: 400,000–500,000 metric tons per year. The production statistics confirm perfectly this figure. At present the average price of the product is ~385 U.S. dollars per metric ton.

   b. CRYOLITE. The exhaustion of natural cryolite reserves, the increasing demand for aluminum, and the need to reduce environmental pollution have led the chemical industry either to prepare synthetic cryolite or to elaborate processes for recovering and reusing fluorine contained in the waste in the form of cryolite. In fact, industrial cryolite often contains two varieties: (1) cryolite ($AlF_3 \cdot 3NaF$) and (2) chiolite ($AlF_3 \cdot \frac{5}{3}NaF$).
   Developed by taking account of the available raw materials and/or the nature of the waste to be treated, the proposed processes are very flexible. They can make use of quite different raw materials: very or not so pure hydrofluoric acid (H_2SiF_6), fluosilicic acid and sodium fluosilicate, soda, various sodium salts (e.g., chloride and sulfate), sodium aluminate, aluminum hydroxide, and various aluminum salts. Almost of these processes are based on the same principle. The only difference is the technology used.

In practice, reagent solutions after more or less extensive purification are mixed in one or two stages in an acid or neutral medium. For example, if we have at our disposal raw materials consisting of $(HF + H_2SiF_6)$ solution, soda solution, and sodium aluminate solution $(Al_2O_3 \cdot Na_2O)$, the following reactions will be carried out:

$$H_2SiF_6 + 2\,NaOH \longrightarrow Na_2SiF_6 + H_2O$$
$$12\,HF + 4\,NaOH + Al_2O_3 \cdot Na_2O \longrightarrow 2\,AlF_3 \cdot 3\,NaF + 8\,H_2O$$

In other words, the solution of impure hydrofluoric acid is cleared of fluosilicic acid by adding an excess of soda and by separating the formed sodium fluosilicate. The solution of purified acid containing sodium is then mixed with a solution of aluminate, the ratios being adjusted to obtain cryolite.

It is difficult to say how much cryolite is produced in the world because of the autoconsumption of many aluminum manufacturers, and it is uncertain how far the scattered published figures reflect the reality.

3. Elemental Fluorine

Even if the production cell technology (Woyteck, 1980) has changed within large limits and can differ from one unit to another, the basic process is practically the same as that developed by Cady in the United States in 1942.

Fluorine is presently produced on an industrial scale by electrolysis of a molten mixture with a composition close to $KF-2\,HF$. This mixture is maintained between 75 and 100°C and can sometimes contain a small percentage of lithium fluoride. All fluorine production cells in use today throughout the world use this technique.

Production requires three inorganic fluorides: calcium fluoride, for the preparation of hydrofluoric acid; potassium fluoride, the basis of the electrolyte; and lithium fluoride, for depolarization.

In the electrolytic process, anhydrous hydrofluoric acid is consumed in a continuous way to produce fluorine at the anode and hydrogen at the cathode. Potassium fluoride indirectly takes part in the electrolysis by making the medium conducting. Despite the excellent basic research carried out by Watanabe and co-workers (1975, 1976, 1978), the role played by lithium fluoride is uncertain; however, its possible influence on anodic polarization explains its use.

Apart from hydrofluoric acid, the consumption of which is directly linked to the production of fluorine at the rate of about 1.1 metric ton hydrofluoric acid per metric ton fluorine, the needs for other fluorides are limited to the first loading of the electrolysis cells and their periodic renewal (from 1 to 1.5 metric tons by cell according to the model) and the compensation for the losses due to mechanical loss.

Indeed, the gases coming out of the anodic and cathodic compartments contain some very fine solid particles rich in potassium fluoride. It is very difficult to eliminate completely this "mist," even with sophisticated mechanical and/or electrical devices.

In addition to electrolysis itself, sodium fluoride is used to purify fluorine destined for special uses in which it is necessary to decrease the concentration of the present hydrofluoric acid below residual content obtained by cooling. This chemical trapping is carried out by putting at the exit of the condensation step at $-80°C$ a tower filled with sodium fluoride in powder or pellet form. As long as the temperature of the reaction medium does not go beyond 50 to 100°C, sodium fluoride fixes the hydrogen fluoride contained in the gases by forming bifluoride according to

$$HF(g) + NaF \longrightarrow NaF \cdot HF(s)$$

This reaction is reversible, and one merely has to heat the bifluoride at 300°C in order to liberate hydrogen fluoride and to regenerate the fluoride reusable for a new absorption cycle.

Almost all of the acid entrained by vapor pressure into the gaseous anodic and cathodic fluxes can be eliminated in this way. This represents a very limited use, however, and a rather insignificant consumption.

Finally, one may mention the possible use of ammonium fluoride as a complete or partial substitute for KF in the electrolyte used for fluorine preparation. The expected advantage is the lower melting point of the $KF/NH_4F/HF$ ternary or the NH_4F/HF binary mixtures; weaker vapor pressures are usually counterbalanced by the corrosion observed at the carbon anode level.

Despite the existence of patents (Société Usines Chimiques, Pierrelatte, 1971), it seems that such a process has so far not been developed beyond the experimental stage.

World fluorine production can be roughly estimated at 15,000 metric tons/year; 75% is used in the fabrication of uranium hexafluoride.

4. Fluorides in Synthesis Chemistry

a. INORGANIC CHEMISTRY. The synthesis of inorganic fluorides is carried out according to the methods discussed in Chapter 2. Industrially, however, inorganic fluorides are prepared by a limited number of methods:

1. Action of hydrofluoric acid in solution or in gaseous phase on an oxide, a hydroxide, or a carbonate:

$$K_2CO_3 + 2\,HF \longrightarrow 2\,KF + H_2O + CO_2$$
$$Al(OH)_3 + 3\,HF \longrightarrow AlF_3 + 3\,H_2O$$
$$UO_2 + 4\,HF \longrightarrow UF_4 + 2\,H_2O$$

In certain cases, the reaction can be carried out just as well in solution as in gaseous phase:

$$NH_3(g) + HF(g) \longrightarrow NH_4F(s)$$
$$NH_4OH(l) + HF(l) \longrightarrow NH_4F(l) + H_2O$$

Depending on the reagent ratios, acid salt can be obtained just as well as neutral salt:

$$NH_3 + 2\,HF \longrightarrow NH_4F \cdot HF \text{ bifluoride}$$

2. Action of fluorine on a metal, an oxide, or a nonmetal:

$$W + 3\,F_2 \longrightarrow WF_6$$
$$UO_2 + 3\,F_2 \longrightarrow UF_6 + 3\,O_2$$
$$S + 3\,F_2 \longrightarrow SF_6$$

The reaction temperatures may be very high (flame), but at the industrial stage they can be easily controlled. Pressure provides more problems, however, and there are practically no examples of industrial processes that use fluorine under high pressure on a large scale.

b. ORGANIC CHEMISTRY. Although it is difficult to speak about industrial uses, it is advisable to mention the use of certain fluorides in organic synthesis. Potassium fluoride, for example, like other fluorides (e.g., CsF) allows a labile chlorine atom to be replaced by a fluorine atom:

1,3-dinitro-5-chlorobenzene $\xrightarrow{\text{KF}}$ 1,3-dinitro-5-fluorobenzene

The higher fluorinating power of the antimony trifluoride SbF_3 allows other chlorine atoms to be replaced by fluorine. The reaction is made all the easier by the reactivity of the chlorine atom. Therefore, if chlorine is activated by the presence of a double bond, reactions of the following type can be carried out:

$$Cl_3C-CH{=}CH_2 \longrightarrow F_3C-CH{=}CH_2$$

However,

$$Cl_3C-CCl_3 \longrightarrow Cl_2FC-CCl_2F$$

Another group of fluorides composed of elements with two possible oxidation states (e.g., cobalt, silver, manganese, mercury, cerium, and lead) allow some delicate hydrocarbon fluorinations to be carried out. These fluorides are solid fluorine carriers. For example, cobalt trifluoride, the most widely used, liberates fluorine according to $2\,CoF_3 \rightarrow 2\,CoF_2 + F_2$ when the temperature is raised to 300 to 400°C.

After utilization, a treatment of CoF_2 by elemental fluorine allows the original fluoride to be regenerated: $2\,CoF_2 + F_2 \rightarrow 2\,CoF_3$.

The use of these fluorides, known as reactive metallic fluorides, made it possible to prepare saturated perfluorinated hydrocarbons (the perfluorocarbons) with the same structure as the original hydrocarbons:

$$C_{16}H_{34} + 34\,CoF_3 \longrightarrow C_{16}F_{34} + 34\,HF + 34\,CoF_2$$

Such syntheses were used mainly during World War II; they have certain disadvantages: (1) high cost due to the use of elemental fluorine; (2) risk of degradation of the organic molecule if the dilution of the hydrocarbon vapor by an inert gas is not sufficient; (3) loss of fluorine, which is partially consumed for producing hydrofluoric acid; and (4) complexity of the obtained mixture, making purification of the final product both difficult and delicate.

Iodine pentafluoride IF_5 is used, on an industrial basis, for the production of perfluoroiodoalkanes, which are intermediates in the synthesis of functional perfluoro compounds by a route other than electrochemical fluorination. Thus, commercial production of alkanecarboxylic acids of the fluorocarbon class can be carried out according to

$$F_2C{=}CF_2$$
$$\text{or} \qquad\qquad \xrightarrow{\;IF_5\;} R_FI$$
$$F_3C{-}CF{=}CF_2$$

$$R_FI + n\,C_2F_4 \longrightarrow R_F{-}(F_2C{-}CF_2)_nI$$

$$R_F{-}(F_2C{-}CF_2)_nI \xrightarrow{\text{oleum}} R_F(F_2C{-}CF_2)_{n-1}{-}CF_2{-}COOH$$

D. FLUORIDES IN THE NUCLEAR INDUSTRY

The contribution of fluorine to this area has, as explained earlier, been crucial to its development.

Figure 6 summarizes the main stages in the nuclear fuel cycle. It illustrates a series of operations by which the fuel is prepared from the ore and reprocessed after utilization in a nuclear reactor in order to recover the reusable elements, uranium and plutonium. Figure 7 indicates in more detail the main utilization points of the fluorine derivatives in this cycle and in particular the contribution of the solid fluorides.

1. Uranium Hexafluoride Produced
for Isotopic Enrichment of Uranium

Whatever the technique used to separate uranium isotopes for industrial or military purposes—gaseous diffusion, nozzles (Becker process), centrifugation, laser, etc.—it is practically impossible to avoid the preparation stage

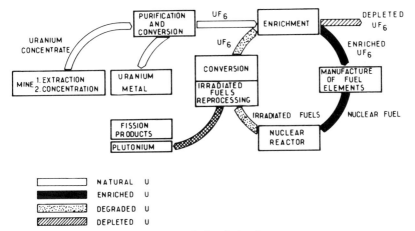

Fig. 6.   Nuclear fuel cycle.

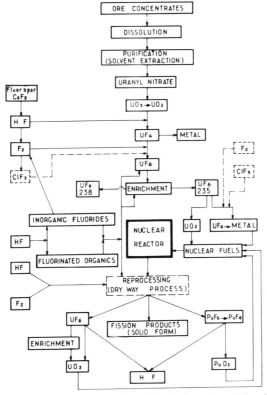

Fig. 7.   Inorganic fluorine derivatives in the nuclear fuel cycle.

of uranium hexafluoride, the only uranium compound that is stable and gaseous at relatively low temperature. One exception, perhaps, is the ion-exchange process (Pesme, 1982), which is being studied in France but the performances of which limit the use to very low enrichments.

Uranium hexafluoride is exclusively prepared by the action of elemental fluorine on uranium tetrafluoride:

$$UF_4 + F_2 \longrightarrow UF_6 \qquad \Delta H_{298} = -59\,\text{kcal/mol}$$

Several techniques have been studied (Galkin et al., 1964) in order to carry out this reaction. It seems that at present the flame reactor (Brater and Smiley, 1958) equips most of the large production units as well as the fluid beds (Vogel et al., 1961).

The preparation of both raw materials, fluorine and tetrafluoride, was the subject of numerous studies during the Manhattan Project (Katz and Rabinovitch, 1951) and, more recently, in the countries that have developed a nuclear industry.

2. Industrial Importance of Uranium Tetrafluoride

The choice of this fluoride over chloride became quickly vital for various reasons, the main one being its superior physicochemical properties: stability, volatility, and purity. Numerous access routes have been proposed and explored (Level, 1962). Various fluorinating agents—hydrofluoric acid, simple or complex fluorides—can be used to perform the transformation of dioxide UO_2 into tetrafluoride by two types of process: wet and dry.

The uranium contained in the original raw material can be tetra- or hexavalent.

The industrial production of UF_4 is carried out almost everywhere in the world according to the dry process using uranium dioxide, which is treated at 500 to 700°C by both gaseous and anhydrous hydrofluoric acid:

$$UO_2 + 4\,HF \rightleftharpoons UF_4 + 2\,H_2O$$

Only the technology used to carry out the reaction between gas and solid varies from one unit to another. Because this reaction is reversible it is usually carried out at countercurrent in a screw reactor, rotary kiln, fluidized bed, or moving (eventually vibrating) bed reactor. Each system has its supporters.

With some equipment, the reaction stage in which uranium trioxide is reduced to dioxide,

$$UO_3 + H_2 \longrightarrow UO_2 + H_2O$$

can be realized in the same reactor as hydrofluorination, which eliminates the risk of the dioxide, an intermediate product that is generally highly pyrophoric, of being reoxidized. The L reactor (Huet, 1960) developed in France by the Commissariat à l'Energie Atomique (Atomic Energy Board) is of this type.

In this reactor the introduced solid (UO_3) flows by gravity, countercurrently to the gases (NH_3 and HF); it is extracted by a horizontal screw, supplied with anhydrous hydrofluoric acid, in which the transformation from UO_2 to UF_4 is achieved.

The quality of the UF_4 is very important, because the eventual presence of untransformed UO_2 or of uranyl fluoride UO_2F_2, resulting from an incomplete reduction of U^{6+} to U^{4+}, interferes in later stages: fluorination to UF_6 or reduction to metal.

Three times more fluorine are necessary to obtain UF_6 from the direct fluorination of UO_2:

$$UO_2 + 3\,F_2 \longrightarrow UF_6 + 3\,O_2$$
$$UF_4 + F_2 \longrightarrow UF_6$$

Generally, the products are of excellent quality and contain more than 97 to 98% UF_4.

Without going into details, it is advisable to point out the existence and sometimes harmful role of a series of solid fluorides with a composition halfway between UF_4 and UF_6: U_4F_{17}, U_2F_9, and UF_5 (α and β). These fluorides are the result either of incomplete fluorination,

$$2\,UF_4 + F_2 \longrightarrow 2\,UF_5$$
$$4\,UF_4 + F_2 \longrightarrow 2\,U_2F_9$$

or of UF_6 reaction on UF_4,

$$UF_4 + UF_6 \longrightarrow 2\,UF_5$$

Such reactions, which inhibit UF_6 preparation and lead to deposits and residues in the fluorination reactors, have been studied extensively, especially in France by Nguyen Hoang Nghi (1961).

3. Production of Uranium Metal

There has been a considerable decline in the production of uranium metal due to the disappearance of graphite gas reactors, which have been replaced by pressurized water reactors, but there is still a certain need for uranium metal outside the military field.

The direct reduction of oxide is thermodynamically possible except with hydrogen. Whichever technique is selected, that is, reduction by carbon or metals (calcium, magnesium, or aluminum) or electrolysis in molten fluorides,[†] it is tricky to carry out the reaction because of metal carburization, alloy formation, and difficult separation of metal and slag. The only industrial method is the reduction of tetrafluoride UF_4 by magnesium, or in special cases by calcium, which eliminates the disadvantages just cited.

The reaction is carried out in batch quantities in a crucible lined with magnesium fluoride (powder form), prepared by a previous operation. After being filled with a mixture in suitable proportions of UF_4 and magnesium chips, the crucible, hermetically sealed with a lid, is heated in a bell-shaped electric furnace. Beyond a certain temperature, the reaction initiates itself:

$$UF_4 + 2\,Mg \longrightarrow U + 2\,MgF_2$$

After a slow cooling, which allows the metal to decant perfectly, the solid mass is removed from the crucible, and then the metal ingot, which apart from the surface is exempt from slag traces, is recovered. The MgF_2 slag is crushed, ground, and washed. It is partly recycled, but its purity makes it an excellent product for the preparation of fluxes in metallurgy. This method allows the elaboration of ingots of different sizes with weights reaching 0.5 metric ton.

4. Uranium Hexafluoride Reduction

In some cases it is worthwhile to convert the uranium contained in the fluxes, which comes out of the enrichment cascades, into a compound other than depleted or enriched UF_6. Some reduction processes of UF_6 using hydrogen or cracked ammonia were developed for this purpose (Patton et al., 1963). The two gases are mixed in a burner at the top of an empty tubular reactor, in which the following reaction is carried out:

$$UF_6 + H_2 \longrightarrow UF_4 + 2\,HF$$

The supply of calories necessary for the reaction can be obtained either from outside of the reactor (hot wall process) or by adding fluorine (or chlorine trifluoride) to UF_6 (cold wall process).

After separation and filtration of the gases, the recovered solid can easily be transformed into metal by reduction or can be stored. This solution was chosen for the long-term storage of depleted uranium, presently stored in containers in the form of UF_6.

5. From Uranium to Plutonium

The similarity in behavior between uranium and plutonium is such that the preceding discussion of uranium also applies to plutonium. This explains the role played by fluorides (Cleveland, 1970) in plutonium chemistry and, in

---

[†] See experiments by the Mallinckrodt Company (Vie, 1966; Stevenson, 1966). The technique is similar to that used for aluminum but is difficult to carry out due to the high melting point and high density of the metal.

particular, the elaboration of this element in metallic form. It is advisable to indicate, however, that hexafluoride is much less stable and therefore has an advantage: with U-Pu there are better separation possibilities. There are also some great disadvantages, namely, decomposition during treatment and undesirable deposits.

6. Reprocessing of Irradiated Fuels

By means of the reprocessing operation after the fuel has been irradiated in the nuclear reactor, the products of uranium fission are separated, and the fissile elements—uranium and plutonium—which have an intrinsic value and can be recycled, are recovered.

Today, this operation is carried out on an industrial basis in large units at La Hague, France, Windscale, United Kingdom, and Tokai Mura, Japan, where highly automatized exploitation is particularly difficult and costly because of the potential risks linked to the fission products.

The process used is a very sophisticated hydrometallurgical technique that makes use of several well-chosen solvents and speculates on the valence states of some principal elements for extraction and/or reextraction. The performances are good enough to ensure a correct decontamination and a suitable separation of uranium and plutonium. Figure 8 summarizes the principles of the process.

For various reasons concerned with the evolution of nuclear reactors, with the increase in their performance and number, and with difficulties in the exploitation of existing reprocessing units, it was not certain, in spite of adaptation efforts, whether this type of process would be able to meet the problems faced in a later stage. These include (1) an increase in irradiation

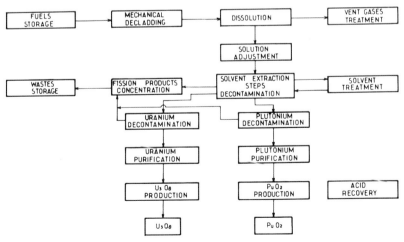

Fig. 8.   Irradiated fuels reprocessing: wet process.

rates, (2) a reduction in the length of the cooling time, (3) the transport of very highly irradiated fuels, (4) the storage of larger and larger quantities of highly radioactive and often liquid waste, and (5) an increase in the financial charges linked to prolonged immobilization of increasingly large amounts of fissile materials.

The possible use of fluorine and fluorine properties (e.g., volatility and capacity to form stable complexes) for the reprocessing of irradiated fuels has given rise to some hopes, serious enough to have generated some important studies (Bourgeois and Cochet-Muchy, 1968) in the United States, France, Belgium, and the Soviet Union.

Although some technical and financial efforts have met with success, these investigations have not yet gone beyond the pilot level; they have justified the basic ideas, however, and the expected large-scale feasibility. This potential use of inorganic fluorides is outlined in Fig. 9.

The dry method, or fluoride volatilization process, involves the following stages: (1) mechanical or chemical decladding; (2) volatilization by fluorination (F_2, ClF_3, BrF_3, etc.) of uranium, plutonium, and all other fission by products whose volatile fluorides (elements whose nonvolatile fluorides are separated at the same time); (3) separation of uranium and plutonium by selective reduction of the plutonium hexafluoride to solid tetrafluoride; and (4) decontamination of uranium and plutonium. The last two steps are based on the volatility difference of the fluorides or their capacity to form

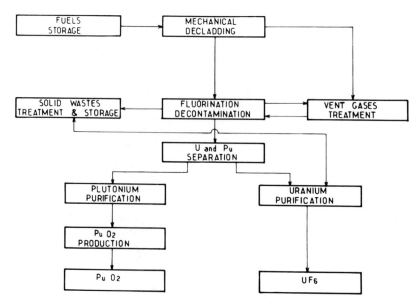

Fig. 9.   Irradiated fuels reprocessing: dry process.

complexes of different stabilities with various solids that are, in general, themselves inorganic fluorides in solid form. As an example, it is possible to separate ruthenium, niobium, and antimony by absorbing them on (or complexing them with) NaF at 400°C and to use MgF_2 in order to eliminate TcF_6 at 100°C.

Moreover, this technique has the advantage of directly producing solid waste.

Finally, at this stage the possibility still remains of combining the wet and dry methods if the entire cycle cannot be completed by the dry method.

Whereas the criticism sometimes voiced against the dry process due to the high temperatures, strongly exothermic reactions, high radioactivity, and corrosion risks does not stand up very well after technical discussion, the expected advantages are certain. They include (1) shorter cycle; (2) absence of radiolysis (no organic products); (3) greatly decreased risks of critical accident (no hydrogenated compounds in condensed phase); (4) entire confinement of the fission products in solid form; (5) obtaining of uranium in the form of UF_6, directly recyclable in an enrichment process; (6) compactness of the units, creating a link between the treatment unit and the reactor, which can be built on the same site, and elimination of the transport problems. We can also expect, apart from these technical advantages, a significant decrease in the cost of the treatment linked to more rapid rotation of the fissile materials and to lower capital investments.

E. FLUORIDES IN METALLURGY

1. Aluminum Production and Refining

The claim by Hall in U.S. patent 400,766 issued in July 1886 was the following: "Process by which aluminum is fabricated by electrolysis, consisting in dissolving alumina in a molten bath composed of aluminum fluoride and one or several metals which are more electropositive than aluminum (sodium, lithium) and in making the electric current pass through the molten mass." The French patent issued to Heroult in February of the same year was roughly the same and mentioned melted cryolite.

These two investigators had simultaneously discovered the contribution of fluorides to the lowering by almost 1500°C of the melting point of pure alumina (2050°C) and the use of a molten medium to carry out its electrolysis. Thus, fluorides act as a flux and even as an alumina solvent.

Almost 40% of hydrofluoric acid production today is used by the aluminum industry in the form of aluminum fluoride or cryolite as well as more modest tonnages of sodium fluoride and fluorspar (acid grade), the latter being used chiefly to adjust the melting point of the bath.

The electrolyte used for the preparation of aluminum has the following average composition: cryolite, 80–85%; aluminum fluoride, 5–7%; fluorspar, 3–5%; and alumina, 2–8%. It can be seen that cryolite plays a prevailing role, which explains, because natural cryolite is disappearing, why it was necessary to prepare a synthetic product in order to ensure replacement. Indeed, cryolite is used both for the renewal of the baths of the old units and for the first loading of the new cells.

Aluminum fluoride is used to stabilize the composition of cryolitic baths. The composition of the electrolyte is disturbed by several factors: (1) the volatilization of $NaAlF_4$; (2) the presence of Na_2O (1%) in the alumina introduced into the cells, which brings about the consumption of aluminum fluoride by the formation of cryolite; and (3) losses by hydrolysis and mechanical removal (dusts).

Another consumption factor is the loss of electrolyte by infiltration into the lining or due to an incident occurring during a run, which can represent 30–40% of the total consumption.

The consumption of fluorides, which theoretically would be zero except for the losses just mentioned, has greatly decreased, changing gradually from 50 to 60 kg/metric ton of aluminum produced to less than 25 kg per metric ton in 1981 for better units. Aluminum fluoride represents about two-thirds of this consumption.

This lowering of the consumption is linked to the considerable effort that was made by the large aluminum producers to make their units nonpolluting. For this purpose, they set up more and more effective devices to trap the gases and dusts and reused all or part of the recovered fluorine in the form of aluminum fluoride or cryolite.

The importance of the gains achieved obviously depends on numerous parameters, including model, age, and size of the cells; nature of the anodes and raw materials; age of the units; efficiency and nature of the trapping systems (washing, absorption on solids); recovery level of gas, dust, and lining of the cells; and the importance of recycling.

The second use of fluorides in the aluminum industry concerns the preparation of flux for metal refining. Thus, for aluminum fabrication, known as second fusion, the metal that comes from the fusion of the scraps is protected from oxidation by a fluoride-based slag. By exchange reactions between this slag and the impurities in the metal in its molten form, it is often possible to purify the metal. In this way, magnesium can be separated by the formation of salts with cryolite or aluminum fluoride.

By adding potassium fluoborate it is possible to eliminate certain impurities (transition metals) by forming borides, which precipitate and can thus be separated. This technique is used to develop high-purity aluminum for electric cables.

As a result of its strong reactivity and its melting point, it is not possible to purify aluminum by controlled oxidation of the impurities or by metal distillation, so electrorefining is required. In this process, known as a three-layer process (Hart and Hills, 1971), the purified aluminum floats at the surface of the electrolyte, made up of fluorides (i.e., barium, sodium, and aluminum), which is above the impure metal situated at the bottom of the cell. The current runs between the anode (at the bottom of the cell) and the floating cathode (purified metal), which is protected from oxidation by a layer of alumina. A wise adjustment of the impure metal composition by the addition of copper and silicon and of the layer of molten-salt composition gives a suitable ratio of density.

Although it is difficult to avoid accidental pollution of the cathode by the impure metal as a result of electromagnetic turbulence, this process gives a 99.99% purity in a single stage.

In order to modify the structure of the metal or to refine the grain, simple or complex fluorides are still used. These include calcium fluoride in the form of fluorspar, sodium and potassium fluorides, alkaline fluoroborates, fluorotitanates, potassium fluosilicate and fluozirconate, and potassium and aluminum double fluorides.

In casting, potassium fluoborates and fluotitanates are used eventually in the ternary alloy form (Ti/B/Al) for grain refining.

2. Iron and Steel Metallurgy

This area of calcium fluoride utilization is a privileged one, the metallurgical-grade fluorspar accounting, as seen earlier, for more than 40% of the extracted tonnages and being used to elaborate the cast iron and the steel as well as in casting processes.

Calcium fluoride has a double role in steel elaboration. It allows (1) a better separation of the metal and the slag by modifying its fluidity and (2) the purification of the molten metal.

The capacity of calcium fluoride to form, with refractory metals, eutectics that have a low melting point is the basis of its first use, which is simultaneously the oldest and the most important. In addition to its influence on the fluidity of the slag, it has a positive action on the dissolution of certain oxides such as lime. Although for this use steel producers prefer fluorspar in gravel or sand form, briquets or pellets are being used more frequently, and the grade and calcium fluoride content of these forms are gradually increasing (25–90%).

The second function of spar is more chemical in nature. The elaboration process of steel essentially consists in eliminating the undesirable impurities (phosphorus, sulfur, silicon, and carbon) from the iron by oxidation. The oxides and/or the complex compounds formed here are eliminated in the form of a slag, the basicity and fluidity of which are stabilized by the presence of fluorspar.

One of the functions of spar is to ensure, in the case of the oxygen elaboration process, a stable emulsion during the refinement stage.

Depending on the process or furnace used the fluorspar consumption varies in a ratio that can range from 1 to 3 or even more, as the following figures show: open-hearth furnace, 3.5 kg per metric ton of steel; basic oxygen process, 6.0 kg per metric ton steel; electric furnace, 2.4 kg per metric ton steel. It also varies according to the rate of production and increases with the productivity of the unit.

According to a publication from the AISI, however, the average specific consumption of fluorspar was relatively stable from 1972 to 1980: about 4–4.5 kg per metric ton. The amount used on a worldwide scale in 1980 was close to 2 million metric tons.

Fluorspar is also a component of the fluxes used for the ferroalloy industry, which consumes between 5 and 90 and even 100 kg per metric ton, depending on the operating conditions and the type of end product aimed at.

In casting, moreover, fluorspar is used because, if it is added to the cupola charge in amounts ranging from 3.5 to 10 kg per metric ton of the metal, it facilitates dissolving of the alloy elements, prevents oxidation of the metal, and gives a more malleable end product, which has a higher tensile strength.

The last important use of fluorspar is the elaboration of high-grade steels. Powder injection processes are used (Scaninject III, 1983). The pulverulent mixture, with a composition adapted to the required aftereffect (e.g., lime + spar for desulfurization) is introduced in the heart of the liquid metal with the help of an injection tube made from refractory material and a current of inert gas (argon). By means of vigorous stirring by injection of argon, washing of the metal by the slag is carried out, allowing the impurities to be separated and to be recovered in the slag above the melt.

For various ecological reasons (freeing of toxic fluorinated products into the atmosphere in gaseous and dust forms) or technoeconomic reasons (e.g., corrosion of the linings, rising prices, and risk of shortage), research work was started in the late 1960s to find a substitution product.

Numerous solutions have been considered: alumina, bauxite, slags containing alumina, colemanite (the most promising one from the technical standpoint), lime, topaz, and ilmenite. None of these products used alone or mixed with others has so far given the same results as fluorspar.

In this way, the specific role played by fluorinated products here, as elsewhere, is confirmed. Furthermore, the economic aspect of their use, which involves very large tonnages, is a determining factor in the choice.

3. Other Metallurgical Uses

With production limited to a few thousand metric tons per year, fluorinated products are used for the preparation of fluxes for refining of copper, chrome, magnesium, nickel, gold, silver, tin, zinc, and antimony, as well as for the preparation of metal-based alloys (i.e., based on the metals listed here or others). In addition, sodium fluoride and sodium fluosilicate are used, in small quantities (3–6%), along with metallic oxides (e.g., iron and manganese) to make exothermic mixtures for casting (steel and aluminum) in order to avoid solidification of the molten metal in the channels of the molds during casting. This is important for maintaining the continuous feeding of these molds.

4. Welding and Brazing

Metal joining is carried out either with an alloy with a melting point lower than those of the metals to be joined (welding and brazing) or with a deposited metal of the same nature (arc welding). In both cases, it is necessary to protect the metallic surfaces that are to be joined, to protect the deposited metal from oxidation, and to remove the metallic oxides formed on the surfaces during heating in order to prevent the surfaces from being damaged. This is the role played by the fluxes, which generally have a high fluorine content.

Depending on the case, it is possible to use fluorspar, potassium fluoride, and fluoroborates with other inorganic halogenated or nonhalogenated salts. Thus, for arc welding, the main components are fluorspar and calcium carbonate. The fluorspar content can reach 50%. For brazing, one more often uses mixtures of alkaline borates and potassium fluorides.

There are several ways to make use of fluxes: (1) The alloy or deposited metal rod can be coated with the flux; (2) the flux can be loaded into the central cavity of a hollow electrode; and (3) the mixture of powder (flux) can be distributed at the point where junction is to be carried out. In the first case a binder must be used, but in both other cases the fine-powder product obtained by grinding of the homogeneous mass, which is the result of the melting of the mixture of components, can be used as it is.

5. Metallic Surface Treatment

There are two different examples of metallic surface treatment: metal pickling and surface finishing treatment, which gives the surface a special appearance ranging from shiny to matt.

Very often, solid inorganic fluorides are used only indirectly in the form of hydrofluoric acid solutions. This is also the case for stainless steel pickling, in which hydrofluoric and nitric acid solutions are used, the role of the latter being to limit and even inhibit the intergranular corrosion.

For nonferrous metals (e.g., aluminum and zinc), the treatment baths contain fluorides, acid fluorides, fluoroborates, fluozirconates, and fluosilicates in various proportions, depending on the formulations and the desired effect. Therefore, to obtain a glazed finish on aluminum surfaces it is possible to use sodium fluoride– and sodium hydroxide–based mixtures. For certain alloys, one can use hydrofluoric acid or ammonium bifluoride. In the chemical processes carried out to obtain luster finishes, alkaline fluorides are used along with strong inorganic acids such as hydrofluoric and nitric acids. In the electrochemical processes carried out to obtain luster finishes, fluoboric acid is almost exclusively used.

These various processes, although widely developed, use only relatively modest amounts of fluorinated compounds. There are two reasons for this: First, the specific consumptions are low; for example, 10 kg of 100% hydrofluoric acid are used for every metric ton of steel in stainless steel pickling. Second, it is possible to recover and recycle most of the products that have not reacted and remain in the polluted treatment baths.

6.   Electrolytic Coatings: Electroplating

The metallic deposits obtained electrolytically are the result of complex processes for which inorganic fluorinated compounds give some interesting results. Fluoboric acid and fluoroborates are often used. Depending on the case, one can obtain thick deposits in a relatively short time, high growth rates even at room temperature, fine-grain deposits, and excellent anodic and cathodic yield (almost 100%). For example, all utilizations requiring lead and tin coatings mainly involve fluoboric acid and fluoroborate solutions.

F.   FLUORIDES IN THE ELECTRICAL INDUSTRIES

Fluorine chemistry also has an important part in the electrical and electronic industries, in the manufacture of cooling fluids for power electronics, welding and etching agents for microelectronics, testing fluids, and dielectric compounds. Inorganic fluorides are not used in solid form, however. Moreover, these uses are relatively new and specific, which explains the modest size of the markets and their relatively experimental character.

Sulfur Hexafluoride SF_6

Although sulfur hexafluoride is a gas, we cannot ignore here its contribution to the development of a new generation of equipment for the distribution of electricity.

a. INDUSTRIAL PRODUCTION.   Almost all the processes described in the technical literature are based on the reaction of elemental fluorine with sulfur in the solid, liquid, or gaseous form, according to $S + 3 F_2 \rightarrow SF_6$. Several techniques are claimed in patents (Kalichemie A.G., 1977; Montedison, 1977a,b; Ugine Kuhlmann, 1982). They range from the use of a simple sulfur melting pot, where the fluorine is distributed by a circular pipe with holes and comes into contact with the surface of the molten sulfur, to more sophisticated equipment derived from the nuclear industry, such as flame reactors. In the last case, the sulfur can be fed in solid or liquid form using a pump linked to a system of spray nozzles.

b. SPECIAL PROPERTIES.   Under the normal conditions of temperature and pressure, sulfur hexafluoride is a gaseous compound. This gas is exceptionally stable; it can be heated to more than 500°C without decomposition. It is nonflammable and practically inert from a chemical and biological standpoint (when it is pure). It is also insoluble in water.

Concerning the origin of its electrical uses, sulfur hexafluoride has a dielectric rigidity that is higher under similar pressure than that of most comparable materials; at 3 or 4 bars it is equivalent to that of oil. It also has a high recombination rate after dissociation by electric discharges (self-regeneration). Finally, its liquefaction temperature is low enough that the equipment can work outside under extremely cold climatic conditions.

Moreover, it has two other interesting properties. It is a good phonic insulating agent; the sound velocity in sulfur hexafluoride is about one-third of that in air. It is a good thermal insulating agent; its conductivity is two times lower than that of air.

c. USES OF SULFUR HEXAFLUORIDE.   Today, there are two important markets for sulfur hexafluoride: circuit breakers for high and very high voltages and armored stations.

For the circuit breakers used in the distribution of electrical energy under high voltage (> 90 kV), sulfur hexafluoride is extremely useful due to the uncommon conjunction of its dielectrical properties and its capacity to trap electrons during arc blowing. No substitution product has been proved to reach similar performances. The situation is different for medium voltages because oil and vacuum are still in competition with sulfur hexafluoride.

In the classical open-air energy distribution stations (open stations), insulators and complementary devices are insulated in open air. Only the circuit breaker contains sulfur hexafluoride.

Several factors—atmospheric pollution, safety, ecology, and aesthetics (site protection)—have led the manufacturers of electrical equipment to build armored stations for high and very high voltage distributions (220 kV). All the elements that form this type of substation (circuit breakers, disconnecting switchers, cable junction boxes, bus bar, and earth switchers) are put inside a closed metallic container, which is tightly sealed and fed with gaseous sulfur hexafluoride under 3 or 4 bars pressure. The decrease in isolation distance allows the volume to be reduced 10 or 12 times compared with that of a conventional substation giving the same performances. Moreover, the armored stations are very reliable, need only a little maintenance, do not disturb broadcasts, are not polluted by the surrounding atmosphere, and are very safe.

The amount of sulfur hexafluoride required for the insulation of an armored station is 30–40 times greater than that used in an open station with the same characteristics.

Another use must be mentioned, because its potential market is important: as a gas-insulated conductor to replace electrical cables, which are insulated with paper and require a forced circulation of oil, for the transportation of high electrical power (3000 MV A and 400 kV). A gas-insulated cable consists of a central hollow conductor surrounded by an aluminum clad, 50 to 100 cm in diameter, fed with sulfur hexafluoride (20 metric tons per kilometer).

Still new, which explains its present price, this technology ought to be an important development in the near future if its cost can be sufficiently reduced. Indeed, it allows passage through difficult or protected sites, as well as discrete penetration into cities. It shows increased performances (specific intensity three to four times higher) and an excellent safety level.

A last but very limited use of sulfur hexafluoride is the insulation of nuclear particle accelerators by this gas. There are examples all over the world.

d. PRICES AND TONNAGES.   In 1980 the world market was estimated to be approximatively 5000–6000 metric tons and was distributed as follows:

Region	Metric Tonnage
North America	1200–1300
Europe	1600–1700
Japan	1300–1400
Soviet Union and socialist countries	100–300
Other countries	800–1000

Seventy-five to 80% of this market is devoted to armored stations and 15–20% to circuit breakers. The average price oscillates between 3.75 and 4.5 U.S. dollars per kilogram.

G. FLUORIDES IN THE GLASS AND CERAMIC INDUSTRIES

1. Glass Industry

Fluoride and related glasses are described in Chapter 7. In this section, only the uses of fluorides in the industry of traditional silicate glasses are discussed.

Fluorides have been used since antiquity for controlled devitrification giving rise to opal glass. The fluorides are dissolved in molten silicates. During the quenching process very fine fluoride crystals precipitate inside the vitreous phase, imparting a milky color to the glass (Scholze, 1963).

Added in small amounts to the melted oxides, fluorides are used for modifying certain properties of commercial glasses (viscosity, linear expansion coefficient, etc.) (Lajarte, 1961).

Strong adhesion of vacuum-evaporated gold to oxide or glass substrates is obtained by using intermediate layers of fluoride (CdF_2, SnF_2, PbF_2, InF_3, or BiF_3) (Zydzik et al., 1977).

2. Ceramic Industry

Fluoride ceramics have useful electrical, refractory, and nuclear applications. Their fabrication processes are described in Chapter 2.

Fluorides are often used as fluxing agents for forming and firing oxide ceramics (Kingery et al., 1976). For instance, CaF_2 and MgF_2 are involved in the sintering process of ceramic dielectrics based on $BaTiO_3$. This method has been improved by the addition of $BaLiF_3$, which gives a solid solution with $BaTiO_3$. The sintering temperature is lowered, and the dielectric constant enhanced (Benziada-Taibi et al., 1984).

H. MISCELLANEOUS APPLICATIONS OF INORGANIC SOLID FLUORIDES

Miscellaneous applications include water fluoridation and dental uses (see Chapter 19), fluxes for metal welding or brazing (Section I,E), and abrasives.

Solid inorganic fluorides are used in relatively small amounts for the manufacture of certain types of abrasives such as "agglomerated abrasives," in which the grain is molded with the help of a binder in the form of blocks or grindstones. Depending on the final utilization, the grain is inserted in a matrix made of ceramic, synthetic resin, rubber, or gumlac. By the addition of fluoride during the manufacturing stage, performances are increased under working conditions.

Natural cryolite was used initially but has been progressively replaced by other simple or complex fluorides such as alkali fluorides or fluoborates, magnesium fluoride, calcium fluoride, potassium and zirconium fluoride (the most commonly used), or potassium and aluminum double fluoride.

Although the role of the fluorides is not yet clearly understood, two possibilities have been suggested: reaction of the fluoride on the removed pieces of metal, which prevents them from being refixed by welding, and fusion of the fluoride due to the high temperature obtained during work on the metal and lubrication of the metallic surface by the melted fluoride.

Some very small tonnages (less than 1000 metric tons) of potassium fluoroborate, double potassium, and aluminum fluoride are also used for the manufacture of "supported abrasives" in which the grain is fixed onto a flat and flexible surface.

Lubrication

Fluorinated polymers (e.g., polytetrafluoroethylene) have self-lubrication properties. These materials are widely used for the manufacture of mechanical parts or as dry lubricants. Polytetrafluoroethylene cannot be used above 250°C.

Inorganic materials (graphite fluorides, alkali and alkaline-earth fluorides) also have lubricant properties, especially at high temperature (Sliney, 1982). Graphite fluoride $(CF_x)_n$ is a member of the family of solid lubricants with layered structures $(MoS_2, WS_2, etc.)$. Such a lattice induces gliding properties. Among those materials, $(CF_x)_n$ is the most stable at high temperature. For instance, the failure temperature is 400°C for MoS_2 but 480°C for $(CF_x)_n$. This behavior is due to the good resistance of graphite fluoride to oxidation.

The fluorides LiF, CaF_2, and BaF_2 are used as lubricants between 500 and 950°C. Extremely oxidation resistant, they are more efficient than soft oxides previously used (PbO and SiO_2).

II. Industrial Developments and Application Prospects

A. MOLTEN FLUORIDES

Despite the fact that this volume is devoted in principle to inorganic solid fluorides, it seems appropriate to deal with some modern aspects of molten fluoride chemistry. Indeed, new developments in molten salt technology will bring about an increased production of fluorides for the benefit of the fluorine industry. In addition, solid-state scientists are becoming more and more interested in the connected liquid state. Knowledge of a liquid is very often useful for explaining and predicting the properties of the corresponding solids. This is obvious in the case of glass research.

The study of the physical properties and structure of molten silicates is important if we take account of their numerous applications. It is worthwhile mentioning the strong analogy between SiO_2 and BeF_2, on one hand, and the couples LiF/MgO, NaF/CaO, and KF/BaO, on the other hand. The phase diagrams of alkali fluoroberyllates and of alkaline-earth silicates largely coincide on a reduced absolute temperature scale. A deeper understanding of molten silicates could be promoted by further experiments on molten fluoroberyllates (Furukawa and Ohno, 1978).

1. Metal Refining

As mentioned previously (Section I,E) a current application of molten fluorides is their use as slag for metal refining. It will most likely be extended in the future to electroslag remelting to produce high-quality metals. One of the reasons for the efficiency of fluorides as components of slags is reduction of the oxygen potential. The requirements for fluorides to be used as slags are (1) appropriate electrical conductivity, (2) stability up to 1500°C, and (3) low vapor pressure. The alkaline-earth fluorides seem to be the most appropriate for this purpose. Fluoride slags with LaF_3 and YF_3 are studied for future applications in refining processes of titanium and zirconium (Society of Molten-Salt Thermal Technology, 1980).

2. Miscellaneous

Inorganic molten fluorides have found use in the field of solar energy as components of molten-salt breeder reactors (see Chapter 20).

B. SOLID FLUORIDES

1. Optical Applications[†]

a. CRYSTALS. The main characteristic of fluorides is their large domain of transparency from the ultraviolet (uv) to the middle infrared (ir) region. The uv transmission is due to the large band gap resulting from the high electronegativity of fluorine (MgF_2 $\Delta E = 12$ eV). The absorption in the ir region is related mainly to the lattice vibration and is therefore connected to the nature of the bonding. The ionic character of the metal–fluorine bonding is favorable. In contrast, the low atomic mass of fluorine limits the transparency. Practically single crystals of alkaline-earth fluorides are widely used from 150 nm to 10 μm (Dubois et al., 1984).

Hot pressed polycrystalline fluoride ceramics with heavy cations are used for military applications.

Another characteristic of fluorides is their low index of refraction. Thin films of MgF_2 are used as antireflection coating on lenses. Fluorides are also appropriate materials for various applications involving luminescence (lasers, cathodoluminescent materials, etc.; see Chapter 13). Lithium fluoride is often used in x-ray technology as a monochromator.

b. FLUOROGLASSES. For information on the application of fluorides in this area, see Chapter 7.

[†] See also Chapter 14.

2. Fluorides in Electronics

Inorganic fluorides are used in silicon-based microelectronics as etching and welding agents (see Section I,F). Solid fluorides have been proposed as dielectric films appropriate for use with III–V semiconductors (Tu et al., 1983).

Group III–V semiconductors (e.g., gallium, arsenide, and indium phosphide) have acquired growing importance because they are the substrate materials for long-wavelength optical communication and high-speed devices. It is not possible to form an insulating film on these materials by simple oxidation as for silicon, because the corresponding oxides do not grow epitaxially. Alkaline-earth fluorides (calcium, strontium, barium, and solid solutions) with fluorite-type structure have been grown as epitaxial films on the surface of zinc blende–type III–V semiconductors.

3. Electrochemical Applications

Electrochromic devices and solid-state batteries are discussed in Chapters 15 and 20, respectively.

a. SPECIFIC ELECTRODES.   The ionic conductivity of fluorides has been used in fluorine ion–specific electrodes for some years (Frant and Ross, 1966). The device contains an LaF_3 crystal, the conductivity of which has been improved by doping with EuF_2. An attempt to replace the costly single crystal by a cheaper and readily prepared $PbSnF_4$ ceramic has been made (Perez, 1980). The titration is limited, however, to 10^{-3} mol/liter compared with 10^5 mol/liter for LaF_3 due to the solubility of the lead tin fluoride.

b. PIEZOELECTRIC GAUGES.   The variation of the ionic conductivity of fluorides has been proposed for measuring pressure. A linear and reversible variation ($\Delta \log \sigma/\Delta P = 0.040$ kbar^{-1}) has been observed, for instance, for $PbSnF_4$ from 20 up to 70 kbars (Matar et al., 1980).

c. GAS SENSORS.   Two types of gas sensors using fluorine ion conductors have been proposed.

Utilization of the ionic conductivity of LaF_3 for gas detection has been proposed (Laroy et al., 1973). Exposure to certain reducibles gases (O_2, CO_2, SO_2, NO_2, and NO) increases sharply the conductivity of a cell composed of a thin film of LaF_3 between two metallic electrodes. For a given gas the increase in conductivity appears at a characteristic voltage.

Other potentiometric gauges have been studied (Birot et al., 1983). They make use of the diffusion of gaseous species on a weak depth at the surface of fluorine ion conductors (e.g., PbF_2 and $PbSnF_4$). Such cells can be used as gas sensors (O_2, H_2, and NH_3) at low temperature.

III. Conclusion

After the outcome of his research on fluorine, Moissan asked himself, "Will fluorine never have any applications? It is very difficult to answer this question." History has answered positively: fluorine has become an important industrial element.

References

Agricola, G. (1950). In "De Re Metallica", (Engl. Transl.). Dover, New York.

Benziada-Taibi, A., Ravez, J., and Hagenmuller, P. (1984). J. Fluorine Chem. 26, 395–404.

Birot, D., Couturier, G., Danto, Y., Portier, J., and Salardenne, J. (1983). Int. Meet. Chem. Sens. p. 357.

Bourgeois, M., and Cochet-Muchy, B. (1968). Energ. Nucl. (Paris) 10, 192.

Brater, D. C., and Smiley, S. H. (1958). In "Process in Nuclear Energy Series III, Process Chemistry" (F. R. Bruce, J. M. Fletcher, and H. H. Hyman, eds.), Vol. 2, pp. 136–148. Pergamon, Oxford.

Buss A.G. (1962). D.B.P. 1,271,086.

Chemical Daily Co., Ltd. (1980). Chem. Ind. Yearb.

Clarke, G. (1980). Ind. Min., June pp. 21–42.

Clarke, G. (1982). Ind. Min., June pp. 25–46.

Cleveland, J. M. (1970). In "The Chemistry of Plutonium," pp. 323–352. Gordon & Breach, New York.

Dubois, B., Portier, J., and Videau, J. J. (1984). J. Optics (Paris) 15, 351.

Frant, M. S., and Ross, J. M. (1966). Science 154, 1553.

Furukawa, K., and Ohno, H. (1978). Trans. Jpn. Inst. Met. 19, 553.

Galkin, N. P., Sudarikov, B. N., Veryatin, U. D., Shishkov, Y. D., and Maiorov, A. A. (1964). In "Technology of Uranium," (A. Aladjem, ed.) Moscow (Engl. Transl.: A. Aladjem, ed. Israel Program for Scientific Translation, Jerusalem, 1966).

Gall, J. F. (1980). Kirk-Othmer Encycl. Chem. Technol., 3rd Ed. 10, 733–753.

Hart, P. F., and Hills, A. W. D. (1971). In "Industrial Electrochemical Processes" (A. T. Kuhn, ed.), p. 247. Am. Elsevier, New York.

Huet, H. (1960). Nucl. Power 5, 48 pp. 130–131.

Kalichemie A. G. (1977). U.S. Patent 4, 108, 967.

Katz, J. J., and Rabinovitch, E. (1951). In "The Chemistry of Uranium," (USAEC, eds.), Natl. Nucl. Energy Ser., Part 1, pp. 355–449. McGraw-Hill, New York.

Kingery, W.D., Bowen H. K., and Whlmann D.R. (1976). In "Introduction to Ceramics,"p. 8. Wiley New York.

Lajarte, S. (1961). Cent. Glass Ceram. Res. Inst. Bull. 8, 81.

Laroy, B. C., Lilly, A. C., and Tiller, C. O. (1973). J. Electrochem. Soc. 12, 120.

Level, A. (1962). Energ. Nucl. (Paris) 4, 279–281.

Matar, S., Réau, J. M., Demazeau, G., Lucat, C., Portier, J., and Hagenmuller, P. (1980). Solid State Commun. 35, 681.

Montedison S. P. A. (1977a). U.S. Patent 4,186,180.

Montedison S. P. A. (1977b). U.S. Patent 4,246,236.

Morlock, J. Y., Nocher, B., and Mollard, P. (1980). Proc. Int. Congr. Phosphorous Compd., 2nd, 1980 pp. 799–808.

Morse, D. E. (1980). *In* "Mineral Facts and Problems," (USDOI, eds.). Bull. 671, pp. 309–310. U.S. Dept. of Interior, Bureau of Mines, Washington, D.C.

Neumark, H. R. (1947). *Trans. Am. Electrochem. Soc.* **91**, 367.

Nguyen Hoang Nghi (1961). CEA, Rep. No. 1976.

Patton, F. S., Googin, J. M., and Griffith W. L. (1963). "International Series of Monographs on Nuclear Energy." Pergamon, Oxford.

Perez, G. (1980). Thesis, University of Montpellier.

Pesme, L. (1982). *Echos CEA* pp. 2–13.

Scaninject III (1983). *Prepr. Conf. Refining Iron Steel Powder Injection, 3rd, 15–17 June 1983,* Lulea, Sweden.

Scholze, H. (1963). "Glass, Natur, Struktur und Eigensschaften." Vieweg, Braunschweig.

Seigneurin, L. (1969). *Chim. Ind., Genie Chim.* **102**, 747–751.

Sliney, H. E. (1982). *Tribol. Int.,* October p. 303.

Société Usines Chimiques, Pierrelatte (1970). French Patent 2,082,366.

Société Usines Chimiques, Pierrelatte (1971). French Patent 2,145,063.

Society of Molten-Salt Thermal Technology. (1980). "Fundamentals of Molten-Salt Thermal Technology." Foundation for Industrial & Economic Research, Tokyo.

Stauffer Chemical Company (1963a,b). French Patents 1,342,890 and 1,342,891.

Stauffer Chemical Company (1966). U.S. Patent 3,257,167.

Stevenson, J. W. (1966). The electrolytic uranium project—terminal progress report—cell operation. Rep. MCW 1514. Mallinckrodt Co., Weldon Spring, Missouri.

Tu, C. W., Sheng, T. T., Read, M. H., Schlier, A. R., Johnson, J. G., Johnston, W. D., Jr., and Bonner, W. A. (1983). *J. Electrochem. Soc.* 2081.

Ugine Kuhlmann (P. C. U. K.) (1982). European Patent 87,338.

U.S. Department of Interior (1980). "Mineral Facts and Problems," Bull. 671. U.S. Bureau of Mines, Washington, D.C.

U.S. Department of Interior (1981). "Mineral Industry Surveys—Fluorspar," Miner. Yearb. U.S. Bureau of Mines, Washington, D.C.

U.S. Department of Interior (1982). "Industrial Minerals," Miner. Yearb. U.S. Bureau of Mines, Washington, D.C.

Vie, J. D. (1966). The electrolytic uranium project—terminal progress report—cell design. Rep. MCW 1513. Mallinckrodt Co., Weldon Spring, Missouri.

Vogel, G., Steunenberg, R., and Sandos, O. (1961). *In* "Progress in Nuclear Energy Series III, Process Chemistry" (F. R. Bruce, J. M. Fletcher, and H. H. Hyman, eds.), Vol. 3, pp. 113–118. Pergamon, Oxford.

Watanabe, N., Imoto, H., and Kakajima, T. (1975). *Bull. Chem. Soc. Jpn.* **48**, 1633.

Watanabe, N., Imoto, H. (1976) *Bull. Chem. Soc. Jpn.* **49**, 1736.

Watanabe, N., Imoto, H., Ueno, K. (1978) *Bull. Chem. Soc. Jpn.* **51**, 2822.

Woytek, A. J. (1980). *In* "Kirk–Othmer Encyclopedia of Chemical Technology," 3rd, ed., Vol. 10, pp. 630–654.

Zydzik, G. J., Van Uitert, L. G., Singh, S., and Kyle, T. R. (1977). *Appl. Phys. Lett.* **31**, 6977.

Suggested for Supplemental Reading

Burgess, G. G. (1974). *Inf. Chim.* **138**, 79–80, 89–91.

Emeleus, H. J. (1969). "The Chemistry of Fluorine and its Compounds." Academic Press, New York.

Fielding, H. C., and Lee, B. E. (1977). *In* "The Modern Inorganic Chemicals Industry" (R. Thompson, ed.), Spec. Publ. No. 31, pp. 149–167. Chemical Society, London.

Massone, J. (1972). *Chem. Ztg.* **96,** 65–75.

Rudje, A. J. (1962). "The Manufacture and Use of Fluorine and its Compounds." Oxford Univ. Press, London and New York.

Ryss, I. G. (1956). "The Chemistry of Fluorine and its Inorganic Compounds," State Publ. House Sci. Tech. Chem. Lit., Moscow. AEC tr 3927. U.S. At. Energy Comm., Oak Ridge, Tennessee (Engl. Transl.: Boston, 1960).

Index

O

Octahedra, rotation or tilting, 136, 139
Odd-composition structure, 162–175
  $BaM^{II}M^{III}F_7$ compounds, 165–168
  $Ba_2M^{II}M^{III}F_9$ and related $(M_2F_9)^{n-}$
    compounds, 166–167, 169–172
  $Ba_2Ni_3F_{10}$, 173, 174
  compounds with noninteger F/M ratio, 162
  $Cs_4Ni_3F_{10}$ and $Cs_4Mg_3F_{10}$, 173, 174
  $Cs_6Ni_5F_{16}$, $Cs_7Ni_4F_{15}$, and $Cs_4CoCr_4F_{18}$,
    173–176
  $M^{II}M^{III}F_5$ compounds, 163, 164
Oligofluoride, 288
Opal glass, 309
  fluorides in, 599
Optical dispersion, fluoride glass, 316
Optical fiber
  applications, 318
  fluorozirconate glasses, 325
  preparation, and viscosity, 323
Optical properties, 477–487, 601
  absorption and emission properties, 479–484
  color centers, 485–486
  dispersion, 315–317
  and ferromagnetism in fluorides, 372
  fluorescence properties, 318–320
  fluoride glass, 315–320
    families, 310
  fluorides, 12–13, 477–487
  fluorine and fluoro compounds in energy
    context, 555
  hexagonal polytypes of perovskite,
    ferrimagnetism, 382
  optical transmission range, 316–318
  refractive index, 478
    and dispersion, 315–317
  transition metal oxyfluorides, 208, 220
  transmission properties, 478–479
Optical spectra, nephelauxetic effect, 8
Optical transmission range, see Transmission
  properties
Organic fluoride, 66–67, 583–584
  resistance to ultraviolet light, 555
  thermal stability, 555
Osmium, OsF_6, high-temperature form, 289
Osteoporosis
  bone apatite, fluoride substitution for
    hydroxyl, 548
  compact bone tissue, effect of fluoride,
    547

Oxidation state
  definition, 276–277
  +2, AgF_2, 279, 280
  +3, copper, silver, and gold, 280–282
    Mn(III), Co(III), and Ni(III), 283–285
  +4, copper and silver, 281, 282
    Cr(IV), Mn(IV), Co(IV), and Ni(IV), 283,
    285, 287
  +5, Cr(V) and Mn(V), 285, 287
    gold, 281, 282
  +6, Cr(VI), 286, 287
  +7, Mn(VII), 286, 287
Oxidative coreagents, 365–366
Oxide–fluoride, see Oxyfluoride
Oxyfluoride, 5–8, 205–258, 287–289
  actinide elements, 218
  compositions, with cubic $KSbO_3$
    structure, 236
    with cubic ReO_3 structure, 226, 227
    spinel structure, 238–241
  compounds, K_2NiF_4 structure, 243, 244
    pyrochlore structure, 246, 247
  elpasolite structure, 112
  families of nonlinear materials, 10
  fluorophosphate glasses, 310, 314–320
  with hexagonal elpasolite structures,
    122
  high oxidation states, 297, 300–301
  hydrated, 159
  ion conductor material, 449
  lanthanide and actinide elements, 218
  mixed, impurity behavior in, 522
  $MO_{3x}F_{3-3x}$, 94
  β-$Na_2Ta_2O_5F_2$, 122
  oxide–fluoride cluster, 444
  oxide–tetrafluorides, 160
  with $PbCl_2$ structure, 221, 223, 224
  pyrochlore structure, 177
  $NaCaNb_2O_6F$ type, 125–126
  scheelite structure, tetragonal, 237
  second moments in rigid-lattice method,
    498–499
Oxygen–fluorine substitution, 5–8

P

Palladium
  Pd_2F_6, electronic conductivity, 14
  -type fluorides, ferromagnetism, 379–381

Y

MATERIALS SCIENCE AND TECHNOLOGY

EDITORS

A. S. NOWICK
Henry Krumb School of Mines
Columbia University
New York, New York

G. G. LIBOWITZ
Solid State Chemistry Department
Materials Research Center
Allied Corporation
Morristown, New Jersey